MW00843559

Patagonian Mesozoic Reptiles

LIFE OF THE PAST
James O. Farlow, editor

Patagonian Mesozoic Reptiles

EDITED BY
Zulma Gasparini,
Leonardo Salgado,
and Rodolfo A. Coria

Indiana University Press
Bloomington & Indianapolis

This book is a publication of
Indiana University Press
601 North Morton Street
Bloomington, IN 47404-3797 USA
http://iupress.indiana.edu
Telephone orders 800-842-6796
Fax orders 812-855-7931
Orders by e-mail iuporder@indiana.edu

The paper used in this publication meets the minimum requirements of American National Standard for Information Sciences–Permanence of Paper for Printed Library Materials, ANSI Z39.48-1984.

Manufactured in the United States of America

Library of Congress Cataloging-in-Publication Data

Patagonian Mesozoic reptiles / edited by Zulma Gasparini, Leonardo Salgado, and Rodolfo A. Coria.
p. cm. — (Life of the past)
Includes bibliographical references and indexes.
ISBN 978-0-253-34857-9 (cloth : alk. paper)
1. Reptiles, Fossil—Patagonia (Argentina and Chile) 2. Paleontology–Mesozoic. I. Gasparini, Zulma. II. Salgado, Leonardo. III. Coria, Rodolfo A.
QE861.P38 2007
567.90982'7—dc22 2007000633

1 2 3 4 5 12 11 10 09 08 07

CONTENTS

Contributors

Adriana Albino, CONICET-Departamento de Biología, Universidad Nacional de Mar del Plata, Funes 3250, (7600) Mar del Plata, Buenos Aires, Argentina

José F. Bonaparte, Museo Argentino de Ciencias Naturales "Bernardino Rivadavia," Ángel Gallardo 1470, (1405) Capital Federal, Argentina

Jorge O. Calvo, Centro Paleontológico "Lago Barreales," Universidad Nacional del Comahue, "Proyecto Dino," Ruta Provincial 51, Km. 65, (8312) Neuquén, Argentina

Andrea V. Cambiaso, Museo "Carmen Funes," (8318) Plaza Huincul, Neuquén, Argentina

Luis M. Chiappe, Department of Vertebrate Paleontology, Natural History Museum of Los Angeles County, 900 Exposition Boulevard, Los Angeles, CA 90007, USA

Laura Codorniú, CONICET-Departamento de Geología, Universidad Nacional de San Luis, Chacabuco y Pedernera, (5700) San Luis, Argentina

Rodolfo A. Coria, CONICET-Museo "Carmen Funes," (8318) Plaza Huincul, Neuquén, Argentina

Marta Fernández, CONICET-Departamento Paleontología Vertebrados, Facultad de Ciencias Naturales y Museo, Universidad Nacional de La Plata, Paseo del Bosque s/n, (B1900TAC), La Plata, Buenos Aires, Argentina

Juan R. Franzese, CONICET-Centro de Investigaciones Geológicas, Facultad de Ciencias Naturales y Museo, Universidad Nacional de La Plata, Calle 1-644, (B1900TAC), La Plata, Buenos Aires, Argentina

Marcelo S. de la Fuente, CONICET-Departamento de Paleontología, Museo de Historia Natural de San Rafael, Parque Mariano Moreno s/n, (5600) San Rafael, Mendoza, Argentina

Zulma Gasparini, CONICET-Departamento Paleontología de Vertebrados, Facultad de Ciencias Naturales y Museo, Universidad Nacional de La Plata, Paseo del Bosque s/n, (B1900TAC), La Plata, Buenos Aires, Argentina

Diego Pol, CONICET-Museo Paleontológico "Egidio Feruglio," Auda. Fontana 140, (9100) Trelew, Chubut, Argentina

Leonardo Salgado, CONICET-Museo de Geología y Paleontología, Universidad Nacional del Comahue, Buenos Aires 1400, (8300) Neuquén, Argentina

Luis A. Spalletti, CONICET-Centro de Investigaciones Geológicas, Facultad de Ciencias Naturales y Museo, Universidad Nacional de La Plata, Calle 1 644, (B1900TAC), La Plata, Buenos Aires, Argentina

Preface

The word *Patagonia* is usually linked to the vast deserts wedged between the South Atlantic and the Andes. The latter separates it from Chile, where at the same latitudes we find some of the most humid forests anywhere. Both the desert or semiarid Argentinean Patagonia and the humid Chilean forests are the consequence of a geologic history shared through millions of years. Much of this history occurs during the Mesozoic.

Patagonia has been distant yet seductive since its discovery by European explorers. Especially since the nineteenth century, it has been one of the most attractive places for explorers and paleontologists. The amazing taxonomic diversity, the temporal and geographic broadness of the fossil record, and the variety of paleoenvironments (from ancient deserts to placid seas) all contribute to making Patagonia a key reference for Gondwanan Mesozoic reptiles.

The first Patagonian geologic and paleontologic studies were mainly conducted on invertebrates and mammals. Other vertebrates did not draw researchers' attention, and there were no detailed analyses until the publication in 1929 of a world-famous monograph by F. von Huene, "Los Saurisquios y Ornitisquios del Cretáceo Argentino" ("The Saurischians and Ornithischians from the Argentinean Cretaceous"). Nonetheless, several decades passed before the discovery of other nonmammalian tetrapods proved that Patagonia held an invaluable part of the evolutionary history of vertebrates.

Today, some 40 years after the first Patagonian discoveries, Patagonia has turned out to be a true "land of dinosaurs." Patagonia has plenty of stories to tell about these icons of the Mesozoic and their closest relatives, the birds, and about other reptiles that lived in Patagonia and its adjacent seas.

The discoveries and subsequent studies of Patagonian Mesozoic reptiles were driven mainly by Rodolfo Casamiquela in the 1960s and 1970s, and later by José Bonaparte. Both provided decisive encouragement for others to carry on the research. Many of those researchers are contributors to the present volume.

Patagonian Mesozoic Reptiles aims to gather, in a single volume and for the first time, our current knowledge about these fossils. The body of information accumulated during the last 100 years has not always been easily available to the international scientific community because of ill-distributed journals, language barriers, or simple ignorance about the existence of the data. This book presents up-to-date information about the evolution of Patagonia and its reptiles, always compared with other faunas from different corners of the planet.

Patagonia provides a distinct and complementary perspective to Gondwana and Laurasia in terms of the evolution of life and the associated landmasses. In turn, the Mesozoic constitutes a key chapter of the geologic history of this region, from both tectonic and paleontologic points of view.

We have included in this volume those taxa that fall within the modern phylogenetic definition of Reptilia ("the most inclusive clade containing *Lacerta agilis* and *Crocodylus niloticus* but not *Homo sapiens*" [Modesto and Anderson 2004]). Although not included in this definition, we have included in Chapter 13 a series of footprints assigned to mammal-like reptiles of the Triassic of Río Negro (*Calibarichnus, Palaciosichnus, Rogerbaletichnus, Gallegosichnus*), but not footprints attributed to mammals from the Jurassic of Santa Cruz (*Ameghinichnus*) (Casamiquela 1964).

Patagonian Mesozoic Reptiles is organized into chapters written by specialists in their respective subjects. Chapter 1 deals with the early history of the studies of fossil reptiles of Patagonia, most of which were known from circumstantial findings and were framed by scientific projects traditionally devoted to the study of Tertiary and Quaternary mammals. The achievements of the researchers who made these early discoveries and conducted the studies are remarkable given the harsh conditions of the operation of working in vast deserts in a developing country. Carlos and Florentino Ameghino, Santiago Roth, Richard Lydekker, John B. Hacther, and Friedrich von Huene (augmented by the myth of a living plesiosaur of Patagonia—the "Nessie" of these latitudes) are part of that colorful and poorly known history. Chapter 2 describes the paleogeographic evolution of Patagonia and the main depositional systems of its sedimentary basins. This chapter provides the paleogeographic and paleoenvironmental frame for the Mesozoic of Patagonia and for the subsequent chapters. In Chapters 3–12 we present a systematic approach to the representative reptile clades of the Patagonian Mesozoic. The type and referred material of each taxon is detailed, with its geographic and chronological provenance. We include diagnoses—in many cases new or modified—together with extensive comments on various topics of the biology or evolutionary history of each group. Differences in these commentaries are due to the main research trends developed by the authors. Chapter 13 includes the information supplied by paleoichnology. Finally, Chapter 14 is an overview on the reptilian

Mesozoic faunal successions, in which some discontinuities, turnovers, and diversifications may be observed, most of which are related to the main geotectonic events that affected Patagonia during the Mesozoic.

All the chapters of this volume have been reviewed by colleagues who generously contributed to the chapters' improvement and to whom the editors are especially indebted: N. Bardet (Musée National d'Histoire Naturelle, France), M. Bond (Museo de La Plata, Argentina), C. Brochu (University of Iowa, USA), J. I. Canudo (Universidad de Zaragoza, Spain), M. Carrano (Smithsonian Institution, USA), I. S. Carvalho (Universidade Federal de Rio de Janeiro, Brazil), J. Clark (George Washington University, USA), E. Gaffney (Natural History Museum, New York, USA), P. Galton (University of Bridgeport, USA), A. Kellner (Universidade Federal de Rio de Janeiro, Brazil), M. Lamanna (Carnegie Museum of Natural History, USA), O. Limarino (Buenos Aires University, Argentina), M. Lockley (University of Colorado at Denver, USA), J. Massare (SUNY College, Brockport, USA), C. A. Meyer (Geologisch-Palaeontologisches Institut Basel, Switzerland), F. Novas (Museo Argentino de Ciencias Naturales, Argentina), K. Padian (University of California, Berkeley, USA), M. J. Parham (University of California, Berkeley, USA), J. C. Rage (Museum National d'Histoire Naturelle, France), S. Sachs (Institut für Paläontologie, Freie Universität Berlin, Germany), J. D. Scanlon (Riversleigh Fossils Centre, Mount Isa, Australia), E. Tonni (Museo de La Plata, Argentina), D. Weishampel (John Hopkins University, USA), and J. Wilson (University of Michigan, Ann Arbor, USA).

Likewise, the editors appreciate the collaboration of the Argentinean artists Jorge González, Carlos Papolio, and Oscar Campos for generously creating the reconstructions of reptiles and environments.

Some volunteer collaborators such as A. Paulina Carabajal, Liliana Rikemberg (Museo "Carmen Funes"), Prebiterio Pacheco (Universidad Nacional del Comahue), Phil Currie (Royal Tyrrell Museum), and Cecilia Deschamps (Museo de La Plata) assisted in the preparation of this volume by checking manuscripts, providing translations into English, mailing and filing, and supplying constant support and enthusiasm.

Finally, we particularly want to thank Indiana University Press; editorial director, Robert J. Sloan; and series editor, James O. Farlow, for their continuous help and assistance.

References Cited

Casamiquela, R. M. 1964. *Estudios icnológicos. Problemas y métodos de la icnología con aplicación al estudio de pisadas mesozoicas (Reptilia, Mammalia) de la Patagonia*. Buenos Aires: Colegio Industrial Pío IX.

Modesto, S. P., and J. S. Anderson. 2004. The phylogenetic definition of Reptilia. *Systematic Biology* 53 (5): 815–821.

Addendum: List of New Species Published

Lepidosauromorpha

Najash rionegrina

Apesteguía, S., and H. Zaher. A Cretaceous terrestrial snake with robust hindlimbs and a sacrum. *Nature* 440: 1037–1040.

Crocodyliformes

Araripesuchus buitrerarensis

Pol, D., and S. Apesteguía. 2005. New *Araripesuchus* remains from the Early Late Cretaceous (Cenomanian-Turonian) of Patagonia. *American Museum Novitates* 3490: 1–38.

Sauropodomorpha

Cathartesaura anaerobica

Gallina, P., and S. Apesteguía. 2005. *Cathartesaura anaerobica* gen. et sp. nov., a new rebbachisaurid (Dinosauria, Sauropoda) from the Huincul Formation (Upper Cretaceous), Río Negro, Argentina. *Revista del Museo Argentino de Ciencias Naturales* 7: 153–166.

Puertasaurus reuili

Novas, F. E., L. Salgado, J. O. Calvo, and F. Agnolín. 2005. Giant titanosaur (Dinosauria, Sauropoda) from the Late Cretaceous of Patagonia. *Revista del Museo Argentino de Ciencias Naturales* 7: 37–41.

Ligabuesaurus leanzai

Bonaparte, J. F., B. González Riga, and S. Apesteguia. 2006. *Ligabuesaurus leanzai* gen. et sp. nov (Dinosauria, Sauropoda), a new titanosaur from the Lohan Cura Formation (Aptian, Lower Cretaceous) of Neuquén, Patagonia, Argentina. *Cretaceous Research* 27: 364–376.

Patagonian Mesozoic Reptiles

1. Patagonia and the Study of Its Mesozoic Reptiles

A Brief History

LEONARDO SALGADO

The Beginnings of the Paleontology in Argentina

Francisco Javier Muñiz (1795–1871), a medical doctor employed by the government of Buenos Aires Province, is usually considered the first Argentinean "paleontologist." In the course of his life, Muñiz unearthed from the *pampas* abundant fossils of large mammals, which years later constituted the basis of the collection of the Public Museum of Buenos Aires, now the Museo Argentino de Ciencias Naturales "Bernardino Rivadavia" (Onna 2000).

Muñiz's time was not marked by a proliferation of knowledge in Argentina. Juan Manuel de Rosas, governor of Buenos Aires since 1829 and in charge of the Argentinean Confederation's foreign relations, had, according to Romero (2000:80), "[destroyed] the foundation of scientific and technological progress." The unpropitious political atmosphere, however, did not prevent Muñiz from becoming an authority on fossils and natural sciences, to the extent that he was the only Argentinean-born naturalist with whom Charles Darwin corresponded.

The 1852 victory of General Justo J. de Urquiza over Rosas in the Battle of Caseros closed one of the most controversial periods in the history of Argentina. In the years that followed, successive governments showed greater interest in public education and in scientific research, particularly in the natural sciences, through which they endeavored to determine the economic potential of the country's natural resources. French naturalist Victor Martin de Moussy's *Description géographique et statistique de la Confédération Argen-*

tine (1860) was requested by President Urquiza for the express purpose of developing the Argentinean Confederation's mineral and other resources (Babini 1949). At the same time, the National Museum was founded in Paraná (then capital city of the Argentinean Confederation), and Alfred Marbais du Graty (who, in turn, authored an historical and geographical description of the Argentinean Confederation) was named its director in 1854, thus demonstrating Urquiza's preoccupation with making up for time lost during the previous period. In 1830, the government of neighboring Chile commissioned another French national, Claude Gay Mouret (1800–1873), to complete a scientific description of the country, which was published in 30 volumes between 1844 and 1870 as *Historia física y política de Chile*.

The Public Museum of Buenos Aires, heir of the former Museum of the Country founded by Bernardino Rivadavia in 1823, had been practically dismantled under the Rosas regime (Babini 1949). The Asociación de Amigos de la Historia Natural was founded in 1854 to rectify the situation. Among its founding members was Francisco Muñiz, just retired from his paleontological excursions.

General Urquiza was defeated by Bartolomé Mitre in the Battle of Pavón, after which the current republican system was instituted. During this period, actions were taken to advance education. In 1862 Mitre appointed the Prussian naturalist Karl Hermann Konrad Burmeister (1807–1892) director of the Museum of Buenos Aires. Some time later, president Domingo F. Sarmiento initiated the organization of the newly founded Academia Nacional de Ciencias de Córdoba, an educational institution of great importance in the professionalization of the natural sciences in Argentina (Tognetti 2000, 2004; García Castellanos 2004). In turn, Sarmiento, as senator, advocated official financial support of Burmeister's treatise *Description physique de la République Argentine* because (among other reasons) enterprises similar to the work of Gay Mouret were underway in Chile (Asúa 1989).

Burmeister's role as director of the Public Museum of Buenos Aires, and as an active member of different societies, substantially changed the societal and governmental perception of scientific activity. Burmeister prepared the ground for the rise of a new generation of naturalists, including Florentino Ameghino (Figs. 1.1, 1.2), Eduardo Holmberg, and Francisco Moreno (Fig. 1.3)—all of them Argentinean.

The First Fossils from Patagonia

During the first years of the republican government, the successive Argentinean presidents (Mitre, Sarmiento, and Nicolás Avellaneda) advocated support of the institutional order and extension of the national dominion southward, toward territories still under the power of Indian leaders. The territorial occupation of Patagonia by the army at the end of the 1870s (during the presidency of

Figure 1.1. Florentino Ameghino (1854–1911).

Avellaneda) brought state institutions to the boundaries of the territory, which were by that time hostile to the national forces. Knowledge, not only topographic but also of the economic possibilities of the new territory, became more valuable (Podgorny 1999). The soldiers that marched to Patagonia in 1879 under the orders of Avellaneda's minister of war, General Julio A. Roca, were accompanied by scientists from the newly founded Academia Nacional de Ciencias de Córdoba (Adolf Doering, zoologist and chemist; and Paul Lorentz, botanist), whose mission was to make a zoological, botanical, and geological survey of the territory being annexed—and, not incidentally, to draw attention away from the military nature of the expedition. The conditions under which the professors undertook their survey were not always optimal. Furthermore, there were complaints about the troops' progress, which was too fast to allow the scientists to take notes on the conquered territories (Podgorny 2004; Tognetti 2004). Nonetheless, the (largely geological) scientific observations made during the military expedition to Patagonia resulted in an extensive "official" document (Doering 1882).

After occupation of Patagonian territory, the soldiers, settled on the shores of the Negro, Neuquén, and Limay Rivers (in the

Figure 1.2. Florentino Ameghino at his place of work. Taken with permission from Tognetti (2004).

present provinces of Río Negro and Neuquén), continued the work begun by the scientists of the Academia de Ciencias. These settlers discovered and collected the territory's first fossil remains—mostly bones and petrified wood. It was thus that Commander Buratowich (in 1882), Captain Rohde (in 1883), and Colonel Romero (in 1887), made their respective contributions to science. Buratowich's fossils were sent to then-president Roca, then shown to the specialist Florentino Ameghino. The bones discovered by Rohde were sent directly to Ameghino; those discovered by Romero were sent to the Museum of La Plata, which had been founded in 1884, and of which Ameghino was subdirector.

Florentino Ameghino (1854–1911) (Figs. 1.1, 1.2) is undoubtedly the leading figure in Argentinean paleontology. In 1876, while still an unknown teacher in an elementary school in Buenos Aires Province, Ameghino contacted the authorities of the Museum of Buenos Aires and of the Sociedad Científica Argentina, offering to evaluate their archaeological and paleontological holdings. A visit to Europe, during which he met such famous paleontologists as Edward D. Cope (to whom he sold part of his collection of fossils [Cabrera 1944:15]), and his publication of *Los Mamíferos Fósiles de la República Argentina* (written in collaboration with the Frenchman Henry Gervais) helped to establish his scientific credentials (González Arrili 1954). Perhaps it is not surprising that the first fossil remains found in Patagonia had ended in the brash young man's hands, even if he did not have an official title.

Although at that time he was unfamiliar with Mesozoic reptiles (he specialized in Tertiary mammals), Ameghino correctly identified the bones sent by Buratowich from Neuquén (and later deposited in President Roca's home), as belonging to dinosaurs. The sensational announcement of the existence of dinosaurs in the ancient geological strata of Patagonia was published in the newspaper *La Nación* on March 23, 1883—that is, before the founding of the Museum of La Plata. This discovery showed beyond all doubt that the geology of the southern part of the continent was more complex than previously thought. In fact, before the Campaña del Desierto (as the military expedition of 1879 is known), Burmeister and other men of science thought that from the north of Argentina to the Strait of Magellan, there were only two geologic formations: the *pampeana* (of Quaternary age) and the *patagónica* (of Pliocene age) (Burmeister 1879:154; Ameghino 1921a:489).

After the unification of the country, Burmeister's museum became the National Museum of Buenos Aires. Francisco P. Moreno (Fig. 1.3), a distinguished member of the Sociedad Científica and former apprentice of Burmeister, persuaded the authorities of the Buenos Aires province to establish a new provincial museum. In 1884, a new museum was founded in a splendid building in La Plata, the new capital city of the province. Moreno, who was named director, invited Florentino Ameghino to work in the institution as subdirector and secretary. Ameghino accepted and moved his private collection of fossils to La Plata, including the bones Rohde collected near the present-day city of General Roca (in Río Negro Province). These few pieces, referable to *Titanosaurus* after Ameghino (1921a:10), together with others sent by Colonel Romero in 1887, were the first dinosaur bones housed in a South American scientific institution.

Figure 1.3. Francisco P. Moreno (1852–1919). With permission of Revista Museo.

The Methodical Exploration of Patagonia

> El desierto ha sido conquistado militar y políticamente; es menester ahora dominarlo para la geografía y la producción.
>
> [The desert has been conquered military and politically; now it is necessary to dominate it for the land and natural resources.]
>
> —Julio A. Roca, president of Argentina, in a speech to the Chamber of Deputies

With the effective presence of the military forces in Patagonia since 1880, the situation became more favorable to naturalists and explorers. With the designation of Florentino Ameghino as subdirector of the Museum of La Plata, his brother Carlos (1865–1936) was designated "traveler naturalist" of that institution. The task given Carlos Ameghino was to collect fossil mammals from localities that had been worked 10 years earlier by Moreno. Carlos made his two first trips to Patagonia in 1887 and 1888. The first expedition was, according to Camacho (2000:18) "el iniciador de las investigaciones geológicas y paleontológicas en la Patagonia" ["the first of the geologic and paleontological investigations in Patagonia"].

In the course of the second expedition, according to his own account, Carlos discovered the bones of a large dinosaur in the area of "Pampa Pelada" in the current province of Chubut. The specimen discovered, a nearly complete skeleton (Camacho 2000), was later excavated by a Museum of La Plata commission without the assistance of Carlos, who was removed from his position in the museum after a fight with Moreno.

In referring to these bones, which were later designated by Lydekker (1893) as holotype of the species *Argyrosaurus superbus*, Florentino (who had left his position in the museum), shed some light on the personal conflict between the Ameghinos and Moreno:

> Desgraciadamente, las personas incompetentes, comisionadas por el Director del Museo de La Plata, para la exhumación de ese esqueleto, sólo consiguieron extraer uno de los miembros, destruyendo lo demás. El viajero que tiene ocasión de cruzar esa región, divisa todavía desde larga distancia, la acumulación de huesos destrozados por esa vandálica expedición.
>
> [Unfortunately, incompetent people commissioned by the director of the Museum of La Plata for the exhumation of that skeleton removed only one of its limbs, destroying all the other pieces. The traveler that has occasion to cross that region still perceives throughout the distance the accumulation of bones destroyed by that vandalistic expedition.] (Ameghino 1921b:9)

The doors of the Museum of La Plata were closed to the Ameghinos. The study of the vertebrate fossils housed in that institution (among these, all the dinosaur remains gathered to that date) was offered to the prominent British paleontologist Richard Lydekker (1849–1915) (Fig. 1.4). Lydekker accepted the invitation and traveled twice to Argentina for that purpose (Colbert 1984; Wilson and Upchurch 2003). His *Contributions to the Study of the Fossil Vertebrates of Argentina* (1893) was published in the journal *Anales del Museo de la Plata* (Lopes and Podgorny 2000). In the volume dedicated to the dinosaurs, Lydekker described remains from northern Patagonia of the genus *Titanosaurus*, which he had identified many years earlier in India (Lydekker 1877) and in England (Lydekker 1888), and a new genus, *Argyrosaurus*, linked to the former (Wilson and Upchurch 2003). The specimen on which the last genus was based was the one collected by the "vandalistic" Museum of La Plata expedition.

Lydekker's discovery was of a diverse family of dinosaurs, the Titanosauridae, to which *Titanosaurus* and *Argyrosaurus* belonged (Fig. 1.5). For Lydekker, "the occurrence of *Titanosaurus* (of that and a closely allied genus) in both India and South America affords one more instance of that remarkable community of type which undoubtedly exists between the faunas of southern continents of the world" (Lydekker 1893:3). It was the first time that a genus of dinosaur was claimed to have a wide austral distribution (Wilson and Upchurch 2003). Lydekker's studies, the first on Patagonian Mesozoic reptiles, reaffirmed the enormous paleontological potential of

Figure 1.4. Richard Lydekker (1849–1915).

Patagonia, up to then corroborated only by its excellent assemblages of Tertiary mammals.

The dispute with Moreno, and the impossibility of accessing the collections of the Museum of La Plata, did not prevent the Ameghinos from continuing their own paleontological expeditions. But Carlos's trips to Patagonia depended on personal funds until Florentino was appointed director of the Museo Nacional de Buenos Aires, and Carlos became its traveler naturalist.

In addition to the Argentinean expeditions, foreign scientific institutions organized explorations during this period. Hugo Zapalowics (1852–1917), a Polish explorer who traveled around the world between 1888 and 1890 (Dembicz 2002), stayed in the Argentinean Patagonia, 80 km from the confluence of the Neuquén and Limay Rivers (in the current Province of Neuquén) in 1889. Zapalowics and his people found vertebral remains that were eventually housed in the geology collection of the University of Vienna. Many years later, these bones were rediscovered and studied by

Figure 1.5. Titanosaurus, *as reconstructed by Á. Cabrera.* Taken from Windhausen (1931, 283).

Ferenc Nopcsa (1877–1933), who tentatively referred them to the genus *Bothriospondylus,* a sauropod from the Jurassic of Europe and Madagascar (Nopcsa 1902).

In 1903, North American paleontologist John B. Hatcher interpreted these bones as belonging to the North American Jurassic sauropod *Haplocanthosaurus* (Hatcher 1903). Later, Friedrich von Huene (1929) interpreted them as remains of "*Titanosaurus*" *australis,* one of the new species of "*Titanosaurus*" described by Lydekker for Argentina. Recently, Calvo and Salgado (1995) reinterpreted these remains as belonging to *Limaysaurus tessonei* (Salgado and Bonaparte this volume).

In light of the enormous paleontological potential of Patagonia, Hatcher organized three Princeton University expeditions (finally undertaken between 1896 and 1899), with the purpose of collecting fossils and establishing a geological stratigraphy to improve or replace the Ameghinos' dubious chronology (Lopes 2001). As will be seen later, the Ameghinos (as well as Santiago [Jacob] Roth) believed that in Patagonia, dinosaurs had coexisted with mammals that were evolutionarily more advanced than their contemporaries on other continents. The Princeton expeditions unearthed a number of dinosaur fossils in the current province of Santa Cruz, but they were unable to transport them, underscoring the difficult conditions that faced foreign explorers and paleontologists who visited Patagonia (Hatcher 1985:132).

Figure 1.6. Santiago Roth (1850–1924).

Vertebrate Paleontology in Argentina between the Centuries

After Florentino Ameghino's removal from the Museum of La Plata, Acides Mercerat and later Santiago Roth (1850–1924) (Fig. 1.6) took charge of its paleontology department. Roth organized a series of field trips to Patagonia between 1895 and 1899. On the first trip to Río Negro and Neuquén, from December 1895 to June 1896, Roth was accompanied by German geologist Walther Schiller (1879–1944), who would be hired by the museum in 1905 (Camacho 2001). The expedition found remains of snakes and crocodiles in the area of the confluence of the rivers Limay and Neuquén (in the actual city of Neuquén); later trips to Chubut, in the surrounding area of Laguna Pelada, recovered some turtle and carnivorous dinosaur remains.

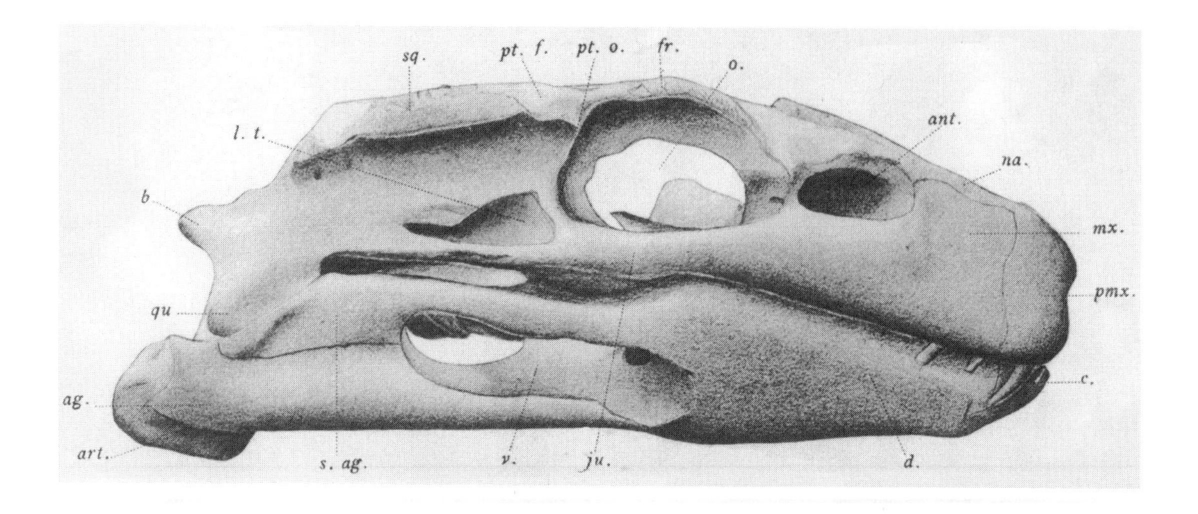

Figure 1.7. Notosuchus terrestris. From Woodward (1896).

Some of these materials were brought to England by Moreno and studied by Arthur Smith Woodward (1864–1944). Woodward identified the Neuquén remains as the *Notosuchus* and *Cynodontosuchus* crocodilians (Woodward 1896; Pol and Gasparini this volume) (Fig. 1.7). Woodward went on to describe *Genyodectes* (the carnivorous dinosaur from Laguna Pelada) (Coria this volume) and *Dinilysia* (a snake from Neuquén) (Albino this volume), together with new remains of the turtle *Niolamia* (a genus described by Ameghino in 1899) (Woodward 1901; de la Fuente this volume) in 1901.

The dinosaur collected by Roth and studied by Woodward was particularly significant in the context within which Roth and the Ameghinos were working, based as it was on the supposed coexistence of dinosaurs and higher mammals. As Roth communicated to Ameghino (in Ameghino 1934:125), the bones of *Genyodctes* had been found above the "Piso Notostylopense" (as named by Ameghino), a mammal-bearing level of ash that was supposed to be Cenomanian in age (Huene 1929:17) (Coria and Salgado 2000). The record of *Genyodectes* seemed to corroborate the idea that the South American Mesozoic mammals were of a higher evolutionary rank than those from other parts of the world. But Roth's field methods were far from reliable, and many doubted the real provenance and stratigraphic position of these materials. Twenty-five years later, and 10 years after Florentino's death, Roth confessed to Friedrich von Huene during a visit to La Plata that he did not pick up the materials from the ground, but "un gaucho que lo acompañaba" ("a gaucho who was with him") (Huene 1929:18).

Carlos Ameghino acted as Florentino's private collector between 1889 and 1902, and as a traveler naturalist for the National Museum of Buenos Aires between 1902 and 1903. These new expeditions were very productive, and the discoveries and observations made during this period allowed to the Ameghinos to frame their controversial stratigraphy of Patagonia. During these new explorations, Carlos discovered many important remains of Mesozoic reptiles. Those of the small dinosaur *Loncosaurus argentinus,* found in the Guaranítica Formation in Río Sehuen, Santa Cruz, were described by Florentino in 1898 (Ameghino 1921c) and interpreted as belonging to a carnivorous dinosaur of the group of Megalosauridae (currently an Ornithopoda; Coria and Cambiaso this volume). Those of *Clasmodosaurus spatula* (Ameghino 1921a:703), found in the same site, were originally interpreted by Florentino as belonging to a sauropod (which coincides with the current interpretation of Powell 2003). Many years later, Huene would suggest that the teeth upon which *Clasmodosaurus* was founded belonged to the same *Loncosaurus,* which in turn was supposed to be a coelurosaurian saurischian (Huene 1929:141).

In both cases, as with Roth's *Genyodectes,* the Ameghinos claimed that all the remains came from levels younger than others bearing notoungulate mammals of the genus *Notostylops*—that is, from the so-called Piso Pyrotheriense (called thus as a result of the supposed presence of the great mammal *Pyrotherium*). From the Pyrotheriense, in turn, Carlos had collected a series of teeth presumed to be from megalosaur dinosaurs, which seemed to be irrefutable proof of the contemporaneity of dinosaurs and mammals (Bond 2000).

The Definitive "Conquest" of Patagonia

In the beginning of the twentieth century, Argentina experienced a period of relative prosperity, signaled by the consolidation of the industrial activity and public investment in infrastructure (Romero 2000). In Patagonia, the settlement of agricultural villages demanded railways (which were constructed between 1897 and 1901) and the construction of irrigation canals. Between 1910 and 1932, the first aqueducts in North Patagonia were built, and the Pellegrini Lake, in the Río Negro, was filled (Nicoletti and Navarro Floria 2000). The fluvial valleys occupied by settlers and farmers cut into ancient continental sediments, and during construction and the geological studies linked to exploration for oil and minerals, numerous vertebrate fossils were discovered (Powell 2003). The 25 years from 1914 to 1939 saw a major transformation in Argentine geology (Camacho 2001). At first, the state focused geological activity on the search for exploitable resources. The discovery of oil in 1907 in the small town of Comodoro Rivadavia, in Chubut, prompted the national government, through its Dirección General de Minas, Geología e Hidrología (DGMH), to hire many foreign geologists. In 1918, thanks to the activity of two of them—Anselmo Windhausen and Juan Keidel—oil was discovered in Plaza Huincul, in Neuquén (Camacho 2001).

A number of geologists of the DGMH showed a special interest in the fossil vertebrates, which were frequent in those areas that were being explored. (So numerous were the fossil remains in Mesozoic levels that Keidel coined the term "Estratos con Di-

Figure 1.8. Richard Wichmann (1881–1930).

nosaurios" to designate the widespread exposures today known as "Grupo Neuquén" and "Grupo Chubut" [Keidel 1917]. Ameghino's old term "Formación Guaranítica" was subsumed under this designation.) One such geologist was Richard Wichmann (Fig. 1.8), who arrived in the country from Germany in 1912 (Camacho 2001). In 1916 Wichmann recorded a large dinosaur found four years earlier in the territory of Río Negro, near the present-day city of General Roca. After a detailed inspection of the bones, Carlos Ameghino, then director of the Buenos Aires Museum, informed Wichmann that the bones did not correspond to *Titanosaurus* (at the time the only genus of dinosaur described for the territory of Río Negro), but to a form similar to *Diplodocus,* from the Jurassic of North America (Wichmann 1918:96). In later publications, however, Wichmann ignored Carlos's identification, speaking of "*Titanosaurus*" (Wichmann 1919:17). The full description of that specimen, which pertained to a titanosaur, was carried out by Huene in 1929 (see Salgado and Bonaparte this volume, in which many opinions with respect to the association are mentioned). A similar case is that of Augusto Tapia, another DGMH geologist, who in 1918 found the mandibles of a dinosaur northwest of Colhué Huapí Lake, in the province of Chubut. One year later, Tapia referred the bone to a new species of ceratopsian (horned dinosaur), *Notoceratops bonarelli* (Coria and Cambiaso this volume), in honor of another geologist employed by DGMH, Guido Bonarelli, who had arrived in 1911 (Tapia 1918). Besides these skull remains, Tapia found other pieces of the same skeleton, but he did not collect them.

In 1922, the Dirección Nacional de Yacimientos Petrolíferos Fiscales was created. This institution provided a strong impetus for geological studies in Argentina until its privatization in the early 1990s. Many geologists linked to Yacimientos Petrolíferos Fiscales made important contributions to the knowledge of the fossil faunas of Patagonia. In 1931, Alejandro Piatnitzky and Egidio Feruglio, through their studies on the stratigraphy and paleontology of the San Jorge Basin (currently in Chubut Province), helped to disprove the theory of coexistence of notoungulates, piroteries, and dinosaurs (Camacho 2001). At about the same time, the Scarritt expeditions of the American Museum of Natural History to Patagonia, led by George Gaylord Simpson between 1930–1931 and 1933–1934, decisively cleared up the controversy. In the Piso Notostylopense deposit, which Ameghino considered part of the Upper Cretaceous Formación Guaranítica (today considered of lower Eocene, mammal-age Casamayorense), Simpson and his team found the skeleton of a large crocodile that had compressed teeth with serrated margins, very similar to those seen in carnivorous dinosaurs (Bond 2000). Isolated teeth of a similar kind, called "megalosaur teeth" in the stratigraphic tables of Ameghino, contributed to this long-sustained misinterpretation.

The Museum of La Plata and a New Period of Exploration

The explorations carried out by the museum stalled as a result of the annexing of the museum to the Universidad Nacional de La Plata and its consequent nationalization and the retirement of Moreno in 1906. (The museum had originally been established as a provincial, not national, museum.)

The years that followed Moreno's departure were difficult, in part because the museum's scientific staff were required both to teach and to conduct research. With the decline in field exploration and in research in general, the museum's administration was anxious to fill the still-empty spaces in the enormous building in the Paseo del Bosque. So in 1912, the director of the Museum of Pittsburgh, William Holland, mounted a magnificent cast of the enormous Jurassic sauropod *Diplodocus carnegii,* whose remains had been originally described by an early visitor to Patagonia, John Bell Hatcher. The cast was a gift from Andrew Carnegie, the North American millionaire for whom the species was named (Podgorny and Plöger 1999).

Figure 1.9. Friedrich von Huene (1875–1969).

The museum's scientific expeditions were suspended until the 1920s, when they were reinitiated under the Hipólito Yrigoyen and Marcelo T. de Alvear (both of the Radical Party) regimes, and thanks to the urging of the new director of the museum, Luis María Torres (1878–1937). This period witnessed the explorations of Santiago Roth to Río Negro and Neuquén in 1921 and again in 1922–1923, during the course of which many dinosaur skeletons were unearthed. It is likely that some of those remains had been discovered by some of the hundreds of workers employed in the construction of the main irrigation canal in Cinco Saltos, Río Negro. During these years, two huge sauropod femora (2.30 m) were discovered at Aguada del Caño, in Neuquén Province. Once in La Plata, the femora were displayed alongside the femora of the cast of *Diplodocus carnegii* donated by Carnegie, which was dwarfed by the Aguada del Caño giant (Podgorny and Plöger 1999). Huene (1929) referred the specimen to *Antarctosaurus giganteus.*

At about the same time, the Marshall Field Paleontological Expedition to Argentina and Bolivia (1922–1924) was carried out under the direction of Elmer Riggs. The expedition made important discoveries of dinosaurs in Sierra de San Bernardo, in the territory of Chubut. These finds included the remains of a hadrosaur, named *Secernosaurus koerneri* by M. Brett-Surman (1979), as well as a titanosaur (a femur 1.77 m long and a tibia 1.24 m long) that Huene referred to ?*Antarctosaurus wichmannianus* (see below).

As result of the work achieved during these years by the Museum of La Plata, a significant number of dinosaur remains was added to its collections, most of which came from the localities of Cerro Policía and, overall, from Cinco Saltos, in Río Negro. More than 50 years had passed since the first paleontological studies had been undertaken. However, in 1920, no Argentinean scientist was competent to initiate an exhaustive study of Mesozoic reptiles. (Florentino Ameghino, to whom the doors of the Museum of La Plata were again opened after reconciliation with Moreno, had died in 1911.) Because of this, Torres decided to invite to a foreign specialist: Friedrich von Huene of Tübingen, Germany, without a doubt the most important Mesozoic reptile specialist of his time (Fig. 1.9).

During his visit to Argentina, Huene traveled to Patagonia with Schiller and Roth with an eye to visiting the most important localities. He was in Senillosa, in Neuquén, and in "Aguada de Córdoba" (possibly near Salitral Moreno) and Cinco Saltos, in Río Negro. In 1924, Huene joined the Field Museum expedition headed by Riggs and witnessed the exhumation of the bones of a large sauropod in the territory of Chubut (Huene 1929:57). The volume of the museum's *Anales, Los saurisquios y ornitisquios del Cretáceo argentino,* that includes his main results is indeed a fundamental publication, mainly as a result of its contribution to the knowledge of the titanosaurian dinosaurs of Argentina. The title of this volume is telling. Unlike Lydekker, Huene does not make reference to dinosaurs. Following the British paleontologist Harry Govier Seeley (who coined the terms *saurischian* and *ornithischian*), the professor from Tübingen thought that the dinosaurs were not a natural single group, but two unrelated groups (Seeley's saurischians and ornithischians) (Colbert 1984:108).

The pages of *Los saurisquios y ornitisquios del Cretáceo argentino* contain virtually all the knowledge on South American dinosaurs to that date. Besides considering the fossils previously studied by Ameghino, Smith Woodward, and Lydekker, Huene described many others, identifying four new saurischian species: *Titanosaurus robustus* (from Cinco Saltos), *Laplatasaurus araukanicus* (from Cinco Saltos, Roca, and Rancho de Ávila), *Antarctosaurus wichmannianus* (from General Roca, mentioned by Richard Wichmann in 1916), and *Campylodon ameghinoi* (a partial sauropod maxilla found by Carlos Ameghino in the Sierra de Bernardo, in Chubut), and one new ornithischian, *Loricosaurus scutatus* (a series of dermal plates from the locality of Cinco Saltos, later attributed to sauropods [Powell 1980]).

Patagonia was revealed as the place where diverse Cretaceous reptiles coexisted: saurischians, ornithischians, crocodiles, snakes, and turtles. The probable connections between South America and the other continents could now be inferred, and thus, the Patagonian Mesozoic reptiles could bring a new perspective to the paleontological history of the planet. Huene's closing words illustrate the enthusiasm that the new discoveries were starting to generate: "El problema íntegro, aquí tratado, es un capítulo importantísimo e instructivo de la historia de la tierra y de sus organismos [The full problem, here exposed, is a very important and instructive chapter of the history of the earth and of its organisms]" (Huene 1929:180).

Hunting Prehistoric Creatures

During the second half of the nineteenth century, Patagonia was still a territory virtually unknown to scientists. This changed once the obstacle represented by hostile aboriginal populations (*Tehuelches* and *Mapuches*) was removed in the 1879 Campaña del Desierto. The scientific literature on Patagonia during this time was rather tendentious; it was devoted to the cause of legitimizing territorial annexations by government troops. Moreno and his *Viaje a la Patagonia Austral* of 1879 are clear examples of this. In this text, Patagonia is shown as an empty, virgin space, full of possibilities, but at the same time a territory that may (and must) be occupied by people (Andermann 2000). The people inhabiting Patagonia before the conquest are represented as living remnants of an ancient world–fossilized humans virtually unchanged during centuries of isolation (Livon-Grosman 2003). Likewise, the desolate Patagonian geography, cut by ancient glacial valleys, with active volcanoes and thousand-year-old *araucarias,* presented a vivid image of a primitive planet.

It is not surprising that well into the twentieth century, it was accepted that creatures extinct on other parts of the planet inhabited Patagonia. The search for a giant sloth constitutes perhaps the most illustrative case of the scientific imagination of the epoch (Podgorny 1999). The beast was known only from fragments of skin and unreliable reports. The explorer Ramón Lista and some aborigines reported sighting the creature, and the well-regarded Florentino Ameghino went as far as naming the unknown beast *Neomylodon listai* (Ameghino 1921d). Santiago Roth, on his own, renamed the mysterious animal as *Jemmisch listai,* affirming that it was an unknown felid rather than a prehistoric relic (*jemmisch* means "tiger of the water" in the language of the *Tehuelches*) (Podgorny 1999). Apparently, Moreno, in spite of having contributed to the idea of a "prehistoric" Patagonia, was opposed to the existence of a living mylodontid. Likewise, British paleontologist Smith Woodward identified the pieces of skin attributed to the extant *Neomylodon* as belonging to the extinct genus *Mylodon* (Camacho 2000).

A similar case, though never supported by direct evidence, concerned the presence of a large aquatic reptile in the cold lakes of western Patagonia. In 1924, a local newspaper spread the report of a strange creature similar to a plesiosaur in the Epuyén Lake, in Chubut Province. As one could imagine, the story of a prehistoric living reptile in Patagonia made international headlines. Although the story was never taken seriously by scientists, Clemente Onelli, explorer of Patagonia and, by that time, director of the Buenos Aires Zoo, took advantage of popular enthusiasm to plan a scientific expedition to the area. But Onelli's true purpose for this trip was merely to acquire specimens for the zoo—and, incidentally, to search for the shy *Neomylodon.* Onelli designated engineer Emilio Frey, a former collaborator of Moreno during his days as *perito* (expert) in the border dispute with Chile (Tonni et al. 2003), as leader of the expedition.

The most famous "witness" to the existence of the incredible plesiosaur was the Texan sheriff Martin Sheffield, a colorful adventurer and prospector who was reported to be on the trail of Butch Cassidy and Sundance Kid in Patagonia. Sheffield, who had served as a guide to Onelli, had written a letter to him about the existence of the creature of Epuyén Lake (Gavirati 2001a).

Frey recounted that the atypical expedition was known to many during that time: "de todas partes me llovían cartas y obsequios entre los que había las cosas más notables: un tango 'el Plesiosaurio,' una caja de cigarrillos marca 'Plesiosaurio,' lápices hechos por los presos con la efigie del presunto monstruo [from everywhere rained letters and gifts, among which there was the most notable things: a 'tango' called "El Plesiosaurio," a box of cigarettes marked "Plesiosaurio," [and] pencils made by the prisoners with the image of the presumed monster]" (Gavirati 2001b:17).

Obviously, neither the living mylodontid nor the plesiosaur was spotted, nor was their existence demonstrated, but the idea of

Figure 1.10. The ichthyosaur Ancanamunia, *as reconstructed by C. Rusconi.* Taken from Rusconi (1967).

Patagonia as a modern Noah's Ark, preserving creatures elsewhere extinct, persisted up to the first decades of the twentieth century.

Florentino Ameghino and Santiago Roth's Legacy

After Carlos Ameghino replaced his brother Florentino as director of the Museum of Buenos Aires in 1911, he aimed to fill the void left by his brother by promoting a group of people, including Lucas Kraglievich, Carlos Rusconi and Lorenzo Parodi. Some years later, under the directorship of Martin Doello Jurado, that group dissolved. Kraglievich moved to Uruguay, and Rusconi, to Mendoza, in 1937, when he was appointed director of the Museum "Juan Cornelio Moyano," in the capital city of this province (Tonni and Pasquali 1999).

Rusconi's work on Mesozoic reptiles included important descriptions. Rusconi named many reptiles from Mendoza, including ichthyosaur *Ancanamunia* (Fernández this volume) (Fig. 1.10) and the crocodile "*Purranisaurus,*" which may be synonymous with *Metriorhynchus* (Pol and Gasparini this volume).

In 1923 the Museum of Buenos Aires was renamed Museo Argentino de Ciencias Naturales "Bernardino Rivadavia" in honor of the first Argentine president and founder of the Museo del País, the former Museo Público de Buenos Aires. After the departure of Rusconi and Kraglievich, the museum hired Noemí Violeta Cattoi,

Figure 1.11. Ángel Cabrera (1879–1960). With permission of Revista Museo.

who from 1948 until her death in 1965 was chief of the Division of Paleozoology of Vertebrates. Cattoi described the first remains of a Jurassic turtle (*Notoemys laticentralis;* Cattoi and Freiberg 1961) on the basis of an incomplete carapace found in a building of Acassuso (Buenos Aires Province), and a new species of *Podocnemis*—a turtle from the Cretaceous—in collaboration with Marcos A. Freiberg, a zoologist of the Museo Argentino de Ciencias Naturales (MACN) (de la Fuente this volume).

Meanwhile in La Plata, Torres had designated the Spanish naturalist Ángel Cabrera (Fig. 1.11) as chief of the Department of Paleontology of the Museum to replace Roth, who died in 1924. Because he lacked a degree in the sciences (he was a licentiate in philosophy and letters), Cabrera had professional problems at the Museum of Natural History of Madrid, where he was an "aggregated naturalist" (Casado 2001). Apparently Torres favored Cabrera because he wanted a Hispanic paleontologist to balance a Nordic and Italian majority (Bond 1998).

As chair, Cabrera described the remains of the first ichthyosaur from the Middle Jurassic of Patagonia (*"Stenopterygius" grandis,* Cabrera 1939; Fernández this volume), the first Jurassic dinosaur from South America (*Amygdalodon patagonicus,* Cabrera 1947; Salgado and Bonaparte this volume), and the first plesiosaur from Argentina (*Aristonectes parvidens,* Cabrera 1941; Gasparini this volume) (Fig. 1.12). The remains of the sauropod had been discovered in Pampa de Agnia, in Chubut, by Tomás Suero, a geologist with the Yacimientos Petrolíferos Fiscales. The first remains of a plesiosaur in South America had been described by Gay a century earlier in *Historia física y política de Chile.* These remains came from the Lirquén area, near Quriquina Island, in the VIII Chilean region.

Ángel Cabrera left his position at the museum and the Universidad de La Plata in 1947 for political reasons, but he remained in Argentina until his death in 1960. A replacement was not hired until 1957, when Rosendo Pascual (Fig. 1.13) became the first Argentinean professor of vertebrate paleontology at the University of La Plata. Pascual would study mostly Tertiary mammals, following Ameghino's and Kraglievich's tradition, but in 1973 he advised the first doctoral thesis on reptiles in the country, written by Zulma Gasparini.

A contemporary of Cabrera, Matilde Dolgopol de Sáez (1901–1957) emerged as an expert on fossil crocodiles (Anonymous 1957). In 1928, she described the unusual *Microsuchus schilleri,* discovered in Cretaceous strata near the confluence of the rivers Limay and Neuquén. Following Huene, Dolgopol de Sáez assigned this crocodile to the family "Goniopholidae" (Dolgopol de Sáez 1928). In 1957, the year of her death, she described a new species of *Notosuchus, Notosuchus lepidus* (Dolgopol de Sáez 1957), which years later would be shown to be synonymous with *Notosuchus terrestris* by Gasparini (1971).

In the meantime, in the Museum "Bernardino Rivadavia," Cat-

Figure 1.12. Aristonectes, *as reconstructed by Á. Cabrera.*

toi was reorganizing the group of vertebrate paleontologists with the incorporation of two ex-partners of the secondary school: Jorge L. Kraglievich (Lucas's son) and Osvaldo Reig (Báez 1992). Reig (Fig. 1.13) had been secretary of the Municipal Museum of Natural Sciences of Mar del Plata in Buenos Aires Province. Galileo Scaglia, son of that museum's founder, had recommended Reig for the post.

In 1958 Reig finally moved from the MACN to the Instituto Miguel Lillo, in Tucumán (northwestern Argentina). He described valuable new fossils, among which were the remains of what was then the oldest known dinosaur: *Herrerasaurus ischiagualastensis* (Reig 1963). After the 1966 military revolution, Reig, like many university professors and researchers, left the country for political reasons. They returned to the country in 1984 when democracy was restored.

Of the paleontologists mentioned in this chapter, Rodolfo M. Casamiquela (Figs. 1.13, 1.14) is the only one born and raised in Patagonia. In his native Ingeniero Jacobacci (Río Negro Province), Casamiquela began a museum that now houses numerous dinosaur and marine reptile remains that he—or people allied with him—found in the surrounding areas. The museum bears the name Jorge Gerhold, after the uncle who aroused his passion for the natural history of Patagonia.

At the beginning of the 1960s, as an adjunct researcher at the Museum of La Plata, Casamiquela brought to that institution some sandstone blocks extracted from a quarry in Los Menucos (south

Figure 1.13. From left to right: R. M. Casamiquela, J. F. Bonaparte, R. Pascual, and O. Reig. Photograph taken in 1968 in Caracas, Venezuela.

of the Río Negro Province). Their surfaces bore curious marks that Rosendo Pascual, now professor of vertebrate paleontology of the Facultad de Ciencias Naturales, identified as fossil footprints. Casamiquela decided to study the fossils, inaugurating a new field of research in Argentina.

The sandstones from Los Menucos are quarried for building material. Casamiquela's field trips were not only to the quarry but also to the places where sandstones with footprints had been discovered; he also removed slabs containing footprints from the walkways of many Rionegrian towns. (Footprints that were not collected can still be seen in those towns.) The sandstones were from the Upper Triassic, and bore an abundant variety of plant imprints of that age. Some time later, Casamiquela also discovered footprints from the Jurassic, in the locality of La Matilde in Santa Cruz Province (Boido and Chiozza 1988–1989; Calvo this volume).

In *Estudios icnológicos,* his most important contribution to paleoichnology (Casamiquela 1964a), Casamiquela gathered all the information on fossil footprints available up to that time (Calvo this volume). He described a variety of new ichnogenera of synapsids from the locality of Los Menucos (*Calibarichnus, Palaciosichnus, Rogerbaletichnus, Gallegosichnus*) and of pseudosuchians (*Shimmelia*). From Laguna Manantiales he described footprints of coelurosaurians (*Wildeichnus, Sarmientichnus, Delatorrichnus*). *Estudios icnológicos,* prepared by the Instituto de Investigaciones Científicas de Río Negro, in Viedma, is the first Patagonian publication dedicated to the study of the Mesozoic faunas of Patagonia.

Besides the Triassic footprints, Casamiquela also discovered the first osseous remains of reptiles of that age in Patagonia; the discovery took place in El Tranquilo, 400 km northwest of Puerto San Julián, in the province of Santa Cruz. These bones belonged to prosauropods (Salgado and Bonaparte this volume) and were collected during three expeditions—in 1960, 1963, and 1964—that were organized by the Museum of La Plata, la Fundación Miguel Lillo of Tucumán, and the Instituto de Investigaciones Científicas de Río Negro. Casamiquela gave only a preliminary report of these bones (Casamiquela 1964b) before leaving the Museum of La Plata. As director of the Centro de Investigaciones Científicas de Río Negro, in Viedma, he authored a second report (Casamiquela 1980), that expanded on the information presented in 1964.

Figure 1.14. Rodolfo M. Casamiquela (b. 1932). Reproduced with permission from *Ciencia Hoy.*

Casamiquela was responsible for the discovery of the first hadrosaur dinosaur from South America, collected in 1949 near Ingeniero Jacobacci (Casamiquela 1964c). At that time, among the ornithischians, only *Notoceratops bonarelli* and *Loricosaurus scutatus* were known. These were the first hadrosaur remains from the Southern Hemisphere, as Casamiquela recognized. The most important aspect of the discovery was the announcement that the fossil came from the Tertiary levels, which seemed to demonstrate that, at least in Patagonia, dinosaurs had survived to the K-T extinction. Later studies cast doubt on this assertion.

At the end of the 1960s, Zulma Gasparini, also at the Museum of La Plata, began studying South American crocodiles. Beginning in the 1970s, she started an intensive search for new marine reptile materials in Chile and Argentina. At that time, studies on marine reptiles were virtually nonexistent, with the exception of the description of *Aristonectes* (Cabrera 1941). Thirty years of continuous exploration and study of Jurassic and Cretaceous marine reptiles, and collaboration with interdisciplinary groups, have made Patagonia the reference point for these reptiles in the Southern Hemisphere. Gasparini, usually accompanied by colleagues and geologists from different institutions, explored the Jurassic rocks of the Neuquina Basin, where she and her team members discovered valuable Jurassic fossils that helped fill gaps in the paleobiogeographic history of Jurassic South America.

During the 1960s and 1970s, the most important advances in research on Patagonian Mesozoic reptiles were in the Triassic and the Jurassic. With respect to the Cretaceous, the contributions of Guillermo del Corro, of the MACN, should be mentioned. In 1965 del Corro collected the bones of a large dinosaur from the province of Chubut; this was the first time dynamite was used in a paleontological excursion in Argentina (del Corro 1975). Five teeth of a meat-eating dinosaur were unearthed, along with the remains of the enormous sauropod. Del Corro assigned the name *Chubutisaurus insignis* to the sauropod (del Corro 1975) and *Megalosaurus inexpectatus* (because of the unexpected finding) to the theropod (del Corro 1966).

A Time of Specialization

The creation of the Consejo Nacional de Investigaciones Científicas y Técnicas (CONICET) in 1958 and of a related research center in 1960 allowed scientists to dedicate themselves full time to research. The creation of CONICET coincided with the restoration of democracy in Argentina, after a military interlude that followed the government of Juan D. Perón. In spite of the difficult economic and political situation, Bernardo Houssay, Nobel Prize winner and founder of CONICET, campaigned for scientific development and the return of exiled scientists. As a result, many Argentinean paleontologists began their scientific careers as members of CONICET (as was the case for Casamiquela and Gasparini) or joined CONICET as established scientists (as did Rosendo Pascual, José Bonaparte, and Ana María Báez).

José Bonaparte (Figs. 1.13, 1.15) was the first Argentinean paleontologist to specialize in fossil reptiles (they constitute the majority of his scientific production—see, for example, Bonaparte 1979, 1986, 1991, 1996). The study of Patagonian Mesozoic reptiles can be divided into two periods: before and after Bonaparte. Until Bonaparte, these studies were rather sporadic and were largely undertaken by nonspecialists, mostly paleomastozoologists. With the arrival of Bonaparte, these studies became systematic.

Bonaparte's life had been linked to paleontology since his youth. In the city of Mercedes, in the province of Buenos Aires, he and others founded the Museo Popular de Ciencias Naturales "Carlos Ameghino." In 1959 he moved to the Instituto Miguel Lillo at the Universidad Nacional de Tucumán, where he worked with Osvaldo Reig and Galileo Scaglia. Three years later, he took a position in the institute's department of vertebrate paleontology. In 1974, he received the honorific title of doctor. He then returned to the MACN as chief of the department of vertebrate paleontology, from which post he continued his expeditions to Patagonia.

Although most of his expeditions were partially supported with funds from CONICET, the National Geographic Society played a crucial role in the continuity of his research projects in Patagonia. Before CONICET was created, official museums and universities were the primary supporters of paleontological research. It was also common for geologists of the Yacimientos Petrolíferos Fiscales to bring material there for identification, for the purpose of completing the biostratigraphical information for those areas. In 1949 geologists Ferrello and Flores delivered a

Figure 1.15. José F. Bonaparte (b. 1928). Photograph courtesy of Carlos Papolio.

carnosaur tooth to the MACN; years later, del Corro would name the carnosaur *Megalosaurus chubutensis* (del Corro 1974). After the creation of CONICET, this organization did much to sustain paleontological expeditions (for instance, the expeditions of Casamiquela to El Tranquilo). But due to Argentina's acute economic instability, National Geographic grants to Bonaparte and Gasparini were crucial in maintaining projects that were beginning to bear fruit.

To discuss all Bonaparte's contributions to the study of Patagonian Mesozoic reptiles is beyond the scope of this book. All of the current paleontologists who focus on Mesozoic reptiles were trained by Bonaparte in some way, and it can be said that much of the information contained in this volume is the result of investigations continued by his students. Bonaparte's work embraces nearly all groups of reptiles, although the studies on dinosaurs are among the best known. Moreover, materials obtained during field trips organized by Bonaparte were studied by other specialists, further extending Bonaparte's contributions to the field of Mesozoic reptiles. Among these materials is the complete skeleton of an adult ichthyosaur, *Caypullisaurus bonapartei* (Fernández this volume).

Bonaparte's activities transcended the boundaries of pure research. Through the years, he mounted several casts of Patagonian dinosaur skeletons he had discovered and studied (including *Amargasaurus, Patagosaurus, Piatnitzkysaurus, Carnotaurus,* and *"Kritosaurus" australis*), a practice that in Argentina had only one antecedent: the skeleton (in fact a mix of casts of gypsum and original pieces) of *"Titanosaurus" australis* mounted by Á. Cabrera and Antonio Castro, a technician of the Museum of La Plata (Bondesio 1977). Bonaparte was thus the first to present these enormous beasts to the public, a task that he completed through the publication of books on paleontology (e.g., Bonaparte 1978) and sustained support for the settlement of young paleontologists in the Argentinean provinces, particularly in Patagonia.

Conclusion

The study of Patagonian Mesozoic reptiles has a relatively short history. For decades, Patagonian paleontology was oriented to the study of the evolution of Tertiary mammals, following the tradition of Burmeister and Ameghino; only recently has knowledge of the diverse forms of Mesozoic reptiles significantly increased. Of course, this increase is not uniform across the field, in part because of differences in the recording conditions and in the number of specialists dedicated to each group of Mesozoic reptiles. However, it is clear that much more is known today than 25 years ago.

Studies of dinosaurs are abundant in relation to other groups. This is undoubtedly because the size of the dinosaur bones caught the attention of collectors at the end of the nineteenth century, and also because of increasing popular interest in this group of reptiles. Other large reptiles, equally well represented, such as ichthyosaurs and other Jurassic marine reptiles, are only now receiving attention. These fossils are usually found in uninhabited areas, close to the Andes, and in rocks that are extremely hard and difficult to prepare. Today the number of professionals focusing on Patagonian Mesozoic reptiles is growing, thanks to the interest promoted by people like Bonaparte and other specialists.

Acknowledgments. Thanks to Eduardo Tonni, Mariano Bond, Jeff Wilson, Kristi Curry, and Pedro Navarro Floria for their comments. Prebiterio Pacheco provided me with many of the illustrations. This work was accomplished within Project 04 H-082 from the Facultad de Humanidades, Universidad Nacional del Comahue, Argentina.

References Cited

Ameghino, F. 1921a. Sinopsis geológico paleontológica de la Argentina. In A. J. Torcelli (comp.), *Obras Completas y Correspondencia Científica,* 12: 485–734. La Plata: Taller de Impresiones Oficiales.

———. 1921b. Notas sobre cuestiones de geología y paleontología argentinas. In A. J. Torcelli (comp.), *Obras Completas y Correspondencia Científica,* 12: 5–34. La Plata: Taller de Impresiones Oficiales.

———. 1921c. Nota preliminar sobre el "*Loncosaurus argentinus.*" In A. J. Torcelli (comp.), *Obras Completas y Correspondencia Científica,* 12: 751–752. La Plata: Taller de Impresiones Oficiales.

———. 1921d. Un sobreviviente actual de los megaterios de la antigua pampa. In A. J. Torcelli (comp.), *Obras Completas y Correspondencia Científica,* 12: 755–760. La Plata: Taller de Impresiones Oficiales.

———. 1934. Las Formaciones sedimentarias del Cretácico Superior y del Terciario de Patagonia, con un paralelo entre sus faunas mastológicas y las del antiguo continente. In A. J. Torcelli (comp.), *Obras Completas y Correspondencia Científica,* 16: 5–747. La Plata: Taller de Impresiones Oficiales.

Andermann, J. 2000. *Mapas de Poder. Una Arqueología Literaria del Espacio Argentino.* Rosario: Beatriz Viterbo.

Anonymous. 1957. Mathilde Dolgopol de Sáez, 1901–1957. *Ameghiniana* 1 (3): 44.

Asúa, M. de. 1989. El apoyo oficial a la "Description Physique de la République Argentine" de H. Burmeister. *Quipu* 6 (3): 339–353.

Babini, J. 1949. *Historia de la Ciencia Argentina.* México: Fondo de Cultura Económica.

Báez, A. M. 1992. Necrológica. Dr. Osvaldo Alfredo Reig. 1929–1992. *Ameghiniana* 29 (2): 191–192.

Boido, G., and E. Chiozza. 1988–1989. Rodolfo M. Casamiquela. El camino de la fascinación. *Ciencia Hoy* 1 (1): 54–61.

Bonaparte, J. F. 1978. *El Mesozoico de América del Sur y sus Tetrápodos.* Opera Lilloana 26. Tucumán.

———. 1979. Dinosaurs: a Jurassic assemblage from Patagonia. *Science* 205: 1377–1378.

———. 1986. History of the terrestrial Cretaceous vertebrates of Gondwana. *Actas del IV Congreso Argentino de Paleontología y Bioestratigrafía* 2: 63–95.

———. 1991. Los vertebrados fósiles de la Formación Río Colorado, de la ciudad de Neuquén y sus cercanías, Cretácico Superior, Argentina. *Revista del Museo Argentino de Ciencias Naturales "Bernardino Rivadavia," Paleontología* 4: 17–123.

———. 1996. Cretaceous tetrapods of Argentina. In G. Arratia (ed.), Contributions of Southern South America to Vertebrate Paleontology. *Münchner Geowissenschaftliche Abhandlungen* (A) 20: 73–130.

Bond, M. 1998. Ángel Cabrera. *Museo* 2 (11): 17–24.

———. 2000. Carlos Ameghino y su obra edita. In S. Vizcaíno (ed.), *Simposio Obra de los Hermanos Ameghino,* 33–41. Luján: Publicación Especial de la Universidad Nacional de Luján.

Bondesio, P. 1977. Cien años de paleontología en el Museo de La Plata. *Obra del Centenario del Museo de La Plata* 1: 75–87.

Brett-Surman, M. K. 1979. Phylogeny and palaeobiogeography of hadrosaurian dinosaurs. *Nature* 277 (5697): 560–562.

Burmeister, H. 1879. *Description Physique de la République Argentine.* II. París: Libraire F. Sabih.

Cabrera, Á. 1939. Sobre un nuevo ictiosaurio del Neuquén. *Notas del Museo de La Plata* 4: 485–491.

———. 1941. Un plesiosaurio nuevo del Cretáceo del Chubut. *Revista del Museo de La Plata* (nueva serie) 2 (8): 113–130.

———. 1944. *El Pensamiento Vivo de Ameghino.* Buenos Aires: Losada.

———. 1947. Un saurópodo nuevo del Jurásico de Patagonia. *Notas del Museo de La Plata* 12 (95): 1–17.

Calvo, J. O., and L. Salgado. 1995. *Rebbachisaurus tessonei,* a new sauropod of the Albian-Cenomanian of Argentina, new evidence on the origin of the Diplodocidae. *Gaia* 11: 13–33.

Camacho, H. 2000. Francisco P. Moreno y su contribución al conocimiento geológico de la Patagonia. *Saber y Tiempo* 9: 5–32.

———. 2001. Las ciencias geológicas en la Argentina, hasta 1939. *Saber y Tiempo* 12: 177–220.

Casado, S. 2001. En la cueva de la Madgalena con Ángel Cabrera. *Quercus* 187: 26–32.

Casamiquela, R. M. 1964a. *Estudios Icnológicos. Problemas y Métodos de la Icnología con Aplicación al Estudio de Pisadas Mesozoicas (Reptilia, Mammalia) de la Patagonia.* Viedma: Ministerio de Asuntos Sociales de la Provincia de Río Negro.

———. 1964b. Sobre el hallazgo de dinosaurios triásicos en la Provincia de Santa Cruz. *Argentina Austral* 35: 10–11.

———. 1964c. Sobre un dinosaurio hadrosáurido de la Argentina. *Ameghiniana* 3 (9): 285–312.

———. 1980. La presencia del género *Plateosaurus* (Prosauropoda) en el Triásico Superior de la Formación El Tranquilo, Patagonia. *Actas del II Congreso Argentino de Paleontología y Bioestratigrafía y I Congreso Latinoamericano de Paleontología* 1: 143–158.

Cattoi, N., and M. Freiberg. 1961. Nuevo hallazgo de chelonia extinguidos en la República Argentina. *Physis* 22 (63): 202.

Colbert, E. H. 1984. *The Great Dinosaur Hunters and their Discoveries.* New York: Dover Publications.

Coria, R. A. and L. Salgado. 2000. Los dinosaurios de Ameghino. In S. Vizcaíno (ed.), *Simposio Obra de los Hermanos Ameghino,* 43–49. Luján: Publicación Especial de la Universidad Nacional de Luján.

Del Corro, G. 1966. Un nuevo dinosaurio carnívoro del Chubut (Argentina). *Comunicaciones del Museo Argentino de Ciencias Naturales "Bernardino Rivadavia"* I, 1: 1–4.

———. 1974. Un nuevo megalosaurio (Carnosaurio) del Cretácico de Chubut (Argentina). *Comunicaciones del Museo Argentino de Ciencias Naturales "Bernardino Rivadavia"* I, 5: 37–44.

———. 1975. Un nuevo saurópodo del Cretácico Superior. *Chubutisaurus insignis* gen. et sp. nov. (Saurischia-Chubutisauridae nov.) del Cretácico Superior (Chubutiano), Chubut, Argentina. *Actas del I Congreso Argentino de Paleontología y Bioestratigrafía.* Tucumán, Argentina. Agosto de 1974. II: 229–240.

Dembicz, A. 2002. Puentes académicos entre Polonia y Chile: experiencias, potencialidades, visiones. *Anales de la Universidad de Chile* 14 (sexta serie). Available at http: //www2.anales.uchile.cl/CDA/an_simple/0,1278,SCID%253D4133%2526ISID%253D271%2526ACT%253D0%2526PRT%253D4108,00.html. Accessed May 22, 2006.

Doering, A. 1882. *Informe Oficial de la Comisión Científica Agregada al Estado Mayor General de la Expedición al Río Negro (Patagonia),* entrega III (Geología). Bs. As.

Dolgopol de Sáez, M. 1928. Un goniofólido argentino. *Anales de la Sociedad Científica Argentina* 106: 287–290.

———. 1957. Cocodrilideos fósiles argentinos. Un nuevo cocodrilo del Mesozoico Argentino. *Ameghiniana* 1 (1–2): 48–50.

García Castellanos, T. 2004. *Sarmiento. Su Influencia en Córdoba.* Córdoba: Academia Nacional de Ciencias.

Gasparini, Z. 1971. Los Notosuchia del Cretácico de América del Sur

como un nuevo infraorden de los Mesosuchia (Crocodilia). *Ameghiniana* 8 (1): 83–103.
Gavirati, M. 2001a. Martín Sheffield: El Gaucho Yanqui. *La Bitácora Patagónica* 16: 22–23.
———. 2001b. Clemente Onelli: Aventurero, explorador, literato, buscador de fósiles y del mítico plesiosaurio. *La Bitácora Patagónica* 15: 16–17.
González Arrili, B. 1954. *Vida de Ameghino*. Buenos Aires: Castellví S. A.
Hatcher, J. B. 1903. Osteology of *Haplocanthosaurus,* with a description of a new species, and remarks on the probable habits of the Sauropoda and the age and origin of the *Atlantosaurus* beds. *Memoirs of the Carnegie Museum* 2 (1): 1–75.
———. 1985. Bone Hunters in Patagonia. Narrative of the Expedition. Connecticut: Ox Bow Press.
Huene, F. v. 1929. Los Saurisquios y Ornitisquios del Cretáceo Argentino. *Anales del Museo de La Plata* vol. 3: 1–194.
Keidel, J. 1917. Über das patagonische tafelland und ihre ziehungen zu den geologischen ercheinnugen in den Argentinischen Anden gebiet und litoral. *Zeitschrift der Deutsche Akademie Wiessenschaft* 3 (5–6): 219–245. Stuttgart.
Livon-Grosman, E. 2003. *Geografías Imaginarias. El Relato de Viaje y la Construcción del Espacio Patagónico*. Rosario: Beatriz Viturbe.
Lopes, M. M. 2001. Viajando pelo campo e pelas coleções: aspectos de uma controvérsia paleontológica. *História, Ciências, Saúde-Manguinhos,* vol. VIII (suplemento): 881–897.
Lopes, M. M., and I. Podgorny. 2000. The shaping of Latin American museums of Natural History, 1850–1990. *Osiris* 15: 1301–1311.
Lydekker, R. 1877. Notices of new and other Vertebrata from Indian Tertiary and Secondary rocks. *Records of the Geological Survey of India* 10: 30–43.
———. 1888. Catalogue of fossil Reptilia and Amphibia in the British Museum. Pt. 1. Containing the orders Ornithosauria, Crocodilia, Dinosauria, Squamata, Rhynchocephalia, and Proterosauria. *British Museum of Natural History.* London.
———. 1893. Contributions to the study of the fossil vertebrates of Argentina. I. The Dinosaurs of Patagonia. *Anales del Museo de La Plata* 2: 1–14.
Nicoletti, M. A., and P. Navarro Floria. 2000. *Confluencias. Una Breve Historia del Neuquén.* Buenos Aires: Editorial Dunken.
Nopcsa, F. 1902. Notizen über cretacische Dinosaurier. 3. Wirbel eines südamerikanischen Sauropoden. *Sitzungsberichte der Kaiserlichen Akademie der Wissenschaften* 111: 108–114.
Onna, A. F. 2000. Estrategias de visualización y legitimación de los primeros paleontólogos en el Río de la Plata durante la primera mitad del siglo XIX: Francisco Javier Muñiz y Teodoro Miguel Vilardebó. In M. Montserrat (comp.), *La Ciencia en la Argentina entre siglos,* 53–70. Buenos Aires: Manantial.
Podgorny, I. 1999. La Patagonia como santuario natural de la ciencia finisecular. *Redes* 7 (14): 157–176.
———. 2004. El Entierro de un perro. *TodaVía* 7: 28–32.
Podgorny, I., and T. Plöger. 1999. El largo viaje al Plata del *Diplodocus carnegii. Ciencia Hoy* 9 (51): 50–55.
Powell, J. E. 1980. Sobre la presencia de una armadura dérmica en algunos dinosaurios titanosáuridos. *Acta Geológica Lilloana* 15: 41–47.

———. 2003. Revision of South American titanosaurid dinosaurs: paleobiological, paleobiogeographical and phylogenetic aspects. *Records of the Queen Victoria Museum* 111: 1–173.

Reig, O. A. 1963. La presencia de dinosaurios saurisquios en los "Estratos de Ischiagualasto" (Mesotriásico Superior) de las provincias de San Juan y La Rioja (República Argentina). *Ameghiniana* 3: 3–20.

Romero, J. L. 2000. *Breve Historia de la Argentina.* Buenos Aires: Fondo de Cultura Económica.

Rusconi, C. 1967. Animales extinguidos de Mendoza y de la República Argentina. Edición Oficial.

Tapia, A. 1918. Una mandíbula de dinosaurio procedente de Patagonia. *Physis* 4: 369–370.

Tognetti, L. 2000. La introducción de la investigación científica en Córdoba a fines del siglo XIX: La Academia Nacional de Ciencias y la Facultad de Ciencias Físico-Matemáticas (1868–1878). In M. Montserrat (comp.), *La Ciencia en la Argentina entre Siglos,* 345–365. Buenos Aires: Manantial.

———. 2004. *La Academia Nacional de Ciencias en el siglo XIX.* Córdoba: Academia Nacional de Ciencias.

Tonni, E. P., and R. C. Pasquali, 1999. El estudio de los mamíferos fósiles en la Argentina. *Ciencia Hoy* 9 (53): 22–31.

Tonni, E. P., M. Bond, and R. C. Pasquali. 2003. El monstruo, el noble, el sheriff y la curiosa historia de una expedición a los lagos del Sur. *Museo* 3 (17): 49–54.

Wichmann, R. 1918. Sobre la constitución geológica del territorio del Río Negro. *1era. Reunión Nacional de la Sociedad Argentina de Ciencias Naturales,* 90–97.

———. 1919. Contribución a la geología de la región comprendida entre el Río Negro y Arroyo Valcheta. *Anales del Ministerio de Agricultura de la Nación. Dirección General de Minas, Geología e Hidrología* XIII (4): 1–45.

Wilson, J. A., and P. Upchurch. 2003. A revision of *Titanosaurus* Lydekker (Dinosauria-Sauropoda), the first dinosaur genus with a "Gondwanan" distribution. *Journal of Systematic Palaeontology* 1 (3): 125–160.

Windhausen, A. 1931. *Geología Argentina.* II. Buenos Aires: Peuser.

Woodward, A. S. 1896. On two Mesozoic crocodilians, *Notosuchus* (genus novum) and *Cynodontosuchus* (gen. nov.) from the sandstone of the territory of Neuquén (Argentina). *Anales del Museo de La Plata* 41–20.

———. 1901. On some extinct reptiles from Patagonia of the genera *Miolania, Dinilysia* and *Genyodectes. Proceedings of the Zoological Society of London* 1901: 169–184.

2. Mesozoic Paleogeography and Paleoenvironmental Evolution of Patagonia (Southern South America)

Luis A. Spalletti and
Juan R. Franzese

Introduction

Patagonia and the southern Pampas (south of the 33°S latitude) extend for more than 2500 km and average 1000 km in width (Fig. 2.1). During Mesozoic times, this region was the scene of paleogeographic changes triggered by different tectonic processes: intracontinental strike-slip deformation along transcrustal megashears, a largely convergent proto-Pacific margin along the west side of South America, and the impact of a series of hot spots or mantle plumes that impinged on or between the cratonic regions and resulted in the Gondwana breakup.

As a result of all these structural and major tectonic controls, several sedimentary basins were formed during the Mesozoic. The main groups of sedimentary basins resulting from these processes in southern South America are isolated Triassic rifts and pull-apart basins, Jurassic-Cretaceous extensional and back-arc basins, Cretaceous rifts (mainly located along the eastern or Atlantic margin of Argentina), and foreland basins developed along the Andes foothills.

The aim of this chapter is to present the paleogeographic evo-

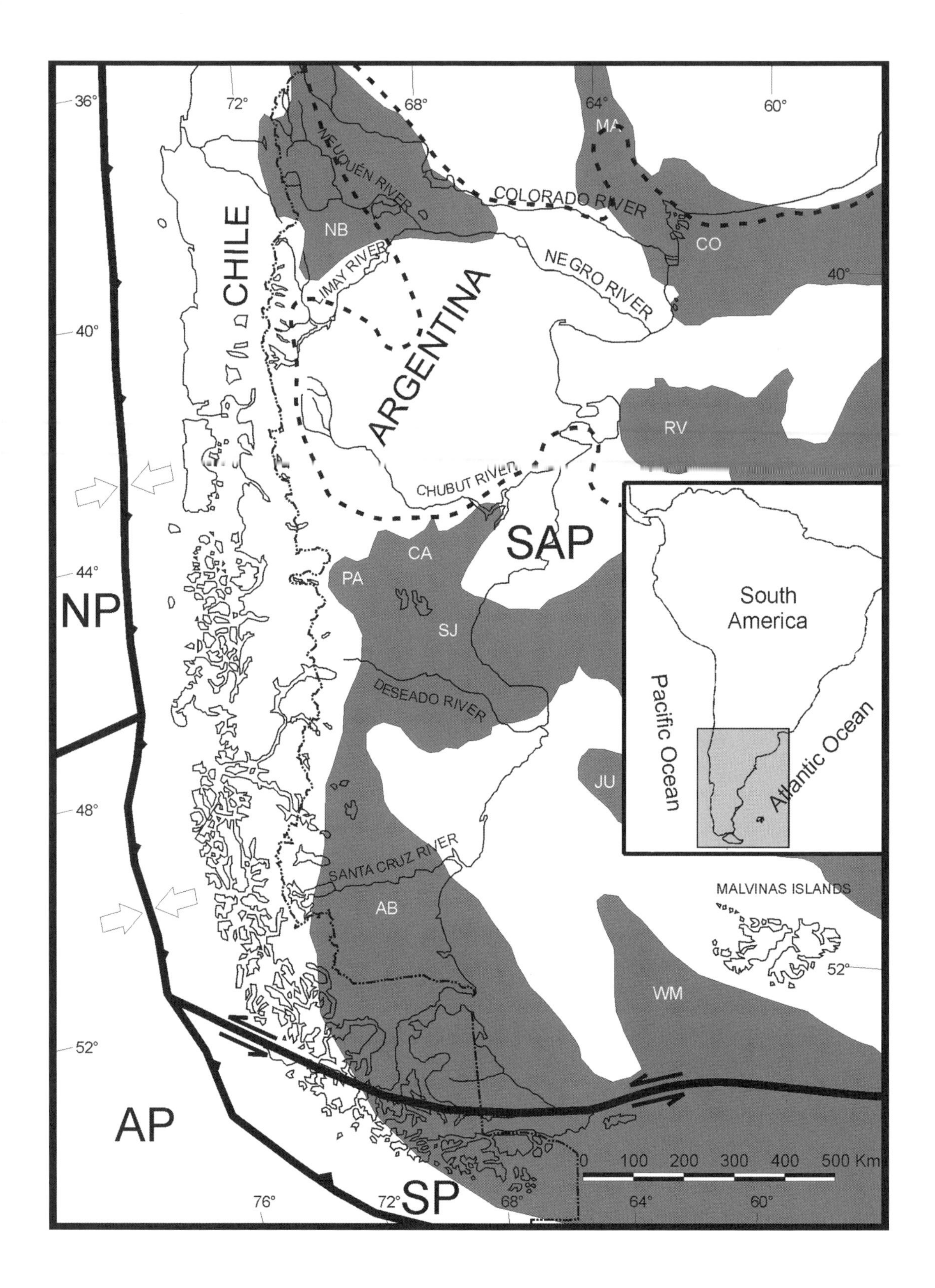
36°
72°
68°
64°
60°
MA
NEUQUÉN RIVER
COLORADO RIVER
NB
CO
CHILE
LIMAY RIVER
NEGRO RIVER
ARGENTINA
40°
40°
RV
CHUBUT RIVER
SAP
CA
PA
44°
NP
SJ
South
America
Pacific Ocean
Atlantic Ocean
DESEADO RIVER
JU
48°
SANTA CRUZ RIVER
MALVINAS ISLANDS
AB
52°
WM
52°
AP
0 100 200 300 400 500 Km
76°
72°
SP
68°
64°
60°

lution of Patagonia and the main depositional systems of its sedimentary basins through a series of maps plotted on a palinspastic plate reconstruction. These maps start in the Middle Triassic and continue in 15-Ma steps through the Mesozoic.

Methods

Different partial paleogeographic reconstructions of southern South America have been presented in previous works by Harrington (1962), Camacho (1967), Riccardi (1987), Uliana and Biddle (1987, 1988), Macellari (1988), Light et al. (1993), Urien et al. (1995), Urien and Zambrano (1996), and Jacques (2003a,b). Despite this, in order to accomplish a better understanding of basin evolution and to achieve the closest fit of the original continental plates, new palinspastic plate reconstruction will be required. In the present case, the base maps were developed by Lawver et al. (1999) and Dalziel et al. (2000).

To achieve a tight-fit reconstruction, South America was split into different tectonic plates along northwest-southeast lineaments and was subdivided into four rigid continental plates (Fig. 2.2), as follows: (1) a north-central Paraná plate, bounded to the south by the Colorado-Huincul fault system (Chernicoff and Zappettini 2003; Ramos et al. 2004); (2) a northern Patagonia plate, bounded to the south by the Gastre fault system (Rapela and Pankhurst 1992); (3) a central Patagonia plate, bounded to the south by the Deseado fault system (Ramos 1996, 2002; Chernicoff and Zappettini 2003); and (4) a southern Patagonia plate (Dalziel et al. 2000). The existence of these major lineations along northwest–southeast fault systems in South America is evident in geological maps of a continental scale (Rapela and Pankhurst 1992; Scotese et al. 1994, 1999; Lawver et al. 1999; Jacques 2003b). Geological fieldwork suggests continental strike-slip faulting during the Late Triassic and Early Jurassic to reconstruct southern South America into its present-day shape. In our pre-195-Ma palinspastic reconstructions, the southern plates were displaced east relative to the northern plates. Before 180 Ma, dextral movement along the Gastre fault and the Deseado fault systems (Rapela et al. 2005) brought the three segments of South America together in their present-day orientation.

These lineaments were later important controls in the orientation of the Atlantic marginal basins of Argentina. Furthermore, reconstruction of the Cretaceous north to south component of extension along these faults solved the problem of the excessive length of the Atlantic margin of South America compared with the West African margin.

The block of crust north of the Río Colorado–Huincul structure (Fig. 2.2) has been rotated to the north along its western edge in order to close the zone of underlap and to distribute extension across a wider area. The southern block (south of the San Jorge Gulf) has been displaced to the west along its southern edge to take

Figure 2.1. (opposite page) Present-day plate tectonic setting of southern South America and sketch map showing the main geographical features of Patagonia. AB = Austral Basin; AP = Antarctic Plate; CA = Cañadón Asfalto Basin; CO = Colorado Basin; JU = San Julián Basin; MA = Macachín Basin; NB = Neuquén Basin; NP = Nazca Plate; PA = Pampa de Agnia Basin; RV = Rawson-Valdés Basin; SAP = South American Plate; SJ = San Jorge Basin; SP = Scotia Plate; WM = Western Malvinas Basin. Dashed lines indicate the extension of the North Patagonian Platform.

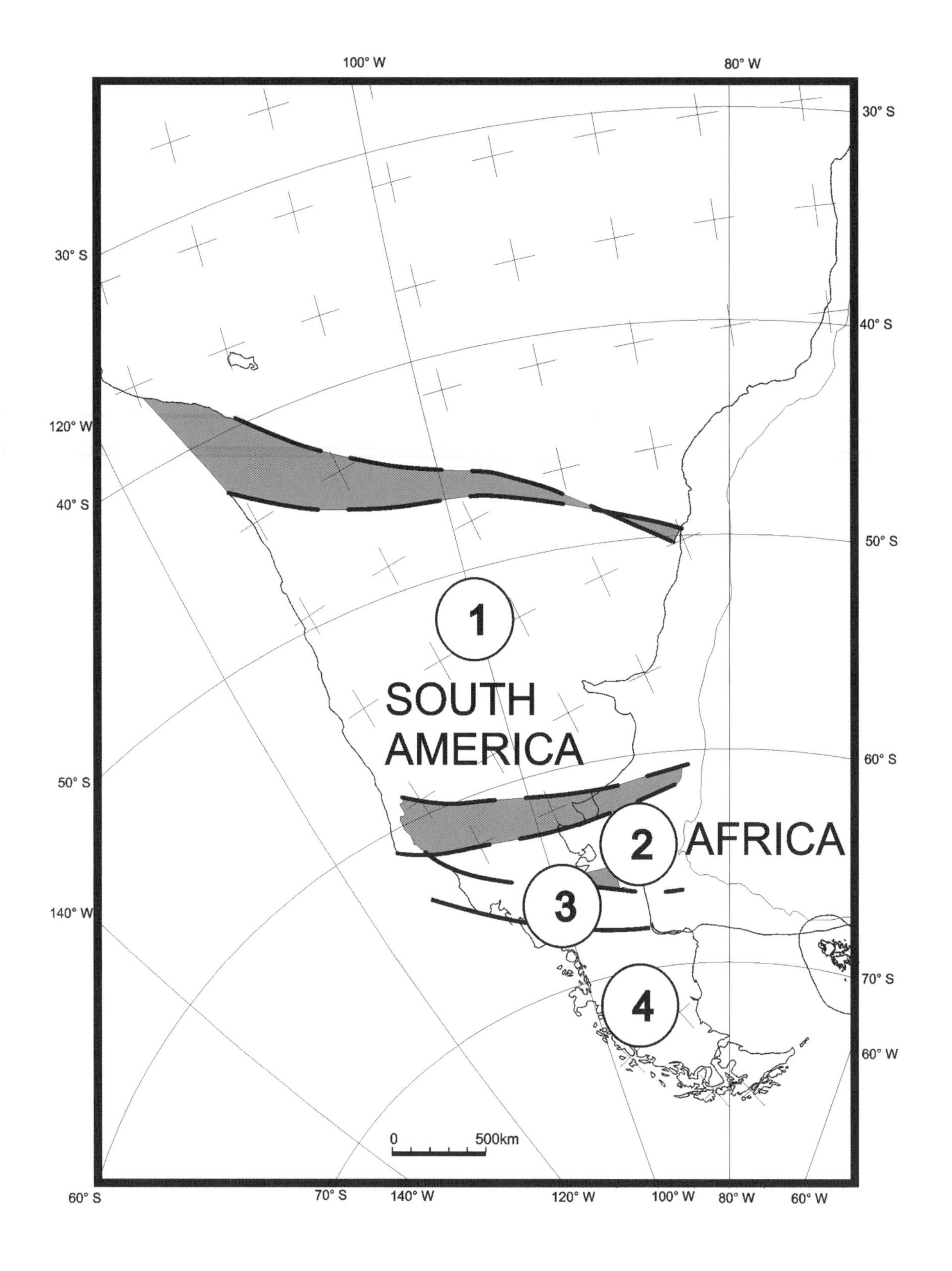
100° W
80° W
30° S
30° S
40° S
120° W
40° S
50° S
1
SOUTH
AMERICA
60° S
50° S
2
AFRICA
3
140° W
70° S
4
60° W
0
500km
60° S
70° S
140° W
120° W
100° W
80° W
60° W

out Late Cretaceous and Tertiary strike-slip movement. Scotia was restored along the transforms by several rectilinear displacements along strike slips and by slight desrotation of neighbor terranes (northern Antarctic Peninsula, and southernmost Patagonia, Tierra del Fuego, Estados, and South Georgia Islands). Every displacement of terranes for these new base maps was made in order to avoid significant changes in paleolatitudinal positioning.

Afterward, the facies and paleoenvironmental information was incorporated on the plate-tectonic reconstruction. We made final adjustments between geophysical reconstructions and geological information, regional correlations, and paleoenvironmental sections. Yet the resulting paleogeographic maps presented here do not contain all of the original data points, but our paleoenvironmental interpretation of them.

The reconstructions of the Patagonian paleogeography consist of 12 maps in 15-Ma steps through the Mesozoic. The following time slices are represented in the reconstructions: (1) Anisian-Ladinian (Middle Triassic, 240 Ma), (2) Carnian–early Norian (Late Triassic, 225 Ma), (3) late Norian–Rhaetian (Late Triassic, 210 Ma), (4) Sinemurian-Pliesbachian (Early Jurassic, 195 Ma), (5) Toarcian-Aalenian (Early–Middle Jurassic, 180 Ma), (6) Bathonian-Callovian (Middle Jurassic, 165 Ma), (7) Kimmerigian-Tithonian (Late Jurassic, 150 Ma), (8) Valanginian-Hauterivian (Early Cretaceous, 135 Ma), (9) Aptian (Early Cretaceous, 120 Ma), (10) Albian (Early Cretaceous, 105 Ma), (11) Cenomanian-Turonian (Late Cretaceous, 90 Ma), and (12) Campanian-Maastrichtian (Late Cretaceous, 75 Ma).

Chronological Review

Anisian-Ladinian (240 Ma) (Fig. 2.3A). Rifting occurred in west-central Argentina and resulted in extensional bimodal volcanism (late stage of the Choiyoi Group) dated around 239–240 Ma, and opening of a series of narrow rifts, such as the San Rafael depocenter, where fluvial-dominated deposits are associated with acidic pyroclastic flow deposits (Puesto Viejo Formation).

As shown in Fig. 2.3A, large areas of Patagonia remained as interbasinal highlands. In northern Patagonia, the intrusion of Gondwanan batholiths (Calvo Granite, Dos Lomas Complex of La Esperanza bimodal magmatism, dated 240 Ma) reveals a new pulse of magmatic activity with alkalic affinities. A subduction system along southern Chile was operating, with accretion and deformation of the Tarlton Limestone and the Denaro complex.

Carnian–early Norian (225 Ma) (Fig. 2.3B). Coeval strike-slip movements at the Gondwana paleo-Pacific margin led to the opening of several rifts (El Quereo–Los Molles and Curepto–Quitacoya Basins) in central Chile, characterized by fluvial and shallow marine facies.

The Los Menucos depocenter in north-central Patagonia was

Figure 2.2. (opposite page) South American tectonic plates. The region was subdivided into four rigid continental plates: (1) north-central Paraná plate, bounded to the south by the Colorado-Huincul fault system; (2) northern Patagonia plate, bounded to the south by the Gastre fault system; (3) central Patagonia plate, bounded to the south by the Deseado fault system; and (4) southern Patagonia plate.

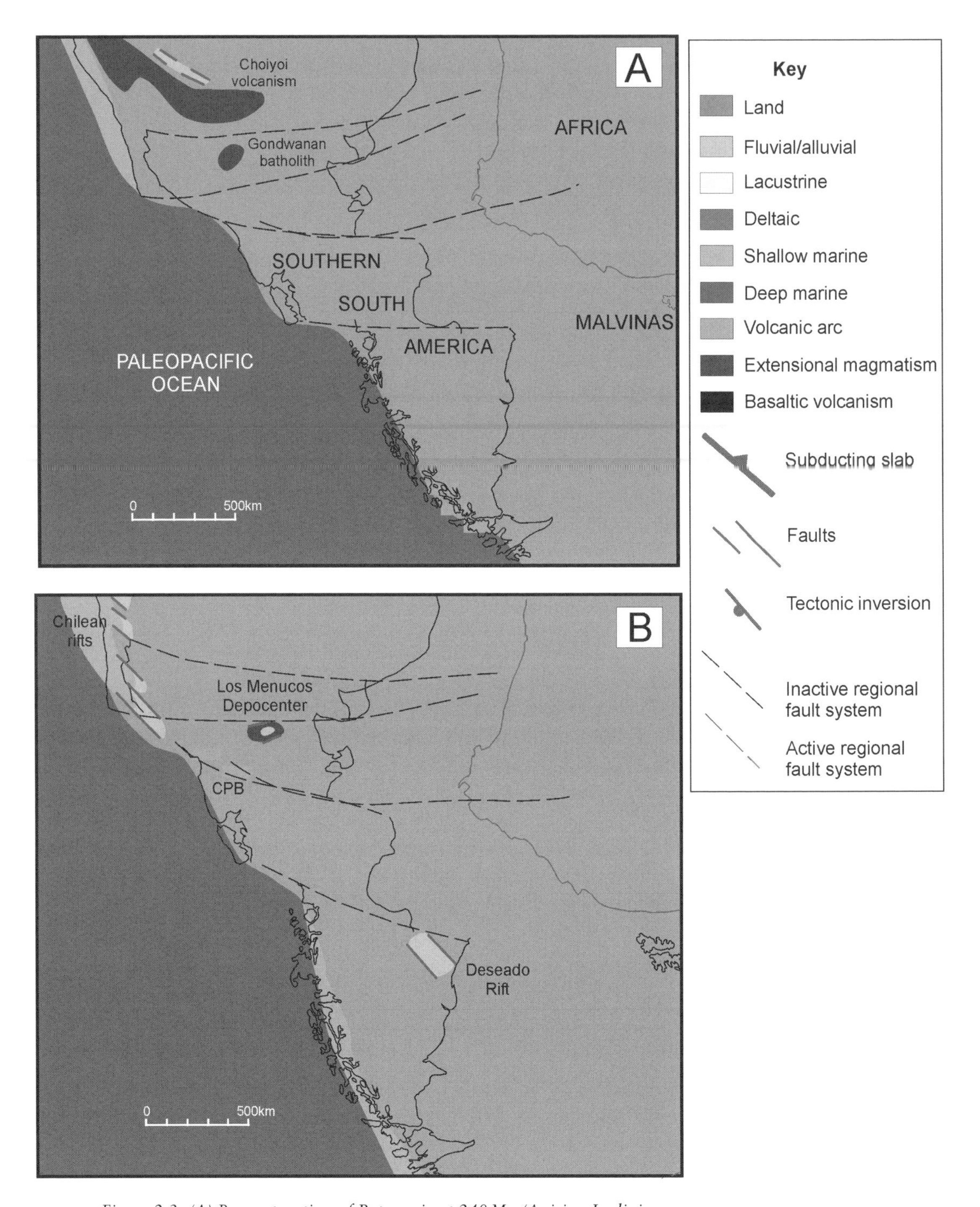

Figure 2.3. (A) Reconstruction of Patagonia at 240 Ma (Anisian-Ladinian, Middle Triassic). (B) Reconstruction of Patagonia at 225 Ma (Carnian–early Norian, Late Triassic). CPB = Central Patagonia batholith. (See Plate 9 in the color section.)

dominated by acidic pyroclastic flow deposits and stream-reworked pyroclastic rocks. The age of these rocks was defined by fossil plants and by radiometric data (223 Ma).

In southern Patagonia, the Deseado Rift is a narrow depocenter with a record of fluvial deposits showing paleocurrent directions to the south and southwest (El Tranquilo Group). An early Late Triassic age for the Deseado Rift is defined by a well-documented paleoflora.

It is likely that the Patagonian western margin was active by these times, although the extent of active volcanism is uncertain because of later overprinting. Subduction operated at least in two areas along the western margin of Gondwana: west-central Patagonia and southern Chile. In west-central Patagonia, the first evidence of calk-alkaline magmatism occurred at about 220 Ma (Central Patagonia batholith; dated 220–207 Ma). The present-day orientation of this batholith is oblique to the continental margin and suggests a very different configuration of the subduction system. Along the continental margin of southern Chile, a subduction-related deep marine succession started (Duque de York Flysch).

Late Norian–Rhaetian (210 Ma) (Fig. 2.4A). In the Late Triassic, the region adjacent to the proto-Pacific margin of Gondwana between 30° and 40°S was subjected to continental extension. As a result, several half-grabens were opened in western Argentina and Chile (e.g., Llantenes-Atuel, Paso Flores–Lapa, Panguipulli, Gomero). The sediments are coarse-grained fluvial deposits (alluvial fan, gravelly braided systems) and local lacustrine deposits. Coeval pyroclastic flow deposits are commonly intercalated. Most of these sediments bear abundant Upper Triassic fossil plants.

For the Late Triassic times, several areas of plutonic rock emplacement were defined in Patagonia: to the north granites of Chasicó-Mencué (210 Ma), slightly south the Gastre-Lipetrén intrusives, and far south the Triassic (202 Ma) La Leona Granite. The Gastre-Lipetrén granites are part of the Central Patagonia batholith. They are characterized by calc-alkaline compositions and have been related to oblique subduction of the proto-Pacific plate beyond western Gondwana.

Convergent tectonic activity along the western Gondwana margin is evidenced by the geochemical signature of the Central Patagonia batholith as well as by the Madre de Dios–Chonos–Chiloé subduction complex. Turbiditic trench deposition is preserved in the Duque de York Flysch, dated 234–195 Ma.

Sinemurian-Pliensbachian (195 Ma) (Fig. 2.4B). Continental extension continued along the northern proto-Pacific margin of Patagonia. The most important expression of this was the development of a system of half-grabens infilled by volcaniclastic and pyroclastic deposits associated with lava flows and plutonic intrusions. Coeval magmatism has a bimodal signature consistent with the interpretation of an extensional tectonic regime. The depocenters located to the west (e.g., Curepto and Lonquimay in Chile,

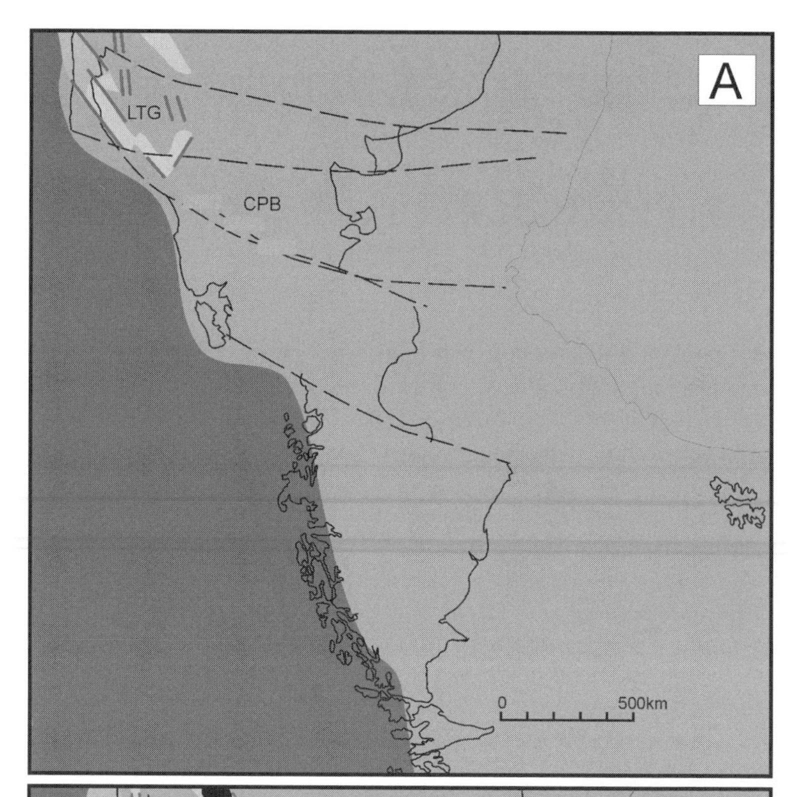

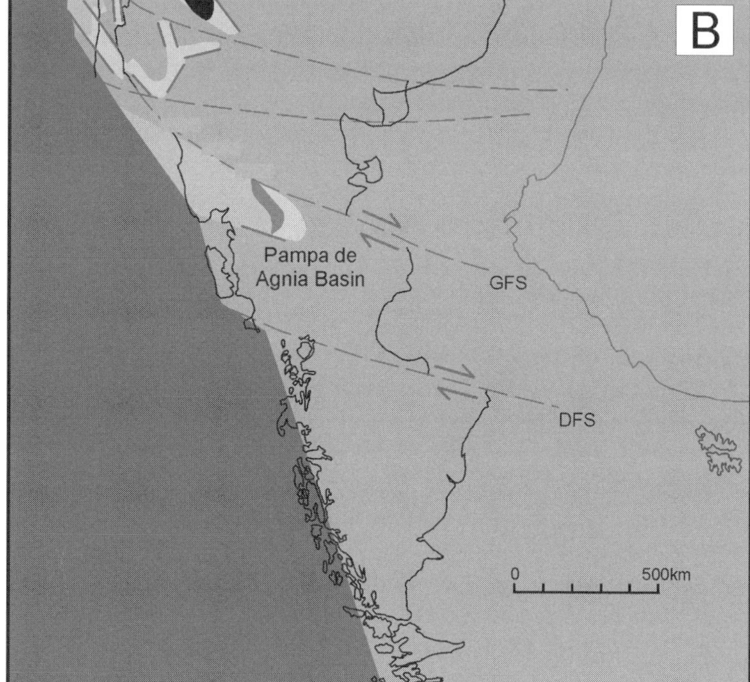

Figure 2.4. (A) Reconstruction of Patagonia at 210 Ma (late Norian–Rhaetian, Late Triassic). (B) Reconstruction of Patagonia at 195 Ma (Sinemurian-Pliesbachian, Early Jurassic). CPB = Central Patagonia batholith; DFS = Deseado fault system; GFS = Gastre fault system; LTG = Late Triassic grabens. (See Plate 10 in the color section.)

Atuel and Malargüe in Argentina) are characterized by marine deposits and/or alternance between continental and marine deposits (El Freno, Puesto Araya Formations, and correlative units), whereas those located toward the Gondwana interior (e.g., Neuquén Embayment, Chachil, Piedra del Águila in Argentina) are entirely dominated by continental (alluvial-fluvial-lacustrine) sedimentation (Lapa, Sañicó Formations). Toward the end of this time slice, the first widespread transgression occurred in the Neuquén Basin (Piedra Pintada, Los Molles Formations).

Patagonia received shallow marine sedimentation from the proto-Pacific Ocean in the Pampa de Agnia Basin, a depocenter oriented northwest-southeast, which coincides with the trace of the Gastre fault system. In the Pampa de Agnia Basin, a mixed clastic-carbonate association (Lepá and Osta Arena Formations) bearing a rich invertebrate fauna is interpreted as a tidal-dominated shallow marine system. Deltaic facies prograded from the north (Piltriquitrón Formation), and toward the southeast a belt of proximal fluvial deposits appears (El Córdoba, Puntudo Alto, and Puesto Lizarralde Formations). Paleocurrent trends from marine sediments suggest a positive area to the west of the Pampa de Agnia Basin.

Along this lapse, Patagonia was affected by a regional strike-slip tectonic regime. Movement along the Gastre fault system started. Magmatism with arc affinities (El Maitén Belt; approximately 190–200 Ma) is defined along a belt located to the south. Further south of the Gastre fault system, large areas of Patagonia remained as interbasinal highlands. The Deseado Rift was closed; even so, a narrow north-south rift, a precursor to the Austral Basin in southernmost Patagonia, is inferred from seismic sections. Deep marine deposition of the youngest Duque de York turbidites and the accretionary prism of Madre de Dios–Chonos–Chiloé suggest a protracted active convergent margin along southwestern Chile. Basic dykes dated about 192 Ma appear at south of the Gran Malvina (West Falkland) Island, as a first manifestation of the Karoo volcanism in South America.

Toarcian-Aalenian (180 Ma) (Fig. 2.5A). The early isolated depocenters located to the northwest of the studied region were integrated in the Neuquén Basin, the largest hydrocarbon producer of western Argentina. During this time slice, a period of generalized subsidence occurred in this basin, and the first proto-Pacific transgression covered large areas of northwestern Patagonia. Turbidites and coeval shallow marine deposits (Los Molles Formation) testified the full extent of the Cuyano transgression. To the south, the deltaic succession of the Piedra Pintada Formation passed to shallow marine siliciclastic deposits and fluvial sandstones and conglomerates. The southern and eastern margins of the Neuquén Basin (Neuquén Embayment) were filled with proximal fluvial deposits (Punta Rosada Formation).

In northern and northeastern Patagonia, a widespread and dominantly acidic volcanism (Marifil Formation) was developed under an extensional tectonic regime. In central Patagonia, this vol-

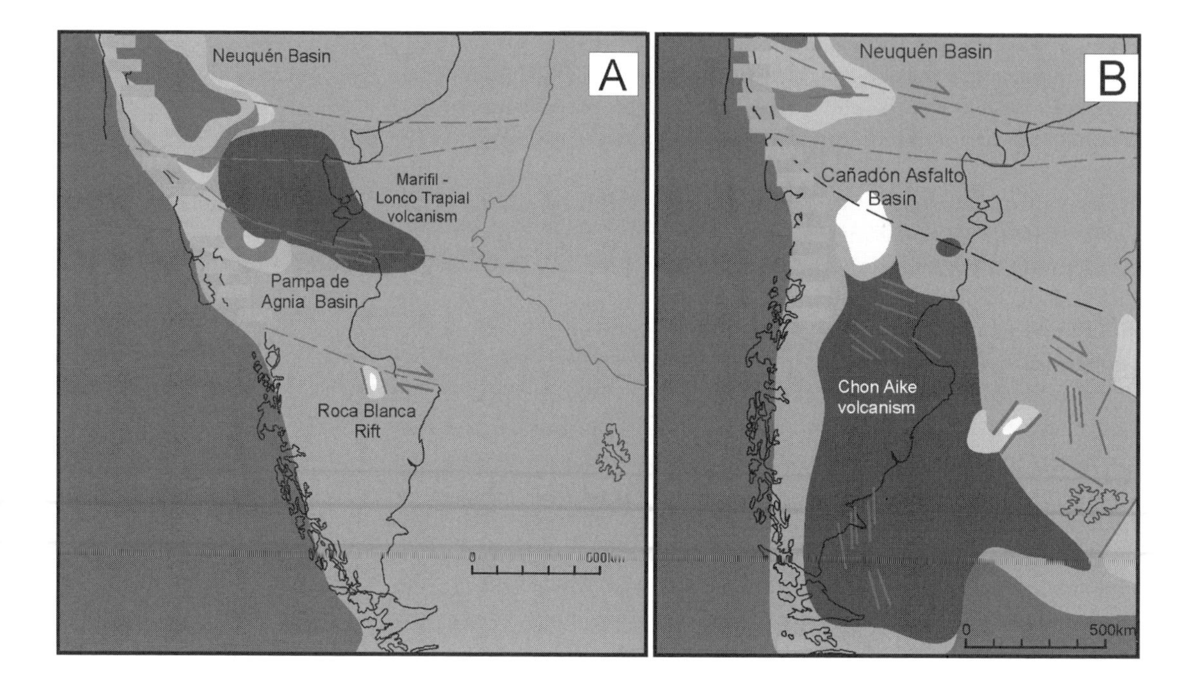

Figure 2.5. (A) Reconstruction of Patagonia at 180 Ma (Toarcian-Aalenian, Early-Middle Jurassic). (B) Reconstruction of Patagonia at 165 Ma (Bathonian-Callovian, Middle Jurassic). (See Plate 11 in the color section.)

canism, known as the Lonco Trapial Group, extends until the eastern margin of the Pampa de Agnia Basin. The Marifil-Lonco Trapial volcanism represents the northernmost and oldest episode of the Chon Aike Magmatic Province and has been dated to about 180 Ma by several Rb/Sr isochrons.

In the Pampa de Agnia depocenter, a transgressive episode (Lepá and Osta Arena Formations) extended to the southeast, and while a prograding deltaic system is located along its northeast side. Pyroclastic and volcaniclastic contributions became frequent in the Pampa de Agnia Basin during the Toarcian-Aalenian interval. The first evidence of the Andean Magmatic Arc appears to the west of the Pampa de Agnia Basin along the Argentina-Chile border between 42° and 44° south latitude (SL), and is represented by the volcanic rocks of the Lago La Plata Formation. Central and southern Patagonia still remained as a huge positive land. Nevertheless, in the Deseado Massif (to the west of the previous Permian and Triassic rifts) a north-south-oriented narrow rift, filled with fluvial and lacustrine siliciclastic facies, was developed (Roca Blanca Formation).

Bathonian-Callovian (165 Ma) (Fig. 2.5B). The Neuquén Basin passed through a critical tectonic phase and paleogeographic reorganization. The Huincul High was formed during this stage, the result of either east–west strike-slip movements or reverse inversion of original normal faults. Major regression and expansion of fluvial and deltaic facies occurred. The southern sector of the basin was dominated by fluvial (Challacó Formation, Punta Rosada Formation) and deltaic to shallow marine deposits, characterized by subtidal and intertidal sand bars, mouth bars, estuarine channels, and interdistributary bay systems (Lajas Formation and top of the Piedra Pintada Formation). Toward the basin center, protracted deep marine deposition of the Los Molles Formation prevailed.

The Pampa de Agnia marine basin evolved to an entirely continental depocenter known as the Cañadón Asfalto Basin because of the growth of the Andean Magmatic Arc. Although this basin is considered mostly extensional, late strike-slip movements along the Gastre fault system were recorded along its northern margin. It was dominated by lacustrine deposits and surrounded (especially toward the south) by fluvial sandstones and conglomerates. The Cañadón Asfalto Formation is mainly volcaniclastic, and in many places, it is closely associated with thick intermediate and basic lavas known as the Taquetrén Formation. The Leleque Granite, dated 165 Ma, seems to be the crustal counterpart of this magmatic episode.

By the end of this time slice, strike-slip movement along the Gastre fault system was completed, and granites postdating its displacement are dated about 165 Ma. The acidic volcanism recorded early in the North Patagonian area ceased. However, to the south of the Gastre fault system Patagonia was almost entirely covered by the rhyolithic lavas and pyroclastic flow deposits of the Tobífera, Bahía Laura, or Chon Aike volcanism. Several radiometric ages are about 165 Ma for this volcanic suite. This volcanism shows bimodal compositions (minor basic lavas) and constitutes the substrate of the Upper Jurassic and Cretaceous successions of southern Patagonia and the Western Malvinas Basin. At the Deseado Massif, since the Middle Jurassic, the Bahía Laura volcanics remained part of a positive land.

The acme of the Chon Aike volcanism coincides with the end of Malvinas-Falkland rotation and migration, as well as with the extensional tectonism that opened the Weddell Sea. It is also the case for the San Julián, Malvinas-Falkland Plateau, and Northern Malvinas–Falkland Basins. Southern Patagonia was also subjected to extension, and a series of gravitational faults defined a topography of alternating horsts and grabens. These fault systems trend west–east in the area to the west of the San Jorge Gulf, and north-northwest–south-southeast to the south of the Deseado Massif.

Kimmeridgian-Tithonian (150 Ma) (Fig. 2.6A). At the beginning of this time slice, a second episode of paleogeographic reorganization occurred in the Neuquén Basin. This episode is related to the growth of the magmatic arc to the west and protracted tectonic inversion of previous extensional structures. Later, a generalized transgression occurred in most of the basin with deposition of anoxic shales (Vaca Muerta Formation, main source rock for the Neuquén oil fields) and marginal carbonate-siliciclastic ramp facies to the south (Carrín Curá Formation, Picún Leufú Formation).

The volcanic arc is completely developed along the western margin of Patagonia. The bimodal volcanism in southeastern Patagonia has ceased; despite this, it is still present in the Andean

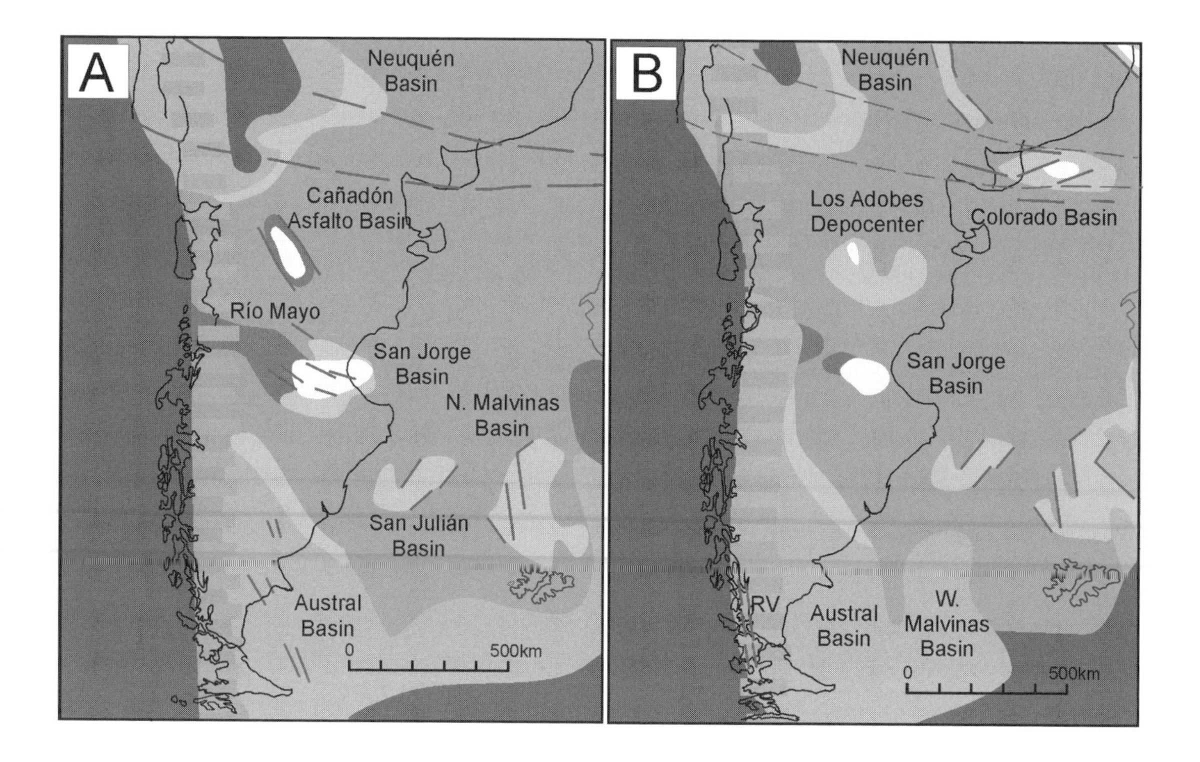

Figure 2.6. (A) Reconstruction of Patagonia at 150 Ma (Kimmerigian-Tithonian, Late Jurassic). (B) Reconstruction of Patagonia at 135 Ma (Valanginian-Hauterivian, Early Cretaceous). (See Plate 12 in the color section.)

area (Ibáñez Group, El Quemado Complex), where it seems to be related to back-arc and intra-arc extension. A submarine rhyolite suite was formed along the active margin of Patagonia between 50° and 55°SL ("Young Tobífera"), and to the east of this volcanic area, the Sarmiento-Tortuga ophiolitic complex (147 Ma) suggests the beginning of back-arc ocean-floor generation. Coevally, several depocenters are opened along western Patagonia, such as the Río Mayo Basin. All these depocenters were associated with explosive volcanism and lava flows, representing the final expression of the Chon Aike volcanism (Pankhurst et al. 2000).

Three important areas of sediment accumulation are defined in Patagonia. From north to south, they are the Cañadón Asfalto Basin, the San Jorge Basin, and the Austral Basin. The San Jorge and the Austral Basins are separated by a permanent positive land located by 48°SL known as the Deseado Massif, which extends southeast as the Dungeness or Río Chico Arch.

The Cañadón Asfalto Basin reduced its size but still generated lacustrine and deltaic deposits prograding from its western and eastern margins. Both the San Jorge and the Río Mayo Basins define a large east-west depocenter at 46°SL. The Río Mayo depression represents the western connection with the paleo-Pacific Ocean. Shallow marine carbonates (Cotidiano Formation) located by the Andean Volcanic Arc can be interpreted as intra-arc and proximal back-arc accumulations. To the east, only periodic marine episodes are recorded in the mixed siliciclastic-carbonate Tres

Lagunas Formation (composed of localized gravitational-flow deposits and patch reefs associated with isolated submarine highs). In Central Patagonia, the early infill of the extensional San Jorge Gulf Basin is represented by lacustrine deposits surrounded by proximal fluvial deposits along both its northern and southern margins (Bajo Grande and Anticlinal Aguada Bandera Formations).

A further analysis of the paleogeographic maps shows an alignment of the San Jorge Basin with the newly opened continental San Julián and Northern Malvinas–Falkland Basins. This lineation would suggest that their initial development could be controlled by strike-slip transtension linked to the Agulhas transform fault. Besides, in the Malvinas-Falkland Plateau, a marine incursion is recorded.

The Austral Basin is the largest depocenter of Patagonia. It was connected to the Pacific Ocean along its southwestern and western margins, where deep marine deposits are associated with basic lavas of the Sarmiento Complex and with the "Young Tobífera" submarine rhyolites. In southern Argentina, Tierra del Fuego Island, and the western Malvinas sector, the basin is an extended shallow marine platform characterized by tidal and deltaic deposits (Lower Springhill Formation). These sediments grade toward the northern and northeastern margins of the basin to a narrow belt of continental deposits. To the south, the Austral or Magallanes Basin is integrated to the Weddell Sea, which is floored with oceanic crust by this time. In fact, the Austral Basin represents the shallow platform of the Weddell Sea.

Valanginian-Hauterivian (135 Ma) (Fig. 2.6B). The opening of the South Atlantic Ocean produced significant paleogeographic changes along the eastern side of Patagonia. As a result of this extension, two huge taphrogenic basins (Salado and Colorado) were developed. The synrift stage in both basins is characterized by fluvial and lacustrine deposits of the Fortín and Río Salado Formations, respectively. The extensional process also affected the continental interior, where narrow and deep rift furrows composed of a thick record of continental red beds were formed (Central Argentinean Rift). This depression is composed of a chain of depocenters known as the Macachín, Levalle, Laboulaye, and Mercedes Basins, connected to the east with the Colorado and Salado Basins.

Continental and marine deposits were accumulated in the Neuquén Basin. To the north, mixed (siliciclastic and carbonate) marine ramp deposits (Agrio Formation) are widely distributed, whereas to the south, these marine deposits are followed by fluvial sandstones associated with lacustrine shales and evaporites (La Amarga Formation).

In Central Patagonia, a continental depocenter dominated by fluvial sandstones and conglomerates (Los Adobes Formation) occupied the area of the Cañadón Asfalto depression, which was enlarged to the east. To the south, the Río Mayo Embayment was drastically reduced. Its sedimentary infill is composed of siliciclastic deltaic facies (Katerfeld Formation) and shallow marine glauconitic sandstones (Apeleg Formation). These deposits suggest an effective connection of the Río Mayo Embayment with the proto-Pacific Ocean to the west, and with the northwest side of the Austral Basin to the south.

The San Jorge Basin was isolated from the proto-Pacific and was reduced to its smallest size. The restricted trough is filled with lacustrine mudstones and deltaic siliciclastic deposits prograding from the west (Cerro Guadal Formation). The Deseado Massif effectively separated the San Jorge Basin form the Austral Basin. The Río Chico Arch became narrower toward its southeastern end. Thus, the Western Malvinas Basin and the Austral Basin seem to be connected only to the south.

Coastal and shallow marine siliciclastic sediments (Upper Springhill Formation) accumulated on the wide platform of the Austral Basin. Fluvial deposits along its northeast margin surrounded it. To the south, in the Austral Basin and the Western Malvinas Basin, outer platform and slope deposits are represented by anoxic fine-grained siliciclastic facies of the lower Río Mayer, or "Favrela bearing beds." These widespread deposits bear testimony to the ample connection between the distal sectors of the Austral–Western Malvinas Basins and the Weddell Sea. To the west, the Andean margin of the Austral Basin is characterized by deep marine turbidites and hemipelagites of the Zapata-Yahgán Formations. They are part of the marginal Rocas Verdes Basin, are floored by oceanic crust, and formed as a result of intense back-arc extension. Connection between the Rocas Verdes Basin and the proto-Pacific ocean occurred across narrow paths within the volcanic chain.

Aptian (120 Ma) (Fig. 2.7A). A progressive transgression occurred in the Atlantic margin of southern South America, while the change from mechanical to thermal subsidence originated the rift/sag unconformity in the taphrogenic basins (e.g., Salado and Colorado). Fluvial deposits that laterally grade to offshore mudstones depict this change in tectonic regime. Coevally, the Macachín and Laboulaye rifts are characterized by fluvial-lacustrine facies associations.

To the west, the Neuquén Basin became a continental-restricted-marine depocenter composed of marginal fluvial red beds and a widespread muddy-evaporitic sabkha (Huitrín Formation, Rayoso Group, Lohan Cura Formation). Since the Aptian, the connection between the Pacific Ocean and the basin was definitively cut.

Between 40° and 47°SL, Patagonia was a continental land separated from the Pacific Ocean by the volcanic chain of the Andean Magmatic Arc. The San Jorge Basin was converted into a large and widespread continental trough, which incorporated the northern (Los Adobes) depocenter and a southern north-south-oriented depression located on the Permian and Triassic Deseado rifts. The marginal areas of the San Jorge Basin are represented by fluvial deposits (Matasiete Formation, upper Los Adobes Formation, and Cerro Barcino Member); deltaic systems prograded from the north-

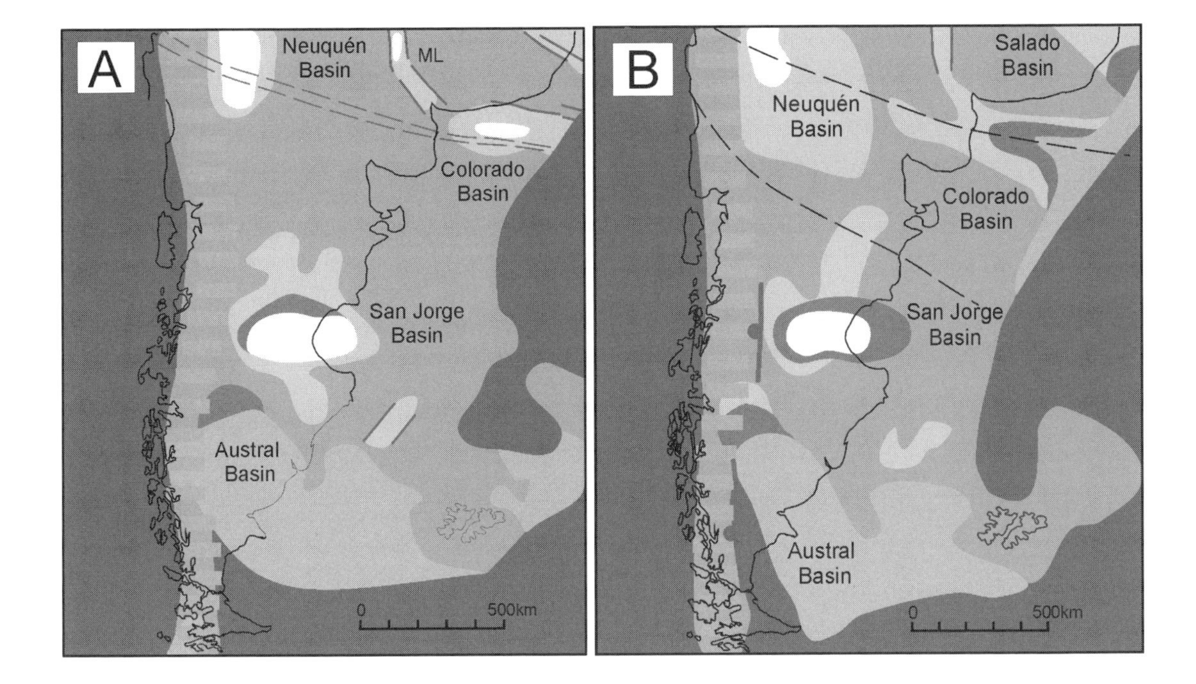

Figure 2.7. (A) Reconstruction of Patagonia at 120 Ma (Aptian, Early Cretaceous). (B) Reconstruction of Patagonia at 105 Ma (Albian, Early Cretaceous). ML = Macachín-Laboulaye rifts. (See Plate 13 in the color secion.)

ern and western margins onto a giant lacustrine system (Pozo D-129 Formation, source rock of the San Jorge oil fields).

Anoxic mudstones were deposited in wide areas of the Weddell Sea and the southern South Atlantic. In the Austral Basin, deep marine facies (interbedded shales and chert) are also widespread and grade into distal platform mudstones and shales to the north ("Lower Inoceramus": lower Palermo Aike and Lago San Martín Formations). Restricted marine conditions are represented by the Río Belgrano and Río Mayer black shales, among which storm sands intercalate. Toward its northwest corner, the Austral Basin is characterized by fluvial deposits (Río Tarde Formation). Along the foot of the Andean Magmatic Arc, black shales and turbidites of the Zapata Formation were accumulated as marginal back-arc and intra-arc deposits. In this region, the magmatic arc was almost submerged, and a tongue of deep-sea deposits "entered" from the proto-Pacific Ocean into the Weddell Sea.

Albian (105 Ma) (Fig. 2.7B). A sag stage, punctuated by widespread transgressive shales, started in the Salado (General Belgrano Formation) and Colorado (Colorado Formation) Basins; the main record of the intracontinental rifts was a thick succession of red beds formed in ephemeral fluvial systems (El Gigante Group and coeval units).

A new tectonic regime started along the convergent margin of southern South America. To the west of the Neuquén Basin, the shallowing of the subducting slab increased compression across the arc and favored the development of epidermic fold and thrust belts in the retroarc, accompanied by accumulation of fluvial red beds (lower Neuquén Group). The Neuquén Basin enlarged its area of fluvial-dominated sedimentation to the south and east.

Southern South America was separated from the Pacific by the topographic volcanic barrier of the Andean Magmatic Arc, and the first extensional plateau basalts were erupted in central Patagonia. In the San Jorge Basin, the southern fluvial domain was reduced, and the whole Deseado Massif became almost a positive land. Tectonic inversion started along the western margin of the basin, and toward the northwest, the Pampa de Agnia–Cañadón Asfalto–Los Adobes depocenter was definitively closed. Yet to the north and northeast, an expansion of the fluvial systems is clearly defined (Chubut Group red beds). At the center of the San Jorge Basin, lacustrine facies dominate (Mina El Carmen Formation), and they are surrounded almost entirely by prograding deltaic and fluvial siliciclastic facies (e.g., Castillo Formation).

The Austral Basin seems to be connected to the ocean through its southern and/or southeastern sides. At the northwest corner, a very active clastic depositional system is composed of fluvial deposits (lower Arroyo Potrancas and lower Puesto El Moro Formations), deltaic sediments (Kachaike Formation), and prodelta shales (upper Río Mayer Formation). Far south, the Palique, upper Palermo Aike, and the classic "Margas Verdes" Formations represent outer shelf shales. As a result of the magmatic arc elevation and the beginning of a compressional tectonic regime (early foreland stage), the marginal Rocas Verdes Basin was closed and the proximal back-arc was characterized by shallow marine sedimentation. Deep marine deposits are restricted to eastern Chile and southernmost Argentina, where an axial north-south-oriented turbidite system was developed (Punta Barrosa Formation).

A very effective connection existed between the Weddell Sea and the South Atlantic Ocean, but integration was less effective between the Pacific and Weddell Oceans. Incipient transforms of the proto-Scotia structure can be traced between southern South America and the Antarctic Peninsula.

Cenomanian-Turonian (90 Ma) (Fig. 2.8A). Fluvial and deltaic progradation upon shallow marine deposits occurred in the Atlantic taphrogenic basins. These deposits are known as the Upper General Belgrano Formation in the Salado Basin, and the Fortín and lower Colorado Formations in the Colorado Basin. The intracontinental rifts of Valdés and Rawson seem to be active during these times, with a continental early rift clastic infill.

Large volumes of granitic rocks (Patagonian Batholith) were intruded within the Andean Magmatic Arc. North and central Patagonia became regions of continental deposition. Thus, the Neuquén and the San Jorge Basins were integrated in a unique depocenter dominated by fluvial red beds, with subordinated shallow lacustrine and playa deposits. These deposits are represented by the red beds of the Neuquén Group in the Neuquén Basin, by the Chubut Group in northern Patagonia, the Bajo Barreal Formation

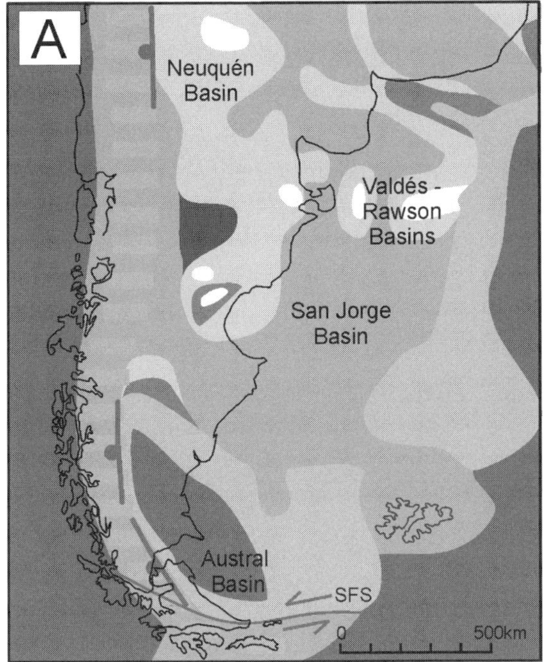

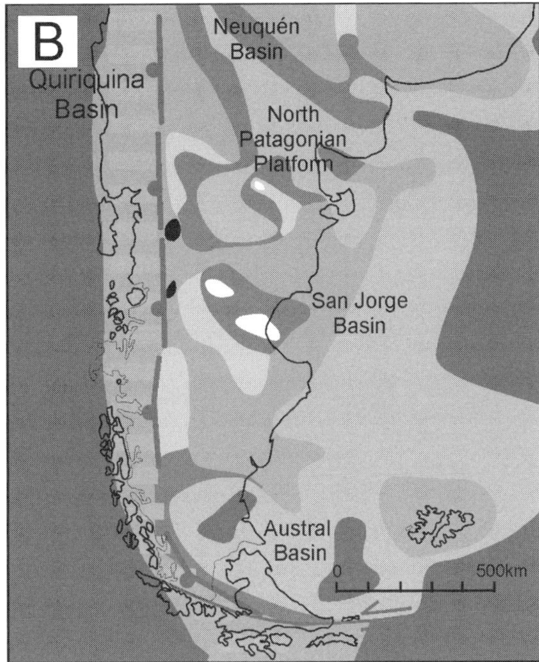

Figure 2.8. (A) Reconstruction of Patagonia at 90 Ma (Cenomanian-Turonian, Late Cretaceous). SFS = Shackleton Fault System. (B) Reconstruction of Patagonia at 75 Ma (Campanian-Maastrichtian, Late Cretaceous). (See Plate 14 in the color section.)

in the western margin of the San Jorge Basin, and the reservoir sandstones of the Comodoro Rivadavia and Cañadón Seco Formations in the subsurface of the San Jorge Basin.

The Austral Basin shows a similar facies pattern when compared with the previous map. However, because of compressional tectonics, an eastward migration of the foredeep occurred. The northwest prograding zone is characterized by fluvial facies (upper Arroyo Potrancas and upper Puesto El Moro Formations), a deltaic siliciclastics system (Piedra Clavada and Shehuen Formations), and shallow marine mudstones and fine sandstones (Lago Viedma and El Álamo Formations in Argentina, and Tres Pasos Formation in Chile). Toward the center of the basin, the Middle Inoceramus, Palermo Aike, and Cabeza de León Formations are composed of mudstones and shales deposited in deeper sectors of the platform and talus slope. North-to-south-oriented axial turbidite systems remained along the western flank of the Austral Basin (lower Cerro Toro and Lago Sofía Formations).

Intense strike-slip displacements along the Shackleton fault system in southernmost South America progressively curved Tierra del Fuego (and neighbor islands) toward the east. This movement caused the development of an east-west-oriented ridge, with partial isolation of the Magallanes–Austral Basin from the Weddell Sea. Instead, to the east, the Austral Basin widely faced the Atlantic Ocean through the western Malvinas–Falkland Basin.

Campanian-Maastrichtian (75 Ma) (Fig. 2.8B). The Atlantic Ocean was in a drift stage, and generalized marine transgression (late sag) occurred in most of the taphrogenic basins facing the ocean. This marine incursion attained its maximum at the uppermost Maastrichtian, probably as a result of the climax of thermal subsidence combined with a global sea-level rise. Siliciclastic shallow marine facies are recorded in the Salado Basin (Chilcas Formation) and in the Colorado Basin (Pedro Luro Formation). To the east of northern Patagonia, two intracontinental rifts (the Valdés and Rawson Basins) received also shallow marine sediments related to the Pedro Luro transgression. To the northwest, the narrow Laboulaye and Macachín rifts were infilled with fluvial and lacustrine deposits (Mariano Boedo Formation).

As a result of an increase in the rate of convergence between the South American and the Pacific plates, a foreland tectonic phase started along most of the length of the retroarc. Large amounts of sediments were delivered to the Argentine Basins from the uplifted Andean chain. A generalized relative sea level rise resulted in connections of Andean Basins with the Atlantic Ocean. Therefore, the Late Cretaceous foreland stage in the Neuquén Basin is associated with shallow marine deposits of the Malargüe Group. The North Patagonian Platform was also dominated by this important transgressive (Atlantic) episode. Transgressive-regressive cycle deposits ranging from siliciclastic fluvial and lacustrine systems to mixed siliciclastic and carbonate shallow marine systems are recorded in north and central Patagonia (e.g., Puntudo Chico, Cerro Bororó, Paso del Sapo, Lefipán, Los Alamitos, Coli Toro, Arroyo Salado Formations). Toward the northeast, the North Patagonian platform was connected to the Colorado Basin.

In central Patagonia, the first plateau basalts, dated approximately 77–81 Ma, are interpreted as extensional back-arc lava flows. The San Jorge Basin is dominated by fluvial sedimentation (Upper Bajo Barreal, Laguna Palacios Formations), although ephemeral lacustrine, deltaic, and fluvial deposits developed toward the center of the depression (upper Yacimiento El Trébol and upper Meseta Espinosa Formations). The Bajo Grande shallow depression located toward the southeast just by the Deseado Massif border received coarse fluvial deposits of the Laguna Palacios Formation.

A general regressive phase is recorded in the Austral Basin, with a huge embayment toward its northwest corner. Fluvial sedimentation is represented by the Cerro Fortaleza, La Anita (La Irene Member), and Chorrillo Formations. To the south, these deposits are replaced by deltaic-estuarine clastics of the Calafate and La Anita Formations. Tidal-dominated coastal-nearshore deposits are documented in marginal areas of the basin (La Anita and upper Cerro Cazador Formations). Distal platform shales and mudstones (known as the "Arcillas Fragmentosas" or upper Cerro Cazador Formation) are widespread in most of the Austral and Western Malvinas Basins. Far south, the sinistral displacement along the Shackleton transform fault zone was active and stretched to southernmost Patagonia and the Antarctic Peninsula.

Final Remarks

The mutual influence between subduction along the western margin of Gondwana and intracontinental extension and opening of the southern Atlantic Ocean governed the evolution of the Mesozoic basins of Patagonia. They may be grouped in strike-slip depocenters, intracontinental rifts, taphrogenic basins, arc-related basins, and large and long-lived polihistory basins.

Intracontinental rifts, preferentially infilled with continental deposits, were formed during the Mesozoic as a response to different extensional processes. During the Triassic, several narrow and isolated depocenters (e.g., the El Tranquilo rift) were opened along a system of northwest-southeast fault inherited from the grain of the basement. During the Jurassic, as a result of the impact of the Karoo Plume and the opening of the Weddell Sea, central Patagonia was subjected to an extensional process and strike-slip movements along major transcurrent zones (Gastre and Deseado fault systems). The continental and volcaniclastic San Jorge and Cañadón Asfalto Basins resulted from this tectonic episode. During the Early Cretaceous, and as a consequence of the opening of the Atlantic Ocean, several elongated rifts were created. Some of them remained as isolated troughs (Valdés, Rawson); others were integrated, forming a long furrow that received continental deposition in north-central Patagonia (Macachín-Levalle and associated rifts).

Strike-slip tectonics affected Patagonia during different stages of the Mesozoic. Despite this, only a few depocenters were entirely related to this tectonic regime. Narrow "en echelon" depocenters located nearby the western margin of northern Patagonia formed during the Lower Triassic as a consequence of a transcurrent regime that precluded the development of the Andean Magmatic Arc. These basins are characterized by shallow marine and fluvial deposits. Some of them were short-lived depocenters, whereas others were lately assimilated to the wider and long-lived Neuquén Basin.

As the opening of the southern Atlantic Ocean proceeded, failed rift arms evolved as taphrogenic basins all along the Cretaceous. The main depocenters are the Salado and Colorado Basins located in the northeastern side of the studied region. These basins started as fault-controlled continental rifts and evolved to a mixed marine and continental sedimentation during the later sag stage.

Coevally with the growth of the Andean Magmatic Arc, a series of intra-arc and back-arc extensional basins related to subduction along western Patagonia were formed. Whereas the Río Mayo Basin was developed upon a continental basement, the Rocas Verdes Basin, located in southernmost Patagonia, was floored by oceanic crust. The intra-back-arc Pampa de Agnia Basin was formed in a more complex scenario that included arc-related extension and strike-slip tectonics along the Gastre fault system.

Two large depocenters characterized by a more complex evolution are recognized in Patagonia: the Neuquén Basin to the north and the Austral Basin to the south. Both basins started at different times during the Jurassic as a series of extensional troughs filled with volcaniclastic continental deposits. During the growth of the Andean Magmatic Arc, these basins passed through a back-arc stage, characterized by starvation and widespread marine sedimentation. Acceleration of plate convergence during the Late Cretaceous produced partial inversion and the development of a retroarc flexural stage, with a progressive change from marine- to continental-dominated sedimentation. All along the Cretaceous, the northeast margin of the Austral Basin remained as a stable marine platform facing the Weddell Sea and the newly opened Atlantic Ocean.

References Cited

Camacho, H. 1967. Las transgresiones del Cretácico Superior y Terciario de la Argentina. *Revista de la Asociación Geológica Argentina* 22: 253–280.

Chernicoff, C., and E. Zappettini. 2003. Delimitación de los terrenos tectonoestratigráficos de la región centro-austral Argentina: evidencias aeromagnéticas. *Revista Geológica de Chile* 30: 299–316.

Dalziel, I., L. Lawver, and J. Murphy. 2000. Plumes, orogenesis and supercontinental fragmentation. *Earth and Planetary Science Letters* 178: 1–11.

Harrington, H. 1962. Paleogeographic development of South America. *American Association of Petroleum Geologists Bulletin* 46: 1773–1814.

Jacques, J. 2003a. A tectonostratigraphic synthesis of the Sub-Andean basins: implications for the geotectonic segmentation of the Andean Belt. *Journal of the Geological Society* 160: 687–701.

———. 2003b. A tectonostratigraphic synthesis of the Sub-Andean basins: inferences on the position of South American intraplate accommodation zones and their control on South Atlantic opening. *Journal of the Geological Society* 160: 703–717.

Lawver, L., L. Gahagan, and I. Dalziel. 1999. A tight fit–Early Mesozoic Gondwana, a plate reconstruction perspective. *Memoir of the National Institute of Polar Research Special Issue* 53: 214–229.

Light, M., M. Keeley, M. Maslanyj, and C. Urien. 1993. The tectonostratigraphic development of Patagonia, and its relevance to hydrocarbon exploration. *Journal of Petroleum Geology* 16: 465–482.

Macellari, C. 1988. Cretaceous paleogeography and depositional cycles of western Sudamerica. *Journal of South American Earth Sciences* 1: 373–418.

Pankhurst, R., T. Riley, M. Fanning, and S. Kelley. 2000. Episodic silicic volcanism in Patagonia and the Antarctic Peninsula: chronology of magmatism associated with the break-up of Gondwana. *Journal of Petrology* 41: 605–625.

Ramos, V. 1996. Evolución tectónica de la plataforma continental. In V. Ramos and M. Turic (eds.), *Geología y Recursos Naturales de la Plataforma Continental Argentina,* 385–404. Relatorio XIII Congreso Geológico Argentino y III Congreso de Exploración de Hidrocarburos.

———. 2002. Evolución tectónica. In M. Haller (ed.), *Geología y Recursos Naturales de Santa Cruz,* 365–387. Relatorio del XV Congreso Geológico Argentino, I-23.

Ramos, V., A. Riccardi, and E. Rolleri. 2004. Límites naturales del norte de la Patagonia. *Revista de la Asociación Geológica Argentina* 59: 785–786.

Rapela, C., and R. Pankhurst. 1992. The granites of northern Patagonia and the Gastre fault system in relation to the break of Gondwana. In B. Storey, T. Alabaster, and R. Pankhurst (eds.), *Magmatism and the Causes of Continental Break-up*, 209–220. Geological Society, Special Publication 68.

Rapela, C., R. Pankhurst, C. Fanning, and F. Hervé. 2005. Pacific subduction coeval with the Karoo mantle plume: the Early Jurassic Subcordilleran belt of northwestern Patagonia. In A. Waughan, T. Leat and R. Pankhurst (eds.), *Terrane Processes at the Margin of Gondwana*, 217–239. Geological Society, Special Publication 246.

Riccardi, A. 1987. Cretaceous paleogeography of southern South America. *Palaeogeography, Palaeoclimatology, Palaeoecology* 59: 169–195.

Scotese, C., J. Golonka, and M. I. Ross. 1994. Phanerozoic paleogeographic and paleoclimatic modelling maps. In A. Embry, B. Beauchamp and D. Glass (eds.), *Pangea, Global Environments and Resources*, 1–47. Canadian Society of Petroleum Geologists, Memoir 17.

Scotese, C., A. Boucot, and W. McKerrow. 1999. Gondwanan paleogeography and paleoclimatology. *Journal of African Earth Sciences* 28: 99–114.

Uliana, M., and K. Biddle. 1987. Permian to late Cenozoic evolution of northern Patagonia: main tectonic events, magmatic activity, and depositional trends. American Geophysical Union, Geophysical Monograph 40: 271–286.

———. 1988. Mesozoic-Cenozoic paleogeographic and geodynamic evolution of southern South America. *Revista Brasileira de Geociencias* 18: 172–190.

Urien, C., J. Zambrano, and M. Yrigoyen. 1995. Petroleum basins of southern South America: an overview. In A. Tankard, R. Suárez and H. Welsink (eds.), *Petroleum Basins of South America*, 63–77. American Association of Petroleum Geologists, Memoir 62.

Urien, C., and J. Zambrano. 1996. Estructura del margen continental. In V. Ramos and M. Turic (eds.), *Geología y Recursos Naturales de la Plataforma Continental Argentina*, 29–25. Relatorio XIII Congreso Geológico Argentino y III Congreso de Exploración de Hidrocarburos. 3. Buenos Aires.

3. Testudines

Marcelo S. de la Fuente

Introduction

Patagonia is an extended continental territory with an ancient and complex geologic history (Ramos 1988; Spalletti and Franzese this volume). During the Jurassic-Cretaceous, it was an area of diversification and evolution of many vertebrate groups, included the chelonians. Extended exposures of Jurassic and Cretaceous age of northwestern, central, and southern Patagonia have yielded numerous fossil turtles. The first references concerning Patagonian Mesozoic (mainly Cretaceous) turtles were documented in traditional paleontologic and geologic papers published in the last centuries. These papers named and described new taxa of turtles (e.g., Ameghino 1899, 1906; Woodward 1901) or merely made reference to chelid turtles as indeterminated chelonians (e.g., Wichmann 1927). Over the last two centuries, the Mesozoic Patagonian record of chelonians has been increased substantially, becoming in the more extended temporal South American record. This is known since Callovian-Oxfordian (Middle–Upper Jurassic) to early Maastrichtian (Upper Cretaceous). The main Jurassic groups are represented by panpleurodires notoemydids and eucryptodires indeterminated; the Cretaceous components are the pleurodires chelids and podocnemidoids, the pancrytodires meiolaniids among the continental turtles, and a cryptodire pancheloniid as a representative of sea turtles.

The aim of this chapter is to report the diversity of chelonians in the late Callovian–early Maastrichtian of Patagonia and to make a brief account of the main Jurassic and Cretaceous turtles. It consists of a diagnostic characterization of the named taxa, with remarks on their taxonomic status and phylogenetic position, with mention of their geographic and stratigraphic occurrence.

Institutional abbreviations. BSP, Bayerische Staatsammlung für Palaeontologie und Historische Geologie, München, Germany; MACN, Museo Argentino de Ciencias Naturales "Bernardino Rivadavia," Buenos Aires, Argentina; MCF, Museo "Carmen Funes," Plaza Huincul, Argentina; MCRN, Museo Provincial "Carlos Ameghino," Cipolletti, Argentina; MLP, Museo de La Plata, Argentina; MOZP, Museo "Profesor Dr; Juan Olsacher," Zapala, Argentina; MPA, Museo Municipal de Ciencias Naturales "Carlos Darwin," Punta Alta, Argentina; Q, Museo Paleontológico del Departamento de Geociencias de la Universidad de Concepción, Chile; QM, Queensland Museum, Brisbane, Australia.

Systematic Paleontology

Testudines Batsch 1788

Jurassic

The oldest known turtles from Patagonia are Jurassic in age and have been recently discovered in continental horizons of the Cañadón Asfalto Formation (Callovian-Oxfordian) at the Queso Rallado site (Rauhut and Puerta 2001; Rauhut et al. 2001) near Cerro Cóndor (Fig. 3.1, locality 24), Central Patagonia (Chubut Province) (Table 3.1). They are pancryptodires, and several specimens are currently in preparation and study (Sterli personal communication). Following a chronological order, pleurodires and cryptodires turtles were found in northwestern Patagonia (Neuquén Province) in a marine Tithonian horizon of the Neuquén Basin (Fig. 3.1, localities 4, 5, 7, 11, and 14), described as pelitic facies of basinal and offshore environments (Gulisano et al. 1984). This basin went through a sag evolutionary stage and behaved as a marginal marine basin from the late Sinemurian to the Albian (Mitchum and Uliana 1985). The basin infill, composed of clastic carbonates and evaporitic deposits, was controlled mainly by eustatic and climatic changes (Gulisano et al. 1984; Legarreta and Gulisano 1989). According to Groeber (1946), two sedimentary cycles have been recognized, Jurásico and Ándico. The Ándico cycle extends from the Tithonian to the Albian and comprises the Mendoza and Rayoso groups.

In most of the Neuquén Basin, the Vaca Muerta Formation is the basal unit of the Mendoza Group, concordantly overlaying the clastic and mostly continental deposits of the Tordillo Formation (Gasparini et al. 1997, fig. 2). The contact marks the beginning of the marine Tithonian transgression (Leanza 1981). This author suggested that the base of the Vaca Muerta Formation is isochronous throughout the basin and corresponds to the early Tithonian (*Virgatosphinctes mendozanus* Zone), but its top is roughly diachronous and includes younger stages from south (middle Tithon-

Table 3.1

Main Patagonian Lithostratigraphic Horizons and Localities Bearing Chelonian Taxa

Formation	Stage	Locality	Taxa
Formation Unknown	Unknown stage, Upper Cretaceous or Eocene	(25) Colhue Huapi area	*N. argentina*
Loncoche	Campanian-Maastrichtian	(1) Ranqui-Có	Chelidae indet.
Allen	Campanian-Maastrichtian	(9) Lago Pelegrini, (13) Salitral Moreno, (17) Trapal-Có, (20) Yaminué (Cerro Blanco, El Abra, El Palomar)	cf. *N. argentina Y. gasparinii* and several unnamed species of chelids
Quiriquina	Campanian-Maastrichtian	(2) Lirquen	*Euclastes* sp.
Los Alamitos	Campanian-Maastrichtian	(22) Ea. Los Alamitos	cf. *N. argentina Pa. patagonicus* and several unnamed species of chelids
La Colonia	Campanian-Maastrichtian	(23) Cerro Bosta–La Colonia	Chelidae indet. Meiolaniidae indet.
Angostura Colorada	Pre-Maastrichtian	(21) Cari-Laufquen Chica	cf. *Yaminuechelys* sp.
Mata Amarrilla	Unknown Upper Cretaceous stage	(27) Río Shehuen	Chelidae indet.
Anacleto	Santonian-Campanian	(6) Embalse Cerros Colorados (Planice Banderitas), (12) Gral. Roca	Podocnemidoidae indet. Testudines indet.
Bajo de la Carpa Río Colorado Subgroup	Santonian	(8) Loma de La Lata	*Lo. neuquina*
Formation Unknown	Santonian-Campanian	(3) Cañadón Amarillo	Podocnemidoidae indet Chelidae indet.
Plottier	Coniacian-Santonian	(19) Agua del Caño	Testudines indet.
Portezuelo	Turonian-Coniacian	(10) Sierra del Portezuelo	*Pr. portezuelae Po. patagonica*
Bajo Barreal	Cenomanian-Turonian	(26) Ea. Ocho Hermanos	Bo. bajobarrealis *Pr. argentinae*
Candeleros	Cenomanian-Turonian	(15) El Chocón, (18) La Buitrera	*Prochelidella* spp.
Lohan Cura	Aptian-Albian	(16) Cerro Leones	*Prochelidella* spp.
Vaca Muerta	Tithonian	(4) Cerro de La Parva, (5) Trincajuera, (7) Las Lajas, (11) Los Catutos, (14) Cerro Lotena	*No. laticentralis, Ne. neuquina,* Testudines indet.
Cañadón Asfalto	Callovian-Oxfordian	(24) Queso Rallado	Pancryptodira indet.

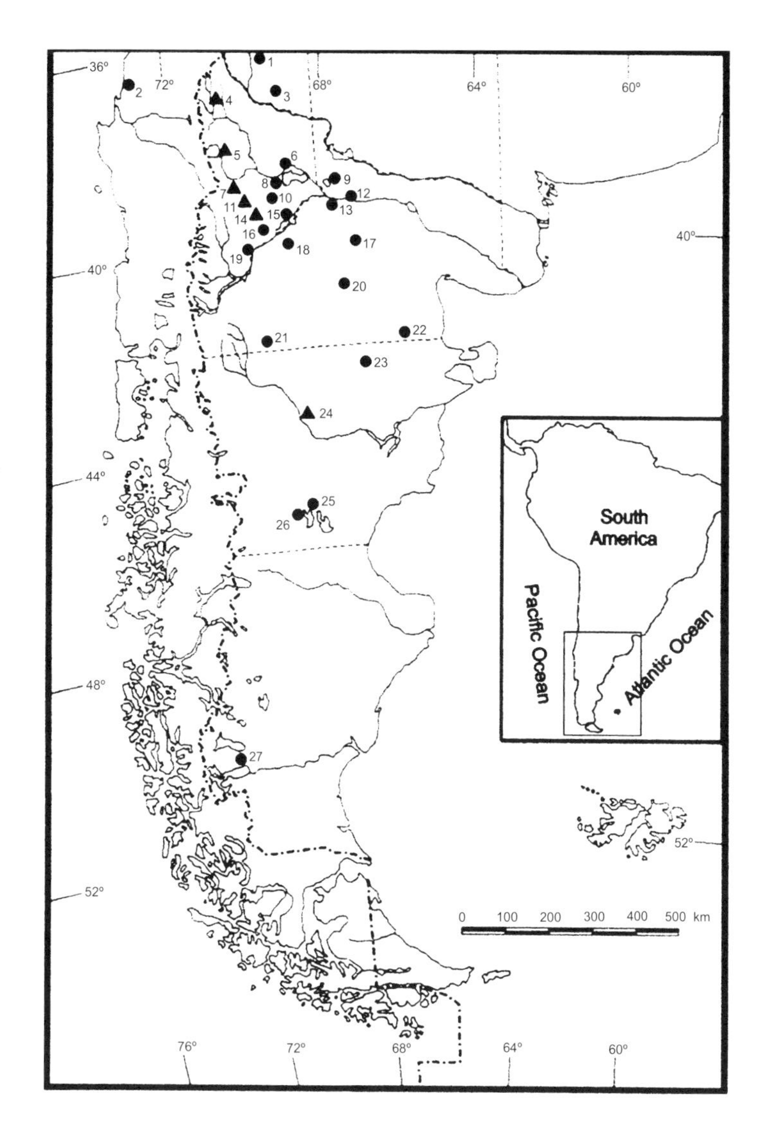

Figure 3.1. Patagonian fossil turtles localities from north to south. Solid triangles = Jurassic sites; solid circles = Cretaceous sites. 1, Ranquil-Có; 2, Lirquen; 3, Cañadón Amarillo; 4, Cerrro de la Parva; 5, Trincajuera; 6, Embalse Cerros Colorados; 7, Las Lajas; 8, Loma de la Lata; 9, Lago Pellegrini; 10, Sierra del Portezuelo; 11, Los Catutos; 12, Gral. Roca; 13, Salitral Moreno; 14, Cerro Lotena; 15, El Chocón; 16, Cerrro Leones; 17, Trapal-Có; 18, La Buitrera; 19, Aguada del Caño; 20, Arroyo Yaminué area (Cerro Blanco, El Abra, El Palomar); 21, Cari-Laufquen Chica; 22, Estancia Los Alamitos; 23, Cerro Bosta (La Colonia); 24, Queso Rallado; 25, Colhue Huapi; 26, Estancia Ocho Hermanos; 27, Río Shehuen.

ian) to north (Valanginian) (Leanza and Hugo 1977). This lithostratigraphic unit yielded the Jurassic panpleurodires and eucryptodires turtles from northwestern Patagonia (Table 3.1).

Panpleurodira Joyce, Parham, and Gauthier 2004

Comments. Instead of Casichelydia (Gaffney 1975), I use Testudines (Batsch 1788), because as proposed by Joyce et al. (2004), the former is a subjective synonymous of the second. As was pointed out by Joyce et al., Cryptodiromorpha (Lee 1995) is here named as Pancryptodira; Polycryptodira (Gaffney and Meylan 1988) is named here Cryptodira (Cope 1868a); and Pleurodiromorpha (Lee 1995) includes the taxa here grouped in Panpleurodira (Joyce et al. 2004).

Family Notoemydidae Broin and de la Fuente 1993

Comments. This family was first named by Broin and de la Fuente (1993) and diagnosed by Fernández and de la Fuente (1994) to include *Notoemys laticentralis,* the oldest Gondwanan pleurodire (Cattoi and Freiberg 1961). A second taxon that might belong to this family is *Caribemys oxfordiensis* form the Oxfordian of Cuba (de la Fuente and Iturralde-Vinent 2001; Lapparent de Broin et al. in press). Recently, Cadena Rueda and Gaffney (2005) referred the species *Caribemys oxfordiensis* to the genus *Notoemys* and included this genus and *Platychelys* (Wagner 1853) in the family Platychelyidae (see below).

Genus *Notoemys* Cattoi and Freiberg 1961

Type species. Notoemys laticentralis (Cattoi and Freiberg 1961; Wood and Freiberg 1977, pls. 1, 2, fig. 1).

Notoemys laticentralis Cattoi and Freiberg 1961
Fig. 3.2A–D

Holotype. MACN 18043, carapace and anterior platral lobe.

Referred specimens. MOZP 2487, carapace, plastron, posterior skull, four anterior cervical vertebrae, right humerus, left radius, ulna and hand, right femur, left tibia and tarsus; MOZP 4040, carapace, a damaged skull, two cervical vertebrae, right humerus, radius and ulna, left scapula, both ilia and femora, and left fibula.

Locality and age. The holotype of *Notoemys laticentralis* was found near Las Lajas, Picunches Department (Fig. 3.1, locality 7); the referred specimens were found in Cerro Lotena (Fig. 3.1, locality 14) and Los Catutos (Fig. 3.1, locality 11), Zapala Department, Neuquén Province, Argentina. These specimens were recovered from the Late Jurassic, lower (Cerro Lotena) and middle levels (Los Catutos), from the Vaca Muerta Formation, lower and middle Tithonian, respectively.

Emended diagnosis. Panpleurodire turtle with relatively low and flattened skull, with the cranioquadrate passage only closed by a ventral expansion of the prootic exceptionally wide up to the area articularis quadrati. The ventral opening of the canal of the stapediotemporal artery is not covered. The shell is hydrodynamic, low, and profiled, with a medial plastral fontanelle; and the carapace is characterized by its anterior neurals, alternately long with more sides, and short with fewer sides. It differs from pelomedusoids and chelids species in having an antrum postoticum not much developed in length and volume, a weak posterior and ventral bony closure of the recessus scalae tympani, absence of a bony wall for a constituted foramen jugulare posterius, fenestra postotica widely open laterally and posteriorly, the vestibulum ventrally incompletely hidden, ventral border of the processus interfenestralis opisthotici nearly completely free, and posterior exoccipital process not enough developed to close the recessus scalae tympani behind the processus interfenestralis. It differs from pelomedusoids in hav-

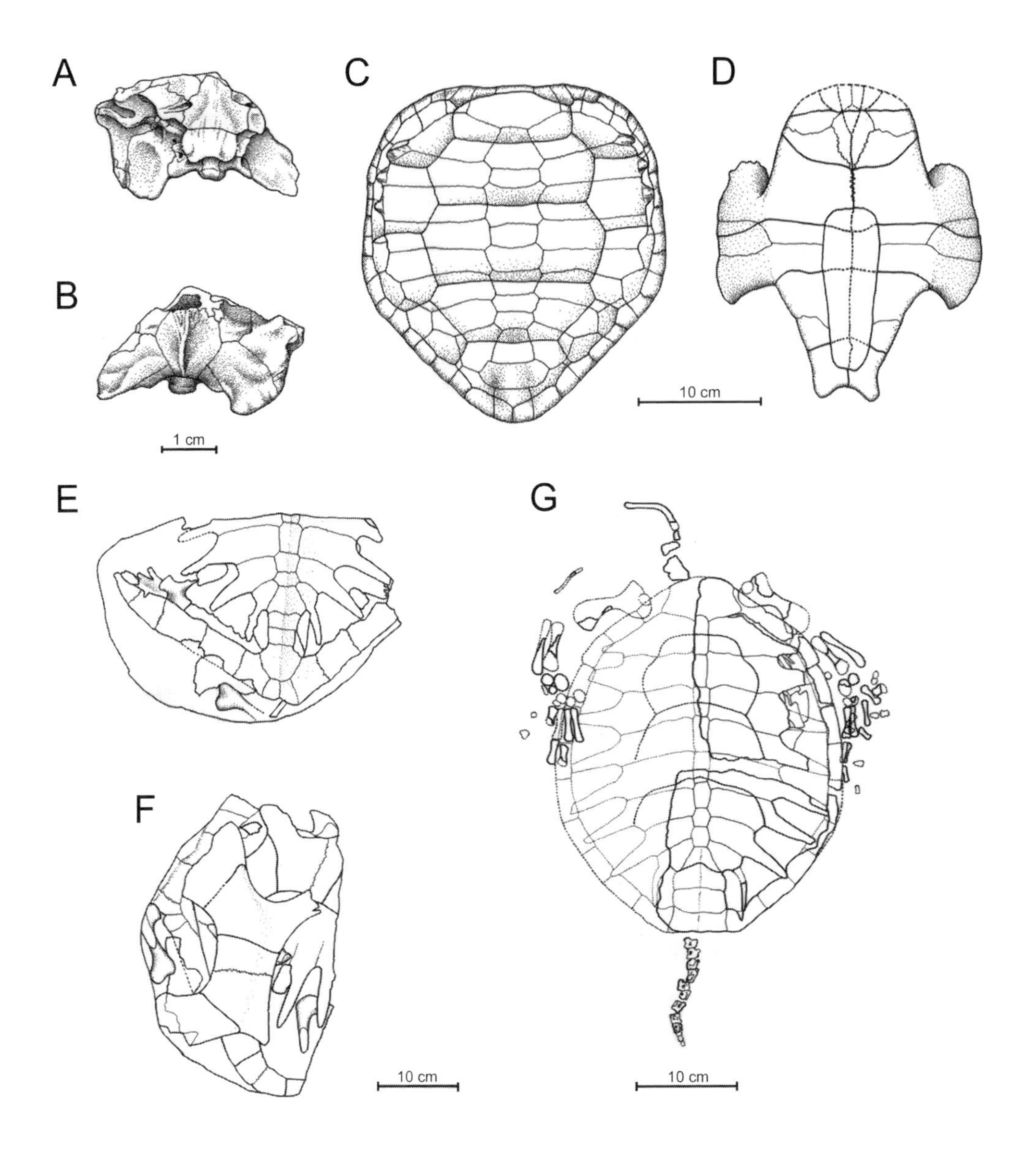

ing the paroccipitalis opisthotic process wide and rounded; the second to fourth cervical vertebrae opisthocoelous, low, and elongated; bearing an anterior triangular transverse apophyses with a transversal anterior border; and having no neural spines developed. It differs from pleurodires in having the prezygapophyses and postzygapophyses of each side widely separated. It differs from *Platychelys oberndorferi* (Wagner 1853) in having a cordiform shape to the carapace, an absence of supramarginal scutes, an absence of carapace protuberances, a rounded anterior lobe, and ilium contacting the suprapygal rather the supapygal and peripheral bones. It differs from *Caribemys oxfordiensis* in having a moderate rather than a large intergular scute, the cordiform shape of the carapace, and the long plastral fenestra. It differs from *Notoe-*

Figure 3.2. Notoemys laticentralis *(MOZP 2487). Skull in (A) ventral and (B) dorsal views. (C) Carapace. (D) Plastron.* Neusticemys neuquina *(Fernández and de la Fuente 1988) (holotype MLP 86-III-30-2). (E) Carapace. (F) Plastron (specimen MLP 92-IV-10-1). (G) Carapace.* (A–D) Modified from Fernández and de la Fuente (1994). (E, F) Modified from Fernández and de la Fuente (1988). (G) Modified from Fernández and de la Fuente (1993).

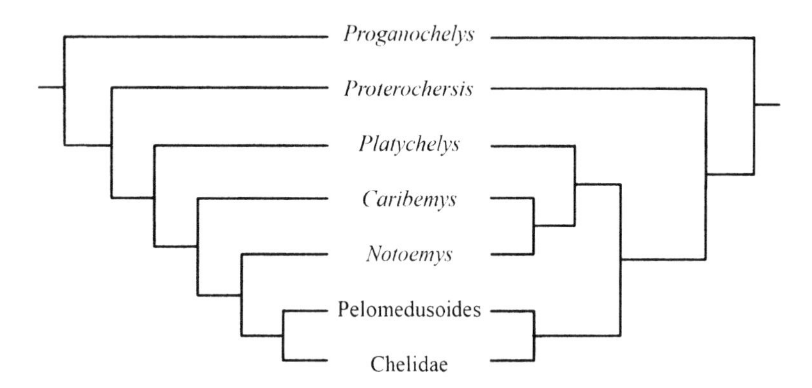

Figure 3.3. Relationships among Notoemys *and close relatives. Left, cladogram proposed by de la Fuente and Iturralde-Vinent (2001, 867). Right, cladogram modified from Cadena Rueda and Gaffney (2005, 18) to reflect the topological change between these two different hypotheses.* Caribemys oxfordiensis *is referred to the genus* Notoemys *by these authors.*

mys zapatocaensis (Cadena Rueda and Gaffney 2005) in having a smooth rather than slightly serrated posterior carapace edge, an absence of pygal notch, wider vertebral scutes, and the first costal contacting the third peripheral.

Comments. Turtles of this species are small, up to 270 mm in length. Three specimens of this species are so far known. *Notoemys laticentralis* was named as a new taxon of the Testudines by Cattoi and Freiberg (1961) on the basis of the carapace and an unidentified fragment of the plastron of the same specimen. This specimen was found by Dr. O. Reig in a slate used as construction material in a building in Acassuso (Buenos Aires Province) (Wood and Freiberg 1977). The quarry and level of the slate could be identified at Las Lajas, Vaca Muerta Formation. Wood and Freiberg (1977) assigned *Notoemys* to the pancryptodires Plesiochelyidae on the basis of a carapace. Later, *Notoemys laticentralis* was fully described (de la Fuente and Fernández 1989; Fernández and de la Fuente 1994), on the better-preserved MOZP 2487 from Cerro Lotena, and corrected its assignment to Pleurodira (= Panpleurodira sensu Joyce et al. 2004). Lapparent de Broin et al. (in press) reviewed the skull and cervical vertebrae of this specimen and confirmed this taxonomic decision on the basis of the study of other pleurodires, including new forms. In addition, the analysis of the evolution of the pleurodiran pelvis (Lapparent de Broin and de la Fuente 1996) showed that the morphology of the sacral area of *Notoemys* is basically panpleurodire because the pelvis is sutured to the carapace and plastron. Some of the new results have been already published (Lapparent de Broin and Murelaga 1999; Lapparent de Broin 2000; de la Fuente and Iturralde-Vinent 2001; de la Fuente 2003) or are in press (Lapparent de Broin et al. in press) The information added in these papers supports the phylogenetic position of *Notoemys laticentralis* or the Notoemydidae (*Notoemys* + *Caribemys*) as the sister taxa of Pleurodira (Chelidae + Pelomedusoides) proposed in previous papers (Fernández and de la Fuente 1994; Meylan 1996). However, this scenario was challenged by Cadena Rueda and Gaffney (2005) when they named a new species of *Notoemys* (*N. zapatocaensis*) from a Valanginian horizon of Colombia. These authors propose the reinterpretation of *Notoemys* and *Platychelys* as sister taxa, including both in the Family Platychelyidae. This interpretation contradicts the anterior phylogenetic proposal (Fig 3.3).

Eucryptodira Gaffney 1975

Comments. After Gaffney and Meylan (1988:177), "The eucryptodires consist of the living cryptodires and their near relatives, plus meiolaniids, plesiochelyids, and other less well-known extinct taxa." These authors and Gaffney (1996) pointed out that the two synapomorphies to test the monophyly of this group are the enclosure of carotid artery by pterygoid and the loss of mesoplastra. Because this second diagnostic condition is recognized in the *Neusticemys* plastron, this taxon is referred as Eucryptodira incertae sedis (see discussion below).

Eucryptodira incertae sedis

Genus *Neusticemys* Fernández and de la Fuente 1993
Type species. ?*Eurysternum neuquinum* Fernández and de la Fuente 1988 (Fernández and de la Fuente 1988, pls. 1, 2).

Neusticemys neuquina Fernández and de la Fuente 1988
Fig. 3.2E–G

Holotype. MLP 86-III-30-2, posterior carapace, left hyo- and hypoplastra, and fragments of appendicular skeleton.

Referred specimens. MLP 86-III-30-1, anterior carapace, both hyo- and hypoplastra and remains of pelvic girdle; MOZP 1106, anterior carapace; MLP 92-IV-10-1, carapace, damaged cervical and caudal vertebrae, right humerus, radius, ulna, and manus; MOZP 6094, carapace, lower jaw, left humerus and femur, and fragments of pelvic girdles.

Locality and age. The holotype and paratypes were found at Cerro Lotena (Fig. 3.1, locality 4), Neuquén Province, Argentina, in the lower levels of the Vaca Muerta Formation. These deposits were referred to the early Tithonian (Leanza and Hugo 1977; Leanza 1980). Two additional referred specimens were found in the middle levels of the Vaca Muerta Formation, assigned to the upper part of the middle Tithonian (Leanza and Zeiss 1990). A sixth specimen of *Neusticemys* was discovered at the northern margin of the Trincajuera Creek (Fig. 3.1, locality 5) and referred by Leanza in Gasparini et al. (1997) to the uppermost Tithonian.

Emended diagnosis. Eucryptodire turtle with a narrow symphysis in the lower jaw indicating the presence of a primitive palate (unknown). Characterized by a flattened and wide carapace of moderate to large size (maximum carapace length 471 mm), a bound keel in the posterior third of the carapace, and anterolateral margins of the carapace that are rectilinear and elongated, resulting in an anteroposterior elongation of the carapace. It has a short and wide nuchal bone, eight neural and three suprapygal bones, vertebral scutes slightly wider than costal, rounded intervertebral and costovertebral sulci, extensive fontanelles between the pleural and peripheral bones from the 3rd to 11th peripherals, large lateral and medial fenestra between the hyo- and hypoplastral bones, and a plastral index between 94 and 101. It differs from panchelonioids in not having a knob on the nuchal that attaches to the eight cervical vertebra. It differs from pancheloniids in that the lateral process of the humerus is not shifted distally. It is also characterized by an elongated anterior limb with flattened carpal element, but it differs from all known testudines in having metacarpal and metatarsal V elongated with hypertrophy of V digit.

Comments. This marine turtle species was named by Fernández and de la Fuente (1988). It was first tentatively assigned to the genus *Eurysternum* from the European Tithonian (Meyer 1839) on the basis of three specimens from the lower levels of the Vaca Muerta Formation, outcropping at Cerro Lotena. These levels belong to the *Virgatosphinctes mendozanus* Zone, which is referred to the early Tithonian (Leanza and Hugo 1977; Leanza 1980). Later, Fernández and de la Fuente (1993) studied two other specimens from the middle levels of the Vaca Muerta Formation referred to the upper part of the middle Tithonian (*Winhauseniceras interniespinosum* Zone, Leanza and Zeiss 1990). When comparing the new remains with the figure of the lost holotype of *Eurysternum wagleri* (Meyer 1839), type species of the genus *Eurysternum,* Fernández and de la Fuente (1993) proposed the relocation of the Neuquén species within a new genus, *Neusticemys.*

The holotype and paratypes of *Neusticemys neuquina* from the Vaca Muerta Formation at Cerro Lotena (Fernández and de la Fuente 1988, pls. 1, 2), as well as other specimens from Los Catutos (Fernández and de la Fuente 1993, pls. 1, 2A, B, and 3) and Trincajuera (Gasparini et al. 1997, figs. 5D, E, 6D, E), share with the species referred to Plesiochelyidae and Eurysternidae (Lapparent de Broin 1994, 2001; Joyce 2000) the width of the carapace, the shortening of the nuchal, and the first pleural bones. But these are primitive characters, not useful when establishing relationships with European taxa (Fernández and de la Fuente 1993). However, *Neusticemys neuquina* shares derived characters of the shell with other specimens such as those unpublished from the Bayerische Staatsammlung für Palaeontologie und Historische Geologie in Munich (BSP-AS-1921 and BSP-1952-1-113) and primitive Panchelonioidea from the Early Cretaceous of Australia as *Notochelone costata* (Owen 1882) (QM F 33511) (Gaffney 1981). Among these characters are the anteroposterior elongation of the carapace, the presence of broad pleuroperipheral fenestrae, the loss of the firm sutural contact between the carapace and the plastron along the bridge, and the development of broad lateral fenestrae and a large central fenestra on the plastron. Although these characters do not suggest a relationship because they characterize shells of marine turtles with pelagic habits (Zangerl 1980), other traits, like the plastral index, are closely similar to conditions recognized in Dermochelyoidae (Hirayama 1998). The tentative assignment of

Neusticemys neuquina and the undescribed Bayern forms to Family Protostegidae or "proto-Protostegidae" by Lapparent de Broin (2001) extends the biochron of this group of sea turtles to the Upper Jurassic. These sea turtles were the dominant panchelonioids from the Albian to the Turonian. They drastically declined after the Campanian (Zangerl 1953; Hirayama 1994) and were well represented in the North American Cretaceous by giant sea turtles such as *Archelon* (Wieland 1896) or *Protostega* (Cope 1871). The oldest record was found in the upper Aptian–lower Albian of Brazil (Santana Formation) (Hirayama 1998). Lapparent de Broin's assignation of *Neusticemys neuquina* to the Protostegidae was justified on the basis of general carapace and plastral morphology. However, considering Hirayama's (1994) diagnosis of this family of panchelonioids, *N. neuquina* is not a protostegid. Likewise, the inclusion of *Neusticemys neuquina* in the Eucheloniodea (Gaffney and Meylan 1988) (=Panchelonioidea, Joyce et al. 2004) has been discussed by Fernández and de la Fuente (1993) because the monophyly of this group was supported by one postcranial synapomorphy (Gaffney and Meylan 1988; Hirayama 1992) that has not been recognized in *Neusticemys neuquina,* such as a humerus with a nearly straight shaft and a lateral process lying more distal to the caput humeri, causing a complete transformation of the forelimbs into paddles relating to a swimming mode of locomotion that is exclusive of this group of sea turtles (Zangerl 1953; Walker 1973). Likewise, in *Neusticemys nueuquina* the elongation of the metacarpals evolved in a different way from the pancheloniods, in which metacarpal 2 is longer than metacarpal 5. This would indicate, as suggested by Fernández and de la Fuente (1993), that the paddles of *Neusticemys* and panchelonioids may be developed in a homoplastic way, as in the case of the extant nonchelonioid *Carettochelys insculpta* (Ramsay 1887), in which many of the same panhelonioid specializations are recognized. Nevertheless, the relationship of *Neusticemys neuquina* to other eucryptodiran turtles is still uncertain as a result of the lack of skulls, which are so important in establishing turtle relationships.

Cretaceous

As for the Jurassic turtles, the main source of Patagonian Cretaceous turtles is the Neuquén Basin (Leanza et al. 2004). However, outcroppings of Cretaceous formations in extra-Andean areas of northeastern, central, and southern Patagonia have yielded several localities with turtle remains (Broin and de la Fuente 1993, fig. 1; Gasparini et al. 2001, fig. 1; Lapparent de Broin and de la Fuente 2001; Goin et al. 2002, fig. 1).

The Cretaceous System in the Neuquén Basin (Fig. 3.1, localities 1, 3, 6, 8, 9, 12, 13, 15, 16, 18, and 19) is situated between marine beds, corresponding to the Mendoza and Malargüe Groups. The lithostratigraphic units of the continental interval are known, from older to younger as the Bajada Colorada, La Amarga, and

Lohan Cura Formations, the Neuquén Group, and the Allen Formation (Leanza 1999 and references therein). The oldest Cretaceous turtle remains in this basin have been recovered from the upper member of the Lohan Cura Formation (Leanza and Hugo 1995) and the Allen Formation (Lapparent de Broin and de la Fuente 2001; Leanza et al. 2004).

The oldest Cretaceous red bed unit in the southern Neuquén Basin is the Bajada Colorada Formation, which in turn is overlain by the marine Agrio Formation through the Catanlilican unconformity. The Initial Miranican unconformity is located between the transition zone (gypsum evaporites and clays) of the Agrio Formation and the fluvial conglomerates of the La Amarga Formation. The Lohan Cura Formation follows next, separated by the Middle Miranican unconformity and overlying the La Amarga, Bajada Colorada, and Agrio Formations. The Neuquén and Malargüe groups, both forming part of the Riograndican Cycle of Groeber (1946), constitute the Upper Cretaceous strata of the Neuquén Basin. They are separated from the previous strata by a regional unconformity known as Main Miranican (Leanza et al. 2004, fig. 2).

The Neuquén Group was deposited along nearly 24 Ma during the Cenomanian–early Campanian. It is composed of a series of wholly continental red beds consisting of conglomerates, sandstones, and claystones corresponding to fluvial, alluvial, and playa lake environments. They are generally arranged in recurrent upward-fining sequences. The stratigraphic division of the Neuquén Group comprises the Candeleros, Huincul, and Cerro Lisandro Formations (Río Limay Subgroup), Portezuelo, and Plottier Formations (Río Neuquén Subgroup), and Bajo de la Carpa and Anacleto Formations (Río Colorado Subgroup).

Such continental sedimentary conditions, together with a large supply of coarse clastics from the west, provide evidence of the final isolation of the basin from the Pacific Ocean. The Huantraiquican unconformity separates the Neuquén Group from the Malargüe Group formed by the Allen, Jagüel, and Roca Formations. The mentioned units and their tetrapod assemblages (turtles included; Table 3.1) have been summarized in Leanza et al. (2004).

In central-northern Patagonia (Fig. 3.1, localities 17, 20–23), additional remains of Upper Cretaceous turtles (Table 3.1) were found together with an Allenian tetrapod assemblage (late Campanian–early Maastrichtian [= Alamitense Bonaparte 1991 = Alamitian SALMA Flynn and Swisher 1995] [see Leanza et al. 2004]) in sedimentary sequences exposed over several wide areas of the slopes of the North Patagonian Massif (= "Macizo de Somún Cura"; see Ramos 1999 and references therein) in north-central Chubut and southern-central Río Negro Provinces (La Colonia, Los Alamitos, and Allen Formations) (Bonaparte et al. 1984; Broin 1987; Broin and de la Fuente 1993; Gasparini and de la Fuente 2000; de la Fuente et al. 2001; Lapparent de Broin and de la Fuente 2001).

The San Jorge Basin is located in central Patagonia, several kilometers south of the region mentioned above (Fig. 3.1, locality 26), and may be divided into west and east sectors (Archangelsky et al. 1994). The strata outcropping in the western part of this basin in San Bernardo Hill ("Estancia Ocho Hermanos") belong to the Chubut Group. This group comprises—from oldest to youngest—the Castillo, Bajo Barreal, and Laguna Palacios Formations (Martínez et al. 1986; Barcat et al. 1989; Bridge et al. 2000), composed of sandstones and mudstones with abundant volcaniclastic material from a western source (Bridge et al. 2000). The upper part of the Lower Member of Bajo Barreal Formation (Martínez et al. 1986, fig. 1) yielded two chelid species (*Bonapartemys bajobarrealis* and *Prochelidella argentinae*) (Lapparent de Broin and de la Fuente 2001) (Table 3.1). Recently, Bridge et al. (2000) reported the Ar-Ar dating of an ignimbrite of Bajo Barreal Formation at Cerro Ballena indicating 91 ± 0.49 Ma, which constrains the Bajo Barreal Formation to the Cenomanian-Turonian. Archangelsky et al. (1994) suggested a late Albian–early Turonian age for the Chubut Group, east of the San Jorge Basin (Santa Cruz Province), on the basis of palynological data obtained at Cañadón Seco Formation, equivalent to the Bajo Barreal Formation.

Finally, fragmentary remains of chelid turtles were recovered from Cretaceous outcrops of the Mata Amarilla Formation at the "Meseta Austral Patagónica" in the Austral Basin, in the southern tip of Patagonia (Fig. 3.1, locality 27) (Goin et al. 2002).

Pleurodira Cope 1865
Family Chelidae Lindholm 1925

Comments. The Family Chelidae is a group of freshwater aquatic to semiaquatic pleurodiran turtles with extant species distributed in South America and Australasia (Pritchard and Trebbau 1984; Iverson 1992), the fossil record of which is known as early as the Albian in Patagonia (Lapparent de Broin and de la Fuente 2001), Paleocene?-Eocene in Australia (Lapparent de Broin and Molnar 2001), and Oligocene in Tasmania (Warren 1969). This extant disjoint geographic range and the early fossil record of this family might be the result of a biogeographic and phylogenetic history developed in Gondwanan landmasses. These turtles are characterized by two main autapomorphic conditions: the formula of articulation of the cervical vertebrae includes—besides the biconcave atlas—three opisthocoelous, two biconvex, one procoelous, and one biconcave vertebrae, and the developed lateral cheek emargination with loss of quadratojugal (Gaffney 1977; Broin and de la Fuente 1993).

Genus *Prochelidella* Lapparent de Broin and de la Fuente 2001

Type species. Prochelidella argentinae Lapparent de Broin and de la Fuente 2001 (Lapparent de Broin and de la Fuente 2001, fig. 3).

Prochelidella argentinae Lapparent de Broin and de la Fuente 2001
Fig. 3.4E

Holotype. MACN-CH-1680, anterior part of the carapace.

Locality and age. Located 6 km north of "Estancia Ocho Hermanos" (Fig. 3.1, locality 26), west of Lake Musters, Sierra de San Bernardo, Sarmiento Department, Chubut Province, Argentina. Upper part of the Lower Member of Bajo Barreal Formation, Upper Cretaceous, Turonian-Campanian (Martínez et al. 1986; Barcat et al. 1989; Leanza 1999), or Cenomanian-Turonian (Bridge et al. 2000).

Emended diagnosis. Carapace low and wide, with a slight cervical notch. Relatively derived within the *Phrynops* group because of the quadrangular neural 1 and moderately narrowed cervical. Primitive with respect to extant *Acanthochelys* species because of the moderate elongation of the anterior border of the carapace, the anteriorly and posteriorly wide nuchal, the less elongated and narrowed cervical, the presence of neurals, and the more anteriorly placed axillary processes. It differs from *Prochelidella portezuelae* in the small shell size, the more anterior axillary buttress, the strong free ribs extremities of the pleurals 1 and 2 in peripherals 2–3 and 3–4, and in the form and proportion of the first and second marginal scutes.

Comments. This species was named by Lapparent de Broin and de la Fuente (2001:466–467, fig. 3) on the basis of a specimen consisting of the anterior margin of a wide and low carapace. These authors also suggested that it might be related to extant species of the genus *Acanthochelys* (Gray 1873) on the basis of the small size and the decoration characterized by a dense microvermiculation with rounded ridges. However, this species retains primitive characters such as a wide and short nuchal bone and cervical scute, the presence of neurals, and the more advanced axillar process, which, as suggested by Lapparent de Broin and de la Fuente (2001), distinguishes it from the extant species of the genus.

Lapparent de Broin and de la Fuente (2001) reported additional remains of small forms similar or very close to *Prochelidella argentinae* from Lower and Upper Cretaceous sites of Neuquén and Río Negro Provinces (Patagonia). These specimens (most of them isolated shell fragments) indicate the presence of several forms referable to *Prochelidella*. Some carapace and plastral characters of these specimens (Broin and de la Fuente 1993) include short pygals (rectangular or trapezoid); pygal bone well overlapped by vertebral 5, pygal posterior border transverse (e.g., *A. radiolata* [Mikan 1820]) or slightly notched or rounded border (e.g., *A. spixii* [Duméril and Bibron 1835] and *A. pallidipectoris* [Freiberg 1945]); and the plastral scute pattern support this assignment.

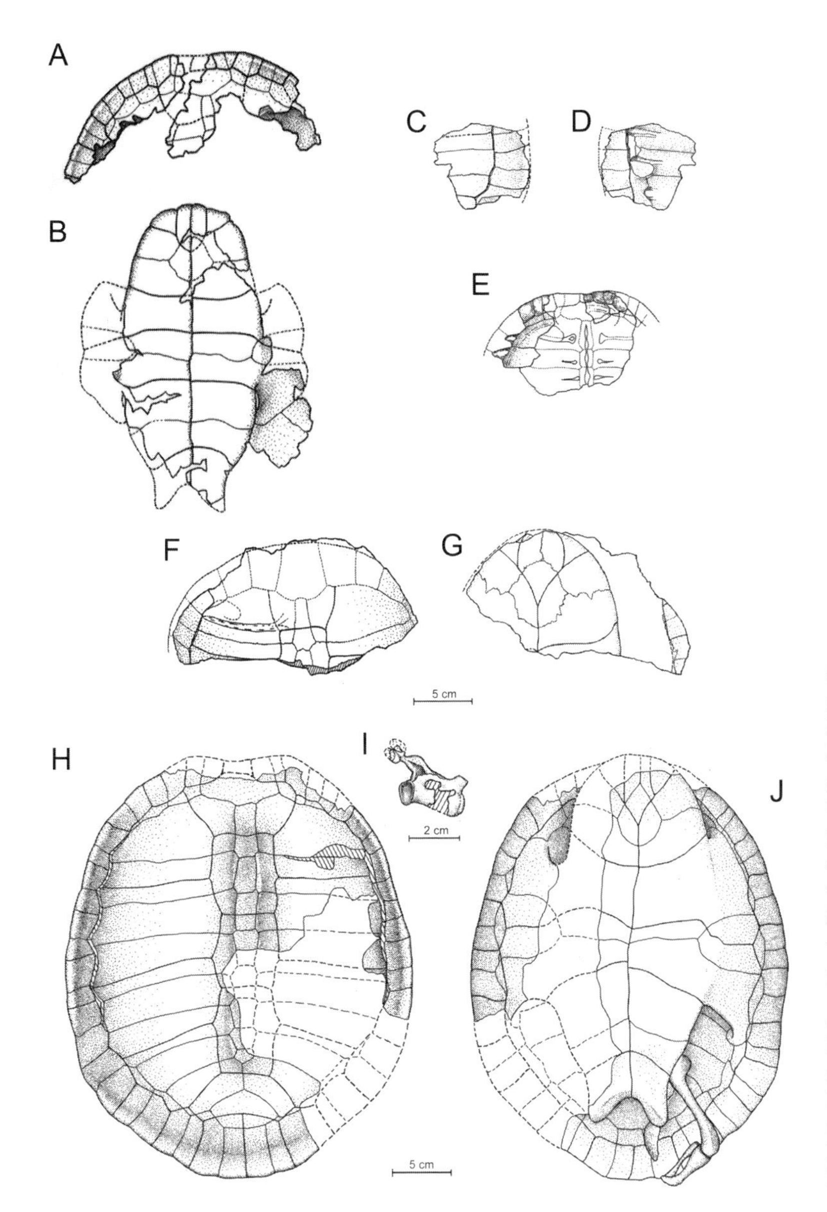

Figure 3.4. Prochelidella portezuelae *(holotype MCF-PVPH-161): (A) Carapace. (B) Plastron.* Prochelidella argentinae *(holotype MACN-CH 1680): (E) Carapace.* Palaeophrynops patagonicus *(holotype MACN-RN 906): (C) Carapace fragment, dorsal view. (D) Carapace fragment, ventral view. (F) Anterior part of the carapace. (G) Anterior part of the plastron. (I) Third or fourth opisthocoelous cervical vertebra.* Bonapartemys bajobarrealis *(MACN-CH 1469): (H) Carapace. (J) Plastron.* (A, B) Modified from de la Fuente (2003). (C–J) Modified from Lapparent de Broin and de la Fuente (2001).

Prochelidella portezuelae de la Fuente 2003
Fig. 3.4A, B

Holotype. MCF-PVPH-161, anterior margin of the carapace and nearly complete plastron, left atlantal arch, and other five cervical vertebrae, both pectoral girdles, left and right centrodistal extremity of the right femur and right tibia.

Locality and age. Sierra del Portezuelo (Fig. 3.1, locality 10), Neuquén Province, Argentina. Portezuelo Formation, Upper Cretaceous (late Turonian–early Coniacian; Hugo and Leanza 2001a; Leanza 1999).

Diagnosis. Short-necked chelid having a carapace with a wide nuchal bone and a wide cervical scale. Contact between first neural and nuchal narrow. Short, wide, laterally placed mesoplastra. Plastral bridge extending from the posterior part of third peripheral bone. Third to eighth cervical vertebrae all slightly longer than high. It differs from *Prochelidella argentinae* in the moderate shell size, in the more posterior axillary buttress, weak free ribs extremities of pleurals 1 and 2 in peripherals 2–3 and 3–4, and in the form and proportion of the first and second marginal scutes. It differs from the extant species of *Acanthochelys* in the presence of postzygapophyseal articular facets broadly expanded and nearly joined in the fifth cervical vertebra, and in the strong development of the ventral keel of the eighth cervical vertebra.

Comments. This species is only known from the holotype MCF-PVPH 161 and was described by de la Fuente (2003) on the basis of an anterior margin of the carapace and nearly complete plastron, left atlantal arch, and five other cervical vertebrae (third or fourth, fifth, sixth, seventh, and eighth), both pectoral girdles, left and right humeri, and mediodistal extremities of the femora (de la Fuente 2003, figs. 1–8). The shell and vertebral morphology of the holotype of *Prochelidella portezuelae* is similar to that of the chelid turtles. The assignment of *Prochelidella portezuelae* to the family Chelidae is based on the suture connection between the pelvic girdle and the shell (pleurodiran condition; see Gaffney and Meylan 1988 and references therein), associated with the presence of a cervical scute, short and wide mesoplastra crossed by humeropectoral sulci, and fifth and eight biconvex cervical vertebrae (also present in Jurassic panpleurodires; see Lapparent de Broin 2000; de la Fuente and Iturralde-Vinent 2001), and loose carapace-plastron and pleuro-peripheral contacts. The inclusion of a second species in the monotypic genus *Prochelidella* (only known from a partial carapace) is proposed because the anterior carapace of the holotype of *Prochelidella portezuelae* fits with the diagnostic character of the genus *Prochelidella* (wide and low carapace with slight nuchal notch, moderate elongation of the anterior border of the carapace, nuchal bone anteriorly and posteriorly wide, neural 1 quadrangular; see Lapparent de Broin and de la Fuente 2001). In contrast to *Prochelidella argentinae,* several traits of *Prochelidella portezuelae* suggest a species-level differentiation between the specimens of the Bajo Barreal and Portezuelo Formations. The latter species includes more posterior axillar process, the different outline and proportions of the first and second marginal scutes, absence of marked growth annuli and the moderate size, and weak free ribs extremities of pleurals 1 and 2 in peripherals 2–3 and 3–4.

The genus *Prochelidella* became a relatively diverse Patagonian chelid taxon because at least two species are named, and other unnamed species are recognized in Albian-Cenomanian horizons (Lohan Cura and Candeleros Formations) and another in Campanian-Maastrichtian horizons (Los Alamitos and Allen Formations). It seems to be the oldest chelid genus worldwide (Broin and de la Fuente 1993; Lapparent de Broin et al. 1997; Lapparent de Broin and de la Fuente 1999, 2001; de la Fuente 2003). *Prochelidella portezuelae* (late Turonian–early Coniacian) is more similar to the older *Prochelidella* unnamed species from the lower Albian–Cenomanian (Cerro Leones, El Chocon sites) than to more recent forms (El Palomar, El Abra localities). The youngest forms (Campanian-Maastrichtian) have narrowed cervical scales and shortened entoplastra (Broin and de la Fuente 1993; Lapparent de Broin and de la Fuente 2001). In the holotype of *Prochelidella portezuelae* (Sierra del Portezuelo), the series of cervical vertebrae is almost complete, but the *Prochelidella* spp. from other localities (e.g., El Chocón, El Abra) are only known from isolated vertebrae (Broin and de la Fuente 1993, pl. 1, fig. 3; Lapparent de Broin and de la Fuente 2001:467). In *Prochelidella portezuelae* the cervical vertebrae are moderately elongated and depressed, the ventral crest being less notched than in the fourth opisthocelous cervical vertebrae of *Palaeophrynops patagonicus* and overall in the extant species *Phrynops* sensu lato (McCord et al. 2001).

Genus *Palaeophrynops* Lapparent de Broin and de la Fuente 2001

Type species. Palaeophrynops patagonicus Lapparent de Broin and de la Fuente 2001 (Lapparent de Broin 2001, fig. 4A–F).

Palaeophrynops patagonicus Lapparent de Broin and de la Fuente 2001

Fig. 3.4C–G, I

Holotype. MACN-RN-906, anterior carapace, right lateral with lateral part of the costals 4 to 6 and peripherals 7 and 8, anterior plastral lobe, fourth cervical vertebra, and left humerus and femur.

Locality and age. West Cerro Colorado, northwestern angle of "Estancia los Alamitos," close to Arroyo Verde, Valcheta Department, Río Negro Province, Argentina (Fig. 3.1, locality 22). Middle part of the Los Alamitos Formation, Upper Cretaceous, upper Campanian–lower Maastrichtian (Bonaparte et al. 1984; Bonaparte 1990; Hugo and Leanza 2001b).

Diagnosis. As for the genus and species by monotypy. Anterior border of the carapace rounded, clearly progressively elongated (although not as much as possible in chelids) from peripherals 3 up to nuchal; the nuchal narrowed, especially anteriorly, long in regard to its width and to the carapace; the vertebrals narrowed from anterior part of vertebral 1 and more at vertebral 2; the dilated intergular, the moderatly elongated and depressed cervical vertebra 4, with oval and close postzygapophyses facets on an undepressed common process. Intergular very large, with lateroposterior border slightly concave, widening much anteriorly with rounded convex borders and then more anteriorly much narrowing, the gular (not contacting entoplastron) much anteriorly widened.

Comments. The holotype of *Palaeophrynops patagonicus* (MACN-RN 906) was described by Lapparent de Broin and de la

Fuente (2001). It is preserved in three shell pieces: an anterior part of a carapace, the morphology of which is mostly seen in the cast, with the anterior border (nuchal, peripherals 1–5 left and right) pleurals 1 and 2, neurals 1 and 2; a right lateral fragment with lateral parts of pleural 4 to 6, and peripherals 7 and 8; and a partial plastral anterior lobe. Also a fourth cervical vertebra, left humerus, and femur are known as part of this holotype. This monotypic species was diagnosed by Lapparent de Broin and de la Fuente (2001:468) on the basis of the rounded and clearly progressive elongated anterior border of the carapace, nuchal narrowed anteriorly and longer than wide, vertebral scutes narrowed, a dilated and very large integular, widening anteriorly with rounded convex border, and then much narrowing anteriorly. The opisthocoelous fourth cervical vertebra is moderately elongated and depressed; the postzygapophyseal facets are oval in shape and very close to each other. These diagnostic conditions are derived in the *Phrynops* group. These authors have suggested that *Palaeophrynops patagonicus* is the first modern representative of the *Phrynops* group because of the elongation of the anterior margin of the carapace (anterior peripherals and nuchal bones) but not of the pleural 1, the narrowed elongated nuchal, and the dilated intergular scale. Lapparent de Broin and de la Fuente (2001) have also suggested that this species might be related to the large extant *Phrynops* species of *Phrynops geoffroanus* complex (Rhodin and Mittermeier 1983) included in the subgenus *Phrynops* (Pritchard and Trebbau 1984) or in the genus *Phrynops* sensu stricto (McCord et al. 2001).

Genus Bonapartemys Lapparent de Broin and de la Fuente 2001

Type species. Bonapartemys bajobarrealis Lapparent de Broin and de la Fuente 2001 (Lapparent de Broin and de la Fuente 2001, figs. 1A–C, 2A, B).

Bonapartemys bajobarrealis Lapparent de Broin and de la Fuente 2001

Figs. 3.4H, J

Holotype. MACN-CH-1469, carapace, plastron, remains of girdles, limbs, and tail.

Locality and age. As for *Prochelidella argentinae.*

Diagnosis. As for the genus and species by monotypy. Large oval carapace, with attenuated decoration, similar to large extant *Phrynops* but with more primitive character. Posterior border of the carapace medially progressively elongated at peripherals 8–11; marginals 12 on the suprapygal, progressively medially elongated, vertebrae 2 to 4 clearly narrowed anteriorly as wide as posteriorly; anteriorly narrowed intergular, elongated bridge. Plastral lobes are prolonged by a ridge at the bridge limit: the anterior lobe ridge particularly marked, up to the mesoplastra. Bridge > posterior plastral lobe > anterior plastral lobe.

Comments. Dr. J. F. Bonaparte found this turtle in the Bajo Barreal Formation (above). Posteriorly, Lapparent de Broin and de la Fuente (2001:465) formally named, described, and diagnosed this specimen as *Bonapartemys bajobarrealis.* The holotype (MACN-CH 1469) is a carapace and plastron with posterolateral and lateral part, lacking part of girdles, limbs, and tail. This species is characterized by a large oval carapace, with attenuated decoration similar to large extant *Phrynops,* but with more primitive traits (see diagnosis and additional characters in Lapparent de Broin and de la Fuente 2001:465–466). These authors pointed out that *Bonapartemys bajobarrealis* was like the large extant *Phrynops* (*Phrynops* sensu stricto, after McCord et al. 2001) with neural bones like those of *Phrynops geoffroanus* (Schweigger 1812) (Pritchard and Trebbau 1984, fig. 18), *Phrynops hilarii* (Duméril and Bibron 1835) (Cabrera 1998, figs. 20, 21), and *Phrynops williamsi* (Rhodin and Mittermeier 1983, figs. 1–6), but with other primitive traits (e.g., wide nuchal bone, little elongated anterior peripheral bones, absence of a dilated intergular scale, with axillary processes at peripherals 2–3).

Genus *Lomalatachelys* Lapparent de Broin and de la Fuente 2001

Type species. Lomalatachelys neuquina Lapparent de Broin and de la Fuente 2001 (Lapparent de Broin and de la Fuente 2001, fig. 5A–D).

Lomalatachelys neuquina Lapparent de Broin and de la Fuente 2001

Fig. 3.5A, B

Holotype. MOZP 5117, carapace, plastron, and remains of pubis and ischia.

Locality and age. Loma de la Lata (75 km northwest from Neuquén City), Confluencia Department, Neuquén Province, Argentina (Fig. 3.1, locality 8). Bajo de la Carpa Formation, Río Colorado Subgroup, Neuquén Group, Upper Cretaceous, Santonian (Bonaparte 1991; Broin and de la Fuente 1993; Leanza 1999; Leanza et al. 2004).

Diagnosis. As for the genus by monotypy. Large quadrangular carapace, slightly narrow, with an elongated anterior border (peripherals 4 to 1, pleural 1, and nuchal bones). Elongated nuchal bone with a narrow anterior border, but with an elongated and narrow cervical scale. Slight notch at middle-peripheral 1 (from seam between marginals 1–2) and nuchal; vertebral 1 narrowed at the nuchal width and vertebral 2, 3, and 4 clearly but progressively narrowed; long marginals 12 on the suprapygal, progressively elongated from their lateral border at suprapygal-peripheral suture; axillary processes at midlength of peripheral 3; rather quadrate anterior lobe with slightly elongated epiplastral symphysis and neither widened nor narrowed intergular. Posterior plastral lobe > anterior plastral lobe > bridge. Posterior wide nuchal border.

Comments. The carapace, plastron, and remains of pelvic girdle are the only preserved elements of the holotype MOZ P 5117. This species was diagnosed by Lapparent de Broin and de la Fuente

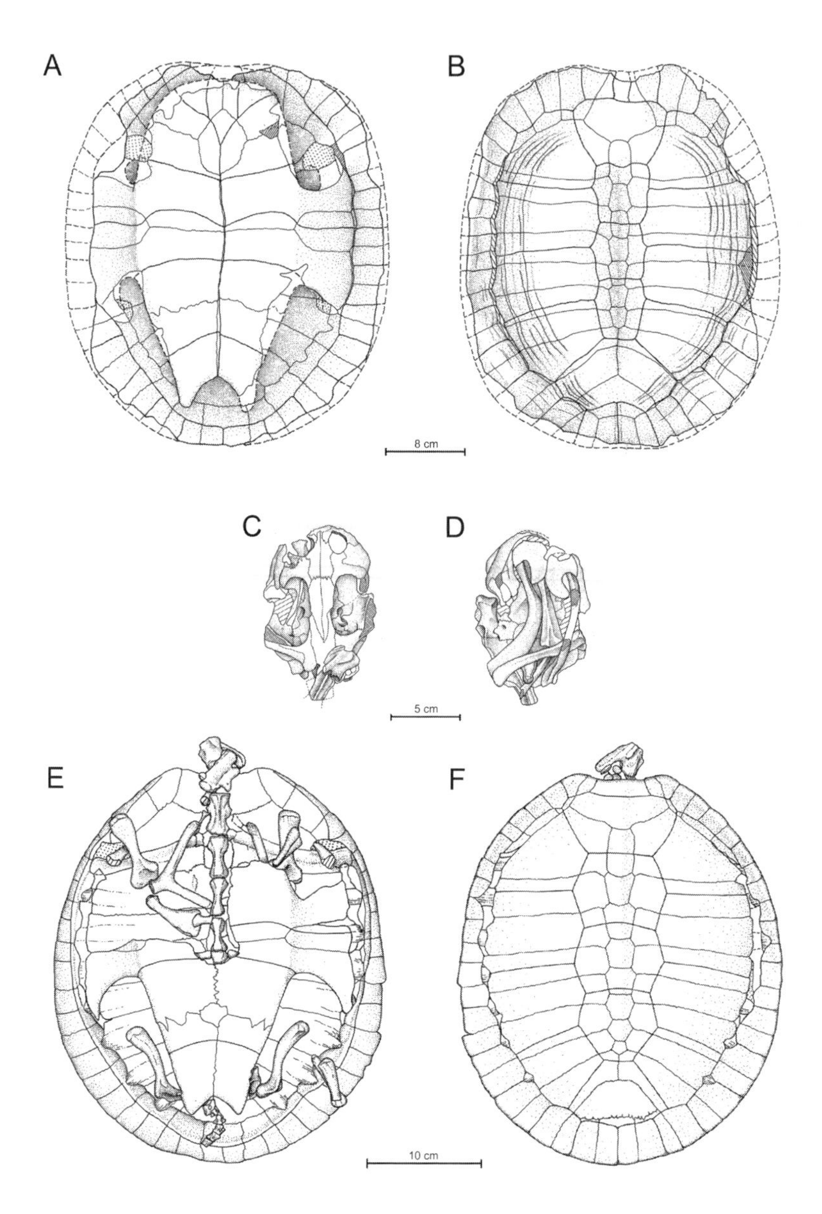

Figure 3.5. Lomalatachelys neuquina *(MOZP-5117): (A) Plastron. (B) Carapace. Modified from Lapparent de Broin and de la Fuente (2001).* Yaminuechelys gasparinii *(MPA 86-86-IC): (C) Dorsal view of the skull and lateral view of atlas and axis. (D) Ventral view of the skull and hyoid apparatus. (E) Plastron, visceral view of the anterior carapace, seventh and eight cervical vertebrae, dorsal vertebrae, and appendicular bones. (F) Carapace.* Modified from de la Fuente et al. (2001).

(2001:469) as a chelid with large quadrangular carapace, slightly narrowed, with elongated anterior margin, and other characters pointed out in the diagnosis (see also additional characters in Lapparent de Broin and de la Fuente 2001:469). These authors stated that *Lomalatachelys neuquina* was similar to *Chelus* (Duméril 1806) in the large size and quadrangular carapace shape, similarly elongated anterior element, scute radiation, and plastral shape. However, the carapace dentation and protuberances of extant *Chelus* are absent. The *Chelus* group is also recorded in Campanian-Maastrichtian horizons (Los Alamitos and Allen Formations) outcropping in other localities of Patagonia (Los Alamitos, El Abra, Trapalcó). These turtles are represented by isolated shell bones and cervical vertebrae (Broin and de la Fuente 1993:185–192, pl. 1, fig. 5; Gasparini et al. 2002). Finally, Lapparent de Broin and de la Fuente (2001) suggested that there are other large forms unidentified below the family level that may or not be related to *Lomalatachelys* in Patagonia at Los Alamitos, Trapalcó and La Colonia, and in southern Mendoza at Ranquilco (Broin and de la Fuente 1993; Lapparent de Broin and de la Fuente 1999; Gasparini and de la Fuente 2000).

Genus *Yaminuechelys* de la Fuente, Lapparent de Broin, and Manera de Bianco 2001

Type species. Yaminuechelys gasparinii de la Fuente, Lapparent de Broin, and Manera de Bianco 2001 (de la Fuente et al. 2001, figs. 1A, B, 2A–G).

Yaminuechelys gasparinii de la Fuente, Lapparent de Broin, and Manera de Bianco 2001

Fig. 3.5C–F

Holotype. MPA 86-86-IC, carapace, plastron, skull, hyoid apparatus, atlas, axis, sixth, seventh, and eight cervical vertebrae, left and right scapular girdle, left and right humeri, right radius and ulna, pelvic girdle, left and right femora, right tibia and fibula, two sacral and seven caudal vertebrae

Locality and age. This turtle was found in a mudstone level belonging to a succession of mudstones and sandstones named as lacustrine Senonian by Wichmann (1927), exposed at Yaminué Creek, Cerro Blanco, Río Negro Province, Argentina (Manera de Bianco 1996) (Fig. 3.1, locality 20). These sediments are similar to those of Lago Pellegrini area, which are referred (Andreis et al. 1974) to the middle member of the Allen Formation, Upper Cretaceous (upper Campanian–lower Maastrichtian; Hugo and Leanza 2001a).

Emended diagnosis. Long-necked (elongated vertebrae and flattened skull) chelid turtle belonging to the *Hydromedusa-Chelodina* group because of the elongated skull and the strongly polygoned decoration of the carapace, and to the *Hydromedusa* subgroup because of the large bony aperturae narium internae. Carapace with a shallow nuchal-anterior peripheral emargination. It differs from *Hydromedusa* (Wagler 1830) in the less emarginated skull. Large postorbital and jugal, more extensive exposition of the parietal, supraoccipital, and squamosal, in dorsal view. The hyoid apparatus with relatively wider and longer components: entoglossa, hyoid body and horns, short and wide (twice as wide as long) nuchal bone, wide and short marginal cervical scute, three times wider than long, presence of lateral mesoplastra, short and wide; two amphicoelous sacral vertebrae, and caudals–two weakly procoelus, one amphicoelous or concavoplatycoelous, two concavoplatycoelous, the other procoelous, and one amphicoelous. It differs from *Yaminuechelys maior* (Staesche 1929) in having a moderate size (418 mm), the shape of the supraoccipital on the dorsal surface forming the lateral posterior edge of the lateral emarginations, the quadrangular pygal bone, the axillar buttresses attached only to the lateral margin of the ventral surface of the first pleural bone, and the straightened lateral margins of the posterior plastral lobe.

Comments. Dr. Teresa Manera de Bianco discovered this turtle in an upper Campanian–lower Maastrichtian horizon outcropping at Yaminué Creek in Patagonia (see above). The holotype (MPA 86-86-IC) is a single specimen with the skull, hyoid apparatus, atlas, axis, sixth, seventh, and eighth cervical vertebrae, scapular girdles, both humeri, right radius and ulna, pelvic girdle, both femora, right tibia and fibula, two sacral vertebrae, and seven caudal vertebrae, carapace, and most of the plastron.

A basic dichotomy in chelid turtles has been recognized (Boulenger 1889) on the extant chelid species with a neck shorter than the dorsal vertebral column (*Pseudemydura, Emydura* group, and *Phrynops* group), and chelids with a neck longer than the dorsal vertebral column (*Chelus* [Duméril 1806], *Chelodina* [Fitzinger 1826], and *Hydromedusa* [Wagler 1830]) by the lengthening of the each vertebrae. *Chelus* differs from the two other genera because it has high postzygapophyseal processes in the cervical vertebrae, whereas *Hydromedusa* and *Chelodina* have lowered postzygapophyses. Because of the elongated and lowered cervicals and the lowered postzygapophyseal processes, *Yaminuechelys gasparinii* is a member of the long-necked chelids, including *Chelodina* and *Hydromedusa*. This basic cluster on the cervical vertebrae supports the chelid phylogenetic relationships proposed by Gaffney (1977) mainly on skull characters of extant genera. Gaffney concluded that the Australasian (*Chelodina*) and South American long-necked chelids (*Chelus* and *Hydromedusa*) form a monophyletic group spanning throughout two continents. In contrast, on the basis of analyses of morphological and serological data, Burdbidge et al. (1974) concluded that all Australian species were more closely related to each other than to any South American species. Coincidentally, recent 12S rRNA and cytocrome b sequencing suggests that the long-necked *Chelodina* are more closely related to the short-necked Australasian genera than to either *Chelus* or *Hydromedusa* (Seddon et al. 1997; Shaffer et al. 1997; Georges et al. 1998). Previously, Pritchard (1984) proposed that the elongated head and neck structure of *Hydromedusa* and *Chelodina* may have arisen not from a close phylogenetic relationship (as proposed by Gaffney 1977) but from parallel evolution as they became specialized. However, this scenario is not supported by either phylogenetic analyses based on morphological data or by recent systematic studies (de la Fuente et al. 2001; Bona and de la Fuente 2001, 2005; Bona 2004).

As was recently argued by de la Fuente et al. (2001) and de la Fuente and Bona (2002) and tested in a phylogenetic context by Bona (2004) and Bona and de la Fuente (2005), *Yaminuechelys* represents the sister group of *Hydromedusa* on the synapomorphic craneal condition of large apertura narium internae with reduced ossification of palatines.

Long-necked chelid remains referred to *Hydromedusa* subgroup and either belonging to *Yaminuechelys* or forms affinis to *Yaminuechelys* have been found in a dozen Upper Cretaceous localities of Patagonia (Broin and de la Fuente 1993; Gasparini and de la Fuente 2000; de la Fuente et al. 2001 and references therein). Isolated plates belong to large Staesche's species that have been found in Punta Peligro (Staesche 1929) and occasional partial shells articulated with cervical vertebrae and skulls recovered from Cerro Hansen (exposures of the Salamanca Formation, lower Paleocene) might be also referred to *Yaminuechelys* (Bona et al. 1998; Bona 2004; Bona and de la Fuente 2001, 2005). The presence of chelid species of *Yaminuechelys* in Campanian–lower Maastrichtian and Paleocene rocks of Patagonia is also direct evidence of continental forms surviving the K/T Great Extinction.

Pelomedusoides Cope 1864
Podocnemidoidea Cope 1868a
Podocnemidoidae Cope 1868a
Genus *Portezueloemys* de la Fuente 2003

Type species. Portezueloemys patagonica (de la Fuente 2003, figs. 9–12).

***Portezueloemys patagonica* de la Fuente 2003**
Fig. 3.6A–D

Holotype. MCF-PVPH-338, carapace, plastron, and skull.

Referred specimens. MCF-PVPH-339, a nearly complete plastron.

Locality and age. As for *Prochelidella portezuelae.*

Diagnosis. A podocnemidoid pleurodiran turtle with a podocnemidoid fossa, enlarged carotid canal, and lacking prolonged pterygoid wing. Foramen jugulare posterius not separated from fenestra postotica, small epiplastral gular scutes, pectoral scutes contacting entoplastron posteriorly but not extending over the epiplastra and mesoplastra. It differs from *Brasilemys* (Lapparent de Broin 2000) in the extensive skull roof formed by enlarged areas of the postorbital, jugal, and quadratojugal, and in the narrow interorbital space; it differs from *Hamadachelys* (Tong and Buffetaut 1996) in having less extended temporal emargination that does not expose the foramen stapedio temporale, and in a dorsoanterior enlargement of the opening in the podocnemidoid fossa; it differs from extant and fossil South American Podocnemidinae in having the pterygoid flange end at the border of the pterygoid on the infratemporal fossa.

Comments. This species was named by de la Fuente (2003:565–570, figs. 9–12) on the basis of the holotype (MCF-PVPH 338, a skull and partially preserved carapace and plastron), and a referred specimen (MCF-PVPH 339, a nearly complete plastron). De la Fuente (2003) also pointed out that *Portezueloemys*

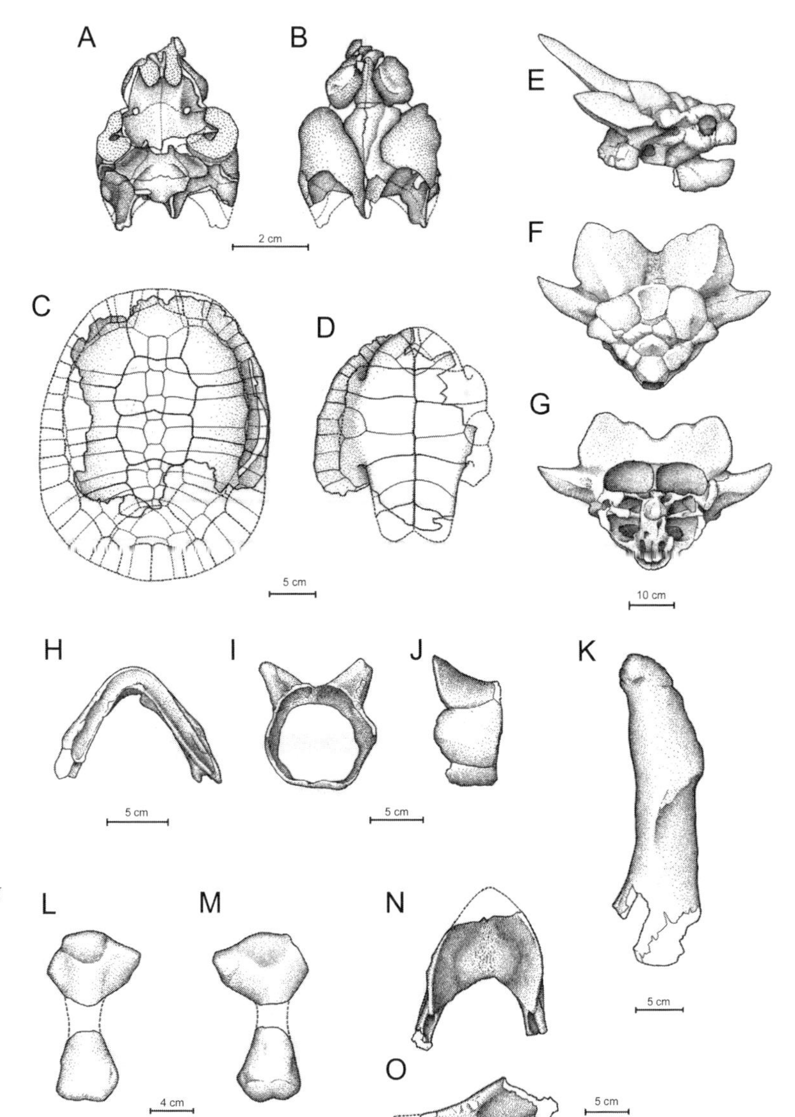

Figure 3.6. Portezueloemys patagonica *(holotype MCF-PVPH-338): Skull in (A) ventral and (B) dorsal views. (C) Carapace. (D) Plastron. Modified from de la Fuente (2003).* Niolamia argentina *(neotype 26-40/41/42 and 44): Skull in (E) lateral, (F) dorsal, and (G) ventral views. (H) Lower jaw. Caudal ring in (I) posterior and (J) lateral views. (K) Left scapula. Modified from Woodward (1901). Cf.* Niolamia *sp. (MACN-RN 36) modified from Broin (1987): left humerus, (L) dorsal and (M) ventral views.* "Osteopygis" *sp. (Q/377): lower jaw in (N) dorsal and (O) lateral views.*

patagonica is a member of Pelomedusoides (sensu Broin 1988) on the basis of the vomer reduced to the anterior interchoanal part and rounded lateral mesoplastra (see characters listed in Lapparent de Broin 2000). Another derived character of this species (e.g., the podocnemidoid fossa becoming the true enlarged carotid canal or pterygoideous channel forming a deeper fossa inside the skull; Gaffney 1979) prompted de la Fuente (2003) to refer this species to the epifamily Podocnemidoidae.

Although the skull and shell morphology of *Portezueloemys patagonica* is *Podocnemic*-like in the general arrangement of bones and scales and with an enlarged podocnemidoid fossa, this suggests that the pterygoid wings stop at the border of the pterygoid on the infratemporal fossa and do not extend posteromedially up to the suture of the basisphenoid as in Podocnemididae. The skull, like that of *Podocnemis,* is rather flat and oblong, with small orbits directed dorsolaterally, and with a narrow interorbital space, but retains some primitive characters (the foramen jugulare posterius confluent with the fenestra postotica and a well-developed paraoccipital opisthotic process). The shell is oval rather than quadrangular, and moderately high with a rounded anterior border. The gular scales are small, and the pectoral scales do not contact the mesoplastra, but only the posterior third of the entoplastron and hyoplastra.

The Lower Cretaceous pelomedusoid turtles from South America include several species such as *Araripemys barretoi,* the unnamed specimens FR 4922, *Brasilemys josei* (Lapparent de Broin 2000), and *Cearachelys placidoi* (Gaffney et al. 2001) (Price 1973; Gaffney and Meylan 1991; Meylan and Gaffney 1991; Lapparent de Broin 2000; Gaffney et al. 2001). All these taxa were recovered from the Romualdo Member of the Santana Formation (Aptian-Albian) in Chapada do Araripe, Brazil. *Portezueloemys patagonica* and *Brasilemys josei* can be differentiated from other Early Cretaceous taxa because they share characters present in the Epifamily Podocnemidoidae (Lapparent de Broin 2000). The phylogenetic analysis proposed by de la Fuente (2003) suggested that *Portezueloemys* shares a common ancestor with its sister group Podocnemididae. This scenario is supported by the following synapomorphies: a dorsoanterior enlargement of the foramen in the podocnemidoid fossa, a processus trochlear pterygoiddeii at a right angle to the skull axis, and pectoral scale not in contact with the mesoplastron (de la Fuente 2003:570, fig. 13).

Podocnemidoidae gen. et sp. indeterminate

Material. MCRN 7049, carapace and plastron.

Locality and age. Planicie Banderitas, Neuquén Province, Argentina (Fig. 3.1, locality 6). Probably Anacleto Formation, Río Colorado Subgroup, Neuquén Group, Upper Cretaceous, Santonian (Broin and de la Fuente 1993; de la Fuente 1993; Leanza 1999; Leanza et al. 2004).

Comments. Another Upper Cretaceous podocnemidoid from northern Patagonia was recently described (de la Fuente 1993). This turtle is represented by a single shell (MCRN 7049) from an unknown horizon (probably the Anacleto Formation) of the Río Colorado Subgroup and was referred to ?Podocnemididae indet. (de la Fuente 1993). This specimen might be considered a close relative of *Portezueloemys patagonica* (de la Fuente 2003). The holotype of this species and the referred specimen have a rounded anterior margin of the carapace and a similar pattern of plastral bone and scales. Minor differences are seen in the pectoroabdominal sulci (which touch the top of mesoplastra) and in the convergence of the lateral margins of the posterior lobes in the specimen from Planicie Banderitas. The incomplete nature of this specimen allows it only to be referred to as Podocnemidoidae incertae sedis.

Pancryptodira Joyce, Parham, and Gauthier 2004
Family Meiolaniidae Günther 1888

Comments. The Family Meiolaniidae is a peculiar and bizarre group of extinct cryptodiran turtles bearing cranial horns and frills and caudal clubs, and is restricted to South America (Upper Cretaceous and Eocene) and Australasia (Oligocene to Holocene) (Simpson 1938; Gaffney 1981, 1983, 1996; Gaffney et al. 1984; Estes and Baéz 1985; Broin 1987; Gaffney et al. 1992; Broin and de la Fuente 1993). The meiolaniid's anatomy is known from Gaffney's monograph series (the skull [Gaffney 1983]; the cervical vertebrae [Gaffney 1985]; and the postcranial morphology [Gaffney 1996]) on *Meiolania platyceps* (Owen 1886), the best-known meiolaniid turtle. The phylogenetic relationships of meiolaniids are controversial because their eucryptodiran condition, proposed by Gaffney (1996, and references therein), is disputed by other authors (Lapparent de Broin et al. 1996; Lapparent de Broin and Murelaga 1999; Hirayama et al. 2000).

Genus *Niolamia* Ameghino 1899

Type species. Niolamia argentina (Ameghino 1899; Woodward 1901:170–176, pls. 15–17, 18, figs. 1–2).

Niolamia argentina Ameghino 1899
Fig. 3.6E–K

Neotype. MLP 26–40/41/42 /44, skull, lower jaw, scapula, caudal ring.

Locality and age. Probably 3 km northwest the Colhue Huapi lake, Chubut Province, Argentina (Marshall et al. 1981, 1983) (Fig. 3.1, locality 25). Unknown horizon, Cretaceous or Eocene (Simpson 1938; Estes and Báez 1985; Broin and de la Fuente 1993).

Diagnosis. After Gaffney (1996:72): "A meiolaniid known only from skull and tail ring, characterized by the unique possession of D scale separate by a large X scale, and A scales significantly larger than in other meiolaniid; within meiolaniids the unique possession of an undivided apertura narium externa and only one accessory ridge on maxillary ridge on maxillary triturating surface; nasal bones not projecting beyond rest of skull; B scale projecting posterolaterally and not recurved as in *Meiolania.* D scale area relatively high in contrast to *Meiolania;* A, B, and C scales form a large shelf at back of skull as in *Ninjemys* but more extensive than in *Ninjemys;* intrapterygoid slit not covering foramen caroticum basisphenoidale; tail ring closed ventrally as in *Ninjemys.*"

Comments. Niolamia argentina was named by Ameghino (1899:10) on the basis of shell and skull fragments neither described nor figured that could not be found in collections. Simultaneously, Moreno (1899) announced that Santiago Roth had discovered the *Miolania* (previous generic name for *Meiolania*). Posteriorly, Woodward (1901) described and figured Roth's material: a complete skull, a scapule, and a tail ring as "*Miolania*" *argentina.* Woodward assumed that this was the material that Ameghino named *Niolamia argentina,* but he assigned this species to the Australian genus. The skull of Roth belongs to the Museo de La Plata collections (MLP 26-40/41/42 /44) and was designed by Simpson (1938) as the neotype of *Niolamia argentina.* When Simpson (1938:242) described *Crossochelys corniger* (a junior synonym of *Niolamia argentina* after Gaffney 1996), he became the first reviewer of *Niolamia argentina* and pointed out several problems involved in this history. Recently Gaffney (1983, 1996) argued that *Niolamia* is a sister taxon to other meiolaniids on the basis of the persistence of the primitive condition of the basicranium (intrapterygoid slit not covering foramen caroticum basisphenoidal) and the primitive condition of the maxillary triturating surface (only one accessory ridge). Additional meiolaniid material (a humerus from the Los Alamitos Formation, caudal vertebrae from the Allen Formation, and a horn from the La Colonia Formation) has been recently discovered from Upper Cretaceous (Campanian-Maastrichtian) outcrops of Patagonia (Río Negro and Chubut Provinces), and could be tentatively referred to *Niolamia argentina* or close relative taxa (Broin 1987; Broin and de la Fuente 1993; Gasparini and de la Fuente 2000).

Eucryptodira Gaffney 1975
Cryptodira Cope 1868a
Panchelonioidea Joyce, Parham, and Gauthier 2004
Pancheloniidae Joyce, Parham, and Gauthier 2004

Comments. The family Cheloniidae is the most diverse group of panchelonioids, and includes five extant genera (*Natator* [McCulloch 1908], *Chelonia* [Brongniart 1800], *Eretmochelys* [Fitzinger 1843], *Caretta* [Rafinesque 1814], and *Lepidochelys* [Fitzinger 1843]) and numerous fossil taxa. After Hirayama (1994, 1997), this family also includes the "toxochelid" turtles, and some Cretaceous and most of the Cenozoic sea turtles. Excluding the extinct genera *Toxochelys* (Cope 1873) and *Ctenochelys* (Zangerl 1953), Parham and Fastovsky (1997) subdivided the Cheloniidae in three groups: Osteopyginae (Zangerl 1953), Eocene stem cheloniines (= "Eochelyinae" of Moody 1968), and Cheloninae. Following this point of view, Hirayama and Tong (2003) have recognized four genera within the subfamily Osteopyginae: *Osteopygis* (Cope 1868b), known from the Upper Cretaceous to Paleocene of North and South America and Lower Tertiary of North Africa; *Erquelinnesia* (Dollo 1887), from the upper Paleocene to lower Eocene of Europe (Zangerl 1971; Groessens Van Dyck and Schleich 1988; Moody 1997); *Rhetechelys* (Hay 1908), from the Late Cretaceous of North America (Zangerl 1953); and *Pampaemys,* from the early Paleocene of Argentina (de la Fuente and Casadío 2000). This traditional point of view was recently modified by Joyce et al. (2004). These authors have restricted Cheloniidae to the crown, and turtles excluded from that group that share a more recent common ancestor with Cheloniidae than Dermochelyoidea are considered stem Cheloniidae, and both (stem Cheloniidae + Cheloniidae) are Pancheloniidae. After the "decapitation" of *Osteopygis emarginatus* and the assignation of the referral skulls of this taxon to pancheloniids and the postcrania to stem-Cryptodira proposed by Parham (2005), Parham suggests to restrict *Osteopygis emarginatus* (Cope 1868b) to postcranial material. This postcranium with plesiomorphic morphology is most likely a Eucryptodira incertae sedis of the "macrobaenid" grade. Parham (2005) also recommends that the name Osteopygine be discarded, and Lynch and Parham (2003) propose to restrict all durophagous osteopygines turtles to oldest generic name *Euclastes* (Cope 1867). This taxonomic decision is followed here.

Genus *Euclastes* Cope 1867

Type species. Euclastes platyops (Cope 1867).

***Euclastes* sp.**
Fig. 3.6N, O

Material. Q/377, lower jaw.

Locality and age. Lirquen, Concepción, Chile (Fig. 3.1, locality 2). Upper part of the Quiriquina Formation (Campanian-Maastrichtian).

Comments. Gasparini and Biro-Bagoczky (1986) referred to *Osteopygis* an almost complete *Osteopygis*-like lower jaw. The morphology of this mandible is coincident with the fairly triturating flat surface with symphyseal swelling seen in the lower jaw described as *Osteopygis emarginatus* by Zangerl (1953, fig. 85). Later, Karl et al. (1998) referred to this genus a skull from the same lithostratigraphic horizon as *Osteopygis* aff. *sculptus* (Staesche 1929). Unfortunately, this specific assignation is uncertain, because the fragmentary material of "*Osteopygis sculptus*" from the upper section of the Salamanca Formation (Danian–lower Paleocene) of Chubut Province described by Staesche has been recently assigned to a long-necked species of the family Chelidae (Broin and de la Fuente 1993; Bona and de la Fuente 2001, 2005; Bona 2004).

Conclusions

Almost half of the records of fossil Argentine turtles (Broin and de la Fuente 1993) are represented by Patagonian Mesozoic turtles. Twenty-seven localities from Callovian (Middle Jurassic) to upper Campanian–lower Maastrichtian (Upper Cretaceous) reported here have yielded several species of turtles.

The oldest Patagonian chelonians have been recently discovered in the Cañadón Asfalto Formation (Callovian-Oxfordian) and are represented by a pancryptodire (Sterli personal communication). The history of Patagonian turtles continues with paratethysean panpleurodires (*Notoemys*) and eucryptodires (*Neusticemys*) from the Tithonian (Upper Jurassic) of the Neuquén Basin. *Notoe-*

mys has been considered the sister taxon of pleurodira (Chelidae + Pelomedusoides) (Fernández and de la Fuente 1994; Meylan 1996; Lapparent de Broin 2000; de la Fuente and Iturralde-Vinent 2001), whereas *Neusticemys* could be referred to Eucryptodira incertae sedis. Cretaceous turtles are represented by pleurodires and cryptodires that Broin (1987, 1988) and Broin and de la Fuente (1993) qualified as south and north Gondwanan according to their biogeographic origin. Among the south Gondwanan, the oldest chelids worldwide (pleurodires) are recorded in Patagonia and in several upper Albian to lower Maastrichtian localities of Neuquén, Chubut, and Río Negro Provinces (Fig. 3.1, localities 1, 3, 6, 8–10, 12–13, 15–23, 26, and 27); and the oldest meiolaniids worldwide (pancryptodires) are recorded in some upper Campanian–lower Maastrichtian localities of Chubut and Río Negro provinces (Fig. 3.1, localities 17, 22–23, and 25?). The five named chelid taxa (*Prochelidella, Bonapartemys, Lomalatachelys, Palaeophrynops,* and *Yaminuechelys*) are close relatives of extant South American generic groups (*Acanthochelys, Phrynops, Chelus,* and *Hydromedusa*). The north Gondwanan turtles are represented only by Pelomedusoides Podocnemidoid of Turonian-Coniacian and Santonian ages (*Portezueloemys*) from a few localities of northwestern Patagonia. Finally, sea turtles are poorly represented by Campanian-Maastrichtian pancheloniids of Concepción area (Chile).

Acknowledgments. I thank editors Zulma B. de Gasparini, L. Salgado, and R. Coria for their kind invitation to contribute to this book; and J. Bonaparte (Buenos Aires), R. Bour (Paris), J. Calvo (Neuquén), R. Coria (Plaza Huincul), E. Gaffney (New York), Ch. Hotton (New York), F. Lapparent de Broin (Paris), T. Manera de Bianco (Punta Alta), R. Molnar (Brisbane), R. Pascual (La Plata), P. Vanzolini (São Paulo), P. Wellnhoffer (München), and R. Wild (Stuttgart) for access to specimens under their care. J. González (Buenos Aires) drew the figures; C. Deschamps (La Plata) and J. Sterli (San Rafael) helped with the preparation of the manuscript. J. Parham (Berkeley) and E. Gaffney (New York) provided helpful commentary. This work was partially supported by the Consejo Nacional de Investigaciones Científicas y Técnicas (PIP 002262/00) (to M. de la Fuente) and by Agencia Nacional de Promoción Científica (PICT 8439) (to Z. B. de Gasparini).

References Cited

Ameghino, F. 1899. *Sinopsis Geológico-Paleontológica.* Suplementos (adiciones y correcciones). La Plata, 1899: 1–13.

———. 1906. Les formations sédimentaires du Crétacé supérieur et du Tertiaire de Patagonie avec un parallèl entre leurs faunes mammalogiques et celles de l'ancien continent. *Anales del Museo Nacional de Buenos Aires* 15 (3): 1–568.

Andreis, R. R., A. M. Iñiguez Rodriguez, J. J. Lluch, and D. A. Sabio. 1974. Estudio sedimentológico de las formaciones del Cretácico superior del Lago Pellegrini (provincia de Río Negro, República Argentina). *Revista de la Asociación Geológica Argentina* 29 (1): 85–102.

Archangelsky, S., E. Bellosi, G. Jelfin, and C. Perrot. 1994. Palynology and alluvial facies from the mid-Cretaceous of Patagonia, subsurface of San Jorge Basin, Argentina. *Cretaceous Research* 15: 127–242.

Barcat, C., J. S. Cortinas, V. A. Nevistic, and H. E. Zucchi. 1989. Cuenca Golfo de San Jorge. In G. Chebbli and L. A. Spalletti (eds.), *Cuencas Sedimentarias Argentinas,* 319–345. Serie Correlación Geológica 6. Trabajos X Congreso Geológico Argentino, San Miguel de Tucumán, 1987. Instituto Superior Correlación Geológica. Universidad Nacional de Tucumán.

Batsch, A. J. G. C. 1788. *Versuch einer Anleitung, zur Kenntniß und Geschichte der Thiere und Mineralien.* Jena: Akademische Buchhandlung.

Bona, P. 2004. *Sistemática y Biogeografía de las Tortugas y Cocodrilos Paleocenos de la Formación Salamanca, Provincia de Chubut, Argentina.* Thesis, Universidad Nacional de La Plata.

Bona, P., and M. S. de la Fuente. 2001. A new long necked chelid turtle of the *Hydromedusa* subgroup from the Lower Paleocene of Patagonia, Argentina. Sixth International Congress of Vertebrate Morphology (Jena, Germany). *Journal of Morphology* 248: 208–209.

———. 2005. Phylogenetic and paleobiogeographic implications of *Yaminuechelys maior* (Staesche, 1929) new. comb., a large long-necked chelid turtle from the early Paleocene of Patagonia, Argentina. *Journal of Vertebrate Paleontology* 25 (3): 569–582.

Bona, P., G. Cladera, and M. S. de la Fuente. 1998. Las tortugas pleurodiras de la Formación Salamanca (Paleoceno inferior) en el área de Cerro Hansen, Provincia de Chubut, Argentina. *Actas del X Congreso Latinoamaericano de Geología y VI Congreso Nacional de Geología Económica* 1: 269–274.

Bonaparte, J. F. 1990. New Late Cretaceous mammals from the Los Alamitos Formation, northern Patagonia. *National Geographic Research* 6: 63–93.

———. 1991. Los vertebrados fósiles de la Formación Río Colorado, de la ciudad de Neuquén y cercanías, Cretácico Superior, Argentina. *Museo Argentino de Ciencias Naturales "Bernardino Rivadavia," Revista (Sección Paleontología)* 4: 15–123.

Bonaparte, J. F., M. R. Franchi, J. E. Powell, and E. G Sepúlveda. 1984. La Formación Los Alamitos (Campaniano-Maastrichtiano) del sudeste de Río Negro, con descripción de *Kritosaurus australis* n. sp. (Hadrosauridae). Significado paleogeográfico de los vertebrados. *Revista de la Asociación Geológica Argentina* 39: 284–299.

Boulenger, G. A. 1889. *Catalogue of the Chelonians, Rhynchocephalians and Crocodiles in the British Museum (Natural History).* London: Trustees British Museum (Natural History).

Bridge, J., G. Jalfin, and S. Georgieff. 2000. Geometry, lithofacies and spatial distribution of Cretaceous fluvial sandstone bodies, San Jorge Basin, Argentina, outcrop analog for the hydrocarbon-bearing Chubut Group. *Journal of Sedimentary Research* 70 (2): 341–359.

Broin, F. de. 1987. The Late Cretaceous fauna of Los Alamitos, Patagonia, Argentina. Part IV, Chelonia. *Revista Museo Argentino de Ciencias Naturales, "Bernardino Rivadavia" Paleontología* 3: 131–139.

———. 1988. Les tortues et le Gondwana. Examen des rapports entre le fractionnement du Gondwana au Crétacé et la dispersion géographique des tortues pleurodires à partir du Crétacé. *Studia Geológica Salmanticensia. Studia Palaeocheloniologica* 2: 103–142.

Broin, F. de, and M. S. de la Fuente. 1993. Les tortues fossiles d'Argentine: synthèse. *Annales de Paléontologie* 79: 169–232.

Brongniart, A. 1800. Essai d'une classifications naturelle des reptiles. *Bulletin des Science par la Société Philomathique de Paris* 2: 81–82, 89–91.

Burdbidge, A. A., J. A. Kirsch, and A. R. Main. 1974. Relationships within the Chelidae (Testudines: Pleurodira) of Australia and New Guinea. *Copeia* 2: 392–409.

Cabrera, M. 1998. *Las Tortugas Continentales de Sudamérica Austral.* Córdoba: Edición del Autor.

Cadena Rueda, E. A., and E. S. Gaffney. 2005. *Notoemys zapatocaensis,* a new side-necked turtle (Pleurodira: Platychelyidae) from the Early Cretaceous of Colombia. *American Museum Novitates* 3470: 1–19.

Cattoi, N., and M. A. Freiberg. 1961. Nuevo hallazgo de Chelonia extinguidos en la Republica Argentina. *Physis* 22 (63): 202.

Cope, E. D. 1865. Third contribution to the herpetology of tropical America. *Proceedings of the Academy of Natural Sciences of Philadelphia* 1865: 185–198.

———. 1867. On *Euclastes,* a genus of extinct Cheloniidae. *Proceedings of the Academy of Natural Sciences of Philadelphia* 1867: 39–42.

———. 1868a. On the origin of the genera. *Proceedings of the Academy of Natural Sciences of Philadelphia* 20: 96–140.

———. 1868b. *Osteopygis emarginatus. Proceedings of the Academy of Natural Sciences of Philadelphia* 1868: 233–242.

———. 1871. [Letter to Professor Lesley giving an account of a journey in the valley of the Smoky Hill River in Kansas.] *Proccedings of the American Philosophical Society* 12: 174–176.

———. 1873. *Toxochelys latiremis. Proceedings of the Academy of Natural Sciences of Philadelphia* 1873: 10.

De la Fuente, M. S. 1993. Un posible Podocnemididae (Pleurodira: Pelomedusoides) en el Cretácico Tardío de la Patagonia. *Ameghiniana* 30: 423–433.

———. 2003. Two new pleurodiran turtles from the Portezuelo Formation (Upper Cretaceous) of northern Patagonia, Argentina. *Journal of Paleontology* 77: 559–575.

De la Fuente, M. S., and P. Bona. 2002. Una nueva especie del género *Hydromedusa* Wagler (Pleurodira Chelidae) del Paléogeno de Patagonia. *Ameghiniana* 39: 77–83.

De la Fuente, M. S., and S. Casadío. 2000. Un nuevo osteopigino (Chelonii: Cryptodira) de la Formación Roca (Paleoceno inferior) de Cerros Bayos, provincia de La Pampa, Argentina. *Ameghiniana* 37 (2): 235–246.

De la Fuente, M. S., and M. S. Fernández. 1989. *Notoemys laticentralis* Cattoi and Freiberg, 1961, from the Upper Jurassic of Argentina: a member of the Infraorder Pleurodira (Cope, 1868). *Studia Geologica Salmanticensis. Studia Paleocheloniologica, Salamanca* 3 (2): 25–32.

De la Fuente, M. S., and M. Iturralde-Vinent. 2001. A new pleurodiran turtle from Jagua Formation (Oxfordian) of Western Cuba. *Journal of Paleontology* 75 (4): 860–869.

De la Fuente, M. S., F. de Lapparent de Broin, and T. Manera de Bianco. 2001. The oldest and first nearly complete skeleton of a chelid, of the *Hydromedusa* sub-group (Chelidae, Pleurodira), from the Upper Cretaceous of Patagonia. *Bullettin de la Société Géologique de France* 172: 237–244.

Dollo, L. 1887. On some Belgian fossil reptiles. *Geological Magazine*, n.s., 4: 392–396.

Duméril, A. M. C. 1806. *Zoologie Analytique ou Méthode Naturelle de Classification des Animaux, Rendue plus Facile á la Áide de Tableaux Synoptiques*. Paris: Allais.

Duméril, A. M. C., and G. Bibron. 1835. *Erpétologie Géneral ou Histoire Naturelle Complète des Reptiles*. Paris: Librairie Encyclopédique de Roret.

Estes, R., and A. Báez. 1985. Herpetofaunas of North and South America during the Late Cretaceous and Cenozoic: evidence for interchange? In F. G. Sthel, and S. D. Webb (eds.), *The Great American Interchange*, 6:139–197. New York: Plenum Press.

Fernández, M. S., and M. S. de la Fuente. 1988. Nueva tortuga (Cryptodira: Thalassemydidae) de la Formación Vaca Muerta (Jurásico, Tithoniano) de la Provincia del Neuquén, Argentina. *Ameghiniana* 25 (2): 129–138.

———. 1993. Las tortugas casiquelidias de las calizas litográficas titonianas del área de Los Catutos, Neuquén, Argentina. *Ameghiniana* 30 (3): 283–295.

———. 1994. Redescription and phylogenetic position of *Notoemys:* the oldest Gondwanian pleurodiran turtle. *Neues Jahrbuch für Geologie und Paläontologie, Abhandlungen* 193 (1): 81–105.

Fitzinger, L. 1826. *Neue Classification der Reptilien nach ihren Natürlichen Verwandtschaften Verlag*. Vienna: J. G. Húbner.

———. 1843. *Systema Reptilium*. Vindobona: Braumüller and Seidel.

Flynn, J. J., and C. C. Swisher. 1995. Cenozoic South American land mammal ages: correlations to a global geochronology. *SEPM Special Publication* 54: 317–330.

Freiberg, M. A. 1945. Tortuga del género *Platemys* Wagler. *Physis* 20: 19–23.

Gaffney, E. S. 1975. A phylogeny and classification of the higher categories of turtles. *Bulletin of the American Museum of Natural History* 155: 389–436.

———. 1977. The side-necked turtle family Chelidae: a theory of relationships using shared derived characters. *American Museum Novitates* 2620: 1–28.

———. 1979. Comparative cranial morphology of recent and fossil turtles. *Bulletin of the American Museum of Natural History* 164: 65–376.

———. 1981. A review of the fossil turtles of Australia. *American Museum Novitates* 2737: 1–22.

———. 1983. The cranial morphology of the extinct horned turtle, *Meiolania platyceps*, from the Pleistocene of Lord Howe Island Australia. *Bulletin of the American Museum of Natural History* 175 (4): 326–479.

———. 1985. The cervical and caudal vertebrae of the cryptodiran turtle, *Meiolania platyceps*, from the Pleistocene of Lord Howe Island, Australia. *American Museum Novitates* 2805: 1–29.

———. 1996. The postcranial morphology of *Meiolania platyceps* and a review of the Meiolaniidae. *Bulletin of the American Museum of Natural History* 229: 1–166.

Gaffney, E. S., and P. A. Meylan. 1988. A phylogeny of turtles. In M. J.

Benton (ed.), *The Phylogeny and Classification of the Tetrapods 1: Amphibians, Reptiles, Birds. Systematic Association,* 35A:157–219. Oxford: Clarendon Press.

———. 1991. Primitive pelomedusid turtle. In J. G. Maisey (ed.), *Santana Fossils, an Illustred Atlas,* 335–339. Neptune City, N.J.: TFH Publications.

Gaffney, E. S, J. C. Balouet, and F. de Broin. 1984. New ocurrence of extinct meiolaniid turtles in New Caledonia. *American Museum Novitates* 2800: 1–6.

Gaffney, E. S., M. Archer, and A. White. 1992. *Warkalania* a new meiolaniid turtle from the Tertiary Riversleigh deposits of Quennsland, Australia. *Beagle* 9 (1): 35–48.

Gaffney, E. S., D. de Almeida Campos, and R. Hirayama. 2001. *Cearachelys,* a new side-necked turtle (Pelomedusoides: Bothremydidae) from the Early Cretaceous of Brazil. *American Museum Novitates* 3319: 1–20.

Gasparini, Z., and L. Biro-Bagoczky. 1986. *Osteopygis* sp. (Reptilia, Testudines, Toxochelyidae) tortuga fósil de la Formación Quiriquina, Cretácico superior, sur de Chile. *Revista Geológica de Chile* 27: 85–90.

Gasparini, Z., and M. S. de la Fuente. 2000. Tortugas y plesiosaurios de la Formación La Colonia (Cretácico Superior) de la Patagonia. *Revista Española de Paleontología* 15: 23–35.

Gasparini, Z., L. Spalletti, and M. S. de la Fuente. 1997. Marine reptiles of a Tithonian transgression, western Neuquén Basin, Argentina: Facies and paleoenvironments. *Geobios* 30: 701–712.

Gasparini, Z., M. S. de la Fuente, M. S. Fernández, and P. Bona. 2001. Reptiles from coastal environments of the Upper Cretaceous of northern Patagonia. VII International Symposium on Mesozoic Terrestrial Ecosystems (Buenos Aires, 1999). *Asociación Paleontológica Argentina Publicación Especial* 7: 101–105.

Gasparini, Z., S. Casadío, M. S. de la Fuente, L. Salgado, M. S. Fernández, and A. Concheyro. 2002. Reptiles acuáticos en sedimentitas lacustres y marinas del Cretácico Superior de Patagonia (Río Negro, Argentina). *Actas del XV Congreso Geológico Argentino,* 1495–1499.

Georges, A., J. Birrell, K. M. Saint, W. McCord, and S. C. Donnellan. 1998. A phylogeny for side-necked turtles (Chelonia: Pleurodira) based on mitochondrial and nuclear gene sequence variation. *Biological Journal of the Linnean Society* 67: 213–246.

Goin, F. J., D. Poire, M. S. de la Fuente A. L. Cione, F. E. Novas, E. S. Bellosi, A. Ambrossio, O. Ferrer, N. D. Canessa, A. Carloni, J. Ferigolo, A. M. Ribereiro, M. S. Sales Viana, M. Reguero, M. G. Vucetich, S. Marenssi, M. F. de Lima Filho, and S. Agostinho. 2002. Paleontología y Geología de los sedimentos del Cretácico Supérior aflorantes al sur del río Shehuen (Mata Amarilla), provincia de Santa Crruz, Argentina. *Actas del XV Congreso Geológico Argentino* 1: 603–608.

Gray, J. E. 1873. Observations on chelonians, with descriptions of new genera and species. *Annals and Magazine of Natural History* 4 (11): 289–308.

Groeber, P. 1946. Observaciones geológicas a lo largo del meridiano 70. I. Hoja Chos Malal. *Revista de la Sociedad Geológica Argentina* 1: 177–208.

Groessens Van Dyck, M-C., and H. H. Schleich. 1988. Zur verbreitung Tertiärer und Quartärer reptilien und amphibien Europas. Belgien,

Dänemark, Niederlande, Schweden. *Studia Geologica Salamanticensia* 3: 113–148.

Gulisano, C. A., A. R. Gutiérrez Pleimling, and R. E. Digregorio. 1984. Análisis estratigráfico del intervalo Tithoniano-Valanginiano (Formación Vaca Muerta, Quintuco y Mulichinco) en el suroeste de la provincia de Neuquén. *Actas del IX Congreso Geológico Argentino* 1: 221–235.

Hay, O. P. 1908. *The Fossil Turtles of North America.* Washington, D.C.: Carnegie Institution.

Hirayama, R. 1992. Humeral morphology of chelonioid sea turtles: its functional analysis and phylogenetic implications. *Bulletin of the Hobetsu Museum* 8: 17–57.

———. 1994. Phylogenetic systematics of chelonioid sea turtle. *The Island Arch* 3: 270–284.

———. 1997. Distribution and diversity of Cretaceous chelonioids. In J. M. Callaway and E. L. Nicholls (eds.), *Ancient Marine Reptiles,* 225–241. San Diego: Academic Press.

———. 1998. Oldest known sea turtle. *Nature* 392: 705–708.

Hirayama, R., and H. Tong. 2003. *Osteopygis* (Testudines: Cheloniidae) from the Lower Tertiary of the Ouled Abdoun Phosphate Basin, Morocco. *Palaeontology* 46: 845–856.

Hirayama, R., D. B. Brinkmann, and I. Danilov. 2000. Distribution and biogeografhy of non-marine Cretaceous turtles. *Russian Journal of Herpetology* 7 (3): 181–198.

Hugo, C. A., and H. A. Leanza. 2001a. *Hoja Geológica 3966–III, Villa Regina, Provincia de Río Negro.* Instituto de Geología y Recursos Naturales. Servicio Geológico y Minero de Argentino (SEGEMAR), Bulletin 309: 1–53.

Hugo, C. A., and H. A. Leanza. 2001b. *Hoja Geológica 3969–IV, General Roca, provincias de Neuquén y Río Negro.* Instituto de Geología y Recursos Naturales. Servicio Geológico y Minero de Argentino (SEGEMAR), Bulletin 308: 1–71.

Iverson, J. B. 1992. *A Revised Checklist with Distribution Maps of the Turtles of the World.* Richmond, Ind.: Privately printed.

Joyce, W. G. 2000. The first complete skeleton of *Solnhofia parsonsi* (Cryptodira, Eurysternidae) from the Upper Jurassic of Germany and its taxonomic implications. *Journal of Paleontology* 74 (4): 684–700.

Joyce, W. G., J. F. Parham, and J. A. Gauthier. 2004. Developing a protocol for the conversion of rank-based taxon names to phylogenetically defined clade names, as exemplified by turtles. *Journal of Paleontology* 78 (5): 989–1013.

Karl, H. V., G. Tichy, and H. Ruschark. 1998. *Osteopygoides priscus* n. gen. n. sp. und die Taxonomie und Evolution der Osteopygidae. *Mittelungen der Geologischen und Paleontologischen Landesmuseum Joaneeum* 56: 329–350.

Lapparent de Broin, F. de. 1994. Turtles from the Chapada do Araripe, Early Cretaceous, Ceará State, Brasil. Boletim do 3° Simpósio sobre o Cretaceo do Brasil, 137–138. Río Branco.

———. 2000. The oldest pre-Podocnemidid (Chelonii, Pleurodira), from the Early Cretaceous, Ceará state, and its environments. *Treballs del Museu de Geología de Barcelona* 9: 43–95.

———. 2001. The European turtle fauna from the Triassic to the Present. *Dumerilia* 4 (3): 155–216.

Lapparent de Broin, F. de, and M. S. de la Fuente. 1996. The analysis of

the character "fused pelvis." *Journal of Vertebrate Paleontology* 16 (Suppl. to 3): 47A.
———. 1999. Particularidades de la fauna continental de tortugas del Cretácico de Argentina. *Ameghiniana* 36: 104.
———. 2001. Oldest world Chelidae (Chelonii, Pleurodira), from the Cretaceous of Patagonia. *Comptes Rendues Académie des Sciences de Paris* 333: 463–470.
Lapparent de Broin, F. de, and R. Molnar. 2001. Eocene chelid turtles from Redbank Plains, southeastern Queensland, Australia. *Geodiversitas* 23 (1): 41–79.
Lapparent de Broin, F. de, and X. Murelaga.1999. Turtles from the Upper Cretaceous of Laño (Iberian Peninsula). In H. Astibia, J. C. Corral, X. Murelaga, X. Xorue-Etxebarria, and X. Pereda-Suberbiola (coord.), *Geology and Palaeontology of the Upper Cretaceous Vertebrate-Bearing Beds of the Laño Quarry (Basque-Cantabrican Region, Iberian Peninsula),* 135–211 Estudios del Museo de Ciencias Naturales de Alava 14 (Número Especial 1).
Lapparent de Broin, F. de, B. Lange-Badré, and M. Dutrieux. 1996. Nouvelles découvertes de tortues dans le Jurassique supérieur du Lot (France) et examen du taxon Plesiochelyidae. *Revue de Paléobiologie* 15: 533–570.
Lapparent de Broin, F. de, M. S. de la Fuente, and J. Calvo. 1997. Presencia de los más antiguos Quélidos (Tortugas pleurodiras) en el Cretácico Inferior de "El Chocón," Provincia del Neuquén, Argentina. *Ameghiniana* 34: 538.
Lapparent de Broin, F. de, M. S. de la Fuente, and M. S. Fernández. In press. *Notoemys* (Chelonii, Pleurodira), Late Jurassic of Argentina: new examination of the anatomical structures and comparisons. *Revue de Paleobiologie.*
Leanza, H. A. 1980. The Lower and Middle Tithonian ammonite fauna from Cerro Lotena, province of Neuquén, Argentina. *Zitteliana* 5: 1–49.
———. 1981. The Jurassic-Cretaceous boundary beds in West Central Argentina and their ammonite zones. *Jahrbuch Geologie und Paläontologie Abhandlungen* 161 (1): 62–92.
———. 1999. The Jurassic and Cretaceous terrestrial beds from Southern Neuquén Basin, Argentina. Field guide. Instituto Superior de Correlación Geológica. INSUGEO. Serie Miscelánea 4: 1–30. San Miguel de Tucumán.
Leanza H. A., and C. A. Hugo. 1977. Sucesión de amonites y edad de la Formación Vaca Muerta y sincrónicas entre los paralelos 35° y 40° lat. S. Cuenca Neuquina-Mendocina. *Revista de la Asociación Geológica Argentina* 32: 248–264.
———. 1995. Revisión estratigráfica del Cretácico inferior continental en el ámbito sudoriental de la Cuenca Neuquina. *Revista de la Asociación Geológica Argentina* 50: 30–32.
Leanza H. A., and A. Zeiss. 1990. Upper Jurassic lithographic limestones from Argentina (Neuquén Basin): stratigraphy and fossils. *Facies* 22: 169–186.
Leanza, H. A., S. Apesteguía, F. E. Novas, and M. S. de la Fuente. 2004. Cretaceous terrestrial beds from Neuquén Basin (Argentina) and their tetrapods assemblages. *Cretaceous Research* 25 (1): 61–87.
Lee, M. S. Y. 1995. Historical burden in systematics and interrelationships of "parareptiles." *Biological Review* 70: 459–547.

Legarreta, L., and C. Gulisano. 1989. Análisis estratigráfico secuencial de la Cuenca Neuquina (Triásico Superior–Terciario Inferior). In G. Chebli and L. Spalletti (eds.), *Cuencas Sedimentarias Argentinas, Serie Correlación Geológica* 6: 221–243.

Lindholm, W. A. 1929. Revidiertes Verzeichnis der Gattungen der rezenten Schildkröten nebst Notizen zur Nomenklatur einiger Arten. *Zoologischer Anzeiger* 81: 275–295.

Lynch, S. C., and J. F. Parham. 2003. The first record of hard-shelled sea turtles (Cheloniidae sensu lato) from the Miocene of California: Including a new species (*Euclastes hutchisoni*) with unusually plesiomorphic character. *PaleoBios* 23 (3): 21–35.

Manera de Bianco, T. 1996. Nueva localidad con nidos y huevos de dinosaurios (Titanosauridae) del Cretácico superior, Cerro Blanco, Yaminué, Río Negro, Argentina. Asociación Paleontológica Argentina, Publicación Especial 4, 59–67.

Marshall, L. G., R. F. Butler, R. E. Drake, and G. H. Curtis. 1981. Calibration of the beginning of the age of mammals in Patagonia. *Science* 214: 43–45.

Marshall, L. G., R. Hoffstetter, and R. Pascual. 1983. Mammals and stratigraphy: geochronology of the continental mammal-bearing Tertiary of South America. *Palaeovertebrata, Mémoire Extraordinaire,* 1–93.

Martínez, R., O. Giménez, J. Rodríguez, and G. Bochatey. 1986. *Xenatarsoaurus bonapartei* nov. gen. et sp. (Carnosauria, Abelisauridae), un nuevo Theropoda de la Formación Bajo Barreal, Chubut, Argentina. *Simposio "Evolución de Vertebrados Mesozoicos," IV Congreso Argentino de Paleontología y Bioestratigrafía* 2: 23–32.

McCord, W. P., M. Joseph-Ouni, and W. W. Lamar. 2001. Taxonomic reevaluation of *Phrynops* (Testudines: Chelidae) with the description of new genera and a new species of *Batrachemys. Revista de Biología Tropical* 49 (2): 715–764

McCulloch, A. R. 1908. A new genus and species of turtle from North Australia. *Records of the Australian Museum* 7: 126–128.

Meyer, H. von. 1839. *Eurysternum Wagleri, Münster. Eine schildkröte aus dem Kalkschiefer von Solnhofen. Beiträge zur Petrefacten-Kunde, Georg, Graf zu Münster.* Buchner'schen Buchhandlung, Bayeruth 1 (11):75–82.

Meylan, P. A. 1996. Skeletal morphology and relationships of the Early Cretaceous side-necked turtle *Araripemys barretoi* (Testudines, Pelomedusoides, Araripemydidae) from the Santana Formation of Brazil. *Journal of Vertebrate Paleontology* 16: 20–33.

Meylan, P. A., and Gaffney, E. S. 1991. *Araripemys* Price, 1973. In J. G. Maisy (ed.), *Santana Fossils: An Ilustrated Atlas,* 326–334. Neptune City, N.J.: TFH Publications.

Mikan, J. C. 1820. *Delectus Florae et Faunae Brasiliensis.* Vienna: Antonii Strauss.

Mitchum, R., and M. Uliana, 1985. Seismic stratigraphy of carbonate depositional sequences, Upper Jurassic–Lower Cretaceous, Neuquén Basin, Argentina. In O. Berg and D. Woolverton (eds.), *Seismic Stratigraphy II, an Integrated Approach,* 255–274. American Association of Petroleum Geologists, Memoir 39.

Moody, R. T. J. 1968. A turtle, *Eochelys crassicostata* (Owen) from the London Clay of the Isle of Sheppey. *Proceeding of the Geological Association of London* 79: 129–140.

———. 1997. The paleogeography of marine and coastal turtle from the North Atlantic and Trans-Saharan regions. In J. M. Callaway and E. L. Nicholls (eds.), *Ancient Marine Reptiles,* 259–280. San Diego: Academic Press.

Moreno, F. P. 1899. Note on the discovery of *Miolania* and of *Glossotherium* (*Neomylodon*) in Patagonia. *Geological Magazine,* n.s. Decade IV 6: 385–388.

Owen, R. 1882. On an extinct reptile (*Notochelys costata* Owen) from Australia. *Quarterly of Journal Geological Society of London* 38: 178–184.

———. 1886. Description of a fossil remains of two species of Megalanian genus (*Meiolania,* Ow) from Howe's Island. *Proceedings of the Royal Society of London* 40: 315.

Parham, J. F. 2005. A reassessment of the referral of sea turtle skulls to the genus *Osteopygis* (Late Cretaceous, New Jersey, USA). *Journal of Vertebrate Paleontology* 25 (1): 71–77.

Parham, J. F., and D. E. Fastovsky. 1997. The phylogeny of cheloniid sea turtle revised. *Chelonian Conservation and Biology* 2: 548–554.

Price, L. I. 1973. Quelonio Amphichelydia no Cretáceo Inferior no Nordeste de Brasil. *Revista Brasileira de Geociencias* 3 (2): 84–96.

Pritchard, P. C. H. 1984. Piscivory in turtles, and evolution of the long-necked Chelidae. *Symposia of the Zoological Society of London* 52: 87–110.

Pritchard, P. C. H., and P. Trebbau. 1984. The turtles of Venezuela. *Contribution to Herpetology, Society Study Amphibians, Reptiles Publications* 2: 1–403.

Rafinesque, C. S. 1814. Prodromo di erpetologie siciliana. *Specchio delle Scienze o Giornale Enciclopedico di Sicilia* 2: 65–66.

Ramos, V. 1988. Late Proterozoic–Early Paleozoic of South America–A collisional history. *Episodes* 2 (3): 168–174.

———. 1999. Las provincias geológicas del territorio argentino. In R. Caminos (ed.), *Geología Argentina,* 41–96. Anales SEGEMAR 29.

Ramsay, E. P. 1887. On a new genus and species of freshwater tortoise from the Fly River, New Guinea. *Proceedings of the Linnean Society of New South Wales* 1: 158–162.

Rauhut, O. W. M., and P. Puerta. 2001. New vertebrate fossils from the Middle-Late Jurassic Cañadón Asfalto Formatión of Chubut, Argentina. *Ameghiniana* 38: 16R.

Rauhut, O. W. M., A. Lopez Arbarello, P. Puerta, and T. Martin. 2001. Jurassic vertebrates from Patagonia. *Journal of Vertebrate Paleontology* 21 (Suppl.): 91A.

Rhodin, A. G. J., and R. A. Mittermeier. 1983. Description of *Phrynops williamsi,* a new species of chelid turtle of the South American *P. geoffroanus* Complex. In A. G. J. Rhodin and K. Miyata (eds.), *Advances in Herpetology and Evolutionary Biology: Essays in Honor of Ernest E. Williams,* 58–73. Cambridge, Mass.: Museum of Comparative Zoology.

Schweigger, A. F. 1812. Prodromus monographiae cheloniorum. *Königsberger Archiv für Naturwissenschaften und Mathematik* 1: 271–368, 406–462.

Seddon, J. M., A. Georges, P. Bavesrstock, and W. McCord. 1997. Phylogenetic relationships of chelid turtles (Pleurodira: Chelidae) based on mitochondrial 12S rRNA gene sequence variation. *Molecular Phylogenetics and Evolution* 7: 55–61.

Shaffer, H. B., P. Meylan, and M. L. Mcknight. 1997. Test of turtle phylogeny: molecular, morphological, and paleontological approaches. *Systematic Biology* 46: 235–268.

Simpson, G. G. 1938. *Crossochelys,* Eocene horned turtle from Patagonia. *Bulletin of the American Museum of Natural History* 74: 221–254.

Staesche, K. 1929. Schildkröten aus der oberen kreide Patagoniens. *Palaeontographica* 72: 103–123.

Tong, H., and E. Buffetaut. 1996. A new genus and species of pleurodiran turtle from the Cretaceous of southern Morocco. *Neues jahrbuch für Geologie und Paläontologie, Abhandlungen* 199: 133–150.

Wagler, J. G. 1830. *Natürliches System der Amphibienmit Vorangehender Classification der Säugethiere und Vögel.* Munich: Cotta'schen.

Wagner, A. 1853. Beschreibung einer fossilen schildkróte und etlicherr anderer reptilienüberreste aus den lithographischen schiefern und dem gründsandsteine von Kelhaim. *Abhanlungen Bayerische Academischer der Wissenschaft. Mathematisch-Physikalische* 7: 239–264.

Walker, W. F., Jr. 1973. The locomotor apparatus of Testudines. In C. Gans and T. S. Parsons (eds.), *Biology of the Reptilia,* 4:1–99. London and New York: Academic Press.

Warren, J. W. 1969. Chelid turtles from the Mid-Tertiary of Tasmania. *Journal of Paleontology* 43: 179–182.

Wichmann, R. 1927. Facies lacustres senonianas de los estratos con dinosaurios. *Boletin de la Academia Nacional de Ciencias* 30: 386–406.

Wieland, G. R. 1896. *Archelon ischyros,* a new gigant cryptodire testudinate from the Fort Pierre Cretaceous of South Dakota. *American Journal of Science* 2 (4): 399–412.

Wood, R. C., and M. A. Freiberg. 1977. Redescription of *Notoemys laticentralis,* the oldest fossil turtle from South America. *Acta Geologica Lilloana* 13 (6): 187–204.

Woodward, A. S. 1901. On some extinct reptiles from Patagonia, of the genera *Miolania, Dinilysia* and *Genyodectes. Proceeding of the Zoological Society of London* 1901: 169–184.

Zangerl, R. 1953. The vertebrate fauna of the Selma Formation of Alabama. Part VI. The turtle of the family Toxochelyidae. *Fieldiana Geology Memoirs* 3: 137–277.

———. 1971. Two toxochelyid sea turtles from the Landenian sands of Erquelinnes (Hainaut) of Belgium. *Institut Royal des Siences Naturelles de Belgique, Mémoires* 169: 1–32.

———. 1980. Pattern of phylogenetic differentiation in the toxochelyid and cheloniid sea turtles. *American Zoologist* 20: 585–596.

4. Lepidosauromorpha

Adriana Albino

Introduction

Among Lepidosauromorpha, only fossil remains of Lepidosauria have been found in South America. Lepidosaurs are by far the most diverse of modern reptiles, including nearly 7000 species of lizards, amphisbaenians, and snakes (Squamata), and two species of tuatara (Rhynchocephalia). Cladistic analyses have shown that lizards do not form a monophyletic group to the exclusion of amphisbaenians and snakes (Estes et al. 1988); hence, the traditional term *lizard* is here used to denote a squamate that is neither an amphisbaenian nor a snake.

The primary dichotomy within Squamata is between Iguania and Scleroglossa; among the latter, clades Gekkota, Scincomorpha, and Anguimorpha are recognized (Estes et al. 1988). The most diverse group of extinct anguimorphs are the aquatic mosasauroids. Recent analyses have placed snakes within Anguimorpha, although their precise relationships continue to be discussed (Caldwell and Lee 1997; Lee 1997, 1998; Lee and Caldwell 1998, 2000; Caldwell 1999; Lee et al. 1999; Zaher and Rieppel 1999, 2002; Rage and Escuillié 2000; Rieppel and Zaher 2000a; Scanlon and Lee 2000; Tchernov et al. 2000; Lee and Scanlon 2002; Rieppel et al. 2003). Living snakes are distributed in two main groups that are considered to be sister groups, Scolecophidia and Alethinophidia (Rieppel 1988). The phyletic position of various fossil snakes is questionable; this is the case of two extinct snake families well represented in the Mesozoic of Patagonia, the Dinilysiidae and the Madtsoiidae.

This chapter discusses the record of Patagonian Mesozoic lepi-

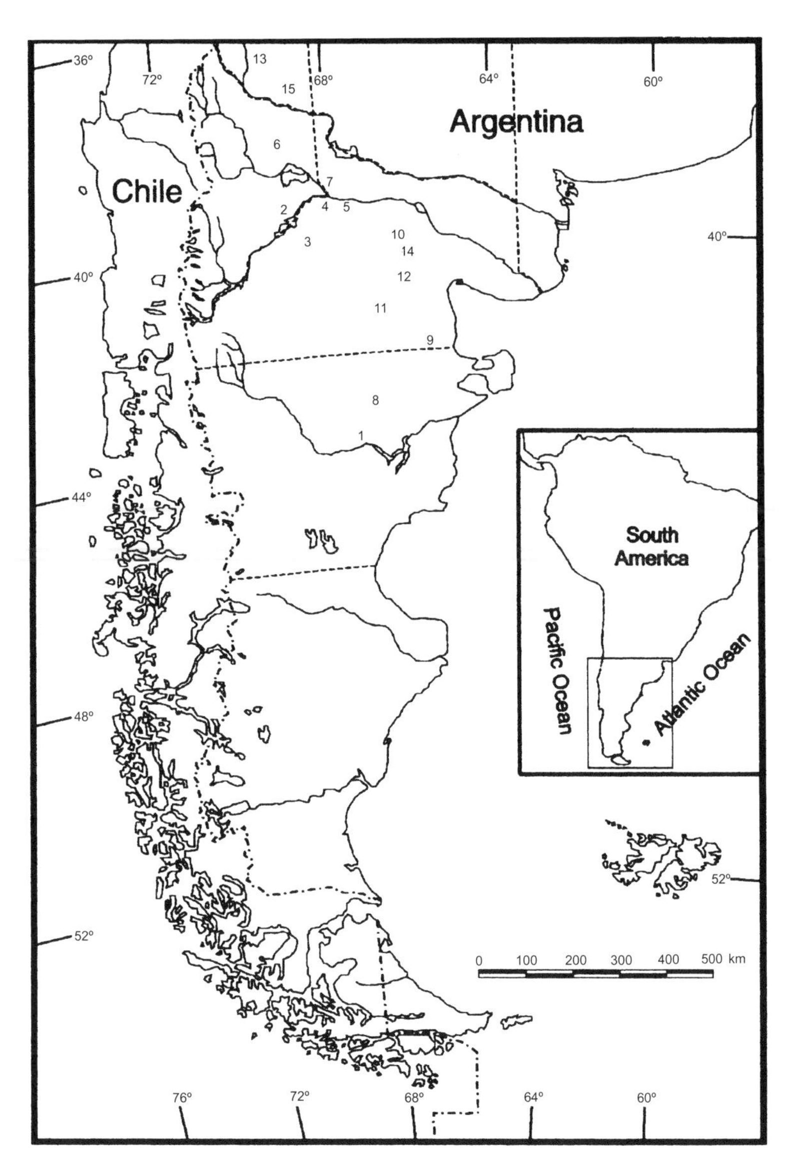

Figure 4.1. Map of the localities cited in the text in chronological order. 1, Cerro Cóndor, Sphenodontia indet.; 2, Araceli site, Kaikaifilusaurus calvoi; *3, La Buitrera,* Kaikaifilusaurus calvoi, *?Iguanidae, Madtsoiidae indet.; 4, Neuquén City,* Dinilysia patagonica; *5, Paso Córdoba,* Dinilysia patagonica; *6, Barreales Norte,* Dinilysia patagonica; *7, Cinco Saltos, Squamata indet.,* Dinilysia patagonica; *8, Ea. La Colonia,* Alamitophis argentinus, *?Madtsoiidae, Serpentes incertae sedis; 9, Ea. Los Alamitos, Sphenodontia indet., Squamata indet., ?Iguanidae,* Alamitophis argentinus, Alamitophis elongatus, Patagoniophis parvus, Rionegrophis madtsoioides, *?Madtsoiidae, ?Boidae; 10, Salinas de Trapalcó,* A. argentinus, A. elongatus; *11, Ea. El Palomar,* A. argentinus; *12, Salitral de Santa Rosa, Sphenodontia indet.,* Alamitophis argentinus, Patagoniophis parvus, *?Madtsoiidae, Madtsoiidae indet.; 13, Ranquil-Có, ?Madtsoiidae; 14, Santa Rosa–Trapalcó area, Mosasaurinae indet.; 15, Liu Malal, Mosasaurinae indet.*

dosaurs. Localities of Neuquén, Río Negro, and Chubut Provinces (Fig. 4.1) provided Mesozoic remains of rhynchocephalians, terrestrial lizards, mosasaurs, and snakes. In addition, remains of mosasaurs and snakes from the Late Cretaceous were found at two extra-Patagonian localities, southern Mendoza Province (Fig. 4.1).

Triassic deposits do not provide lepidosaur remains at the moment, and Patagonian Jurassic lepidosaurs are represented by a few undescribed rhynchocephalian remains. In contrast, the Cretaceous Patagonian record is large and diverse; hence, the treatment of the subject in this chapter follows a systematic arrangement. Phylogenetic hypothesis about the families Dinilysiidae and Madtsoiidae are briefly discussed. The geographic distribution of fossil lepidosaurs in the South American Mesozoic and its paleobiogeographical implications are especially considered.

Institutional abbreviations. MACN-Pv, Colección Nacional de Paleovertebrados, Museo Argentino de Ciencias Naturales "Bernardino Rivadavia," Buenos Aires, Argentina; MCF-PVPH, Museo "Carmen Funes," Paleontología de Vertebrados, Plaza Huincul, Neuquén, Argentina; MEGyP, Museo Educativo de Geología y Paleontología, General Roca, Río Negro, Argentina; MLP, Museo de La Plata, La Plata, Buenos Aires, Argentina; MPCA, Museo Provincial "Carlos Ameghino," Cipolletti, Río Negro, Argentina; MPCHv, Museo "Ernesto Bachmann," El Chocón, Neuquén, Argentina; MPEF, Museo Paleontológico "Egidio Feruglio," Trelew, Chubut, Argentina; MUCPV, Museo de Geología y Paleontología, Universidad Nacional del Comahue, Neuquén, Argentina.

Systematic Paleontology

Lepidosauria Haeckel 1866

Rhynchocephalia Günther 1867

Sphenodontia Williston 1925

Opisthodontia Apesteguía and Novas 2003

Eilenodontinae Rasmussen and Callison 1981

Kaikaifilusaurus Simón and Kellner 2003

Kaikaifilusaurus calvoi Simón and Kellner 2003

Fig. 4.2

Holotype. MPCHv 4, left lower jaw lacking the posterior region.

Referred specimens. MPCA 300, a juvenile incomplete skeleton (Fig. 4.2); MPCA 302, postcranial isolated remains; MPCA 303, a partially articulated adult skeleton.

Locality and age. The type specimen was found in the Araceli site, located close to National Route 237, near Villa El Chocón, about 80 km southeast of Neuquén City, Neuquén Province, Argentina (Fig. 4.1, locality 2) (Simón and Kellner 2003), Candeleros Formation, early Cenomanian (Leanza et al. 2004). The other specimens were found in sediments of the same formation at "La Buitrera" fossil quarry, Cerro Policía, Río Negro Province, Argentina (Fig. 4.1, locality 3) (Apesteguía and Novas 2003).

Diagnosis. Robust skeleton with adults up to 1 m long; single, sharp, beaklike structure on the rostral end of the skull; bulging nasals; frontals rostrally straight; jugals dorsoventrally deep; suborbital fenestrae closed by transversely oriented ectopterygoids; maxillary and palatine tooth rows elongate, straight, and parallel; palatal roof narrow; teeth closely packed, with imbricate labial and lingual flanges; no caniniform teeth and no dental regionalization; and square, distally expanded unguals (Apesteguía and Novas 2003, modified from Simón and Kellner 2003) (Fig. 4.2).

Comments. Kaikaifilusaurus calvoi was described on the basis of an isolated lower jaw from the Araceli site (Simón and Kellner 2003,

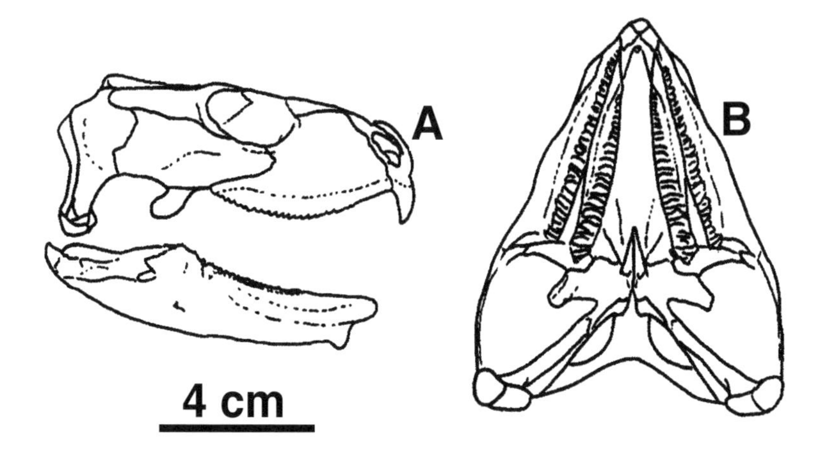

Figure 4.2. Kaikaifilusaurus calvoi *(MPCA 300): (A) Skull and mandible in lateral view. (B) Skull in ventral view.* Modified from Apesteguía and Novas (2003).

accepted July 21). Apesteguía and Novas (2003, accepted July 23) described the new species *Priosphenodon avelasi* on the basis of some partially articulated adult and juvenile skeletons of sphenodontians collected at "La Buitrera." Also, Apesteguía and Novas (2003) presented a phylogenetic analysis of the Sphenodontia in which Opisthodontia emerges as a new taxon (Fig. 4.3). In contrast, Simón and Kellner (2003) only suggested autapomorphies of the lower jaw of *Kaikaifilusaurus*, but did not present a formal phylogeny. Descriptions and illustrations do not evidence dissimilarities between *Kaikaifilusaurus* specimens and material studied by Apesteguía and Novas (2003). In addition, the lower jaw of *Kaikaifilusaurus* shows all synapomorphies mentioned by Apesteguía and Novas (2003). Thus, *Priosphenodon* is here considered a synonym of *Kaikaifilusaurus*. The skull of *Kaikaifilusaurus* reaches 150 mm, with total adult body length slightly surpassing 1 m—making this the largest terrestrial sphenodontian yet recorded (Apesteguía and Novas 2003). Eilenodontine opisthodontians increased in size throughout their evolutionary history. *Kaikaifilusaurus* represents an extreme of an evolutionary trend toward a large size (Apesteguía and Novas 2003).

Sphenodontia gen. et sp. indeterminate

Undescribed sphenodontians from the Cañadón Asfalto Formation (Middle Jurassic) of Cerro Cóndor, Chubut Province, Argentina (Fig. 4.1, locality 1) (Rougier personal communication 2003; Rauhut personal communication 2003), constitute the earliest record of lepidosaurs in Patagonia. Sphenodontian remains of Upper Late Cretaceous age found in the Los Alamitos Formation at Ea. Los Alamitos (Fig. 4.1, locality 9) (Apesteguía and Novas 2003) and in the Allen Formation at Salitral de Santa Rosa (Fig. 4.1, locality 12) (Martinelli and Forasiepi 2004), both in the Río Negro Province (Argentina), represent the latest record of sphenodontians outside New Zealand.

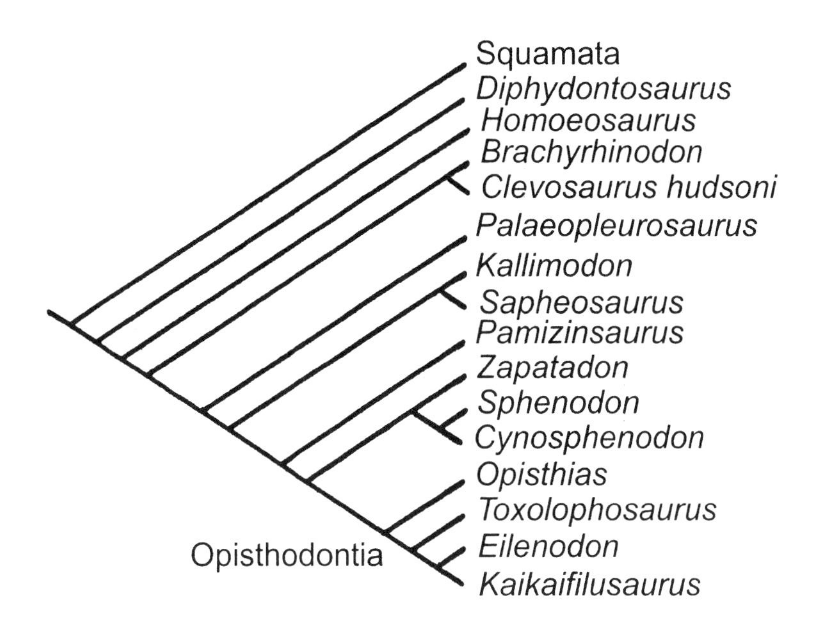

Figure 4.3. Phylogenetic relationships among Sphenodontia (Apesteguía and Novas 2003).

Squamata Oppel 1811
Gen. et sp. indeterminate

Undescribed remains were identified as unquestionably continental lizards from the Patagonian Mesozoic. One incomplete dentary was found in the Anacleto Formation at Cinco Saltos, Río Negro Province, Argentina (Fig. 4.1, locality 7) (Albino 2002), lower Campanian (Leanza et al. 2004). It resembles the midregion of the dentary of the Late Cretaceous South American lizard *Pristiguana brasiliensis;* however, the specimen is not preserved well enough to make a better comparison possible. Maxillae and dentaries of at least three noniguanian lizard taxa also come from Los Alamitos Formation (Campanian-Maastrichtian) at Ea. Los Alamitos (Fig. 4.1, locality 9) (personal observation on uncataloged specimens). They include probable teiid specimens.

Iguania Cope 1864
Iguanidae (sensu Schulte et al. 2003)
Gen. et sp. indeterminate

The incomplete fused frontals of an iguanian lizard were found at "La Buitrera" (Fig. 4.1, locality 3), Candeleros Formation (lower Cenomanian, Leanza et al. 2004) (Apesteguía et al. 2004, 2005). They are strongly constricted between the orbits, a derived character state of Iguania (reversed in Chamalaeonidae) (Estes et al. 1988; Frost and Etheridge 1989). The presence of dorsal rugosities differs from the condition in Agamidae (sensu Schulte et al. 2003). Apesteguía et al. (2005) said that the frontals are moderately rugose, resembling those of *Liolaemus,* a living genus of iguanid. This primitive condition contrasts with the strongly rugose frontals of the extinct Priscagaminae (Frost and Etheridge 1989) and supports the assignment of the remains to the Iguanidae. In addition, uncataloged specimens from Los Alamitos Formation (Campanian-Maastrichtian) at Ea. Los Alamitos (Fig. 4.1, locality 9) have pleurodont teeth, not fused to underyling bone, with deeply tricuspidate flared crowns resembling those of some iguanid taxa (personal observation).

Scleroglossa Estes, de Queiroz, and Gauthier 1988
Anguimorpha Fürbringer 1900
Mosasauridae Gervais 1853
Mosasaurinae Gervais 1853
Gen. et sp. indeterminate

Mosasauridae were large predatory lizards that inhabited warm epicontinental seas during the Late Cretaceous. Nearly 20 genera are recognized, the largest specimens of which exceed 10 m (Carroll 1988; Bell 1997). The skull is basically similar to that of modern varanoid lizards, but the closest relationship is with the aigialosaurs, a group from the Cretaceous with more limited specializations toward an aquatic way of life (Carroll 1988). Within the family, the mosasaurines constitute a clearly differentiated clade (Novas et al. 2002).

Hemal arches fused to the vertebral centra, cervical vertebrae with circular condyles, short premaxillary rostrum, and teeth with strongly convex lingual faces permit assignment of the fossils from the Jagüel Formation (late Maastrichtian) at the Salitral de Santa Rosa–Salinas de Trapalcó area (Fig. 4.1, locality 14) and Liu Malal (Fig. 4.1, locality 15) to the subfamily Mosasaurinae (Gasparini et al. 2001a, 2002). None of these remains contributes to the knowledge of mosasaurine interrelationships.

Serpentes Linnaeus 1758
Dinilysiidae Romer 1956
***Dinilysia* Woodward 1901**
***Dinilysia patagonica* Woodward 1901**
Fig. 4.4

Holotype. MLP 26-410, a nearly complete skull and partial postcranium.

Referred specimens. MACN-Pv RN 1013, isolated fragmentary skull (Fig. 4.4); MACN-Pv RN 1014, MACN-Pv RN 1015, MLP 79-II-27-1, MLP 71-VII-29-1, MUCPV 38, isolated fragmentary skulls; MPCA-PV 527, single skull with associated partial postcranium; MACN-Pv RN 1011, isolated maxilla; MACN-Pv RN 1016, MACN-Pv RN 1017, MACN-Pv RN 1018, MACN-Pv RN 1021, MACN-Pv N 26, MCF-PVPH-517, MEGyP 145, MLP 79-II-27-2 to 79-II-27-17, MPCA-PV 228, MPCA-PV 229, MUCPV 39, 40, 98-102, 104-116, 119, 121, isolated vertebrae and articulated strings of vertebrae.

Locality and age. The type specimen and the other specimens housed at the MLP collection were found at Boca del Sapo, at the

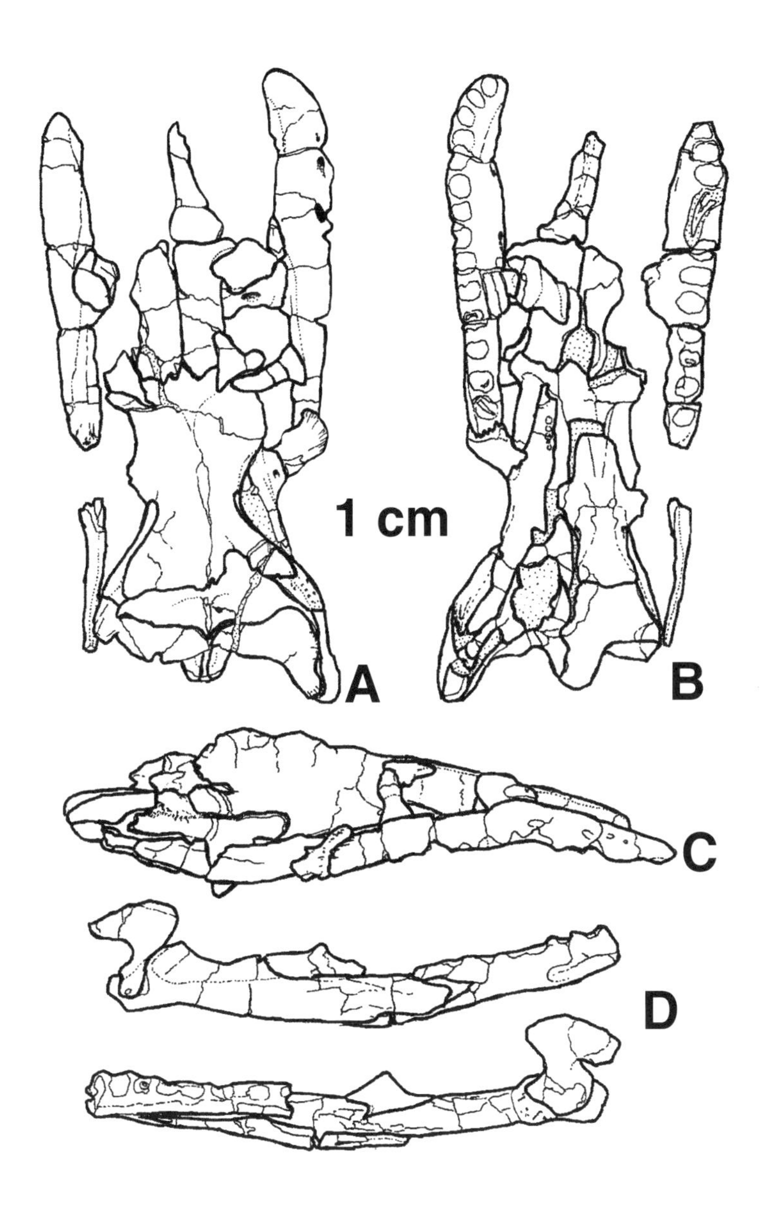

Figure 4.4. Dinilysia patagonica *(MACN-Pv RN 1013): Skull in (A) dorsal, (B) ventral, and (C) lateral views. (D) Lateral and medial views of right mandible.* Modified from Caldwell and Albino (2003).

confluence of the Neuquén and Limay Rivers, currently located in Neuquén City (Fig. 4.1, locality 4), Neuquén Province, Argentina (Woodward 1901; Estes et al. 1970; Hecht 1982; Rage and Albino 1989). The specimens housed at MACN-Pv N and MUCPV collections were found in the vicinities of the Universidad Nacional del Comahue, also located in Neuquén City (Fig. 4.1, locality 4) (Caldwell and Albino 2003). The referred specimens housed in the MACN-Pv RN and MEGyP collections and specimen MPCA-PV 527 were found near Paso Córdoba (Fig. 4.1, locality 5), Río Negro Province, Argentina (Caldwell and Albino 2003). The specimen from the MCF-PVPH collection was found in the locality Barreales Norte, approximately 45 km south of the Auca Mahuida volcano (Fig. 4.1, locality 6). All these sediments were referred to the Bajo de la Carpa Formation, Río Colorado Subgroup (Hugo and Leanza 2001). This formation is currently considered Santonian (Hugo and Leanza 2001; Leanza and Hugo 2001; Leanza et al. 2004). The specimens MPCA-PV 228 and 229 were found in Cinco Saltos (Fig. 4.1, locality 7), Río Negro Province, Argentina (Heredia and Salgado 1999), from the Anacleto Formation (Río Colorado Subgroup), considered early Campanian in age (Leanza and Hugo 2001; Hugo and Leanza 2001; Leanza et al. 2004).

Diagnosis. Medium-sized snake uniquely possessing complex, interdigitating, frontal-parietal suture; maxilla with deep, anterolaterally directed trough on suborbital surface; palatine and prefrontal forming broad ventral facet for articulation with maxilla and also forming choanal groove; ventromedial process of coronoid contacting angular; triradiate postfrontal overlapping frontal-parietal suture; postorbital closing posterior orbital margin with distinct "foot" articulating with maxillary trough; large question-mark-shaped quadrate; stapes robust with extremely large and expanded stapedial footplate; extracolumella-intercalary element contacting quadrate suprastapedial process; unfused intercentra present on hypapophyses of third-fourth precloacals; fused hypapophyses/intercentra to at least 10th precloacal; anterior third of precloacals bear prominent ventral hypapophyseal "keels" (amplified by Caldwell and Albino 2003) (Fig. 4.4).

Comments. Dinilysia was a large-bodied, cylinder-shaped snake (approximately 1.5 m in length), with a relatively large head (approximately 10 cm in length), and large, dorsally exposed orbits (Albino and Caldwell 2003). Postcranial skeleton was constituted by more than 120 relatively large precloacal vertebrae (Albino and Caldwell 2003). Frazzetta (1970) hypothesized that *Dinilysia* might be an aquatic snake on the basis of the dorsally exposed orbits (i.e., dorsally exposed eyes) (Fig. 4.4). Although a comparison of orbit orientation and aspects of postcranial osteology in diverse ecological types of snakes does not clearly resolve life modes and habits, *Dinilysia* was interpreted as a partially terrestrial snake whose morphology may have been adaptable to semiaquatic (seasonal lagoons and streams) or semifossorial habits (dune fields and interdune basin deposits) (Albino and Caldwell 2003). The large body size of *Dinilysia* is more compatible with a semiaquatic lifestyle. Also, the depositional environment of the Bajo de la Carpa Formation was described as high-energy channels in a braided stream, fluvial-dominated environment (Cazau and Uliana 1972; Chiappe and Calvo 1994; Leanza and Hugo 2001). These environmental conditions are adequate for a semiaquatic snake. Alternatively, Caldwell and Albino (2001) suggested a dune-interdune environment for the Bajo de la Carpa Formation at the locality of Paso Córdoba on the basis of supposed ichnological evidence of burrowing insects and rooting plants. Although aeolic sediments are recognized by other authors in outcrops of the Bajo de la Carpa Formation (Heredia and Calvo 1997), the presence of insect burrow systems cannot be supported at present (Genise personal communication 2002). Specimens of *Dinilysia* from the Anacleto Formation enlarge the record of this snake to the early Campanian.

Madtsoiidae Hoffstetter 1961
Alamitophis Albino 1986

Type species. Alamitophis argentinus (Albino 1986).

Geographical distribution. North Patagonia, Argentina, Campanian-Maastrichtian (Albino 1986, 1987a, 1994, 2000); southeast Queensland, Australia, early Eocene (Scanlon 1993, 2005).

Diagnosis. Small trunk vertebrae; neural arch moderately elevated and wide; neural spine well developed; zygosphene relatively thin, not or only slightly wider than cotyle; lateral and subcentral foramina present; prezygapophyses located at a high level and inclined above horizontal; neural canal relatively large; centrum narrow, with subcentral ridges clearly defined, not widened anteriorly, slightly to moderately elongate; distinctive hemal keel, well delimited laterally by deep depressions; cotyle and condyle wider than high, condyle slightly oblique; diapophysial and parapophysial areas slightly distinctive (modified from Albino 1986, 1987a) (Fig. 4.5).

Comments. A relatively elongate centrum is infrequent in midtrunk vertebrae of madtsoiid genera. Large and medium-sized madtsoiids have short vertebrae, with the centrum midline length significantly shorter than the minimum width of the neural arch (probable plesiomorphic state). Long centra are present in small madtsoiid species like *Nanowana godthelpi* (Scanlon 1997), *Patagoniophis parvus* (Albino 1986), and *Herensugea caristiorum* (Rage 1996), as well as in the genus *Alamitophis*.

Alamitophis argentinus Albino 1986
Fig. 4.5A

Holotype. MACN-Pv RN 27, fragmentary trunk vertebra (Fig. 4.5A).

Referred specimens. MACN-Pv RN 28, MACN-Pv RN 1053, MLP 88-III-31-2, MLP 88-III-31-3, MLP 88-III-30-2, MPEF-PV 643, fragmentary trunk vertebrae; MACN-Pv RN 29 to 31, MACN-Pv RN 37, MACN-Pv RN 53 to 55, MLP 88-III-31-4, MLP 88-III-31-5, vertebral centra.

Locality and age. The type specimen and most specimens belonging to the MACN-Pv RN collection were found in sediments of the Los Alamitos Formation at Ea. Los Alamitos, Río Negro Province, Argentina (Fig. 4.1, locality 9) (Albino 1986, 1989). The specimen MACN-Pv RN 1053 was found in levels of the Allen Formation at Salitral de Santa Rosa (Fig. 4.1, locality 12) (Martinelli and Forasiepi 2004). The specimens MLP 88-III-31-2 to 5 are from Salinas de Trapalcó (Fig. 4.1, locality 10), whereas MLP 88-III-30-2 is from Ea. El Palomar (Fig. 4.1, locality 11), both in sediments tentatively assigned to the Los Alamitos Formation, in the Río Negro Province, Argentina (Albino 1989, 1994). The spec-

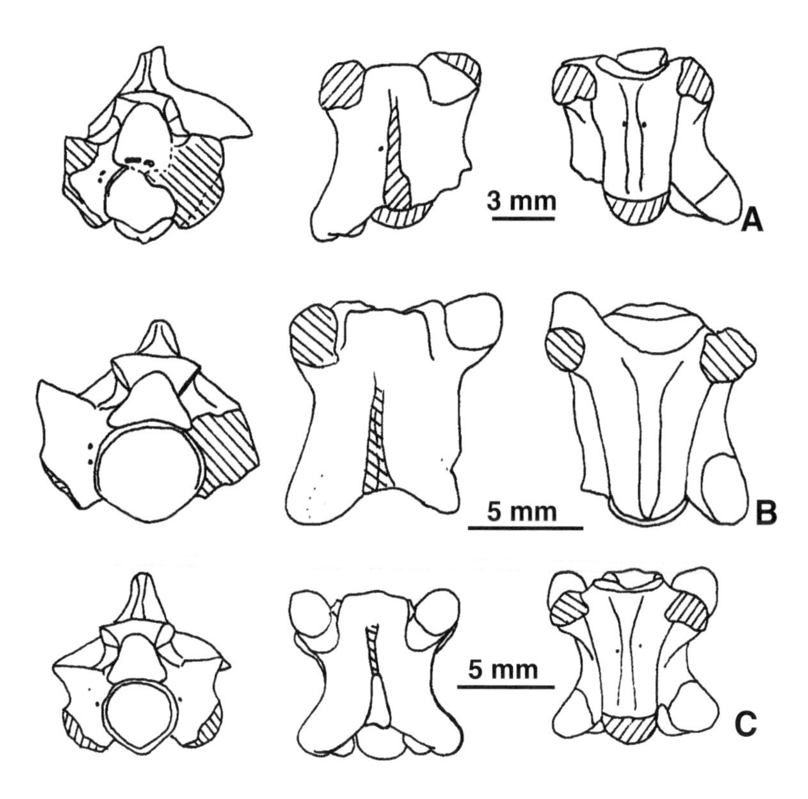

Figure 4.5. (A) Alamitophis argentinus *(MACN-Pv RN 27, holotype). Modified from Albino (1986).* A. elongatus: *(B) MACN-Pv RN 38, holotype. (C) MLP 88-III-31-1. From left to right, anterior, dorsal, and ventral views.* Modified from Albino (1994).

imen MPEF-PV 643 was found in the La Colonia Formation at Ea. La Colonia, Chubut Province, Argentina (Fig. 4.1, locality 8) (Albino 2000). The age of all these units is Campanian-Maastrichtian (Leanza et al. 2004).

Diagnosis. Anterior border of the zygosphene rectilinear, without any distinct prominence; centrum only slightly elongate (the centrum midline length is shorter or almost equal than the minimum width of the neural arch); paradiapophyses reaching the maximum width across prezygapophyses (modified from Albino 1986, 1987a) (Fig. 4.5A).

Comments. The rectilinear zygosphene of *Alamitophis argentinus* is present in most madtsoiids (probable plesiomorphic state in the Madtsoiidae family), except the species *Alamitophis elongatus* (Albino 1994) (Fig. 4.5).

Alamitophis elongatus Albino 1994
Fig. 4.5B

Holotype. MACN-Pv RN 38, fragmentary trunk vertebra (Fig. 4.5B).

Referred specimens. MACN-Pv RN 39, fragmentary caudal vertebra; MLP 88-III-31-1, fragmentary trunk vertebra.

Locality and age. The type specimen and the specimen MACN-Pv RN 39 were found in the Los Alamitos Formation at Ea. Los Alamitos (Fig. 4.1, locality 9) (Albino 1994), Campanian-

Maastrichtian (Leanza et al. 2004). The specimen MLP 88-III-31-1 was found in levels tentatively assigned to the Los Alamitos Formation outcropping in the southeast of Salinas de Trapalcó (Fig. 4.1, locality 10) (Albino 1994).

Diagnosis. Anterior border of the zygosphene distinguished by a median prominence; centrum elongate (the centrum midline length is longer than the minimum width of the neural arch); paradiapophyses nearly surpassing width across prezygapophyses (modified from Albino 1994).

Comments. Alamitophis elongatus is clearly distinguished from *A. argentinus* (and species of all other madtsoiid genera) by the presence of a median prominence on the zygosphene (Fig. 4.5). On the basis of the figures of *Herensugea* (Rage 1996), *Alamitophis elongatus* shares with it the presence of a clearly elongate vertebral centrum. The projected paradiapophyses of *A. elongatus* are also present in other madtsoiid genera, especially *Madtsoia.* The zygapophyses are strongly inclined above the horizontal in the holotype of *Alamitophis elongatus,* contrasting with the slightly inclined zygapophyses in other madtsoiid genera and in a tentatively referred specimen of the species (Fig. 4.5) (Albino 1994). Eocene Australian specimens referred to a new species of *Alamitophis* have strong affinities with *A. elongatus* (Scanlon 1993, 2005).

Patagoniophis Albino 1986

Type species. Patagoniophis parvus (Albino 1986).

Geographical distribution. North Patagonia, Argentina, Campanian-Maastrichtian (Albino 1986, 1987a); southeast Queensland, Australia, early Eocene (Scanlon 1993, 2005).

Patagoniophis parvus Albino 1986
Fig. 4.6A

Holotype. MACN-Pv RN 33, fragmentary trunk vertebra (Fig. 4.6A).

Referred specimens. MACN-Pv RN 34, MACN-Pv RN 41, MACN-Pv RN 1049, MACN-Pv RN 1056, fragmentary trunk vertebrae; MACN-Pv RN 60, vertebral centrum.

Locality and age. The type specimen and the specimens MACN-Pv RN 34, MACN-Pv RN 41, and MACN-Pv RN 60 were found in sediments of the Los Alamitos Formation at Ea. Los Alamitos (Fig. 4.1, locality 9) (Albino 1986, 1989). The specimens MACN-Pv RN 1049 and MACN-Pv RN 1056 occurred in the Allen Formation at Salitral de Santa Rosa (Fig. 4.1, locality 12) (Martinelli and Forasiepi 2004). The age is Campanian-Maastrichtian (Leanza et al. 2004).

Diagnosis. Very small trunk vertebrae; neural arch relatively low, long, and wide; neural spine well developed but low, long, and thin; zygosphene thin, slightly wider than cotyle; lateral and subcentral foramina present; prezygapophyses located at a high level and inclined above horizontal; neural canal relatively large; centrum relatively elongate, narrow, with subcentral ridges defined;

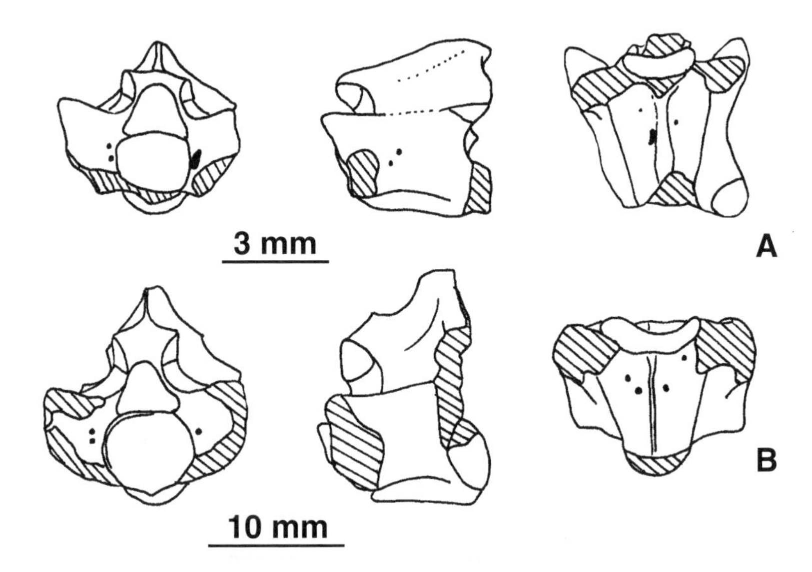

Figure 4.6. (A) Patagoniophis parvus *(MACN-Pv RN 33, holotype). (B)* Rionegrophis madtsoioides *(MACN-Pv RN 32, holotype). From left to right, anterior, lateral, and ventral views.* Modified from Albino (1986, 1987a).

distinctive hemal keel but not well delimited laterally by deep depressions; cotyle and condyle wider than high, condyle slightly oblique; paradiapophyses projected laterally, reaching width across prezygapophyses; diapophysial and parapophysial areas slightly distinctive (modified from Albino 1986, 1987a).

Comments. Specimens referred to *Patagoniophis* sp. cf. *P. parvus* coming from the early Eocene of Australia (Scanlon 1993) belong to a new species (Scanlon 2005).

Rionegrophis Albino 1986
Rionegrophis madtsoioides Albino 1986
Fig. 4.6B

Holotype. MACN-Pv RN 32, fragmentary trunk vertebra (Fig. 4.6B).

Referred specimens. MACN-Pv RN 226, fragmentary trunk vertebra.

Locality and age. All materials assigned to *Rionegrophis* come from sediments of the Los Alamitos Formation at Ea. Los Alamitos, Río Negro Province, Argentina (Fig. 4.1, locality 9), Campanian-Maastrichtian (Albino 1986).

Diagnosis. Medium-sized and robust trunk vertebrae; neural arch vaulted, wide, and short; neural spine well developed; zygosphene thick and narrower than the cotyle; subcentral foramina present; neural canal small, triangular shaped; vertebral centrum short and wide; subcentral ridges well defined and anteriorly divergent; hemal keel distinctive, thin; cotyle and condyle wider than high; condyle slightly oblique (modified from Albino 1986, 1987a).

Comments. Albino (1986) referred *Rionegrophis* only tentatively to the Madtsoiidae because of the badly preserved posterior neural arch, which did not permit observation of the parazygantral foramina, the unique vertebral synapomorphy of the family (Hoffstetter 1959). Preparation and observation of a new trunk vertebra (MACN-Pv RN 226) corroborated the presence of parazygantral foramen and madtsoiid affinities of *Rionegrophis* (Martinelli and Forasiepi 2004).

Madtsoiidae gen. et sp. indeterminate

Indeterminate madtsoiids have been recognized in Late Cretaceous beds of the Río Negro Province, Argentina (Fig. 4.1, localities 3 and 12), associated with sphenodontian lepidosaurs (Apesteguía personal communication 2002; Martinelli and Forasiepi 2004). Specimen from La Buitrera includes articulated vertebrae of the posterior precloacal region bearing parazygantral foramina, and possibly fragments of hind limbs (Apesteguía personal communication 2002). Unfortunately, the specimen could not be studied by the author, and therefore, the presence of true hind limbs in madtsoiids cannot be confirmed at present. It represents the earliest South American snake (early Cenomanian).

?Madtsoiidae gen. et sp. indeterminate

Possible madtsoiid specimens were found in sediments of the Los Alamitos, Allen, La Colonia, and Loncoche Formations (Fig. 4.1) (González Riga 1999; Albino 2000; Martinelli and Forasiepi 2004), all of Campanian-Maastrichtian age. The parazygantral foramina are not visible in all these specimens because the posterior neural arch is broken. Therefore, the remains are tentatively referred to as madtsoiids because of the general constitution and the absence of prezygapophysial processes. The medium-sized ?madtsoiid trunk vertebra from the Allen Formation at Salitral de Santa Rosa could be a new taxon of snake (Martinelli and Forasiepi 2004). The specimens from the Loncoche Formation at Ranquil-Có (Fig. 4.1, locality 13) were badly illustrated (González Riga 1999). They are poorly preserved and need reexamination to clarify the assignation within snakes.

Alethinophidia Nopcsa 1923
Macrostomata Müller 1831
?Boidae Gray 1825
Gen. et sp. indeterminate

It has not yet been established that clear synapomorphies are present in isolated vertebrae of boids. Some nonboid snakes have vertebrae with a boidlike overall morphology (Rage 1998); hence, taxonomic resolutions are frequently based only on comparisons of partially preserved specimens. Fragmentary vertebrae with high and wide neural arches without parazygantral foramina (resembling boine vertebrae) and depressed neural arches with thin zygosphenes (resembling erycine vertebrae) have been observed among the snake assemblage of the Los Alamitos Formation at Ea. Los Alamitos (Fig. 4.1, locality 9) (Albino 1990). Hence, at least two taxa of probable boids were present in the Late Cretaceous of Patagonia.

Serpentes incertae sedis
Gen. et sp. indeterminate

Only one snake vertebra (MPEF-PV 642) distinct from dinilysiid, madtsoiid, and boid trunk vertebrae has been recorded in Late Cretaceous Patagonian beds. The vertebra was found in sediments of the La Colonia Formation at Ea. La Colonia, Chubut Province, Argentina (Fig. 4.1, locality 8) (Albino 2000), Campanian-Maastrichtian (Leanza et al. 2004). It is small, narrow, and rather depressed and elongate (Fig. 4.7). It has no marked median notch in the posterior border of the neural arch. This is a plesiomorphic state in snakes (Albino 1996). The roof of the neural arch is depressed in posterior view. The neural spine is long, inclined backward, and posteriorly thick. The zygapophysial plane is located at a high level, and the prezygapophysial facets are clearly slanting, as in primitive snakes. There is not a true prezygapophysial process (plesiomorphic state, Rage 1998; Lee and Scanlon 2002; Rieppel et al. 2003), although there is an expansion under the prezygapophysial articular facets and the vertebra is wider at that level. The postzygapophyses are at a higher level than the prezygapophyses. The thin zygosphene is narrower than the cotyle, it is moderately elevated, and its anterior border is not notched. The subcentral ridges of the vertebral centrum are not well defined, and the centrum is not significantly widened anteriorly. There is no marked hemal keel. The cotyle is large and wider than high, and there are no paracotylar foramina. The combination of characters in the fossil has not been found in any known snake (Albino 2000).

Phylogenetic Relationships

Until recently, the fossil record of snakes has not contributed much to the understanding of the phylogenetic relationships of major lineages. Most fossil specimens consist only of isolated vertebrae that present relatively few phylogenetically valuable characters. However, some well-preserved specimens studied during the last few years have created new debate about the origins and early evolution of snakes (Caldwell and Lee 1997; Lee 1997, 1998; Lee and Caldwell 1998; Caldwell 1999; Lee et al. 1999; Greene and Cundall 2000; Rage and Escuillié 2000; Rieppel and Zaher 2000a,b; Scanlon and Lee 2000; Tchernov et al. 2000; Lee and Scanlon 2002; Rieppel et al. 2003). The Late Cretaceous Patagonian snake *Dinilysia* was postulated to be the sister group of all modern snakes (Scolecophidia and Alethinophidia, Lee and Scanlon 2002) or the sister group of the Alethinophidia (Kluge 1991; Rieppel et al. 2003) (Fig. 4.8). Both proposed hypotheses place *Dinilysia* as a relatively basal snake, prior to the macrostomatan origin. The evaluation of the phylogenetic relationships of

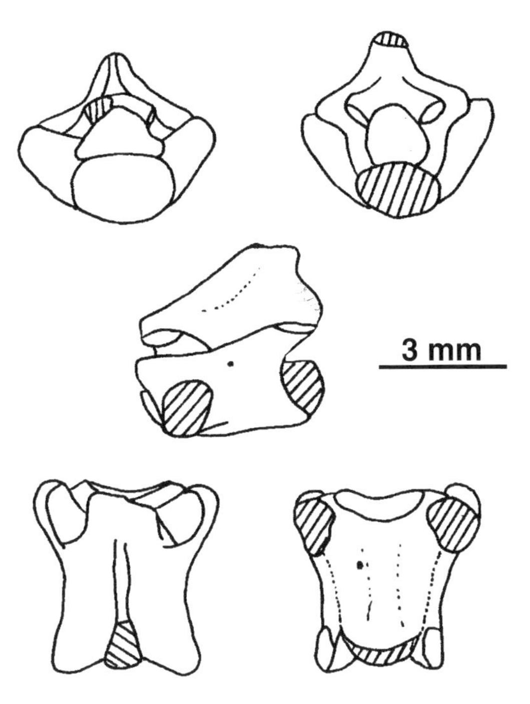

Figure 4.7. Serpentes incertae sedis, MPEF-PV 642. From left to right, anterior, posterior, lateral, dorsal, and ventral views. Modified from Albino (2000).

Dinilysia exceeds the scope of this contribution; however, its impact in the interpretation of the early evolution of snakes may be explained.

Diet has repeatedly been mentioned as an important factor in the evolution of snakes (Greene 1983). Some authors have proposed that innovations of the feeding apparatus of macrostomatans allowed these snakes to eat heavier and bulkier prey than more basal groups of extant snakes (scolecophidians and anilioids) (Greene 1983). Morphological changes, which include acquisition of elongate, more movable quadrate and supratemporal bones, could then have allowed macrostomatans to utilize a broad range of prey types (Gans 1961; Greene 1983). Most primitive living snakes (scolecophidians and anilioids) include small snakes (maximum length nearly 1 m) that inhabit soil, litter, or wet tropical muck, displaying a fossorial mode of life (Greene 1997) and feeding on small or elongate prey items (Rodríguez-Robles et al. 1999; Cundall and Greene 2000). Cundall and Greene (2000) assumed that early snakes were gape-limited predators like living scolecophidians and basal alethinophidians, and that the evolution of feeding in snakes commenced with a hypothetical ancestor to these two very divergent clades. According to that view, it would be expected that all premacrostomatan snakes were small and fossorial, with a limited gape size, and that their prey items were small, elongate, and limbless, and relatively small in diameter. However, the phylogenetic position of *Dinilysia* contradicts this expectation. *Dinilysia,* as has been established, was a large-bodied terrestrial snake, with a large, broad, posteriorly expanded head, and elon-

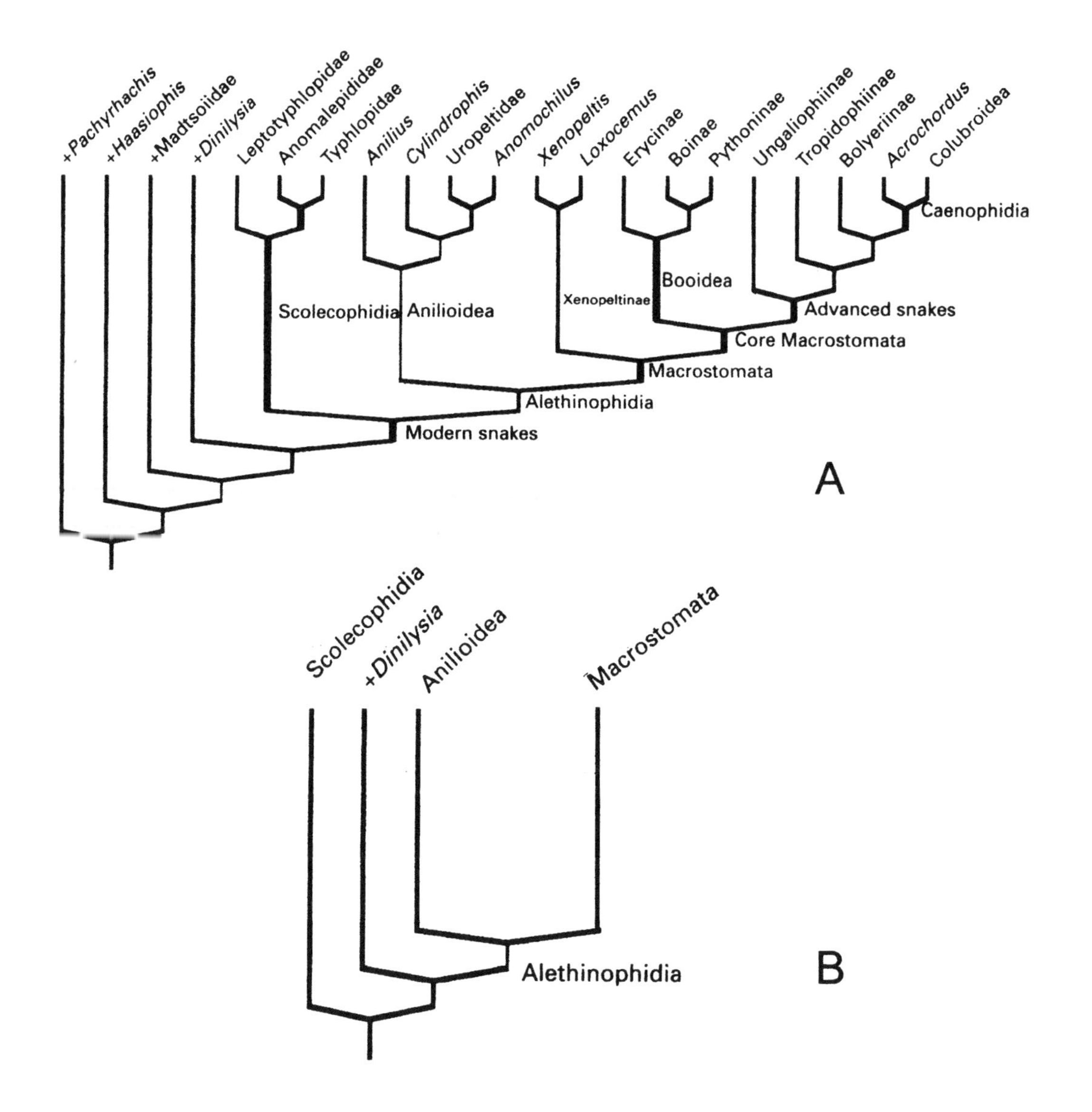

Figure 4.8. Phylogenetic relationships among snakes. (A) Lee and Scanlon (2002). (B) Rieppel et al. (2003).

gated supratemporal bones, well adapted for swallowing relatively large-diameter prey. Although *Dinilysia* had a relatively consolidated skull, its body size and head size were larger than those of primitive living snakes, and the elongation of the supratemporal that carries the quadrate posteriorly distinguished this snake from all living basal forms. These characters increased its gape size and improved the possibility of its preying on items larger than those possible for scolecophidians and basal alethinophidians. In addition, Frazzetta (1970) noted that the skull of *Dinilysia* was kinematically similar to that of living booids, although the absolute amount of motion may have been less in *Dinilysia*. Among the associated vertebrate fauna found in the fossil localities that produced *Dinilysia*, the notosuchian crocodiles, the theropod dinosaurs and the flightless birds were small- and medium-sized terrestrial adaptive types that may have been among the prey items selected by *Dinilysia* (Albino and Caldwell 2003). Hence, diverse ecological morphotypes of snakes would have developed in the earliest evolution of the group, aquatic (*Pachyrhachis, Haasiophis,* and others) and terrestrial (Madtsoiidae, *Dinilysia*) (Lee and Scanlon 2002), or only fossorial (Scolecophidia) (Rieppel et al. 2003) (Fig. 4.8). Nonfossorial snakes with large bodies that may have exploited prey of various shapes and sizes (represented by *Dinilysia*) would have appeared early in snake phylogeny (premacrostomatan) in both hypotheses.

On the other hand, the family Madtsoiidae was for a long time placed within the Boidae as a subfamily (Hoffstetter 1961). McDowell (1987) ranked it as a family of the Cholophidia, a paraphyletic group of primitive fossil snakes. Later, only on the basis of vertebral characters, madtsoiids were interpreted as a monophyletic group of basal alethinophidians (Scanlon 1992, 1993; Albino 1996). At the moment, madtsoiids include at least 16 species allocated to 9 genera (Albino 1994; Rage 1998; Rage and Werner 1999; Scanlon 2005). Until a few years ago, they were known mainly by vertebrae, where only one character may be considered autapomorphic: the presence of a parazygantral foramen opening in a small fossa on either side of the zygantrum (Hoffstetter 1961). Cranial elements preserved on the Australian genus *Wonambi* were used to regard Madtsoiidae as basal snakes, outside the modern radiation or crown clade (Scanlon and Lee 2000; Lee and Scanlon 2002) (Fig. 4.8). Other interpretations have questioned madtsoiid monophyly and considered that *Wonambi* and fossil snakes with relatively well-developed hind limbs (*Haasiophis, Pachyrhachis* and *Eupodophis*), previously supposed to be primitive (Rage and Escuillié 2000, 2002; Scanlon and Lee 2000), must be included in the Macrostomata (Rieppel et al. 2003) (Fig. 4.8). The doubtful characters of madtsoiids are the presence of parazygantral foramina on the vertebrae, and two or three mental foramina on the dentary (Rieppel et al. 2003). Vertebrae of large Australian madtsoiids (*Wonambi* and *Yurlunggur*) frequently have the parazygantral foramina divided into two or three on each side, rather than the single foramen consistently found in the other genera (Rage 1998). Nevertheless, Rieppel et al. (2003) did not state any reason to doubt the structures were homologous. In addition, the supposed absence of this character in an isolated vertebra of *Madtosia* sp. (Albino 1986) cannot be corroborated because this specimen belongs to a boid (Albino 1993). According to Hoffstetter (1961), madtsoiids were also characterized by the presence of two or three mental foramina on the dentary. Because the dentary of *Wonambi* shows only a single foramen (not the pair reconstructed by Scanlon and Lee 2000), Rieppel et al. (2003) also questioned this character. Later, Lee and Scanlon (2002) coded Madtsoiidae as polymorphic for this character, after the confirmation of the presence of only one mental foramina in *Wonambi*. Dentaries with at least two mental foramina are known in the South American genus *Madtsoia* (Hoffstetter 1959; Rage 1998; Scanlon 2005), the Australian genera *Nanowana* and *Yurlunggur* (Scanlon 1997, 2005), and Australian species of *Patagoniophis* and *Alamitophis* (Scanlon 2005).

Although madtsoiids are frequent in continental Patagonian deposits of Cretaceous age, the phylogenetic information they provide is limited because the remains mostly consist of isolated and fragmentary vertebrae. Nevertheless, some primitive character states are found in madtsoiid vertebrae. The lack of prezygapophyseal processes is considered plesiomorphic (Rage 1998; Lee and Scanlon 2002; Rieppel et al. 2003), contrasting with the state found in *Dinilysia,* Scolecophidia, and most Alethinophidia. The great width through the diapophyses (which approaches or exceeds the width through the prezygapophyses) is more plesiomorphic than in Scolecophidia and Alethinophidia (McDowell 1987; Albino 1996). Additionally, a well-developed neural spine is considered a plesiomorphy by Lee and Scanlon (2002), in opposition to the absent or poorly developed neural spine of scolecophidians and basal alethinophidians (anilioids, *Xenopeltis, Loxocemus*). Occurrence of paracotylar foramina on most vertebrae was alternatively considered as a derived character state (Albino 1996), primitive (Lee and Scanlon 2002), or of uncertain polarity (Rage 1998). The strong inclination of the zygapophyses above the horizontal may be interpreted as more plesiomorphic than in Scolecophidia and Alethinophidia (Scanlon 1992), but this character shows a great variability all along the vertebral column, within madtsoiid species (Albino 1994), and among madtsoiid genera (Scanlon 1993). In conclusion, the precise relationships of the Madtsoiidae are still not definitely established (Fig. 4.8). Monophyly of the family needs to be confirmed to use *Wonambi* and *Yurlunggur* cranial elements in a phylogenetic interpretation of madtsoiid relationships. Within madtsoiids, a *Madtsoia-Gigantophis-Yurlunggur-Wonambi* assemblage is distinguished from the other forms (*Alamitophis-Patagoniophis-Rionegrophis-Nanowana-Herensugea*) by their large size.

Biogeographical Considerations

Mesozoic Lepidosaur Record in South America

Sphenodontia is a group of rhynchocephalian lepidosaurs well documented from the Late Triassic to the Early Cretaceous of North America, Mexico, Europe, China, Africa, and India (Reynoso 1997; see Evans et al. 2001). Until recent discoveries of abundant remains of sphenodontians in the Late Triassic of Brazil (Ferigolo 1999, 2000) and the Late Cretaceous of Argentina (Apesteguía and Novas 2003; Simón and Kellner 2003), they were unknown in South America. Jurassic sphenodontian remains were collected recently in Patagonia (Rougier personal communication 2003; Rauhut personal communication 2003), suggesting that this lepidosaurian lineage inhabited southern South America since at least Jurassic times. The South American sphenodontian record suggests that the real taxonomic diversity of Mesozoic sphenodontians has not yet been well documented in the southern continents.

On the other hand, *Tijubina pontei* from the Crato Formation (Early to Middle Cretaceous of northeastern Brazil) has recently been defined as the sister group of the crown group Squamata (Scleroglossa + Iguania) (Bonfim and Avilla 2002). Assuming the above relationships are correct, they suggest an early unknown evolutionary history of this clade in southern continents. *Tijubina* was previously cited as a teiid lizard (Bonfim 1999). Supposed squamate remains from the Jurassic of Patagonia (Rauhut et al. 2001; Rauhut and Puerta 2001; Evans 2003) could not be corroborated (Rauhut personal communication 2003).

Cretaceous terrestrial lizards are infrequent and doubtful in South America. Huene (1931) cited possible lizards for the Early Cretaceous of northwestern Argentina, but the specimens are missing at present and doubtfully belong to this group (Estes 1983). Gayet et al. (1992) and Albino (1996) reported Valencia et al. (in press at that time), who mentioned the presence of a teiid in the Early Cretaceous of Chile. Because Valencia et al. was not published, there is no definitive evidence of teiids in the Early Cretaceous of South America. More recently, Evans and Yabumoto (1998) described a new lizard species (*Olindalacerta brasiliensis*) of uncertain taxonomic affinities from the Crato Formation (Early to middle Cretaceous of Brazil). *Pristiguana brasiliensis* from the late Maastrichtian of Brazil (Estes and Price 1973) is the best-known lizard from the South American Mesozoic. It was assigned to the Iguania by Estes and Price (1973), and to the Teiidae by Borsuk-Białynicka and Moody (1984). Late Cretaceous outcrops of Patagonia provided an undescribed lizard dentary fragment of early Campanian age (Albino 2002). In addition, some fragmentary maxillae and dentaries from the latest Cretaceous of Patagonia (Campanian-Maastrichtian) probably belong to diverse lizard taxa, including iguanids and teiids (personal observation). A fragmentary frontal from the Cenomanian of Patagonia has recently been assigned tentatively to the Iguanidae (Apesteguía et al. 2004, 2005). All these remains suggest that Patagonia may have been an unsuspected territory for Mesozoic lizard evolution, as it was for other reptiles.

South American mosasaurs have been discovered in the Late Cretaceous of Colombia, Chile, Peru, Brazil, Patagonia (reviewed by Gasparini et al. 2001a,b, 2002), and especially Antarctica (reviewed by Martin et al. 2002). Antarctic and Patagonian mosasaur remains are mostly limited to isolated teeth and fragmentary bones assigned to mosasaurines, plioplatecarpines, and tylosaurines (Martin et al. 2002). The basal tylosaurine *Lakumasaurus antarcticus* from the latest Campanian–early Maastrichtian is the most complete mosasaur yet recorded in Antarctica (Novas et al. 2002). Patagonian mosasaurines are restricted to the late Maastrichtian, inhabiting the seas of the first Atlantic transgression over the north part of this region (Gasparini et al. 2001a).

Concerning the snakes, South America appears to have played an important role in the evolutionary history of the group. They were present in South America since the early Cenomanian. Dinilysiidae is represented by the unique species *Dinilysia patagonica* from the Santonian to early Campanian of northern Patagonia (Woodward 1901; Estes et al. 1970; Hecht 1982; Rage and Albino 1989; Caldwell and Albino 2003; this work). Madtsoiidae have been recovered from early Late Cretaceous to Pleistocene deposits in South America, Africa, Madagascar, India, and Australia (reviewed by Albino 1994; Rage 1998; Rage and Werner 1999; Scanlon and Lee 2000; Gayet et al. 2001; Scanlon 2005; this work), with non-Gondwanan records in Spain (Rage 1996, 1999) and perhaps France (Sigé et al. 1997). Late Cretaceous South American madtsoiids presently include four described species from Patagonia (Albino 1986, 1987a, 1994). Indeterminate madtsoiids were also mentioned from two Late Cretaceous localities of Patagonia (Apesteguía personal communication 2002; Martinelli and Forasiepi 2004), and probable madtsoiids appeared in Late Cretaceous deposits of Argentina (González Riga 1999; Albino 2000; Martinelli and Forasiepi 2004), and Bolivia (Gayet et al. 2001). In addition, Zaher et al. (2003) have recently reported the earliest snake specimens of Brazil (Turonian-Santonian), assigned to the Anilioidea. This is the oldest South American record of this probably paraphyletic snake group. The records of Boidae, a group of macrostomatan snakes, in the Cretaceous of South America is still doubtful. Poorly preserved Campanian-Maastrichtian remains found in Patagonia were tentatively referred to this family (Albino 1990, 1996). They are clearly not madtsoiid or dinilysiid vertebral remains. Primitive snakes (probable prealethinophidians) of uncertain affinities are also present in Campanian-Maastrichtian beds of Patagonia (Albino 2000). In conclusion, the snake record suggests a great diversity of snakes since the early Late Cretaceous of South America.

Biogeographic Evolution of Lepidosaurs in South America

Although the tuataras (genus *Sphenodon*) of New Zealand are the only surviving rhynchocephalians, during Triassic and Early Jurassic times, this group had a relatively good record showing considerable diversity and a worldwide distribution (Evans 2003). In contrast, the earliest record of squamates is from the Early-Middle Jurassic, although indirect evidence predicts that they had evolved by at least the Middle Triassic and had diversified into existing major lineages before the end of this period (Evans 2003). The Jurassic and Early Cretaceous are poorly sampled in comparison to the Late Cretaceous (Evans 2003).

The record of fossil sphenodontians in Patagonia suggests that they were an important component of the Mesozoic terrestrial ecosystems of South America. Although in Laurasia this lineage declined in the Early Cretaceous, probably because of the rapidly diversifying lizards, the occurrence of sphenodontians in the latest Cretaceous of Patagonia suggests that they persisted longer in Gondwana than on other landmasses (Apesteguía and Novas 2003). According to Apesteguía and Novas (2003), the postulated ecological replacement of sphenodontians by terrestrial lizards would have been delayed in Gondwana with respect to the northern continents.

The Mesozoic squamate record is relatively well known in Laurasia, contrasting with the very poor Gondwanan record. Exceptions to this pattern are provided by the mosasaurs and the snakes. Mosasaurs are relatively abundant in deposits of both northern and southern continents, whereas snakes are conspicuously more dominant in squamate assemblages of Gondwanan territories (Evans 2003; Krause et al. 2003). Patagonian Mesozoic representatives of all these groups contribute considerably to the understanding of lepidosaurian evolution in the southern continents. Although we know that lizards were present in Gondwana during the Jurassic and Cretaceous, the richness of species is still substantially lower than in Laurasia (Krause et al. 2003), and most of the material is too fragmentary to be assigned to known lineages (Evans 2003). This situation makes it premature to reconstruct centers of origin or dispersal of lizard lineages on the basis of fossil data alone (Evans 2003). Nevertheless, because basal squamates and probable iguanids and teiids were present in South America since Cretaceous times, Gondwana seems to have been involved in the early history of major lizard lineages.

The diversity of pelagic marine reptiles during the Late Cretaceous of northern Patagonia was higher than previously supposed (Gasparini et al. 2001a). Occurrences of indeterminate mosasaurine mosasaurs in Patagonia are still scarce, contrasting with the known mosasaur fauna of the Antarctic Peninsula, which includes at least six taxa (Martin et al. 2002; Novas et al. 2002). The mosasaurs from Antarctica appear to have a relatively cosmopolitan distribution, probably facilitated by the rather close position of continental masses and connecting marine corridors during the latest Cretaceous (Martin et al. 2002). Novas et al. (2002) suggested that Late Cretaceous mosasaur faunas from the Antarctic seas do not completely conform to the coeval northern paleogeographic pattern. Although knowledge of marine vertebrates from the Late Cretaceous of northern Patagonia is limited, especially when compared with palynomorphs, microinvertebrates, and macroinvertebrates, some coincidences may be observed with respect to the paleogeographic distribution pattern (Gasparini et al. 2002). Camacho (1992) argued that the association of Weddellian and Tethyian elements in the northern part of this region defines a transitional biogeographic area. At the moment, neither Weddellian nor Tethyian affinities of the Late Cretaceous Patagonian mosasaurines can be established.

Concerning the snakes, the oldest records in the world suggest that the origin of snakes antedates the earliest Late Cretaceous

(Rage and Werner 1999). The basal *Dinilysia patagonica* from the Late Cretaceous of Patagonia, the Gondwanan distribution of the primitive madtsoiids, and the distribution of extant primitive alethinophidians in Gondwanan areas suggest that a great diversity of primitive snakes lived in Gondwana (Underwood and Stimson 1990). The Cretaceous-Paleogene South American record of snakes (Simpson 1933; Hoffstetter 1959; Albino 1990, 1996, 2000; Rage 1998; Zaher et al. 2003; this work) supports this opinion. In addition, the earliest snake assemblage, including advanced forms (colubroids), from the Cenomanian of Africa (Rage and Werner 1999) and the findings of madtsoiids in the Cenomanian of Patagonia (Apesteguía personal communication 2002) suggest that the earliest radiation of terrestrial snakes probably occurred in Gondwana. The relatively high species richness of snakes in Late Cretaceous Gondwanan vertebrate assemblages (Krause et al. 2003) relative to Laurasia may be explained under this consideration. Recent discoveries of anilioid grade snakes from the uppermost Albian/lowermost Cenomanian of North America (Gardner and Cifelli 1999) demonstrated that alethinophidian snakes were distributed in Laurasia and Gondwana by the early Late Cretaceous. Until recently, the known record of anilioids suggested that these snakes originated in Gondwana and migrated from South America to North America during the latest Cretaceous (Rage 1981, 1998; Estes and Báez 1985; Bonaparte 1986; Gayet et al. 1992). However, the oldest known anilioids come from both Laurasia (Gardner and Cifelli 1999) and Gondwana (Rage and Werner 1999); hence, the continent where these snakes originated remains unresolved (Gardner and Cifelli 1999). The anilioid vertebrae found in the Upper Cretaceous of Brazil (Zaher et al. 2003) and the known anilioid record in the Paleocene deposits of Peru and Brazil (Rage 1981, 1998; Albino 1990, 1996) suggest a large Cretaceous-Paleocene distribution of these snakes in South America.

On the other hand, although the presence of boids in the Late Cretaceous of South America cannot be definitely supported, they were probably present since the latest Cretaceous of North America (Holman 2000) and in Paleocene deposits of Argentina (Albino 1993), Brazil (Albino 1990; Rage 1998), and Bolivia (Rage 1991). According to geodynamic and faunal data, there would have been some kind of connection between South and North America that permitted continental faunal dispersion in both directions around the end of the Cretaceous or the beginning of the Tertiary (Bonaparte 1984, 1986; Estes and Báez 1985; Ross and Scotese 1989; Gayet et al. 1992; Parrish 1993). This connection could probably have been used by boid snakes to disperse from South America into North American territory (Rage 1981, 1998).

According to the record, both madtsoiids and boids were common in South American Cretaceous-Paleogene land communities (Albino 1996, 2000; Rage 1998; Gayet et al. 2001). Also, the most diverse and abundant fauna of madtsoiid species from the Cretaceous to the Eocene comes from Patagonia, including *Alamitophis*,

Patagoniophis, Rionegrophis, and *Madtsoia* genera (Simpson 1933; Hoffstetter 1959; Albino 1986, 1987a,b, 1993, 1994, 1996, 2000). Among the four Cretaceous madtsoiid species coming from the Late Cretaceous of Patagonia, *Alamitophis* and *Patagoniophis* are also recorded from the early Eocene of Australia (Scanlon 1993, 2005). Also, a rib fragment referred as cf. *Madtsoia* sp. was found in the early Eocene of Australia (Scanlon 2005). This distribution evidences a geographical continuity of these lineages between Australia and southern South America across Antarctica between the latest Cretaceous and early Eocene (Scanlon 1993; Albino 2000). Among reptiles, both meiolaniid and chelid turtles have been recorded in Patagonia (Cretaceous-Paleogene) as well as in Australasia (Eocene-Holocene), also suggesting a close biogeographic relationship between Patagonia and Australia (Gasparini et al. 2001b). A dispersal event of terrestrial vertebrates between South America and Australia through Antarctica by the end of the Maastrichtian to the earliest Tertiary is supported by additional evidence provided by mammals (Woodburne and Case 1996).

Acknowledgments. I thank the editors for the invitation to contribute to this volume. I am indebted to Dr. J. Bonaparte (MACN) and Dr. R. Pascual (MLP) for letting me study specimens collected during field trips conducted by them. Analysis of snake fossil material was possible thanks to L. E. Ruigomez and R. Cúneo (MPEF), M. Reguero and S. Bargo (MLP), J. Calvo (MUCPV), R. Coria (MCF), and L. Salgado. I also thank A. Martinelli and J. Scanlon for providing access to their respective unpublished manuscripts. The criticisms and suggestions of the two reviewers, J. D. Scanlon and J. C. Rage, made for a much improved manuscript. The fieldwork in Paso Córdoba, where the specimen MPCA-PV 527 was found, was supported by a National Geographic grant to a project in collaboration with M. Caldwell.

References Cited

Albino, A. M. 1986. Nuevos Boidae Madtsoiinae en el Cretácico tardío de Patagonia (Formación Los Alamitos, Río Negro, Argentina). *Actas del IV Congreso Argentino de Paleontología y Bioestratigrafía* 2: 15–21. Mendoza.

———. 1987a. The Late Cretaceous fauna of Los Alamitos, Patagonia, Argentina. Part V. The Ophidians. *Revista del Museo Argentino de Ciencias Naturales "Bernardino Rivadavia," Paleontología* 3: 141–146.

———. 1987b. Un nuevo Boidae (Reptilia: Serpentes) del Eoceno temprano de la Provincia del Chubut, Argentina. *Ameghiniana* 24: 61–66.

———. 1989. Los Booidea (Reptilia, Serpentes) extinguidos del territorio argentino. Ph.D. thesis, Universidad Nacional de La Plata, La Plata, Argentina.

———. 1990. Las serpientes de São José de Itaboraí (Edad Itaboraiense, Paleoceno medio), Brasil. *Ameghiniana* 27: 337–342.

———. 1993. Snakes from the Paleocene and Eocene of Patagonia (Argentina): paleoecology and coevolution with mammals. *Historical Biology* 7: 51–69.

———. 1994. Una nueva serpiente (Reptilia) del Cretácico tardío de Patagonia, Argentina. *Pesquisas* 21: 58–63.

———. 1996. The South American fossil Squamata (Reptilia: Lepidosauria). In G. Arratia (ed.), *Contributions of Southern South America to Vertebrate Paleontology* (A) 30: 9–72. Münchner Geowissenschaftliche Abhandlungen.

———. 2000. New record of snakes from the Cretaceous of Patagonia (Argentina). *Geodiversitas* 22: 247–253.

———. 2002. El lagarto más antiguo de la Argentina. *Resúmenes I Congreso "Osvaldo A. Reig" de Vertebradología básica y evolutiva e Historia y Filosofía de la Ciencia,* 21. Buenos Aires.

Albino, A. M., and M. Caldwell. 2003. Hábitos de vida de la serpiente cretácica *Dinilysia patagonica* Woodward. *Ameghiniana* 40: 407–414.

Apesteguía, S., and F. E. Novas. 2003. Late Cretaceous sphenodontian from Patagonia provides insight into lepidosaur evolution in Gondwana. *Nature* 425: 609–612.

Apesteguía, S., F. L. Agnolin, and G. Lío. 2004. Shy ghosts: iguanid lizards at "La Buitrera"? (Candeleros Formation, Cenomanian Turonian). *Ameghiniana* 4 (Suppl.): 34R.

Apesteguía, S., F. L. Agnolin, and G. Lío. 2005. An early Late Cretaceous lizard from Patagonia, Argentina. *Comptes Rendus Palevol* 4: 311–315.

Bell, G. 1997. A phylogenetic revision of North American and Adriatic Mosasauroidea. In J. Callaway and E. Nicholls (eds.), *Ancient Marine Reptiles,* 293–332. San Diego: Academic Press.

Bonaparte, J. F. 1984. Late Cretaceous faunal interchange of terrestrial vertebrates between the Americas. In E. Reif and F. Westphal (eds.), *Third Symposium on Mesozoic Terrestrial Ecosystems, Short Papers,* 19–24. Tübingen.

———. 1986. History of the terrestrial Cretaceous vertebrates of Gondwana. *Actas del IV Congreso Argentino de Paleontología y Bioestratigrafía* 2: 63–95. Mendoza.

Bonfim, F. 1999. A dentição de *Tijubina pontei* Bonfim Jr. & Marques 1997 (Squamata, Sauria, Teiidae) Formação Santana, Cretáceo Inferior do bacia do Araripe, Nordeste do Brasil. *Paleontologia em Destaque, Boletím Informativo da Sociedade Brasileira de Paleontologia* 26: 67.

Bonfim, F., and L. Avilla. 2002. Phylogenetic position of *Tijubina pontei* Bonfim & Marques, 1997 (Lepidosauria, Squamata), a basal lizard from the Santana Formation, Lower Cretaceous of Brazil. *Journal of Vertebrate Paleontology, Abstracts* 22: 37A.

Borsuk-Białynicka, M., and S. Moody. 1984. Priscagaminae, a new subfamily of the Agamidae (Sauria) from the Late Cretaceous of the Gobi Desert. *Acta Palaeontologica Polonica* 29: 51–81.

Caldwell, M. W. 1999. Squamate phylogeny and the relationships of snakes and mosasauroids. *Zoological Journal of the Linnean Society* 125: 115–147.

Caldwell, M. W., and A. M. Albino. 2001. Palaeoenvironment and palaeoecology of three Cretaceous snakes: *Pachyophis, Pachyrhachis,* and *Dinilysia.* In S. F. Vizcaíno, R. A. Fariña, and C. Janis (eds.), *Biomechanics and Palaeobiology of Vertebrates, Acta Palaeontologica Polonica* 46: 71–86.

———. 2003. Exceptionally preserved skeletons of the Cretaceous snake

Dinilysia patagonia Woodward, 1901. *Journal of Vertebrate Paleontology* 22: 861–866.
Caldwell, M. W., and M. S. Y. Lee. 1997. A snake with legs from the marine Cretaceous of the Middle East. *Nature* 386: 705–709.
Camacho, H. H. 1992. Algunas consideraciones acerca de la transgresión marina paleocena en la Argentina. *Academia Nacional de Ciencias de Córdoba, Miscelánea* 85: 1–41.
Carroll, R. 1988. *Vertebrate Paleontology and Evolution.* New York: W. H. Freeman.
Cazau, L. B., and M. A. Uliana. 1972. El Cretácico Superior continental de la Cuenca Neuquina. *Actas del V Congreso Geológico Argentino* 3: 131–163.
Chiappe, L. M., and J. O. Calvo. 1994. *Neuquenornis volans,* a new Late Cretaceous bird (Enantiornithes: Avisauridae) from Patagonia, Argentina. *Journal of Vertebrate Paleontology* 14: 230–246.
Cundall, D., and H. Greene. 2000. Feeding in snakes. In K. Schwenk (ed.), *Feeding: Form, Function, and Evolution in Tetrapod Vertebrates* 9: 293–333. San Diego: Academic Press.
Estes, R. 1983. Sauria terrestria, Amphisbaenia. In P. Wellnhofer (ed.), *Handbuch der Paläoherpetologie,* 249. Stuttgart: Gustav Fischer Verlag.
Estes, R., and A. M. Báez. 1985. Herpetofaunas of North and South America during the Late Cretaceous and Cenozoic: evidence for interchange? In F. Stehli and S. Webb (eds.), *The Great American Biotic Interchange,* 139–195. New York: Plenum Press.
Estes, R., and L. Price. 1973. Iguanid lizard from the Upper Cretaceous of Brazil. *Science* 180: 748–751.
Estes, R., T. H. Frazzeta, and E. E. Williams. 1970. Studies on the fossil snake *Dinilysia patagonica* Woodward: Part I. Cranial morphology. *Bulletin of the Museum of Comparative Zoology* 140: 25–73.
Estes, R., K. de Queiroz, and J. Gauthier. 1988. Phylogenetic relationships within Squamata. In R. Estes and G. Pregill (eds.), *Phylogenetic Relationships of the Lizard Families: Essays Commemorating Charles L. Camp,* 119–281. Stanford, Calif.: Stanford University Press.
Evans, S. E. 2003. At the feet of the dinosaurs: the early history and radiation of lizards. *Biological Review* 78: 513–551.
Evans, S. E., and Y. Yabumoto. 1998. A lizard from the Early Cretaceous Crato Formation, Araripe Basin, Brazil. *Neues Jahrbuch für Geologie und Paläontologie Monatshefte* 6: 349–364.
Evans, S. E., G. V. R. Prasad, and B. K. Manhas. 2001. Rhynchocephalians (Diapsida: Lepidosauria) from the Jurassic Kota Formation of India. *Zoological Journal of the Linnean Society* 133: 309–334.
Ferigolo, J. 1999. South American first record of a sphenodontian (Lepidosauria, Rhynchocephalia) from the Late Triassic–Jurassic of Rio Grande do Sul State, Brazil. In H. A. Leanza (ed.), *International Symposium on Mesozoic Terrestrial Ecosystems, VII, Abstracts,* 24–25. Buenos Aires: Museo Argentino de Ciencias Naturales.
———. 2000. Esfenodontídeos do Neo-triássico/?Jurássico do Estado do Rio Grande do Sul, Brasil. In M. Holz, and L. F. De Ros (eds.), *Paleontologia do Rio Grande do Sul,* 236–245. Porto Alegre: CIGO/UFRGS.
Frazzetta, T. H. 1970. Studies on the fossil snake *Dinilysia patagonica* Woodward. Part II. Jaw machinery in the earliest snakes. *Forma et Functio* 3: 205–221.

Frost, D. R., and R. Etheridge. 1989. A phylogenetic analysis and taxonomy of iguanian lizards (Reptilia: Squamata). *University of Kansas, Museum of Natural History, Miscellaneos Publications* 81: 1–65.

Gans, C. 1961. The feeding mechanism of snakes and its possible evolution. *American Zoologist* 1: 217–227.

Gardner, J. D., and R. L. Cifelli. 1999. A primitive snake from the Cretaceous of Utah. *Special Papers in Palaeontology* 60: 87–100.

Gasparini, Z., S. Casadío, M. Fernández, and L. Salgado. 2001a. Marine reptiles from the Late Cretaceous of northern Patagonia. *Journal of South American Earth Sciences* 14: 51–60.

Gasparini, Z., M. de la Fuente, M. Fernández, and P. Bona. 2001b. Reptiles from Late Cretaceous coastal environments of northern Patagonia. *VII International Symposium on Mesozoic Terrestrial Ecosystems, Asociación Paleontológica Argentina, Publicación Especial* 7: 101–105.

Gasparini, Z., S. Casadío, M. de la Fuente, L. Salgado, M. Fernández, and A. Concheyro. 2002. Reptiles acuáticos en sedimentitas lacustres y marinas del Cretácico superior de Patagonia (Río Negro, Argentina). *Actas del XV Congreso Geológico Argentino*, 495–499. El Calafate.

Gayet, M., J. C. Rage, T. Sempéré, and P. Gagnier. 1992. Modalités des échanges de vertébrés continentaux entre l'Amérique du Nord et l'Amérique du Sud au Crétacé Supérieur et au Paléocène. *Bulletin de la Société Géologique de France* 163: 781–791.

Gayet, M., L. Marshall, T. Sempéré, F. Meunier, H. Cappetta, and J. C. Rage. 2001. Middle Maastrichtian vertebrates (fishes, amphibians, dinosaurs and other reptiles, mammals) from Pajcha Pata (Bolivia). Biostratigraphic, palaeoecologic and palaeobiogeographic implications. *Palaeogeography, Palaeoclimatology, Palaeoecology* 169: 39–68.

González Riga, B. 1999. Hallazgo de vertebrados fósiles en la Formación Loncoche, Cretácico Superior de la Provincia de Mendoza, Argentina. *Ameghiniana* 36: 401–410.

Greene, H. 1983. Dietary correlates of the origin and radiation of snakes. *American Zoology* 23: 431–441.

———. 1997. *Snakes: The Evolution of Mystery in Nature*. Berkeley: University of California Press.

Greene, H., and D. Cundall. 2000. Limbless tetrapods and snakes with legs. *Science* 287: 1939–1941.

Hecht, M. K. 1982. The vertebral morphology of the Cretaceous snake *Dinilysia patagonica* Woodward. *Neues Jahrbuch für Geologie und Paläontologie* 9: 523–532.

Heredia, S., and Calvo, J. 1997. Sedimentitas eólicas en la Formación Río Colorado (Grupo Neuquén) y su relación con la fauna del Cretácico Superior. *Ameghiniana* 34: 120–121.

Heredia, S., and L. Salgado. 1999. Posición estratigáfica de los estratos supracretácicos portadores de dinosaurios en Lago Pellegrini, Patagonia septentrional, Argentina. *Ameghiniana* 36: 229–234.

Hoffstetter, R. 1959. Un dentaire de *Madtsoia* (serpent géant du Paléocène de Patagonie). *Bulletin du Muséum National d'Histoire Naturelle*, ser. 2, 31: 379–386.

———. 1961. Nouveaux restes d'un serpent Boidé (*Madtsoia madagascariensis*) dans le Crétacé supérieur de Madagascar. *Bulletin du Muséum National d'Histoire Naturelle*, ser. 2, 33: 152–160.

Holman, J. A. 2000. *Fossil Snakes of North America: Origin, Evolution, Distribution, Paleoecology*. Bloomington: Indiana University Press.

Huene, F. 1931. Verschiedene mesozoische Wirbeltierreste aus Südamerica. *Neues Jahrbuch für Mineralogie, Geologie und Paläontologie* 66: 181–198.

Hugo, C. A., and H. A. Leanza. 2001. *Hoja Geológica 3969–IV, General Roca, Provincias del Neuquén y Río Negro.* Instituto de Geología y Recursos Naturales, SEGEMAR 308: 1–71.

Kluge, A. 1991. Boine snake phylogeny and research cycles. *Museum of Zoology, University of Michigan, Miscellaneous Publications* 178: 1–58.

Krause, D. W., S. E. Evans, and K. Gao. 2003. First definitive record of Mesozoic lizards from Madagascar. *Journal of Vertebrate Paleontology* 23: 842–856.

Leanza, H. A., and C. A. Hugo. 2001. Cretaceous red beds from southern Neuquén Basin (Argentina): age, distribution and stratigraphic discontinuities. *VII International Symposium on Mesozoic Terrestrial Ecosystems, Asociación Paleontológica Argentina, Publicación Especial* 7: 117–122.

Leanza, H. A., S. Apesteguía, F. E. Novas, and M. S. de la Fuente. 2004. Cretaceous terrestrial beds from the Neuquén Basin (Argentina) and their tetrapod assemblages. *Cretaceous Research* 25: 61–87.

Lee, M. S. Y. 1997. The phylogeny of varanoid lizards and the affinities of snakes. *Philosophical Transactions of the Royal Society of London, Series B* 352: 53–91.

———. 1998. Convergent evolution and character correlation in burrowing reptiles: towards a resolution of squamate phylogeny. *Biological Journal of the Linnean Society* 65: 369–453.

Lee, M. S. Y., and M. W. Caldwell. 1998. Anatomy and relationships of *Pachyrhachis*, a primitive snake with hindlimbs. *Philosophical Transactions of the Royal Society of London, Series B* 353: 1521–1552.

———. 2000. *Adriosaurus* and the affinities of mosasaurs, dolichosaurs, and snakes. *Journal of Paleontology* 74: 915–937.

Lee, M. S. Y., and J. D. Scanlon. 2002. Snake phylogeny based on osteology, soft anatomy and ecology. *Biological Review* 77: 333–401.

Lee, M. S. Y., G. L. Bell, and M. W. Caldwell. 1999. The origins of snake feeding. *Nature* 400: 655–659.

Martin, J., G. Bell, J. Case, D. Chaney, M. Fernández, Z. Gasparini, M. Reguero, and M. Woodburne. 2002. Late Cretaceous mosasaurs (Reptilia) from the Antarctic Peninsula. *Royal Society of New Zealand Bulletin* 35: 293–299.

Martinelli, A. G., and A. M. Forasiepi. 2004. Late Cretaceous vertebrates from Bajo de Santa Rosa (Allen Formation), Río Negro province, Argentina, with the description of a new sauropod dinosaur (Titanosauridae). *Revista del Museo Argentino de Ciencias Naturales "Bernardino Rivadavia"* 6: 257–305.

McDowell, S. B. 1987. Systematics. In R. A. Siegel, J. T. Collins, and S. S. Novak (eds.), *Snakes: Ecology and Evolutionary Biology,* 3–50. New York: Macmillan.

Novas, F. E., M. Fernández, Z. Gasparini, J. Lirio, H. Núñez, and P. Puerta. 2002. *Lakumasaurus antarcticus,* n. gen. et sp., a new mosasaur (Reptilia, Squamata) from the Upper Cretaceous of Antarctica. *Ameghiniana* 39: 245–249.

Parrish, J. T. 1993. The Palaeogeography of the opening South Atlantic. In W. George and R. Lavocat (eds.), *The Africa–South America Connection,* 8–27. Oxford: Clarendon Press.

Rage, J. C. 1981. Les continents péri-atlantiques au Crétacé supérieur: migrations des faunes continentales et problèmes paléogéographiques. *Cretaceous Research* 2: 65–84.

———. 1991. Squamate Reptiles from the Early Paleocene of the Tiupampa Area (Santa Lucia Formation), Bolivia. *Revista Técnica de Yacimientos Petrolíferos Fiscales Bolivianos* 12: 503–508.

———. 1996. Les Madtsoiidae (Reptilia, Serpentes) du Crétacé supérieur d'Europe: témoins gondwaniens d'une dispersion transtéthysienne. *Comptes Rendus de l'Académie des Sciences de Paris* 322: 603–608.

———. 1998. Fossil snakes from the Paleocene of São José de Itaboraí, Brazil. Part II. Boidae. *Palaeovertebrata* 30: 111–150.

———. 1999. Squamates (Reptilia) from the Upper Cretaceous of Laño (Basque Country, Spain). *Estudios del Museo de Ciencias Naturales de Alava* 14: 121–133.

Rage, J. C., and A. M. Albino. 1989. *Dinilysia patagonica* (Reptilia: Serpentes): matériel vertébral additionnel du Crétacé supérieur d'Argentine. Etude complémentaire des vertébrés, variations intraspécifiques et intracolumnaires. *Neues Jahrbuch für Geologie und Paläontologie, Montsh* 7: 433–447.

Rage, J. C., and F. Escuillié. 2000. Un nouveau serpent bipède du Cénomanien (Crétacé). Implications phylétiques. *Comptes rendues de l' Academie du Science de Paris* 330: 513–520.

———. 2002. *Eupodophis,* new name for the genus *Podophis* Rage and Escuillié, 2000, an extinct bipedal snake, preoccupied by *Podophis* Wiegmann, 1834 (Lacertilia, Scincidae). *Amphibia-Reptilia* 23: 232–233.

Rage, J. C., and C. Werner. 1999. Mid-Cretaceous (Cenomanian) snakes from Wadi Abu Hashim, Sudan: the earliest snake assemblage. *Palaeontologia Africana* 35: 85–110.

Rauhut, O., and P. Puerta. 2001. New vertebrate fossils from the Middle-Late Jurassic Cañadón Asfalto Formation of Chubut, Argentina. *Ameghiniana* 38 (Suppl.): 16R.

Rauhut, O., A. López Arbarello, and P. Puerta. 2001. Jurassic vertebrates from Patagonia. *Journal of Vertebrate Paleontology, Abstracts,* 91A.

Reynoso, V. H. 1997. A "beaded" sphenodontian (Diapsida: Lepidosauria) from the Early Cretaceous of central Mexico. *Journal of Vertebrate Paleontology* 17: 52–59.

Rieppel, O. 1988. A review of the origin of snakes. *Evolutionary Biology* 22: 37–130.

Rieppel, O., and H. Zaher. 2000a. The intramandibular joint in squamates, and the phylogenetic relationships of the fossil snake *Pachyrhachis problematicus* Haas. *Fieldiana* 43: 1–69.

———. 2000b. The braincases of mosasaurs and *Varanus* and the relationships of snakes. *Zoological Journal of the Linnean Society* 129: 489–514.

Rieppel, O., A. G. Kluge, and H. Zaher. 2003. Testing the phylogenetic relationships of the Pleistocene snake *Wonambi naracoortensis* Smith. *Journal of Vertebrate Paleontology* 22: 812–829.

Rodríguez-Robles, J. A., C. J. Bell, and H. W. Greene. 1999. Gape size and evolution of diet in snakes: feeding ecology of erycine boas. *Journal of Zoology* 248: 49–58.

Ross, M., and C. Scotese. 1989. A hierarchical tectonic model of the Gulf of Mexico and Caribbean region. In C. Scotese, and W. Sager (eds.), *Mesozoic and Cenozoic Plate Reconstructions, Tectonophysics* 155: 139–168.

Scanlon J. D. 1992. A new large madtsoiid snake from the Miocene of the Northern Territory. *The Beagle, Records of the Northern Territory Museum of Arts and Sciences* 9: 49–60.

———. 1993. Madtsoiid snakes from the Eocene Tingamarra Fauna of eastern Queensland. *Kaupia: Darmstädter Beiträge zur Naturgeschichte* 3: 3–8.

———. 1997. *Nanowana* gen. nov., small madtsoiid snakes from the Miocene of Riversleigh: sympatric species with divergently specialised dentition. *Memoirs of the Queensland Museum* 41: 393–412.

———. 2005. Australia's oldest known snakes: *Patagoniophis, Alamitophis,* and cf. *Madtsoia* (Squamata: Madtsoiidae) from the Eocene of Queensland. *Memoirs of the Queensland Museum* 51: 215–235.

Scanlon, J. D., and Lee, M. 2000. The Pleistocene serpent *Wonambi* and the early evolution of snakes. *Nature* 403: 416–420.

Schulte, J. A., J. P. Valladares, and A. Larson. 2003. Phylogenetic relationships within Iguanidae inferred using molecular and morphological data and a phylogenetic taxonomy of Iguanian lizards. *Herpetologica* 59: 399–419.

Sigé, B., A. D. Buscalioni, S. Duffaud, M. Gayet, B. Orth, J. C. Rage, and J. L. Sanz. 1997. Etat des données sur le gisement Crétacé Supérieur continental de Champ-Garimond (Gard, Sud de la France). *Münchner Geowissenschaftliche Abhandlungen (A)* 34: 111–130.

Simón, M. E., and A. W. A. Kellner. 2003. New sphenodontid (Lepidosauria, Rhynchocephalia, Eilenodontinae) from the Candeleros Formation, Cenomanian of Patagonia, Argentina. *Boletim do Museu Nacional, Nova Série, Geologia* 68: 1–12.

Simpson, G. G. 1933. A new fossil snake from the *Notostylops* beds of Patagonia. *Bulletin of the American Museum of Natural History* 67: 1–22.

Tchernov, E., O. Rieppel, H. Zaher, M. Polcyn, and L. L. Jacobs. 2000. A fossil snake with limbs. *Science* 287: 2010–2012.

Underwood, G., and A. F. Stimson. 1990. A classification of pythons (Serpentes, Pythoninae). *Journal of Zoology* 221: 565–603.

Woodburne, M. O., and J. A. Case. 1996. Dispersal, vicariance, and the Late Cretaceous to early Tertiary land mammal biogeography from South America to Australia. *Journal of Mammalian Evolution* 3: 121–161.

Woodward, A. S. 1901. On some extinct reptiles from Patagonia of the genera *Miolania, Dinilysia* and *Genyodectes. Proceedings of the Zoological Society of London* 1901: 169–184.

Zaher, H., and O. Rieppel. 1999. The phylogenetic relationships of *Pachyrhachis problematicus,* and the evolution of limblessness in snakes (Lepidosauria, Squamata). *Comptes Rendus de l'Académie des Sciences de Paris* 329: 831–837.

———. 2002. On the phylogenetic relationships of the Cretaceous snakes with legs, with special reference to *Pachyrhachis problematicus* (Squamata, Serpentes). *Journal of Vertebrate Paleontology* 22: 104–109.

Zaher, H., M. Cardoso Langer, E. Fara, I. S. Carvalho, and J. T. Arruda. 2003. A mais antiga serpente (Anilioidea) brasileira: Cretáceo superior do Grupo Bauru, General Salgado, SP. *Paleontologia em Destaque, Boletím Informativo da Sociedade Brasileira de Paleontologia* 44: 50–51.

5. Crocodyliformes

Diego Pol and Zulma Gasparini

Introduction

Crocodyliformes is a large group traditionally referred as Crocodylia. Currently, the latter term is restricted to the crown group (i.e., those forms closely related to extant species of crocodiles, gavials, and alligators). The first crocodyliform record dates from the Late Triassic, concurrent with those of the major groups of amniotes that diversified and dominated the megafauna of continental ecosystems during the Mesozoic and Cenozoic (i.e., dinosaurs, pterosaurs, turtles, lepidosaurs, and mammaliaforms). The earliest record of this group is a small and gracile form, *Hemiprotosuchus leali* (Bonaparte 1971), found in the Upper Triassic beds of Los Colorados Formation (northwest Argentina). Later, during the Jurassic and Cretaceous, the major crocodyliform clades diversified, and this group achieved its peak of morphological diversity. This wide range of morphologies included the bauplane typical of crocodiles and alligators with amphibious habits, as well as completely terrestrial forms (some of them reported as possible omnivorous and herbivorous taxa), and forms presumably with exclusive marine habits (with completely modified paddlelike forelimbs for aquatic locomotion). Most of this diversity disappeared by the end of the Cretaceous, leaving the Cenozoic record exclusively formed by crown-group crocodylians and a few relict occurrences of non-crocodylian crocodyliforms that diversified in some regions during the Tertiary (e.g., sebecosuchians, dyrosaurids).

The major diversity of South American Mesozoic crocodyliforms is recorded in Northern Patagonia in Argentina (Fig. 5.1). Jurassic and Cretaceous forms include some of the most extreme cases of morphological diversity. The Jurassic record of Patagonian crocodyliforms is exclusively formed by marine forms found in the

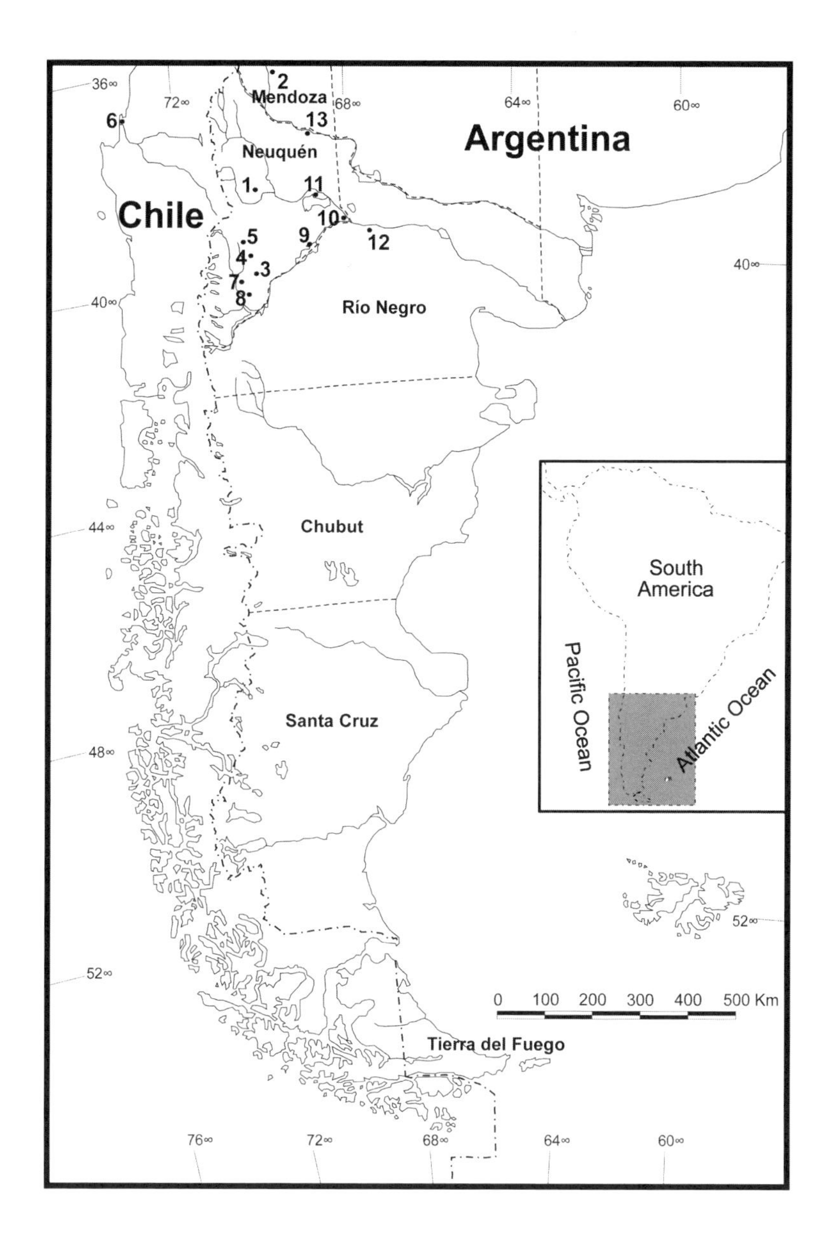

Figure 5.1. Geographic location of Mesozoic crocodyliform localities of Patagonia, 1, Chacay Melehue; 2, Catan Lil; 3, Cerro Lotena; 4, Los Catutos; 5, Yesera del Tromen; 6, Lo Valdéz; 7, Chacaico Sur; 8, La Amarga; 9, El Chocón; 10, Neuquén City (Boca del Sapo); 11, Loma de la Lata; 12, Paso Córdoba; 13, Cañadón Río Seco.

Neuquén Basin, principally in the Tithonian levels of the Vaca Muerta Formation referred to the Tithonian-Berriasian (Gasparini 1981, 1996; Gasparini and Fernández 2005). The presence of crocodyliform remains in the Jurassic beds was early recognized by several researchers during the early twentieth century (Leeds 1908; Huene 1927), but these remains were exceedingly fragmentary and lacked diagnostic features that would justify the creation of formal taxonomic names. Rusconi (1948a,b) studied the first metriorhynchid from South America, which he named *Purranisaurus potens*. Although this author referred *Purranisaurus potens* to Plesiosauria, other authors considered this form as related to *Metri-*

orhynchus (Gasparini 1981). Later, two new marine crocodyliforms found in the Late Jurassic Vaca Muerta Formation were studied and named *Geosaurus araucanensis* and *Dakosaurus andiniensis* (Gasparini and Dellapé 1976; Vignaud and Gasparini 1996; Gasparini et al. 2006; see below). These metriorhynchid crocodyliforms were referred to previously known genera from the Jurassic of Europe because of their remarkable similarities.

The Cretaceous record of Patagonia is notably rich and diverse, and it is also mainly recorded in continental sediments of the Neuquén Basin (except for some fragmentary remains found in central Patagonia [Lamanna et al. 2003]). The first Cretaceous crocodyliforms to be named were the Late Cretaceous *Notosuchus terrestris* and *Cynodontosuchus rothi* (Woodward 1896). After this initial study, research on Cretaceous crocodyliforms from Patagonia was almost absent from the literature, until the revisions of these forms within the context of other South American taxa by Gasparini (1971, 1981, 1984). More recently, a considerable amount of research has been conducted on the anatomy and phylogenetic relationships of Patagonian crocodyliforms, and several new forms have been described (Chiappe 1988; Bonaparte 1991; Gasparini et al. 1991; Ortega et al. 2000; Martinelli 2003; Pol 2005). Currently, a large and diverse assemblage of crocodyliforms is known from Cretaceous beds of Patagonia, ranging from the Early to the Late Cretaceous (see below).

The Jurassic and Cretaceous crocodyliforms from Patagonia are extremely important for understanding the evolutionary history and diversification patterns of this large clade of archosaurs. Additionally, because of their respective relationships with crocodyliforms from Europe, Africa, and Madagascar, these forms play a significant role in Mesozoic biogeographical studies, as has been repeatedly noted by several authors (Buffetaut 1981, 1982; Gasparini 1981, 1996; Bonaparte 1986, 1996; Gasparini and Fernández 1996; Buckley et al. 2000; Ortega et al. 2000; Martinelli 2003; Sereno et al. 2003; Turner 2004).

Here, we review the current knowledge and systematic status of Mesozoic Patagonian crocodyliforms, focusing on their outstanding morphological diversity, phylogenetic relationships, and biogeographical significance.

All diagnoses given below are based on unique combination of characters; autapomorphic conditions are indicated with an asterisk.

Institutional abbreviations. DGM, Departamento de Produção Mineral, Rio de Janeiro, Brazil; MACN, Museo Argentino de Ciencias Naturales "Bernardino Rivadavia," Buenos Aires, Argentina; MAU-PV-CRS, Museo Municipal "Argentino Urquiza," Rincón de los Sauces, Neuquén, Argentina; MHNSR, Museo de Historia Natural de San Rafael, Mendoza, Argentina; MLP, Museo de Ciencias Naturales de La Plata, La Plata, Argentina; MOZ, Museo "Prof. Dr. Juan Olsacher," Zapala, Neuquén, Argentina; MPCA, Museo Provincial "Carlos Ameghino," Cipolletti, Río Negro, Argentina;

MUCPv, Museo de Geología y Paleontología, Universidad Nacional del Comahue, Neuquén, Argentina.

Systematic Paleontology

Crocodyliformes Hay 1930
Mesoeucrocodylia Whetstone and Whybrow 1983
Thalattosuchia Fraas 1902
Metriorhynchidae Fitzinger 1843
***Metriorhynchus* Meyer 1830**
Metriorhynchus* aff. *M. brachyrhynchus
Fig. 5.2B

Material. MOZ 6913PV, fragment of skull, mandible, and vertebral centra (Fig. 5.2B).

Locality and age. Chacay Melehue, located 27 km northwest of Chos Malal, Neuquén Province, Argentina (Fig. 5.1, locality 1). Middle to upper section of Los Molles Formation, upper Bathonian (Gasparini et al. 2005).

Comments. The specimen MOZ 6913PV is referred to Metriorhynchidae because of the presence of an elongated antorbital fossa, oriented obliquely on the lateral surface of the rostrum; lachrymal exclusively exposed laterally; upper margin of the orbit strongly excavated; and absence of external mandibular fenestra. Furthermore, the specimen shares with *Metriorhynchus* the presence of a squamosal posteriorly inclined and the lack of a marked neck on the occipital condyle. The anterior part of the snout is wide and bears large teeth, suggesting the presence of a short rostrum. These characters and the presence of an ornamented dorsal surface of the rostrum are shared with *M. brachyrhynchus* (Deslongchamps) from the Callovian-Oxfordian of Europe. The specimen MOZ 6913PV represents the first metriorhynchid from western South America in which the rostrum is unambiguously ornamented.

Metriorhynchus sp.

Leeds (1908) noted the presence of *Metriorhynchus brachyrhynchus* in Upper Jurassic beds of Neuquén Province; and Ameghino (1906) mentioned the presence of crocodyliform specimens with remarkable long and narrow snout. However, none of the above-mentioned authors explored Jurassic rocks of Neuquén or Mendoza Province, and it is evident that long-snouted crocodyliforms were relatively abundant in the extensive Jurassic outcrops of northwest Patagonia. Rusconi (1948a,b) described *Purranisaurus potens* from the Late Jurassic of Mendoza Province (west Argentina). As noted above, although Rusconi identified this taxon as a plesiosaur, several authors have agreed in considering this taxon as a metriorhynchid (Kälin 1955; Wenz 1968). Furthermore, the validity of *Purranisaurus* as a different genus has been questioned (Kuhn 1968; Gasparini 1973) or rejected, synonymiz-

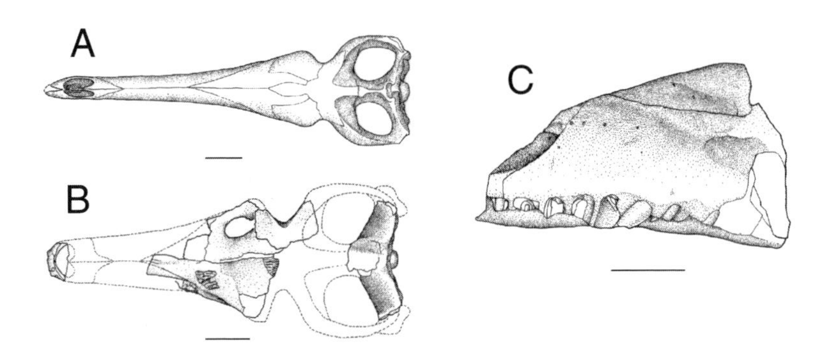

Figure 5.2. Jurassic crocodyliforms from Patagonia. (A) Geosaurus araucanensis *(MLP 72-IV-7-1) in dorsal view. (B)* Metriorhynchus *aff.* M. brachyrhynchus *(MOZ 6913PV) in dorsal view. (C)* Dakosaurus andiniensis *(MHNSR PV 344) in left lateral view. Scale bar = 5 cm.*

ing it with *Metriorhynchus* (*M. potens;* Gasparini 1985; Gasparini and Fernández 1997). Although the validity of *Purranisaurus* as a separate genus was widely rejected, it is not currently clear whether the type specimen of "*Purranisaurus*" *potens* (MJM-PV 2060, a partially prepared skull associated with cervical vertebrae) represents a valid species (i.e., *Metriorhynchus potens*) or whether it should be considered as nomen dubium. This specimen certainly deserves further preparation and study that would reveal more precise information about its taxonomic status and affinities.

Dakosaurus Quenstedt 1856
Dakosaurus andiniensis Vignaud and Gasparini 1996
Fig. 5.2C

Holotype. MHNSR PV 344, an isolated rostrum slightly eroded (Fig. 5.2C).

Locality and age. Catan Lil, Barranca River, Malargüe, Mendoza Province, Argentina (Fig. 5.1, locality 2). Vignaud and Gasparini (1996, 247) incorrectly mentioned that this specimen was originally found in Cari-Lauquen. The specimen MHNSR PV 344 was found in Tithonian sediments of the Vaca Muerta Formation, Mendoza Group.

Diagnosis (from Vignaud and Gasparini 1996). Rostrum very short and stout, in which the separation between the posterior part of the premaxillae and the antorbital fossa is shorter than the height of the rostrum measured in front of the antorbital fossa. The anterior tips of the nasals nearly meet the premaxillae. The teeth, incomplete, are stout and less compressed than in *Dakosaurus maximus* (Plieninger).

Comments. The only known specimen of *Dakosaurus andiniensis* is rather fragmentary; however, it shows a set of diagnostic characters that are only found in *D. maximus,* from the lower Tithonian beds of Germany (Fraas 1902)—in particular, the presence of a notably high rostrum, the great lateral development of the nasal and the short longitudinal development of the dorsal part of the maxillae. *D. andiniensis* differs from *D. maximus* in having less compressed teeth (at the base of the tooth crowns) and a higher and shorter rostrum. More material of this Patagonian taxon was recently studied (Gasparini et al. 2006).

Geosaurus Cuvier 1824
Geosaurus araucanensis Gasparini and Dellapé 1976
Fig. 5.2A

Holotype. MLP 72-IV-7-1, skull and lower jaws (Fig. 5.2A) associated with postcranial remains (scapular girdle, forelimb, vertebrae, and ribs).

Referred specimens. MLP 72-IV-7-2, fragmentary skull including rostral region up to the prefrontals, associated with right mandibular ramus; MLP 72-IV-7-3, skull lacking occipital region, associated with mandibular remains; MLP 72-IV-7-4, complete skull with rostral region preserved up to the antorbital openings; MLP 86-IV-25-1, dentary remains; MLP 86-XI-5-7, skull fragment; MACN-N 95, skull and lower jaws in articulation with cervicodorsal vertebrae; MACN-N 64, anterior tip of rostrum.

Locality and age. The type specimen and most of the referred material was found in Cerro Lotena (Neuquén Province, Argentina; Fig. 5.1, locality 3), in the Vaca Muerta Formation, early Tithonian. Other specimens found in the Neuquén Province and referred to *Geosaurus* were found in the mid-Tithonian locality of Los Catutos (Gasparini et al. 1995) and the late Tithonian locality of Yesera del Tromen (Pampa Tril) (Spalletti et al. 1999, fig. 1).

Geographical distribution. Specimens from this taxon were found in Cerro Lotena, Los Catutos, and Yesera del Tromen in Neuquén Province, Argentina (Fig. 5.1, localities 3–5) (Gasparini and Dellapé 1976; Gasparini et al. 1995; Spalletti et al. 1999). Additionally, a cervical vertebra identical to those of *G. araucanensis* was found in Tithonian rocks in Lo Valdéz, Chile (Fig. 5.1, locality 6) (Gasparini 1985).

Diagnosis (modified from Gasparini and Dellapé [1976] and Vignaud [1995]). Skull proportionally small and gracile, and lacking any signs of ornamentation. Temporal region subquadrangular with large supratemporal fenestrae. Infratemporal openings small and subtriangular. Antorbital fossa deep, anteroposteriorly elongated, and dorsoventrally low. Small, circular antorbital fenestra located on the posterior end of the antorbital fossa, enclosed exclusively by the lacrimal posteriorly and the nasal anteriorly*. Large orbital openings facing laterally and with well-developed sclerotic ring. Completely septated external nares. Long and acute rostrum, subcircular in cross section. Lack of contact between nasals and premaxillae. Laterally expanded prefrontals. Anterior process of frontals lanceolate shaped, extending between the nasals*. Acute teeth, slightly recurved and lacking carinae. Upper dentition consists of 30 to 34 teeth and lower dentition of 29 teeth. Choana suboval and enclosed between palatines and pterygoids. External mandibular fenestra completely obliterated. Amphyplatyan vertebral centra. Radius and ulna disc shaped.

Comments. This taxon is the best-known Jurassic crocodyliform from South America and shares numerous derived characters with *Geosaurus suevicus* (Fraas 1902) from the Tithonian of Switzerland (Buffetaut 1982; Vignaud 1995). These forms are among the most derived taxa of marine crocodyliforms. Their exclusive marine habits are inferred on the basis of remarkable transformations on its postcranial anatomy (e.g., discoidal radius, ulna, and hypocercal tail) as well as the presence of hypertrophied salt glands (Fernández and Gasparini 2000). The close relationships of *Geosaurus* taxa from Europe and western South America indicate the presence of a seaway connection through the proto-Caribbean or Spanish Corridor (Gasparini 1985, 1992). More recently, the presence of *Geosaurus* sp. in the Oxfordian of Cuba (Gasparini and Iturralde-Vinent 2001) and *Geosaurus vignaudi* in the Tithonian of central-east Mexico (Frey et al. 2002) provided additional support to this biogeographic connection between Europe and western South America.

Thalattosuchia gen. et sp. indeterminate

Huene (1927) described *Steneosaurus gerthi* on the basis of two dorsal vertebrae found in the Portezuelo Ancho locality, the Lower Jurassic of Mendoza Province (west Argentina). This specimen is interesting because it represents the oldest crocodyliform record from Patagonia, although it lacks any lower-level diagnostic characters and therefore its taxonomic status can only be determined as a thalattosuchian crocodyliform.

In the area of Chacaico Sur, southwest of the Neuquén Basin (Fig. 5.1, locality 7), a thalattosuchian vertebra has been reported from sediments of the Los Molles Formation (early Bajocian). This element is associated with other marine reptiles such as pliosaurs and ichthyosaurs (Spalletti et al. 1994; Fernández this volume; Gasparini this volume). Finally, in rocks of the same age of Central Chile, the oldest record of *Metriorhynchus* sp. was reported (Gasparini et al. 2000).

Mesoeucrocodylia incertae sedis
Amargasuchus Chiappe 1988
Amargasuchus minor Chiappe 1988

Fig. 5.3C

Holotype. MACN-N 12, isolated incomplete right maxilla (Fig. 5.3C).

Locality and age. La Amarga, 70 km south of Zapala City, Neuquén Province, Argentina (Fig. 5.1, locality 8). Puesto Antigual Member, La Amarga Fm., early Barremian (Leanza et al. 2004).

Diagnosis (modified from Chiappe 1988). Small form with narrow, elongated, and moderately high snout. Presence of antorbital fenestra, probably of small extension. Shallow and reduced anterior region of the antorbital fossa. In dorsal aspect, the maxillary alveolar edge is straight. Maxillary teeth laterally compressed and subequal in size (none of them is hypertrophied). The largest

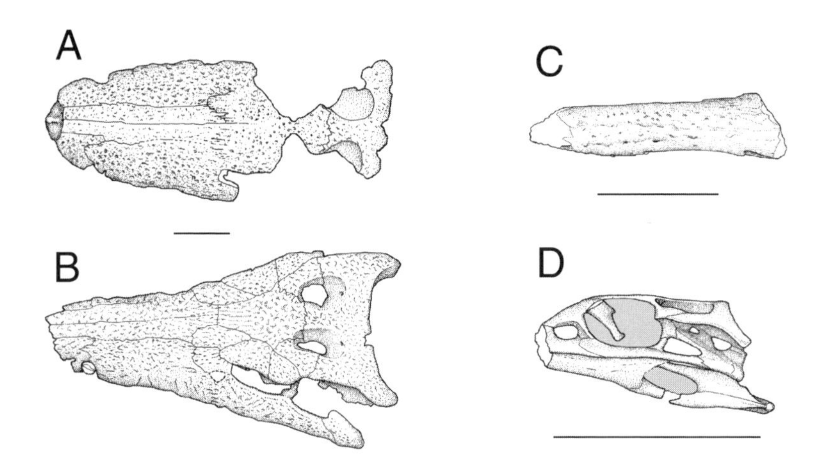

Figure 5.3. Cretaceous crocodyliforms from Patagonia I. (A) Peirosaurus torminni *(MOZ P 1750) in dorsal view (modified from Gasparini et al. 1991). (B)* Lomasuchus palpebrosus *(MOZ P 4084) in dorsal view (modified from Gasparini et al. 1991). (C)* Amargasuchus minor *(MACN-N 12) in left lateral view. (D)* Araripesuchus patagonicus *(MUCPv 269) in left lateral view (modified from Ortega et al. 2000). Scale bar = 5 cm.*

preserved alveoli are the fifth and sixth. Alveolar row extending backward beyond the anterior margin of the suborbital fenestra. The ventral margin of the maxilla is slightly festooned. The lateral surface of the maxilla is strongly ornamented on its dorsal half, whereas the ventral half is smooth and pierced by several large neurovascular foramina.

Comments. Originally, Chiappe (1988) referred this form to Trematochampsidae, a family of crocodyliforms that traditionally clustered ziphodont and false-ziphodont (sensu Prasad and Broin 2002) forms with neosuchian affinities (e.g., *Trematochampsa taqueti* [Buffetaut 1974], *Trematochampsa oblita* [Buffetaut and Taquet 1979], and *Hamadasuchus reboulli* [Buffetaut 1994]). Trematochampsidae has been a problematic group because of the lack of a clear diagnosis, the incompleteness of some of its members (e.g., *T. oblita*), and the unclear association of the many isolated elements that have been referred to *T. taqueti.* Furthermore, several of these forms were never analyzed within a phylogenetic framework. However, most cladistic analyses have corroborated the monophyly of a clade that includes *Trematochampsa* along with peirosaurids (*Peirosaurus* and *Lomasuchus;* Gasparini et al. 1991) and some other Cretaceous forms from Africa and Madagascar (e.g., *Mahajangasuchus insignis* [Buckley and Brochu 1999]; *Stolokrosuchus lapparenti* [Larsson and Gado 2000]). The limited information available from *Amargasuchus minor* suggests that it might be related to these forms (e.g., high and long rostrum, reduced antorbital opening). However, more information is clearly needed from this taxon to thoroughly test its affinities. Furthermore, more research is necessary to thoroughly test the phylogenetic relationships of all these crocodyliforms with an exhaustive taxon sampling regime because their monophyly and phylogenetic position is still debated (Broin 2002; Sues and Larsson 2002). Because of the temporal and geographical distribution of these forms, a thorough understanding of their evolutionary relationships will add critical information to our understanding of the biogeographic relationships of Gondwanan landmasses during the Cretaceous.

Notosuchia Gasparini 1971
***Araripesuchus* Price 1959**
***Araripesuchus patagonicus* Ortega, Gasparini, Buscalioni, and Calvo 2000**
Fig. 5.3D

Holotype. MUCPv 269, anterior half of an articulated skeleton with skull and mandibles (Fig. 5.3D).

Referred specimens. MUCPv 267, anterior half of an articulated skeleton including skull and lower jaws; MUCPv 268, postcranial articulated remains; MUCPv 268b, distal end of tibia and fibula; MUCPv 270, distal end of femur articulated with tibia and fibula; MUCPv 283, anterior region of snout.

Locality and age. El Chocón (Lake Ezequiel Ramos Mexía), Neuquén Province, Argentina (Fig. 5.1, locality 9). These specimens were found in the lower section of the Candeleros Formation, Río Limay Subgroup (Hugo and Leanza 2001). Calvo (1991) considered that the section where these crocodyliforms were found is likely to be Albian, whereas Leanza et al. (2004) referred this section to the early Cenomanian.

Comments. The six known specimens of *Araripesuchus patagonicus* are probably juvenile individuals, as denoted by the absence of well-developed ornamentation in the skull, the relative size of the orbits (occupying 30% of the skull length), and the poor ossification at the proximal and distal ends of the limb elements. Despite their early ontogenetic stage, they show characters that clearly differentiate them from other species of *Araripesuchus*. This genus has been described from the Early Cretaceous of northeast Brazil (*A. gomesii* [Price 1959]) and Niger ("*A.*" *wegeneri* [Buffetaut 1981]). Additionally, new specimens from the Late Cretaceous of Madagascar and Niger have been referred as related forms to this genus (Buckley et al. 1997; Sereno et al. 2004; Turner 2004). There is a strong consensus that these taxa form a monophyletic group (except for the fragmentary "*A.*" *wegeneri*, which may not belong to this group; see Ortega et al. 2000). However, there is some disagreement regarding the position of this clade as more closely related to neosuchian (Buckley and Brochu 1999; Ortega et al. 2000; Sues and Larsson 2002) or to notosuchian crocodyliforms (Pol 1999; Sereno et al. 2003; Pol and Norell 2004).

The widespread distribution of members of this clade during the Cretaceous of Gondwana provides critical information to test competing paleobiogeographical hypotheses regarding the faunal affinities of South America, Africa, and Madagascar (Buckley et al. 2000; Sereno et al. 2003, 2004; Turner 2004). Although further study is needed, currently available information suggests a closer relationship of the Patagonian taxon with the Brazilian forms than with the African and Malagasy taxa (Turner 2004).

Notosuchus Woodward 1896
Notosuchus terrestris Woodward 1896
Fig. 5.4A

Lectotype. MLP 64-IV-16-5, skull and mandible (Fig. 5.4A). This material was designated the lectotype by Gasparini (1971).

Referred specimens. MLP 64-IV-16-1, weathered skull, lacking most of posterior palatal region; MLP 64-IV-16-6, isolated rostrum; MLP 64-IV-16-7, skull and mandible (juvenile); MLP 64-IV-16-8, skull and mandible (juvenile); MLP 64-IV-16-10, braincase and temporal region of skull; MLP 64-IV-16-11, anterior region of skull and mandible; MLP 64-IV-16-12, isolated left maxilla and premaxilla; MLP 64-IV-16-13, anterior region of right mandibular ramus; MLP 64-IV-16-14, anterior region of right mandibular ramus; MLP 64-IV-16-15, fragmentary right side of rostrum; MLP 64-IV-16-16, partially preserved rostrum; MLP 64-IV-16-17, isolated basisphenoid; MLP 64-IV-16-18, basioccipital and basisphenoid; MLP 64-IV-16-19, partially skull remains associated with postcranial remains (probably belonging to more than one individual); MLP 64-IV-16-20, incomplete left mandibular ramus; MLP 64-IV-16-21, anterior region of skull; MLP 64-IV-16-22, partially preserved rostrum; MLP 64-IV-16-23, anterior region of skull and mandible; MLP 64-IV-16-24, anterior region of skull and mandible (juvenile); MLP 64-IV-16-25, fragment of right lower jaw associated with ectopterygoid, pterygoid flanges, and supratemporal region; MLP 64-IV-16-27, isolated dorsal vertebrae; MLP 64-IV-16-30, braincase and supratemporal region of the skull; MLP 26-IV-16-28, isolated distal tibia, calcaneum, and astragalus; MACN-N 22, isolated braincase; MACN-N 23, quadrate distal body and articular region of lower jaw in articulation; MACN-N 24, rostral region of a skull; MACN-RN 1037, skull associated with cervical and dorsal vertebrae, pectoral girdle, and forelimb; MACN-RN 1042, humerus proximal end, radius distal end; MACN-RN 1043, isolated osteoderms; MACN-RN, 1044 fragmentary skull associated to cervicodorsal vertebrae, scapulae, and pelvic girdle elements; MACN-RN, 1027 fragmentary skull associated to osteoderms; MACN-RN, 1024 coracoid proximal end; MPCA-PV, 249 dorsolumbar, sacral, and caudal vertebrae in articulation with right ilium, femur, and proximal end of tibia; MPCA-PV 250, forelimb associated with skull and lower jaws; MUCPv 118, skull and lower jaws of juvenile specimen; MUCPv 287, associated postcranial remains including vertebral series, scapular girdle, forelimb, pelvis, and hind limb.

Locality and age. The lectotype and referred material described by Woodward (1896) and Gasparini (1971) were reported to be found on the left margin of the Pichi Picún Leufú River, at its contact with the Limay River (Neuquén Province, Argentina). Some of these elements, however, are labeled in the MLP collection information as coming from Boca del Sapo (a historic locality placed at the confluence of the Limay and Neuquén Rivers, currently located in Neuquén City). The specimens housed at MUCPv and MACN-N collections were found in the vicinity of the Universidad Nacional del Comahue, Neuquén City, Neuquén Province, Argentina (Fig. 5.1, locality 10). The referred specimens housed at the MACN-RN and MPCA collections were found in the Paso Córdoba locality, Gral. Roca city area, Río Negro Province (Fig. 5.1, locality 12). All these sediments are referred to Bajo de La Carpa Formation, Río Colorado Subgroup (Hugo and Leanza 2001). This unit has been considered either Coniacian or Santonian (Legarreta and Gulisano 1989; Bonaparte 1991; Leanza et al. 2004).

Diagnosis (modified from Woodward [1896], Gasparini [1971], Bonaparte [1991], and Pol [2005]). Snout short and high, with anteriorly located external nares. Well-developed perinarial depressions. Well-developed bulge on the maxilla, anteroventrally located from the small antorbital fossa*. Ectopterygoid extending along the lateral margin of the choanal opening*, contacting the palatine and excluding the pterygoid from the suborbital fenestra. Extremely elongate articular facet for the quadrate condyles, being approximately three times as long as the condyles*, and lacking a posterior buttress. Five premaxillary teeth, being the fourth element notably hypertrophied. Seven maxillary teeth, subtriangular in cross section with major axis oriented posteromedially. Anteroposteriorly short axial neural spine. Three sacral vertebrae, two of which are fused to each other*.

Comments. In addition to being the first described crocodyliform from Patagonia, *Notosuchus terrestris* is the best known Mesozoic crocodyliform from South America. Currently, there are over 20 relatively complete skulls, including some extremely young individuals (e.g., skull length of 5 cm in MUCPv 118). Furthermore, it is one of the few Mesozoic crocodyliforms for which the postcranial anatomy is known (Pol 2005). In some outcrops of the Bajo de La Carpa Formation, *Notosuchus terrestris* is the most abundant vertebrate (e.g., Paso Córdoba, Neuquén City; see Fig. 5.1), being between five and ten times more abundant than other taxa (depending on the locality).

Because of the early recognition of this form and the remarkable quality of some specimens, *Notosuchus terrestris* occupies a pivotal place for understanding the evolution of South American crocodyliforms. In particular, the cranial anatomy of *Notosuchus terrestris* has an interesting combination of crocodyliform plesiomorphies and autapomorphic characters that early distinguished this taxon (and allies) from the rest of the Crocodyliformes (Gasparini 1971). In particular, the most interesting set of characters of *N. terrestris* is related to its feeding apparatus, such as the degree of heterodonty (e.g., small anterior teeth, hypertrophied premaxillary caniniforms, and subtriangular maxillary tooth crowns oriented obliquely) and the craniomandibular articulation located ventrally to tooth-row level. Most importantly, the extensive anteroposterior length of the articular facets for the quadrate condyles would have allowed anteroposterior movements of the lower jaw (Bonaparte

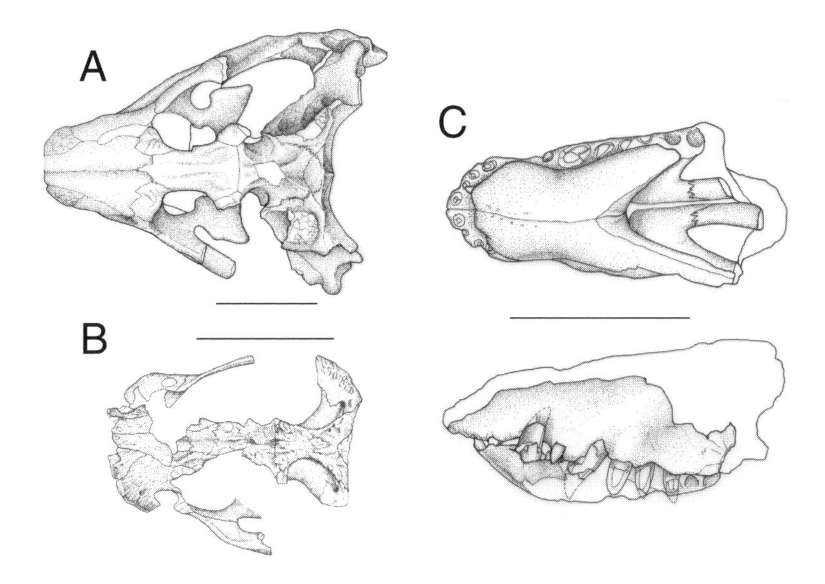

Figure 5.4. Cretaceous crocodyliforms from Patagonia II. (A) Notosuchus terrestris *(MLP 64-IV-16-5) in dorsal view. (B)* Comahuesuchus brachybuccalis *(MOZ P 6131) in dorsal view (modified from Martinelli 2003). (C)* Cynodontosuchus rothi *(MLP 64-IV-16-25) in ventral (top) and left lateral (bottom) views. Scale bar = 5 cm.*

1991). Some of these characters, once thought to be autapomorphies of *N. terrestris,* are now known to be common derived features of some clades within Notosuchia, a conspicuous clade of Cretaceous crocodyliforms (see below).

Comahuesuchus Bonaparte 1991
Comahuesuchus brachybuccalis Bonaparte 1991
Fig. 5.4B

Holotype. MUCPv 202, partial skull with articulated lower jaws of a subadult individual.

Referred specimens. MACN-N 30, incomplete rostrum and anterior region of the lower jaws of a subadult individual; MACN-N 31, incomplete rostrum and anterior region of the lower jaws of a subadult individual; MOZ P 6131, partial skull with articulated lower jaws of a mature individual.

Locality and age. The specimens MUCPv 202, MACN-N 30, and MACN-N 31 were from the vicinities of the Universidad Nacional del Comahue in Neuquén City (Fig. 5.1, locality 10), Neuquén Province, Argentina (Bonaparte 1991). The specimen MOZ P6131 (Fig. 5.4B) recently described by Martinelli (2003), was found in the Paso Córdoba locality, Río Negro Province (Fig. 5.1, locality 12). These sediments have been referred to the Bajo de La Carpa Formation, Río Colorado Subgroup (Hugo and Leanza 2001). This unit has been considered Santonian (Bonaparte 1991; Leanza et al. 2004).

Diagnosis (modified from Bonaparte [1991] and Martinelli [2003]). Extremely short and wide snout, with subcircular outline in dorsal view*. Only five maxillary teeth, with the third element hypertrophied*. Eleven dentary teeth, being the anterior ones procumbent. Paracanine fossa in the posterior maxillary region for the reception of a hypertrophied posterior dentary tooth*. Ectopterygoid contacts palatine, excluding the pterygoid from the suborbital fenestra. Descending ramus of lachrymal acute and tapering ventrally. Symphyseal region of dentary low and transversally broad, extending to the level of the ninth dentary tooth.

Comments. *Comahuesuchus* is one of the most bizarre crocodyliforms known, showing unusual modifications in the skull morphology (Bonaparte 1991; Martinelli 2003). Its mode of life and feeding habits are intriguing because of its peculiar snout morphology and remarkably reduced and small dentition (with the exception of the enlarged maxillary and dentary caniniforms). Several paleobiological aspects of this taxon are still unknown, such as its postcranial anatomy. In part, this lack of knowledge on *Comahuesuchus* arises from its rarity in the Bajo de La Carpa beds. This taxon has been found in the same horizons and localities as *Notosuchus terrestris*, although the latter is remarkably more abundant (e.g., some localities have an abundance ratio of 10:1 between these two taxa).

Its remarkably wide and short snout vaguely resembles the morphology of *Simosuchus clarki* (Buckley et al. 2000); however, their similarities are restricted to superficial similarities. Recently, Sereno et al. (2003) described a new form from the Early Cretaceous of Niger (*Anatosuchus minor*) as the sister taxon of *Comahuesuchus*. This grouping was justified on the presence of an edentulous anterior region of the premaxilla and dentaries, upper tooth row offset labially and ventrally from dentary tooth row, and external nares inset posteriorly from the snout anterior margin. The condition for the first character in *Comahuesuchus* is ambiguous, while the second and third characters seem to be more widespread among notosuchian crocodyliforms. Furthermore, *Anatosuchus* seems to lack many of the derived characters supporting the relationships of *Comahuesuchus* with *Mariliasuchus* and *Notosuchus* (see below). Thus, the affinities of *Comahuesuchus* with *Anatosuchus* deserve a critical revision.

Sebecosuchia Colbert 1946
***Cynodontosuchus* Woodward 1896**
***Cynodontosuchus rothi* Woodward 1896**
Fig. 5.4C

Holotype. MLP 64-IV-16-25, anterior end of rostrum and lower jaw (Fig. 5.4C).

Locality and age. The only data available for the type locality of this enigmatic taxon refer to Boca del Sapo, located at the confluence of the Neuquén and Limay Rivers, Neuquén City, Neuquén Province, Argentina (Fig. 5.1, locality 10). These remains were probably found in a similar horizon to the type specimen of *Notosuchus terrestris*, a taxon currently known exclusively from the Bajo de La Carpa Formation, Río Colorado Subgroup (Hugo and Leanza 2001). These sediments were considered either Coniacian or Santonian (Legarreta and Gulisano 1989; Bonaparte 1991; Leanza et al. 2004).

Geographical distribution. This crocodyliform is only known from the poorly preserved holotype, for which there are imprecise collection data. Its geographical precedence is referred to Boca del Sapo, Neuquén Province (northwest Patagonia), Argentina, in the MLP collections. A fragmentary specimen found in the El Molino Formation (Bolivia) was referred to *Cynodontosuchus* (Buffetaut 1980). Although the Bolivian specimen shares some derived characters with *C. rothi*, it clearly represents a different taxon. Furthermore, the rocks where this specimen was found are currently considered to be Paleocene (Ortiz Jaureguizar and Pascual 1989; Bonaparte 1990; Gasparini 1996).

Diagnosis. Snout narrow, high, and unsculpted, with spatulated anterior end*. Large and shallow perinarial depressions. Notch at the premaxilla-maxilla contact for the reception of enlarged anterior dentary tooth is anteroposteriorly enlarged but dorsoventrally low. Palatines sutured to each other forming an extensive and remarkably narrow secondary palate between enlarged suborbital fenestra. Palatine-maxilla suture transversally oriented, and suture located just posteriorly to the suborbital fenestra anterior margin. Maxillary dentition composed of seven (or possibly more) teeth. The first maxillary tooth is reduced and procumbent*. Second maxillary tooth caniniform. Posterior maxillary teeth reduced in size and lateromedially compressed.

Comments. No remains of *Cynodontosuchus rothi* have been found since the discovery of the type specimen. Despite its incompleteness, the holotype of this taxon shows clear similarities with *Baurusuchus pachecoi* from the Late Cretaceous of Brazil (Price 1945) that early indicated the affinities between these two taxa. Furthermore, it has been suggested that these two taxa might be considered as cogeneric (Gasparini et al. 1991; Gasparini 1996). *Cynodontosuchus rothi,* however, is distinguished from *Baurusuchus pachecoi* and related forms from the Adamantina Formation of the Bauru Basin (e.g., *Baurusuchus salgadoensis* [Carvalho et al. 2005], *Stratiotosuchus maxhechti* [Campos et al. 2001; Riff 2003]) because it lacks the following characters: a hypertrophied posterior premaxillary tooth and the well-developed rugosities on the palatal branches of the maxilla (lateral to the suborbital fenestra). Furthermore, the snout of *Cynodontosuchus rothi* is much shorter and dorsoventrally lower than those of the three Brazilian taxa mentioned above (although this could be caused by ontogenetic differences). The maxilla-palatine suture of *C. rothi* is transversally oriented and located close to the anterior margin of the suborbital opening, while this suture is V-shaped and located much more posteriorly in *B. pachechoi.* Finally, *C. rothi* has the second maxillary tooth distinctly enlarged, while *B. pachecoi* (Price 1945) and *B. salgadoensis* (Carvalho et al. 2005) have the second and third elements of the maxilla remarkably larger than the rest of the maxillary teeth. An increasing number of taxa referred to this group have been recently discovered and described (Campos et al. 2001; Wilson et al. 2001; Riff 2003; Carvalho et al. 2005), although their phylogenetic relationships are poorly understood. Future analyses will certainly reveal interesting aspects on the evolutionary and biogeographic history of this enigmatic group of carnivorous crocodyliforms.

As mentioned above, a fragmentary specimen found in the Paleocene beds of El Molino Formation from Bolivia was referred to *Cynodontosuchus* (Buffetaut 1980). These two taxa may be related to each other, as suggested by the presence of a spatulate anterior end of the snout (i.e., the premaxilla projects anteriorly, tapering dorsoventrally and is not vertically oriented as in *Baurusuchus*) and an anteroposteriorly long and dorsoventrally low notch at the premaxilla-maxilla contact. Despite these derived similarities, the Bolivian specimen seems to represent a different taxon. The holotype of *C. rothi* is slightly more robust and lacks any signs of ornamentation, while the notably smaller Bolivian specimen bears a well-developed pitted ornamentation on its rostrum and mandible. In contrast to the condition of *C. rothi,* in the Bolivian specimen, the second to last premaxillary tooth is notably enlarged (as in *B. pachecoi*), the first maxillary teeth are not procumbent (instead of the procumbent condition also present in *Stratiotosuchus maxhechti* [Riff 2003] and *Pabwehshi pakistanensis* [Wilson et al. 2001]), and the enlarged maxillary tooth is the third element (instead of the second as in *C. rothi*). In sum, although the evidence is fragmentary and more complete specimens are certainly needed, these incomplete specimens may represent the only available evidence of a subclade of small-bodied baurusuchids that diversified in the Late Cretaceous of South America.

Pehuenchesuchus Turner and Calvo 2005
Pehuenchesuchus enderi Turner and Calvo 2005

Holotype. MAU-PV-CRS-440, an isolated right dentary.

Locality and age. The holotype was found in the Cañadón Río Seco site, 2 km north of Rincón de los Sauces, Neuquén Province, Argentina (Fig. 5.1, locality 13). The sediments in which this specimen was found are considered part of the Río Neuquén Subgroup (Turner and Calvo 2005), although it is not clear whether the specimen comes from the Portezuelo Formation or the overlying Plottier Formation. The age of this subgroup has been considered Turonian-Coniacian (Hugo and Leanza 2001; Leanza et al. 2004).

Diagnosis (taken from Turner and Calvo 2005). Narrow and deep lower jaw. Sixteen teeth in dentary, the first and the fourth larger than the remaining ones, with the first slightly procumbent. Tooth row sigmoidal in dorsal view, with lateral surface of dentary bearing a longitudinal depression anterior to the external mandibular fenestra. Differs from all other sebecosuchians by having laterally compressed teeth with carinae lacking serrations*.

Comments. In the recent description of *Pehuenchesuchus enderi,* Turner and Calvo (2005) have identified derived characters shared by this specimen and most sebecosuchians (e.g., sigmoidal tooth row in dorsal view, longitudinal depression on lateral surface of the dentary). The phylogenetic analysis presented by these authors depicted *Pehuenchesuchus enderi* as the most basal sebecosuchian (because of the absence or serrations in the mesial and distal margins of dentary tooth crowns). This suggests that, although fragmentary, the *Pehuenchesuchus* remains can offer some information on the early evolutionary history of Sebecosuchia. Interestingly, this also represents the oldest record of a sebecosuchian crocodyliform (Turonian-Coniacian) because previous Mesozoic sebecosuchians (i.e., baurusuchids) were restricted to Campanian and Maastrichtian beds of Argentina (*Cynodontosuchus rothi*), Brazil (*B. pachecoi, B. salgadoensis, Stratiotosuchus maxhechti* [Price 1945; Riff 2003; Carvalho et al. 2005]), and possibly Pakistan (*Pabwehshi pakistanensis* [Wilson et al. 2001]).

Peirosauridae Gasparini 1982
***Lomasuchus* Gasparini, Chiappe, and Fernández 1991**
***Lomasuchus palpebrosus* Gasparini, Chiappe, and Fernández 1991**
Fig. 5.3B

Holotype. MOZ P 4084PV, nearly complete skull found in association with the lower jaws (Fig. 5.3B).

Locality and age. Loma de la Lata, Neuquén Province, Argentina (Fig. 5.1, locality 11). Bajo de La Carpa Formation, Santonian (Hugo and Leanza 2001; Leanza et al. 2004). However, recent sedimentological studies on this locality revealed that the type specimen of *Lomasuchus palpebrosus* comes from the Portezuelo Formation (A. Garrido personal communication 2004; Poblete and Calvo 2005). These sediments are currently considered late Turonian–early Coniacian (Leanza et al. 2004).

Diagnosis (modified from Gasparini et al. 1991). Snout moderately narrow and high, with a large notch at the premaxilla-maxilla contact that receives an enlarged mandibular tooth. Maxilla with an anterior wedgelike process at its contact with the premaxilla. Maxillary and dentary teeth lateromedially compressed and with serrated margins. Small antorbital fenestra. Anterior and posterior palpebral sutured to each other and to the lateral margin of the frontal, covering the entire dorsal margin of the orbit. Lateral margin of the squamosal is markedly convex and sharply downturned, producing an internally concave roof of the otic recess. Otic notch subrectangular in shape. Robust posteroventral end of the quadratojugal.

Comments. Lomasuchus palpebrosus is particularly interesting because of its unique combination of characters. Its skull morphology resembles the neosuchian condition in some characters, although its snout is remarkably high (contrasting with the platyrostral condition of most neosuchians). Additionally, *Lomasuchus* has a peculiar combination of plesiomorphic and apomorphic characters in its dentition, with an enlarged anterior dentary tooth, ziphodont teeth (with well-developed and individualized denticles), and variation of tooth size in two "waves" (as in neosuchians).

As noted above, *Lomasuchus palpebrosus* has been postulated as being closely related to *Peirosaurus torminni* (Gasparini et al. 1991). This group has been commonly referred to as Peirosauridae, the monophyly of which has never been questioned (Gasparini et al. 1991; Buckley and Brochu 1999; Larsson and Gado 2000; Carvalho et al. 2004). However, the most interesting question regarding *Lomasuchus* relationships concerns which taxa are more closely related to Peirosauridae. Several hypotheses have been proposed: the long-snouted *Stolokrosuchus lapparenti* from the Early Cretaceous of Niger (Larsson and Gado 2000), *Trematochampsa* (Gasparini et al. 1991), *Uberabasuchus terrificus* (Carvalho et al. 2004), or the enigmatic *Mahajangasuchus insignis* from the Campanian of Madagascar (Buckley and Brochu 1999). These hypotheses entail important biogeographic consequences regarding the faunal affinities of South America, Africa, and Madagascar. The precise pattern of relationships among these forms (simultaneously included all the available evidence) is yet to be established and deserves further research. However, a broad consensus exists regarding the strong affinities of *Lomasuchus* and peirosaurids with other Cretaceous Gondwanan crocodyliforms, underscoring their relevance for understanding the biogeographical relationships of Gondwanan landmasses during this period.

Peirosaurus Price 1955
Peirosaurus torminni Price 1955
Fig. 5.3A

Holotype. DGM 433-R. The specimen consists of a left premaxilla bearing five teeth, isolated maxillary and dentary teeth, a left palpebral bone, radius, ulna, pubis, ischium, some presacral and a single caudal vertebrae, associated ribs, chevrons, and osteoderms.

Referred Specimen. MOZ P 1750PV. Incompletely preserved skull (Fig. 5.3A).

Locality and age. The type specimen comes from the famous Peirópolis Site (Price 1955; Candeiro et al. in press), Minas Gerais State, Brazil. These sediments are considered to belong to the Serra de Galga Member, Marilia Formation, Baurú Group (Fernandes and Coimbra 1996; Candeiro et al. in press). This unit has been recently considered as Maastrichtian (Dias-Brito et al. 2001). The referred Patagonian specimen comes from Loma de la Lata, Neuquén Province, Argentina (Fig. 5.1, locality 11). It was originally reported as found in sediments of the Río Colorado Subgroup (Gasparini et al. 1991). Later, Hugo and Leanza (2001) noted that this specimen was actually found in the underlying Portezuelo Formation (as the type of *Lomasuchus palpebrosus*). However, recent stratigraphic work on this area identified the horizon where MOZ P 1750 was found as belonging to the Plottier Formation of the Río Neuquén Subgroup (A. Garrido personal communication 2004;

Poblete and Calvo 2005). These sediments are currently considered to be Coniacian (Leanza et al. 2004).

Geographical distribution. On the basis of the two available specimens, this taxon is one of the most widespread crocodyliform taxa, ranging from the Baurú deposits in Southern Brazil to the Neuquén Province in northwest Patagonia (Fig. 5.1).

Diagnosis (modified from Price [1955] and Gasparini et al. [1991]). Snout wide and high, with a notch at the premaxilla-maxilla contact for the reception of a mandibular tooth. External nares subvertically oriented at the anterior end of the snout, with a noticeably smooth perinarial region*. Five conical premaxillary teeth. Posterior maxillary and dentary teeth with short lateromedially compressed spatulate crowns and bearing serrated margins. Osteoderms with pitted ornamentation and low keels.

Comments. Although *Peirosaurus torminni* remains have been known for half a century (Price 1955), its anatomy is incompletely known. To date, no additional remains of this taxon have been found in the Brazilian Baurú Group after the initial finding of the fragmentary type specimen. However, the recent finding of *Uberabasuchus terrificus* (Carvalho et al. 2004), another peirosaurid from the Marilia Formation (Baurú Group), suggests that the diversity of this group was greater than previously thought. The Patagonian remains of this taxon are certainly more complete (Gasparini et al. 1991), although its preservation is not perfect and many details in its skull morphology cannot be determined. Moreover, knowledge on the postcranial anatomy of *Peirosaurus torminni* (and the related *Lomasuchus palpebrosus*) is extremely limited. More material of this taxon would be highly beneficial in achieving a better understanding of peirosaurid anatomy and relationships. The diversity, geographical, and temporal distribution of this group in Patagonia might also be significantly wider than currently known because of the presence of several fragmentary remains that possibly belong to this group (e.g., postcranial remains from Bajo de La Carpa [MUCPv 27] reported by Bonaparte [1991], fragmentary lower jaw from Bajo Barreal in Central Patagonia [Lamanna et al. 2003]).

Phylogenetic Relationships

The phylogenetic relationships of Patagonian taxa are considered here on the basis of a cladistic analysis within the context of all major crocodyliform lineages (modified from Gasparini et al. 2006). In the most parsimonious hypothesis of this analysis (Fig. 5.5), it can be seen that Patagonian crocodyliforms belong to three major lineages: the Jurassic forms related to the marine radiation of thalattosuchians also recorded in Europe, the Late Cretaceous peirosaurids depicted as basal neosuchians, and the diverse forms grouped in the Notosuchia clade, including *Araripesuchus*, as basal members and baurusuchids as derived forms.

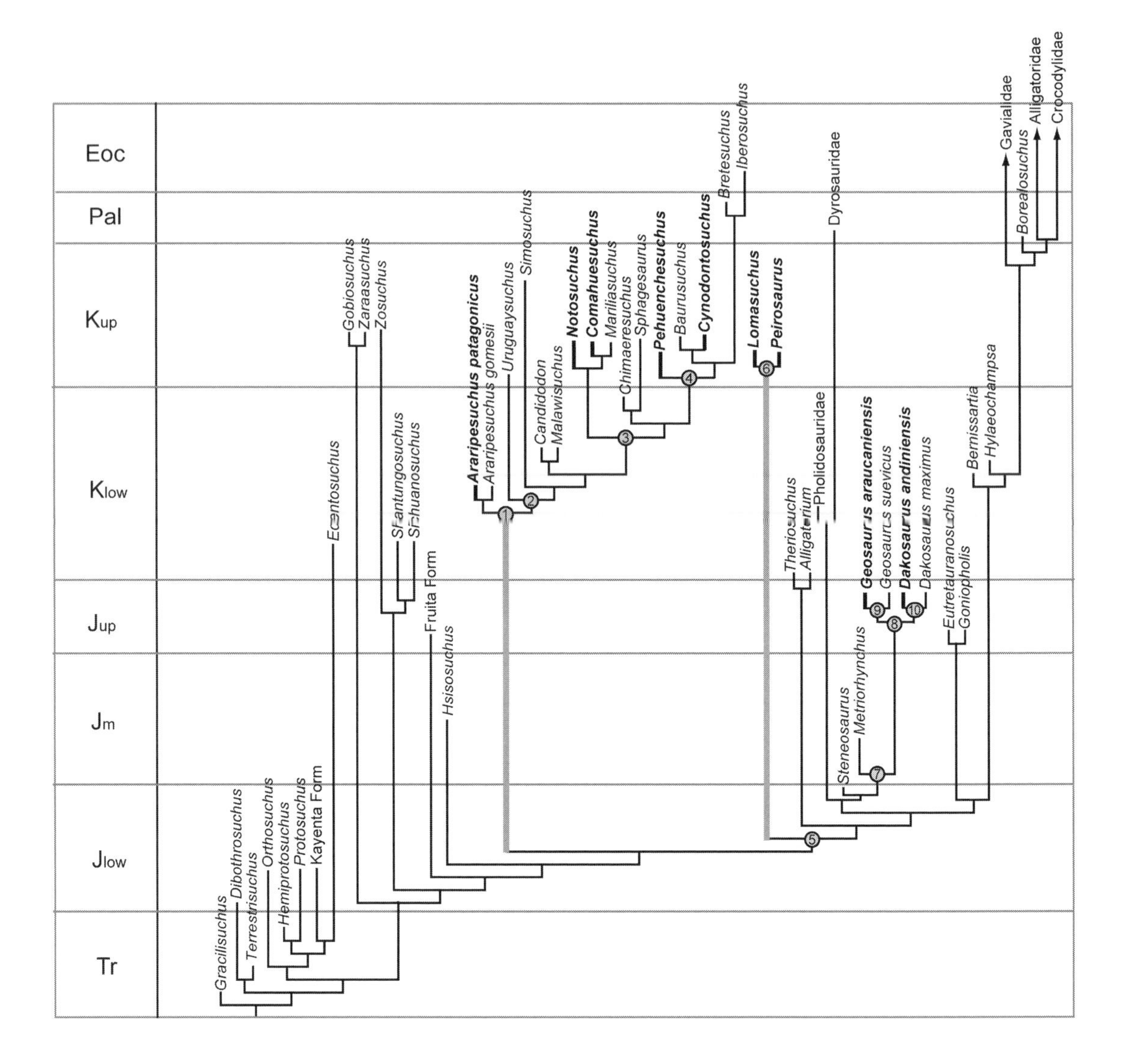

Figure 5.5. Phylogenetic relationships of Crocodyliformes plotted against geologic time. Phylogeny modified from Gasparini et al. (2006). Patagonian taxa are indicated with bold taxon names and branches. Gray branches indicate extensive ghost lineages of clades with Patagonian members as implied by this phylogenetic hypothesis.

Jurassic Taxa

Geosaurus araucanensis and *Dakosaurus andiniensis* are the only two Patagonian taxa with enough preserved information to be included in the phylogenetic analysis. These two taxa are depicted nested well within the Metriorhynchidae (node 7; Fig. 5.5), a group diagnosed by the presence of the following derived characters: elongated and low antorbital fenestra oriented obliquely; nasals descending on lateral surface of skull with extensive participation on antorbital fossa and fenestra; enlarged prefrontals; squamosal with large, rounded, flat surface facing posterolaterally; enlarged foramen for the internal carotid artery; large and deep groove on lateral surface of surangular and posterior region of dentary (below alveolar margin); and closed mandibular fenestra. Furthermore, *Geosaurus* and *Dakosaurus* are depicted as closely related (node 8; Fig.

5.5) because of the presence of the following derived characters: jugal entering antorbital fossa, postorbital bar completely formed by descending process of postorbital, surangular groove ends anteriorly into a large foramen in the dentary. The monophyly of the *Geosaurus* clade, clustering *G. araucanensis* and *G. suevicus* (node 9; Fig. 5.5), is supported by the presence of external nares retracted and completely divided by a bony premaxillary septum, and subrounded prefrontals. Finally, *Dakosaurus andiniensis* is located as the sister taxon of *D. maximus* (node 10; Fig. 5.5) because they are the only metriorhynchids with a short and high rostrum.

Cretaceous Taxa

As noted above, the Cretaceous Patagonian taxa belong to two different clades, the large and diverse Notosuchia and the peirosaurid clade. Notosuchia (node 1; Fig. 5.5) is diagnosed by the following characters: absence of dentary extension beneath mandibular fenestra; rounded retroarticular process, projected posteroventrally and facing dorsomedially; quadrate major axis ventrally directed; absence of posterior ridge of articular glenoid fossa; postzygapophyseal lamina in dorsal vertebra; depression on lateral surface of neural arch, between base of neural spine and postzygapophyses; ventral margin of postacetabular process of ilium located ventrally than acetabular roof.

Within Notosuchia, *Araripesuchus patagonicus* and allies occupy a basal position as a result of the absence of the following characters that diagnose more advanced notosuchians (node 2; Fig. 5.5): jaw joint level below tooth row; dorsal osteoderms rounded or ovate; parasagittal dorsal osteoderm keel lateromedially centered; humeral deltoid crest low and thin; dorsal margin of postacetabular process of ilium projected posteroventrally.

Notosuchus, Comahuesuchus, and more derived notosuchians share the presence of the following derived characters (node 3; Fig. 5.5): depression posterior to choana wider than palatine bar; elongated and narrow concavity lateral to deltoid crest; caudal margin of nasals V-shaped; palatine projecting posterolateral processes towards the ectopterygoids; ectopterygoid widely extended covering approximately the lateral half of the ventral surface of the pterygoid flanges.

The node recognized as Sebecosuchia (node 4; Fig. 5.5) is diagnosed by the following synapomorphies: deep and anteriorly convex mandibular symphysis in lateral view; lateral surface of dentaries below alveolar margin flat, sculpted, and continuous with the ventral region of the lateral surface of the dentaries (at anteroposterior midpoint of tooth row); longitudinal depression in lateral surface of dentary. *Cynodontosuchus* and other sebecosuchians share the presence of the following derived characters absent in notosuchian condition are ambiguously optimized because of the unknown skull morphology of *Pehuenchesuchus:* rostrum higher than wide; large ventrally opened notch at premaxilla-maxilla contact;

denticulate carinae on tooth crown margins; broad perinarial depression facing anteriorly and extending anteroventrally down to the alveolar margin.

Peirosaurids are depicted as more closely related to neosuchians than to notosuchians because of the presence of the following derived characters (node 5; Fig. 5.5): external nares dorsally concealed from the anterior edge of rostrum; reduced antorbital fenestra; five premaxillary teeth; basioccipital and ventral region of otoccipital subvertically oriented, facing posteriorly; absence of a distinct smooth region in dentary below tooth row; maxillary teeth implanted in discrete alveoli. The sister group relationship of *Lomasuchus palpebrosus* and *Peirosaurus torminni* (node 6; Fig. 5.5) is supported by the presence of wedgelike process of premaxilla on lateral surface of premaxilla-maxilla contact (Gasparini et al. 1991).

Conclusions

The Jurassic record of Patagonian crocodyliforms is mainly composed of marine Tithonian thalattosuchians closely related to several taxa from the Callovian-Oxfordian of Europe. As depicted in the phylogenetic analysis, these relationships support the taxonomic assignment previously recognized for the Patagonian taxa (Gasparini and Dellapé 1976; Vignaud and Gasparini 1996). Furthermore, the close faunal relationship between these two distant areas of the globe provides support to the biogeographical connection through the proto-Caribbean Corridor (Gasparini 1985, 1992). The addition of relevant thalattosuchian taxa from the Jurassic of Cuba and Mexico (Gasparini and Iturralde-Vinent 2001; Frey et al. 2002) to the phylogenetic analysis would be critical to further test the above mentioned biogeographical hypothesis.

The faunal composition of Cretaceous crocodyliforms from Patagonia is clearly dominated by two groups of basal mesoeucrocodylians (Fig. 5.5). Most of the diversity is concentrated in a large and diverse clade: Notosuchia. The morphological diversity within this clade (Fig. 5.5) is remarkably large, although all of them depart significantly from the characteristic bauplane of modern crocodylians. The South American Cretaceous taxa developed a wide diversity in tooth morphologies, including multicusped molariform teeth (e.g., *Candidodon*), extensive wear facets (e.g., *Notosuchus, Sphagesaurus*), and theropodomorph carnivorous teeth (e.g., *Baurusuchus, Cynodontosuchus*). These characters are usually present along with modifications in the cranio-mandibular articulation that would probably imply mandibular movements unique among Crocodyliformes and numerous modifications in the rostral anatomy (e.g., *Notosuchus, Sphagesaurus;* Bonaparte 1991; Pol 2003).

The crocodyliform record in other regions of Gondwana shows remarkable similarities in its faunal composition with the South American record (e.g., notosuchians and *Araripesuchus* in Africa and Madagascar), which suggests that this particular faunal assemblage extended to other regions of Gondwana, such as Antarctica. In contrast, the crocodyliform record of Laurasian landmasses during the Cretaceous is mainly composed of neosuchian taxa (e.g., basal Eusuchia) and basal crocodyliforms (e.g., *Gobiosuchus, Zosuchus, Sichuanosuchus*), with the remarkable exception of some isolated records (e.g., *Chimaerasuchus* from the Early Cretaceous of China).

Additionally, crocodyliform taxa in some Patagonian deposits are outstandingly abundant in comparison with other tetrapod groups. Crocodyliforms form 40% of the known taxa from Candeleros Formation, and 70% of the known taxa in the Bajo de La Carpa Formation. Interestingly, this degree of abundance and diversity is also present in some Cretaceous units in other regions of Gondwana (e.g., Baurú Group, Brazil; Gadoufaua, Niger; Maevarano Formation, Madagascar; Buckley et al. 1997).

Acknowledgments. We thank Dr. J. Bonaparte (MACN), S. Cocca (MOZ), and J. O. Calvo (MUC) for their help through the years and for letting us study published and unpublished specimens housed at the collections under their care. We are grateful to J. Clark (GWU) and C. Brochu (UI) for providing thoughtful suggestions as reviewers of this chapter. Jorge González drew the specimen figures. Financial support to D. P. was provided by the Department of Earth and Environmental Sciences of Columbia University and the American Museum of Natural History. Financial support to Z. G. for related projects was provided by Agencia Nacional de Promoción Científica y Tecnológica, Argentina (PICT 8439 and PICT 25276). Comparisons with other crocodyliforms were possible thanks to J. Maisey and M. Norell (AMNH), A. Milner and S. Champman (BMNH), M. Moser (BSP), M. Maisch (GPIT), X. Xu (IVPP), J. M. Clark (GWU), E. Gomani (MAL), D. Unwin (MB), L. E. Ruigomez and R. Cuneo (MEF), M. Reguero (MLP), F. L. de Broin (MNHN), A. Kellner (MNUFRJ), H. Zaher (MZUSP), J. Powell (PVL), C. Cartelle (RCL), A. Chinsamy (SAM), R. Wild (SMNS), D. Krause and G. Buckley (UA), A. Buscalioni and F. Ortega (UAM), and I. S. Carvalho (UFRJ).

References Cited

Ameghino, F. 1906. Les formations sédimentaires du Crétacé supérieur et du Tertiaire de Patagonie avec un paraléle entre leus faunes mammalogiques et celles de l'ancien continent. *Anales del Museo Nacional de Buenos Aires* 3: 1–568.

Bonaparte, J. F. 1971. Los tetrápodos del sector superior de la Formación Los Colorados, La Rioja, Argentina. *Opera Lilloana* 22: 1–183.

———. 1986. History of the terrestrial Cretaceous vertebrates of Gondwana. *Actas Primer Congreso Argentino de Paleontología y Bioestratigrafía* 4: 63–95.

———. 1990. New Late Cretaceous mammals from the Los Alamitos Formation, Northern Patagonia. *National Geographic Research* 6: 63–93.

———. 1991. Los vertebrados fósiles de la Formación Río Colorado, de la ciudad de Neuquén y sus cercanías, Cretácico superior, Argentina. *Re-*

vista del Museo Argentino de Ciencias Naturales "Bernardino Rivadavia," Paleontología 4: 17–123.

———. 1996. Cretaceous tetrapods of Argentina. In G. Arratia (ed.), *Contributions of Southern South America to Vertebrate Paleontology. Münchner Geowissenschaftliche Abhandlungen* (A) 20: 73–130.

Broin F. L. de. 2002. *Elosuchus,* a new genus of crocodile from the Lower Cretaceous of the North of Africa. *Comptes Rendus Palevol* 1: 275–285.

Buckley, G. A., and C. A. Brochu. 1999. An enigmatic new crocodile from the Upper Cretaceous of Madagascar. In D. M. Unwin (ed.), *Special Papers in Palaeontology* 60: 149–175. London: Palaeontological Association.

Buckley, G. A., C. A. Brochu, and D. W. Krause. 1997. Hyperdiversity and the paleobiogeographic origins of the Late Cretaceous crocodyliforms of Madagascar. *Journal of Vertebrate Paleontology* 17 (Suppl. to 3): 35A.

Buckley, G. A., C. A. Brochu, D. W. Krause, and D. Pol. 2000. A pugnosed crocodyliform from the Late Cretaceous of Madagascar. *Nature* 405: 941–944.

Buffetaut, E. 1974. *Trematochampsa taqueti,* un crocodilian nouveau du Sénonien inférieur du Niger. *Comptes Rendus de l'Académie des Sciences Série D* 279: 1749–1752.

———. 1980. Histoire biogéographique des Sebecosuchia (Crocodylia, Mesosuchia): un essai d'interprétation. *Annales de Paléontologie (Vertébrés)* 66: 1–8.

———. 1981. Die biogeographische geschichte der krokodilier, mit beschreibung einer neuen Art, *Araripesuchus wegeneri. Geologische Rundschau* 70: 611–624.

———. 1982. Radiation evolutive, paleoécologie et biogéographie des crocodiliens mesosuchiens. *Memoires de la Societe Géologique de France* 60: 1–88.

———. 1994. A new crocodilian from the Cretaceous of southern Morocco. *Comptes Rendus de l'Académie des Sciences Série II* 319: 1563–1568.

Buffetaut, E., and P. Taquet. 1979. Un nouveau crocodilian mésosuchien dans le Campanien de Madagascar, *Trematochampsa oblita,* n. sp. *Bulletin de la Société Géologique de France* 21: 183–188.

Calvo, J. O. 1991. Huellas fósiles de dinosaurios en la Formación Río Limay (Albiano-Cenomaniano), Provincia de Neuquén, Argentina. *Ameghiniana* 28: 241–253.

Campos, D. A., J. M. Suarez, D. Riff, and A. W. A. Kellner. 2001. Short note on a new Baurusuchidae (Crocodyliformes, Metasuchia) from the Upper Cretaceous of Brazil. *Boletim do Museu Nacional Geologia* 57: 1–7.

Candeiro, C. R. A., L. P. Bergqvist, L. C. B. Ribeiro, and S. Apesteguia. In press. The Late Cretaceous fauna and flora of the Peirópolis site (Minas Gerais State, Brazil). *Journal of South American Earth Sciences.*

Carvalho, I. S., L. C. B. Ribeiro, and L. S. Avilla. 2004. *Uberabasuchus terrificus* sp. nov., a new crocodylomorpha from the Bauru Basin (Upper Cretaceous), Brazil. *Gondwana Research* 7: 975–1002.

Carvalho, I. S., A. C. Arruda Campos, and P. H. Nobre. 2005. *Baurusuchus salgadoensis,* a new Crocodylomorpha from the Bauru Basin (Cretaceous), Brazil. *Gondwana Research* 8: 11–30.

Chiappe, L. M. 1988. A new trematochampsid crocodile from the Early Cretaceous of north-western Patagonia, Argentina and its palaeobiogeographical and phylogenetic implications. *Cretaceous Research* 9: 379–389.

Dias-Brito, D., E. A. Musacchio, J. C. Castro, M. S. A. S. Maranhão, J. M. Suárez, and R. Rodrigues. 2001. Grupo Bauru: uma unidade continental do Cretáceo do Brasil-concepções baseadas em dados micropaleontológicos, isotópicos e estratigráficos. *Revue de Paléobiologie* 20: 245–304.

Fernandes, L. A., and A. M. Coimbra. 1996. A Bacia Bauru (Cretáceo Superior, Brasil). *Anais da Academia Brasileira de Ciências* 68: 195–205.

Fernández, M., and Z. Gasparini. 2000. Salt glands in a Tithonian metriorhynchid crocodyliform and their physiological significance. *Lethaia* 33: 269–276.

Fraas, E. 1902. Die meer crocodilier (Thalattosuchia) des oberen Jura unter spezieller Berücksichtigung von *Dacosaurus* und *Geosaurus*. *Palaeontographica* 49: 1–72.

Frey, E., M.-C. Buchy, W. Stinnesbeck, and J. López Oliva. 2002. *Geosaurus vignaudi* n. sp. (Crocodyliformes: Thalattosuchia), first evidence of metriorhynchid crocodilians in the Late Jurassic (Tithonian) of central-east Mexico (State of Puebla). *Canadian Journal of Earth Sciences* 39: 1467–1483

Gasparini, Z. 1971. Los Notosuchia del Cretácico de América del Sur como un nuevo Infraorden de los Mesosuchia (Crocodylia). *Ameghiniana* 8: 83–103.

———. 1973. Revisión de "*?Purranisaurus potens*" Rusconi 1948 (Crocodylia, Thalattosuchia). Los Thalattosuchia como un nuevo Infraorden de los Crocodylia. *Actas del V Congreso Geológico Argentino* 3: 423–431.

———. 1981. Los Crocodylia fósiles de la Argentina. *Ameghiniana* 18: 177–205.

———. 1984. New Tertiary Sebecosuchia (Crocodylia: Mesosuchia) from Argentina. *Journal of Vertebrate Paleontology* 4: 85–95.

———. 1985. Los reptiles marinos jurásicos de América del Sur. *Ameghiniana* 22: 23–34.

———. 1992. Marine reptiles of the Circum-Pacific region. In G. E. G. Westermann (ed.), *The Jurassic of the Circum-Pacific, World and Regional Geology*, 361–364. Cambridge: Cambridge University Press.

———. 1996. Biogeographic evolution of the South American crocodilians. In G. Arratia (ed.), *Contributions of Southern South America to Vertebrate Paleontology. Münchner Geowissenschaftliche Abhandlungen (A)* 30: 159–184.

Gasparini, Z., and D. Dellapé. 1976. Un Nuevo cocodrilo marino (Thalattosuchia, Metriorhynchidae) de la Formación Vaca Muerta (Jurásico, Tithoniano) de la provincia de Neuquén. *Actas del I° Congreso Geológico Chileno:* C1–C21. Santiago.

Gasparini, Z., and M. Fernández. 1996. Biogeographic affinities of the Jurassic marine reptile fauna of South America. Proceedings of the IV International Congress of Jurassic Stratigraphy and Geology, Mendoza, 1994. *GeoResearch Forum* 1–2: 443–450.

———. 1997. Tithonian marine reptiles of the Eastern Pacific. In J. Callaway and E. Nichols (eds.), *Ancient Marine Reptiles*, 435–450. San Diego: Academic Press.

———. 2005. Jurassic marine reptiles in the Neuquén Basin. In G. Veiga,

L. Spalletti, J. Howell, and E. Schwarz (eds.), *The Neuquén Basin: A Case Study in Sequence Stratigraphy and Basin Dynamics,* Special Publication 252: 279–294. London: Geological Society of London.

Gasparini, Z., and M. Iturralde-Vinent. 2001. Metriorhynchid crocodiles (Crocodyliformes) from the Oxfordian of Western Cuba. *Neues Jahrbuch für Geologie und Paläontologie Monatshefte* 2001: 534–542.

Gasparini, Z., L. M. Chiappe, and M. Fernández. 1991. A new Senonian peirosaurid (Crocodylomorpha) from Argentina and a synopsis of the South American Cretaceous crocodilians. *Journal of Vertebrate Paleontology* 11: 316–333.

Gasparini, Z., M. de la Fuente, and M. Fernández. 1995. Sea reptiles from the lithographic limestones of the Neuquén Basin, Argentina. *II International Symposium on Lithographic Limestones,* 81–84. Cuenca, Spain.

Gasparini, Z., P. Vignaud, and G. Chong. 2000. The Jurassic Thalattosuchia (Crocodyliformes) of Chile: a paleobiogeographic approach. *Bulletin de la Société Géoogique de France* 171: 657–664.

Gasparini, Z., M. Cichowolsky, and D. Lazo. 2005. First record of *Metriorhynchus* (Reptilia: Crocodyliformes) in the Bathonian (Middle Jurassic) of the Eastern Pacific. *Journal of Paleontology* 79: 801–805.

Gasparini, Z., D. Pol, and L. Spalletti. 2006. An unusual marine crocodyliform from the Jurassic-Cretaceous boundary of Patagonia. *Science* 311 (5757): 70–73.

Huene, F. von. 1927. Beitrag zur kenntnis mariner mesozoicher wirbeltiere in Argentinien. *Centralblatt für Mineralogie, Geologie und Paläontologie B* 1: 22–29.

Hugo, C. A., and H. A. Leanza. 2001. Hoja Geológica 3969–IV, General Roca, Provincias del Neuquén y Río Negro. *Instituto de Geología y Recursos Naturales, SEGEMAR* 308: 1–71.

Kälin, J. A. 1955. Crocodilia. In J. Piveteau (ed.), *Traite de Palaeontologie* 5: 695–784. Paris: Masson et Cie.

Kuhn, O. 1968. *Die Vortzeitlichen Krokodile.* Munich: Verlag Oeben, Krailing.

Lamanna, M., M. Luna, G. Casal, R. D. Martinez, L. Ibiricu, and J. C. Sciutto. 2003. New crocodyliform and dinosaur discoveries from the Upper Cretaceous (Campanian-?Maastrichtian), Upper Member of the Bajo Barreal Formation, Southern Chubut Province, Argentina. *Journal of Vertebrate Paleontology* 23 (Suppl. to 3): 70A.

Larsson, H. C. E., and B. Gado. 2000. A new Early Cretaceous crocodyliform from Niger. *Neues Jahrbuch für Geolgie und Paläontologie Abhandlungen* 217: 131–141.

Leanza, H. A., S. Apesteguía, F. E. Novas, and M. S. de la Fuente. 2004. Cretaceous terrestrial beds from the Neuquén Basin (Argentina) and their tetrapod assemblages. *Cretaceous Research* 25: 61–87.

Leeds, E. 1908. On *Metriorhynchus brachrhynchus* (Desl.) from the Oxford Clay near Peterborough. *Quarterly Journal of Geological Society, London* 64: 345–357.

Legarreta, L., and C. A. Gulisano. 1989. Análisis estratigráfico secuencial de la Cuenca Neuquina (Triásico superior-Terciario inferior, Argentina). In G. Chebli and L. Spalletti (eds.), *Cuencas Sedimentarias Argentinas,* 6: 221–243. Serie Correlación Geológica 6. Tucumán, Argentina: Universidad Nacional de Tucumán.

Martinelli, A. G. 2003. New cranial remains of the bizarre notosuchid *Comahuesuchus brachybuccalis* (Archosauria, Crocodyliformes) from the Late Cretaceous of Río Negro Province (Argentina). *Ameghiniana* 40: 559–572.

Ortega, F., Z. Gasparini, A. Buscalioni, and J. O. Calvo. 2000. A new *Araripesuchus* (Crocodylomorpha, Lower Cretaceous) from northwestern Patagonia. *Journal of Vertebrate Paleontology* 20: 57–76.

Ortiz Jaureguizar, E., and R. Pascual. 1989. South American land-mammal faunas during the Cretaceous-Tertiary transition: evolutionary biogeography. *Contribuciones de los Simposios sobre el Cretácico de América Latina* A: 231–252.

Poblete, J. F., and J. O. Calvo. 2005. A crocodyliform tooth and the age of peirosaurids in Neuquén, Patagonia, Argentina. *Boletim de Resumos II Congreso Latino-Americano de Paleontología de Vertebrados,* 206–207.

Pol, D. 1999. Basal mesoeucrocodylian relationships: new clues to old conflicts. *Journal of Vertebrate Paleontology* 19 (Suppl. to 3): 69A.

———. 2003. New remains of *Sphagesaurus huenei* (Crocodylomorpha, Mesoeucrocodylia) from the Late Cretaceous of Brazil. *Journal of Vertebrate Paleontology* 23: 817–831.

———. 2005. Postcranial remains of *Notosuchus terrestris* (Archosauria: Crocodyliformes) from the Upper Cretaceous of Patagonia, Argentina. *Ameghiniana* 42: 21–38.

Pol, D., and M. A. Norell. 2004. A new crocodyliform from Zos Canyon, Mongolia. *American Museum Novitates* 3445: 1–36.

Prasad, G. V. R., and F. L. de Broin. 2002. Late Cretaceous crocodile remains from Naskal (India): comparisons and biogeographic affinities. *Annales de Paléontologie* 88: 19–71.

Price, L. I. 1945. A new reptil [*sic*] from the Cretaceous of Brazil. *Notas Preliminares e Estudos, Ministerio de Agricultura, Divisao Geologia e Minería* 25: 1–9.

———. 1955. Novos crocodilideos dos Arenitos da Série Bauru, Cretáceo do estado de Minas Gerais. *Anais Academia Brasileira de Ciencias* 27: 487–498.

———. 1959. Sobre um crocodilideo notossúquido do Cretácico Brasileiro. *Boletim Divisão de Geologia Mineralogia do Brasil* 188: 1–55.

Riff, D. 2003. *Descrição Morfológica do Crânio e Mandíbula de Stratiotosuchus maxhechti (Crocodylomorpha, Cretáceo Superior do Brasil) e seu Posicionamento Filogenético.* Ph.D. dissertação de mestrado, Universidade Federal do Rio de Janeiro.

Rusconi, C. 1948a. Nuevo plesiosaurio, pez y langosta de mar jurásico de Mendoza. *Revista del Museo de Historia Natural* 2: 3–12.

———. 1948b. *Plesiosaurios del Jurásico de Mendoza y de la Argentina.* Mendoza: Impr. Oficial.

Sereno, P. C., C. A. Sidor, H. C. E. Larsson, and B. Gado. 2003. A new notosuchian from the Early Cretaceous of Niger. *Journal of Vertebrate Palaeontology* 23: 477–482.

Sereno, P. C., J. A. Wilson, and J. L. Conrad. 2004. New dinosaurs link southern landmasses in the Mid-Cretaceous. *Proceedings of the Royal Society of London B* 271: 1325–1330.

Spalletti, L., Z. Gasparini, and M. Fernández. 1994. Facies, ambientes y reptiles marinos de la transición entre las Fm Los Molles y Lajas (Jurásico medio). Cuenca Neuquina. Argentina. XIII Congresso Paleontológico Brasileiro e I Simpósio Paleontológico do Cone Sul (Sao Leopoldo, September 19–26, 1993). *Acta Geologica Leopoldensia* 39: 329–344.

Spalletti, L. A., Z. Gasparini, G. Veiga, E. Schwarz, and M. Fernández. 1999. Facies anóxicas, procesos deposicionales y herpetofauna de la rampa marina titoniano-berriasiana en la Cuenca Neuquina (Yesera del Tromen), Neuquén, Argentina. *Revista Geológica de Chile* 26: 109–123.

Sues, H. D., and H. Larsson. 2002. Cranial structure and phylogenetic relationships of the enigmatic crocodyliform *Hamadasuchus rebouli* from the Cretaceous of Morocco. *Journal of Vertebrate Paleontology* 22 (Suppl. to 3): 56A.

Turner, A. 2004. Crocodyliform biogeography during the Cretaceous: evidence of Gondwanan vicariance from biogeographical analysis. *Proceedings of the Royal Society of London B* 271: 2003–2009.

Turner, A., and J. O. Calvo. 2005. A new Sebecosuchian crocodyliform from the Late Cretaceous of Patagonia. *Journal of Vertebrate Paleontology* 25 (1): 87–98.

Vignaud, P. 1995. *Les Thalattosuchia, Crocodiles Marins du Mesozoique: Systématique Phylogénétique, Paleoécologie, Biochronologie et Implications Paléogéographiques.* Ph.D. thesis, Université de Poitiers.

Vignaud, P., and Z. Gasparini. 1996. New *Dakosaurus* (Crocodylomorpha, Thalattosuchia) in the Upper Jurassic of Argentina. *Comptes Rendues de l'Académie de Sciences* 322: 245–250.

Wenz, S. 1968. Contribution a l'étude du genre *Metriorhynchus* Crane et moulage endocranien de *Metriorhynchus superciliosus. Annals de Paléontologie* 54: 148–191.

Wilson, J. A., M. S. Malkane, and P. D. Gingerich. 2001. New crocodyliform (Reptilia, Mesoeucrocodylia) from the Upper Cretaceous Pab Formation of Vitakri, Balochistan (Pakistan). *Contributions from the Museum of Paleontology, University of Michingan* 30: 321–336.

Woodward, A. S. 1896. On two Mesozoic crocodilians, *Notosuchus* (genus novum) and *Cynodontosuchus* (gen. nov.) from the red sandstones of Territory of Neuquén (Argentina). *Anales del Museo de La Plata* 4: 1–20.

6. Pterosauria

Laura Codorniú and
Zulma Gasparini

Introduction

Pterosaur remains have been found on every continent except Antarctica, and they have ranged from the Late Triassic (Norian) to the latest Cretaceous (Maastrichtian) (Wellnhofer 1991; Kellner 2003; Unwin 2003). As with other groups of Mesozoic reptiles, both continental and marine, their fossils have been found mainly in the Northern Hemisphere. In the Southern Hemisphere, and particularly in South America, discoveries have been scarce (Kellner 2001) compared with the propitious extent of Mesozoic exposures.

Most South American pterosaurs have been found in two extensively prospected areas. The first is in northeastern Brazil (Araripe Basin, Aptian-Albian) (Kellner and Tomida 2000; Kellner 2003), where the highest pterosaur diversity on the continent has been recorded; the other is in Central Argentina (San Luis, Aptian) (Bonaparte 1970; Chiappe et al. 2000; Codorniú and Chiappe 2004), where hundreds of specimens have been collected, most of which seem to belong to a single species, the filter-feeding *Pterodaustro guinazui* (Bonaparte 1970). The discoveries in the rest of South America belong to very incomplete specimens from Venezuela (Kellner and Moody 2003), Peru (Bennett 1989), Chile (Casamiquela and Chong Díaz 1980; Bell and Padian 1995; Martill et al. 2000; Rubilar et al. 2002), and southern Argentina.

Except for the pterosaurs from San Luis, an area outside Patagonia, systematic surveys of pterosaur fossils have never been conducted in Argentina. However, the stratigraphic range of these Argentinian records extends from the Callovian to the Coniacian (164.7 to 85.8 Ma) and geographically to the southern tip of the

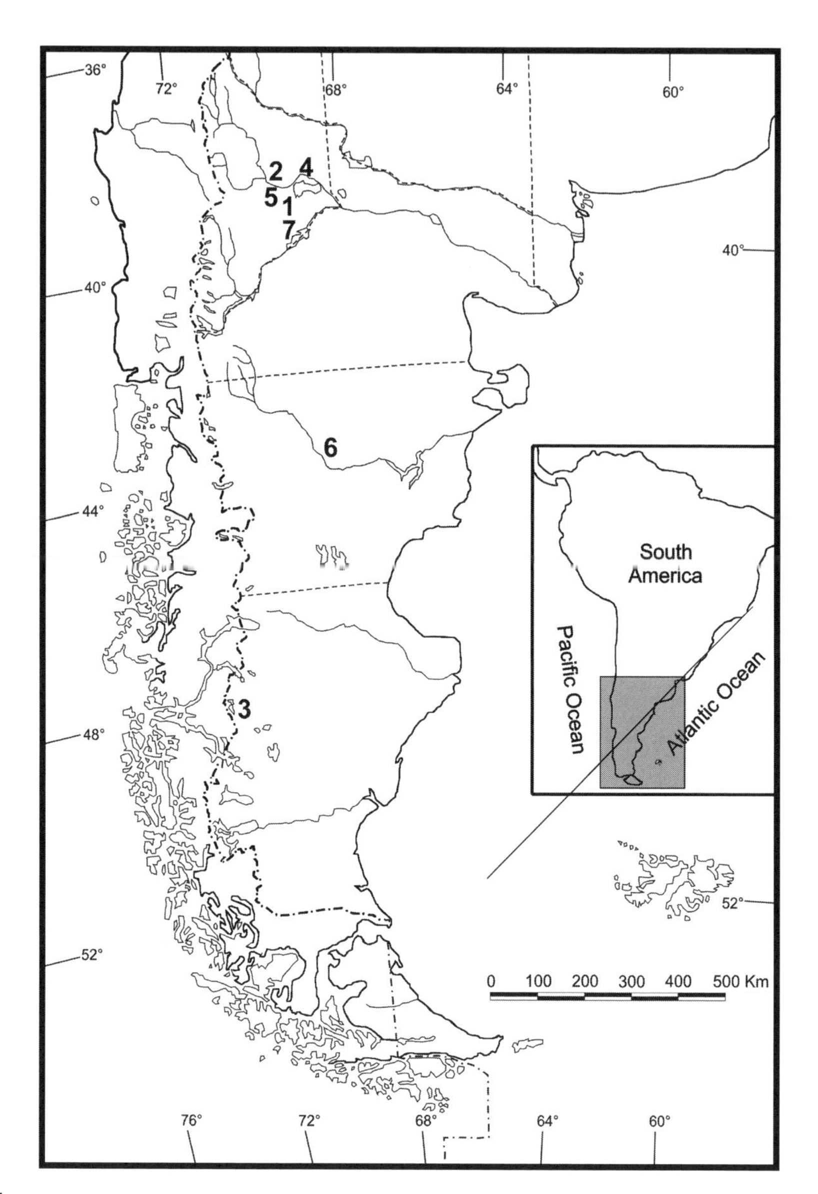

Figure 6.1. Map of Patagonia. (1) Herbstosaurus pigmaeus, *Vaca Muerta Formation, Tithonian, Neuquén. (2) MOZ 3625P and MOZ 2280P, Vaca Muerta Formation, upper-middle Tithonian, Neuquén. (3) MACN-SC 3617, Río Belgrano Formation, lower-middle Barremian, Santa Cruz. (4) MUCPv 358, Lago Barreales, El Portezuelo Formation, upper Turonian–lower Coniacian, Neuquén. (5) MACN-N 02, La Amarga Formation, Hauterivian-Barremian, Neuquén. (6) Pterosaurs from the Cañadón Asfalto Formation, Callovian-Oxfordian, Chubut. (7) ?Pterosaur tracks from the Candeleros Member (cf.* Pteraichnus *ichnosp. indet.) of Río Limay Formation, Albian-Cenomanian, Neuquén.*

continent (Fig. 6.1). This provides proof of their wide and early distribution in southwestern Gondwana. In addition, pterosaur tracks have been registered from the Candeleros Formation (lower Cenomanian) (Calvo this volume), Neuquén Province (Fig. 6.1, locality 7). These are the first pterosaur tracks reported for Gondwana (Calvo and Lockley 2001), but they are disputed (Padian 2003).

The study of pterosaurs began more than 200 years ago, but only in the last decades of the twentieth century did some attempt begin to apply cladistic methodology to investigate pterosaur phylogeny (Kellner 1996). Those works have been published by Howse (1986), Bennett (1989, 1994), and Unwin (1995). Recently, Kellner

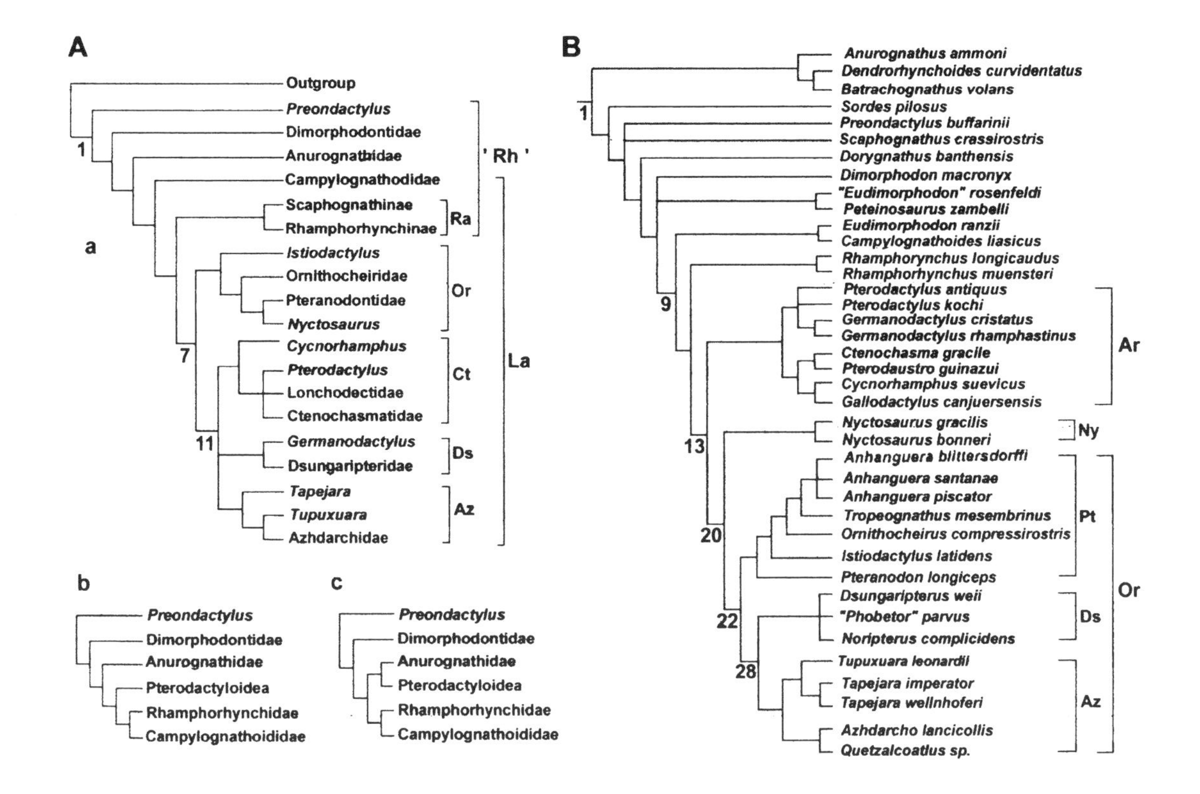

Figure 6.2. (A) a, strict consensus tree of six trees resulting from PAUP analysis; b and c, alternative phylogenies (according to Unwin 2003). (B) Strict consensus cladogram of the 80 most parsimonious cladograms recovered in the cladistic analysis (according to Kellner 2003). Ar = Archaeopterodactyloidea; Az = Azhdarchoidea; Ct = Ctenochasmatoidea; Ds = Dsungaripteroidea (in A) and Dsungaripteridae (in B); La = Lonchognatha; Ny = Nyctosauridae; Or = Ornithocheiroidea; Pt = Pteranodontoidea; Ra = Rhamphorhynchidae; 'Rh' = "Rhamphorhynchoidea." 1, Pterosauria; 7 and 13, Pterodactyloidea; 9, Novialoidea; 11, Lophocratia; 20, Dsungaripteroidea; 22, Ornithocheiroidea; 28, Tapejaroidea.

(2003, table 1, fig. 1) and Unwin (2003, table 1, fig. 7) synthesized previous analyses, including substantial new information, into a new phylogenetic analysis of the Pterosauria. The proposals of these authors yielded strikingly different tree topologies because they used different taxa, and as a consequence, different outgroups and characters (Fig. 6.2).

In this review, new anatomical information and taxonomic interpretations of all Patagonian pterosaurs are provided, and the reported ages of some of them are modified according to recent stratigraphic studies. Likewise, we include *Pterodaustro guinazui*, which, although recorded outside Patagonia, is the most relevant South American pterosaur. New observations of this taxon are reported, and the first known embryo of this species is mentioned (Chiappe et al. 2004; Codorniú et al. 2004).

Institutional abbreviations. CTES-PZ, Laboratorio de Paleozoología, Departamento de Biología, Facultad de Ciencias Exactas, Naturales y Agrimensura, Universidad Nacional del Nordeste, Corrientes, Argentina; MACN, Museo Argentino de Ciencias Naturales "Bernardino Rivadavia," Buenos Aires, Argentina; MHIN-UNSL-GEO V, Museo de Historia Natural, Universidad Nacional de San Luis, San Luis, Argentina, Departamento de Geología, Vertebrados; MMP, Museo Municipal de Ciencias Naturales "Galileo Scaglia," Mar del Plata, Argentina; MOZ, Museo de la Dirección Provincial de Minería, "Prof. Dr. Juan Olsacher," Zapala,

Neuquén, Argentina; MUCPv, Museo de la Universidad Nacional del Comahue, Neuquén, Argentina, Colección de Paleontología de Vertebrados; PVL, Instituto "Miguel Lillo," Tucumán.

Systematic Paleontology

Pterosauria Kaup 1834
Pterodactyloidea Pleininger 1901
Ctenochasmatidae Nopcsa 1928
Ctenochasmatinae (sensu Unwin 2003)
***Pterodaustro guinazui* Bonaparte 1970**
Figs. 6.3–6.6

***Puntanipterus globosus* Bonaparte and Sánchez 1975**

Previous comment. Although this taxon was found north of Patagonia, it is included in this synthesis because it is the best represented taxon of South America. As a contribution to the knowledge of the flying reptiles of the "southern cone," recent advances in the study of *Pterodaustro guinazui* are reported, as well as a short mention of an embryo of this spectacular ctenochasmatid pterodactyloid (Chiappe et al. 2004). The previously proposed synonymy of *Puntanipterus globosus* (La Cruz Formation) with *Pterodaustro guinazui* (Lagarcito Formation) (Chiappe et al. 1998a; Dávila et al. 1999) is here confirmed.

Holotype. PVL 2571, a right humerus (Fig. 6.3) and a few other postcranial elements as hypodigm (Bonaparte 1970); PVL 3860 (hypotype) (Sánchez 1973).

Locality and age. A small quarry named Loma del Pterodaustro (Chiappe et al. 1995) at the Quebrada de Hualtarán (32° 29.65'S, 66° 59.38'W) 4 km west of National Road 147 (from San Luis to San Juan) and Perfil Quebrada Larga-Agüero in the Parque Nacional Sierra de Las Quijadas, San Luis. Lagarcito Formation (Díaz 1946; Flores 1969), Albian (Chiappe et al. 1998a,b), and La Cruz Formation, Aptian (Yrigoyen 1975; Rivarola 1999).

Comments. The material of *Pterodaustro guinazui* was collected during fieldwork from 1963 (Bonaparte 1970, 1971; Sánchez 1973) to the last decade of the twentieth century (Chiappe et al. 1995, 1998a,b, 2000). Approximately 250 specimens have already been prepared, and many still unprepared are available. However, only two complete and articulated skeletons with crania (PVL 3860 and MMP 1089) and another four without crania have been preserved. The rest of the collection includes a dozen isolated, complete skulls (Fig. 6.4); mandibles with very unusual and specialized filtering dentition; wings; pelvic and scapular girdles; hind limbs; and a large number of isolated bones from different parts of the skeleton.

It is noteworthy that these articulated skeletons are very different in size, from 27 to 160 cm in wingspan, thus suggesting the presence of early juveniles, subadults, and adults in the sample. However, the estimated wingspan of the new isolated specimens

Plate 1. Reconstruction of the marine Tithonian herpetofauna of the Neuquén Basin. The ichthyosaur *Caypullisaurus* and the crocodyliforms *Geosaurus* (small forms) and *Dakosaurus* (huge form). The testudine *Neusticemys*, at right.

Illustration © by Jorge González.

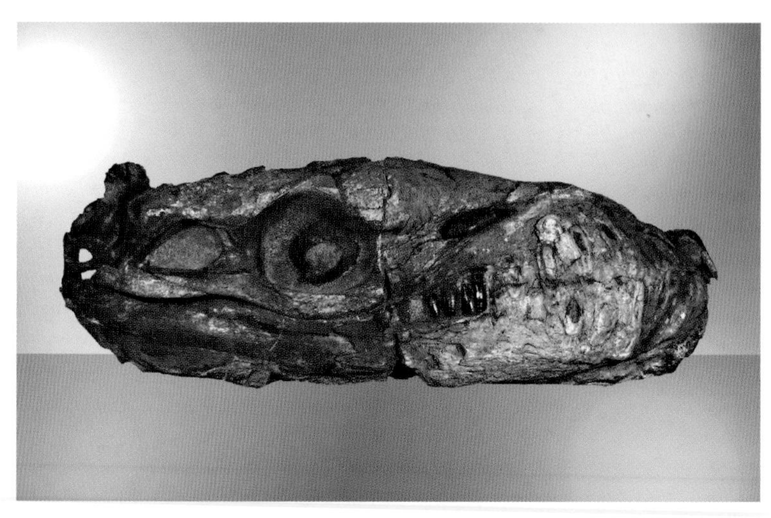

Plate 2. The splendid specimen of the huge thalattosuchian crocodyliform *Dakosaurus andiniensis* MOZ 6146P, popularly known as "Godzilla."

Plate 3. Reconstruction of the Portezuelo Formation herpetofauna at Portezuelo locality, Neuquén Province, 90 million years ago. A large carcharodontosaurid (bottom left), observes a megaraptor and two *Unenlagia* (all non-avian theropods), disputing the carcass of a large sauropod. One crocodyliform peirosaurid and one chelid (*Prochelidella*) complete the scene.

Illustration © by Jorge González.

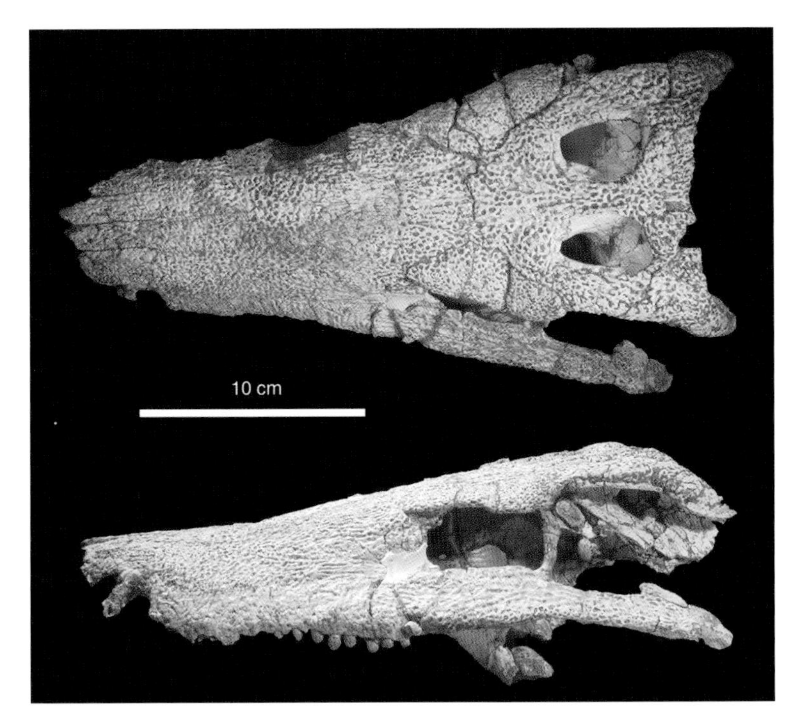

Plate 4. Skull of the peirosaurid crocodyliform *Lomasuchus palpebrosus*. MOZ P 4084.

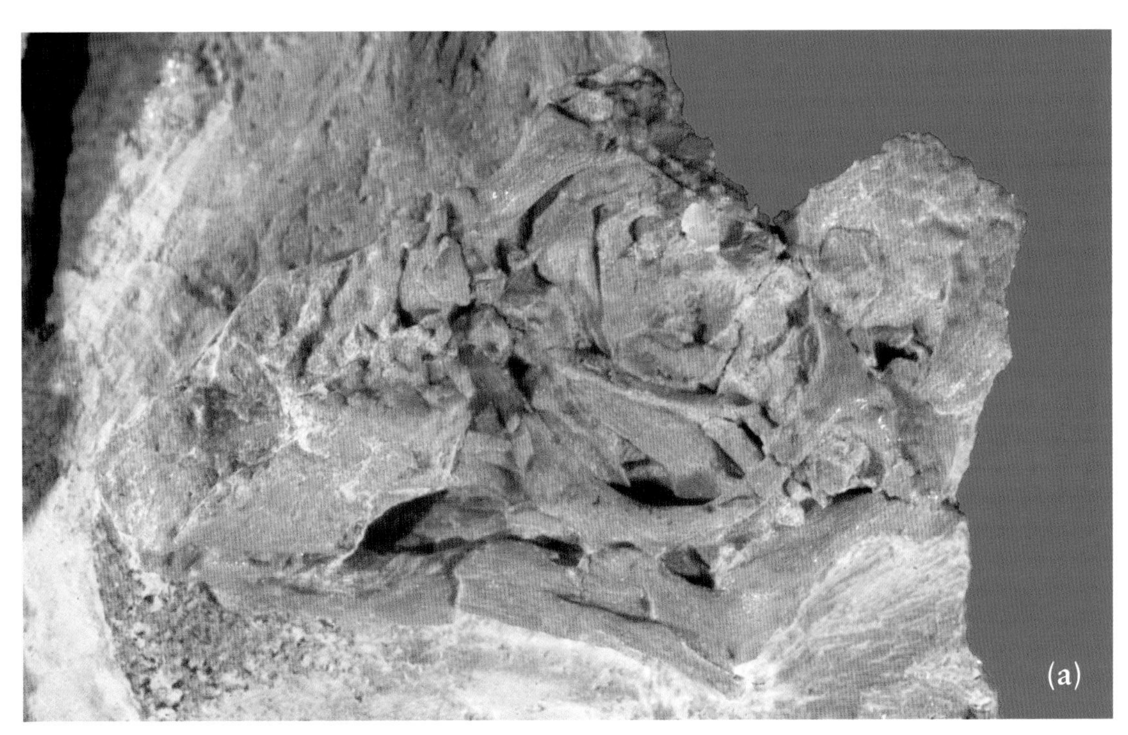

Plate 5. Embryonic sauropod skulls from Auca Mahuevo, Neuquén Province. This Late Cretaceous locality has brought thousands of exquisitely preserved dinosaur eggs and embryos. MCF-PVPH-272 (a) and MCF-PVPH 263 (b).

Plate 6. Reconstruction of the enantiornithine bird *Neuquenornis volans*.
Illustration © by Carlos Papolio.

Plate 7. Reconstruction of the ornithuromorph bird *Patagopteryx deferrariisi*.

Illustration © by Carlos Papolio.

Plate 8. Several Late Cretaceous reptiles from the Rio Colorado Subgroup (Bajo de la Carpa and Anacleto Formations). From the upper-left to the lower-right corner: *Gasparinisaura*, *Patagopteryx*, *Dinylisia*, *Notosuchus* and *Alvarezsaurus*.

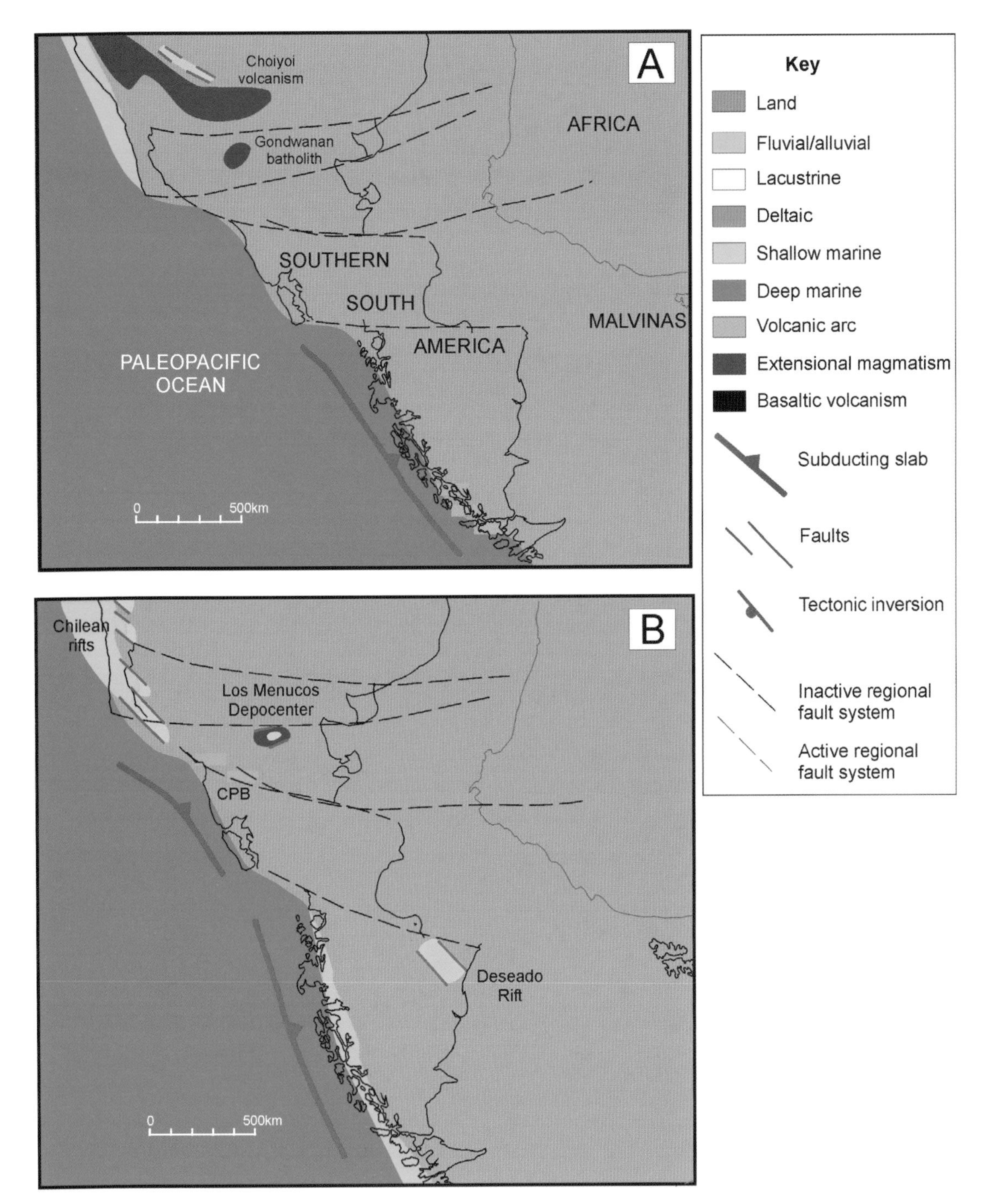

Plate 9. (A) Reconstruction of Patagonia at 240 Ma (Anisian-Ladinian, Middle Triassic). (B) Reconstruction of Patagonia at 225 Ma (Carnian–Early Norian, Late Triassic). CPB = Central Patagonia batholith.

Plate 10. (A) Reconstruction of Patagonia at 210 Ma (Late Norian–Rhaetian, Late Triassic).

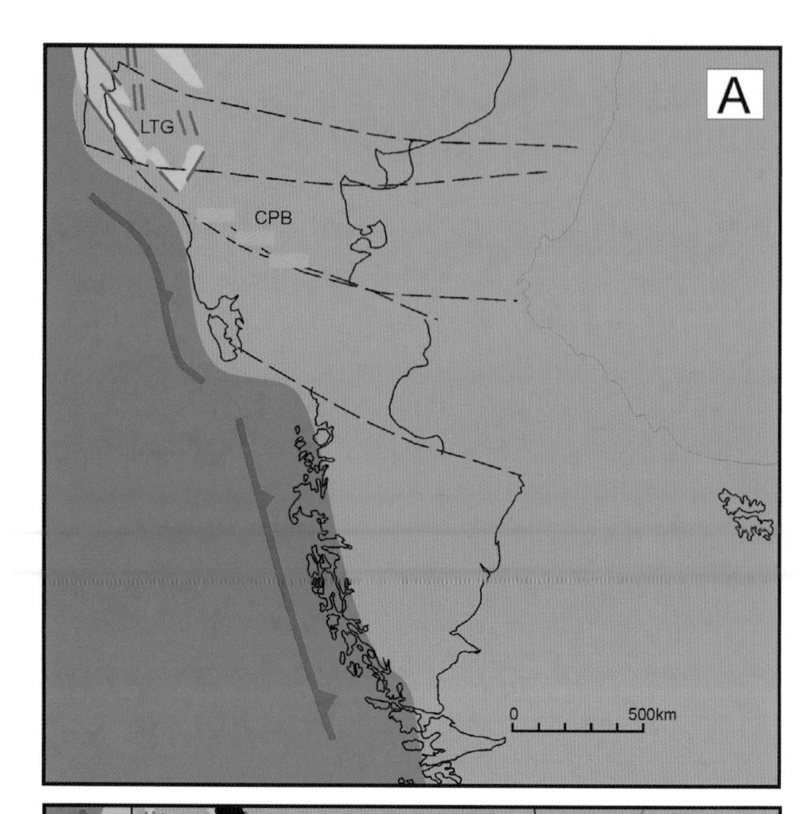

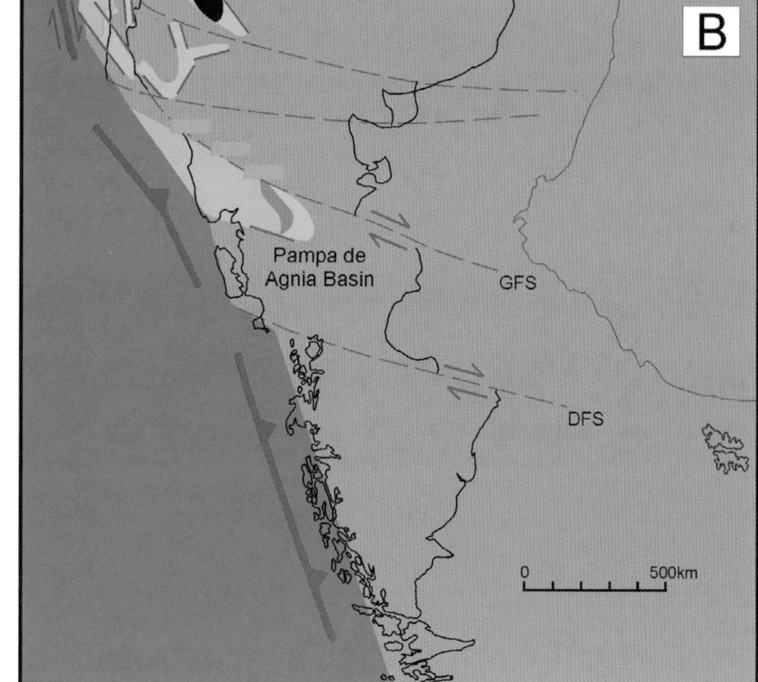

(B) Reconstruction of Patagonia at 195 Ma (Sinemurian-Pliesbachian, Early Jurassic). CPB = Central Patagonia batholith; DFS = Deseado fault system; GFS = Gastre fault system; LTG = Late Triassic grabens.

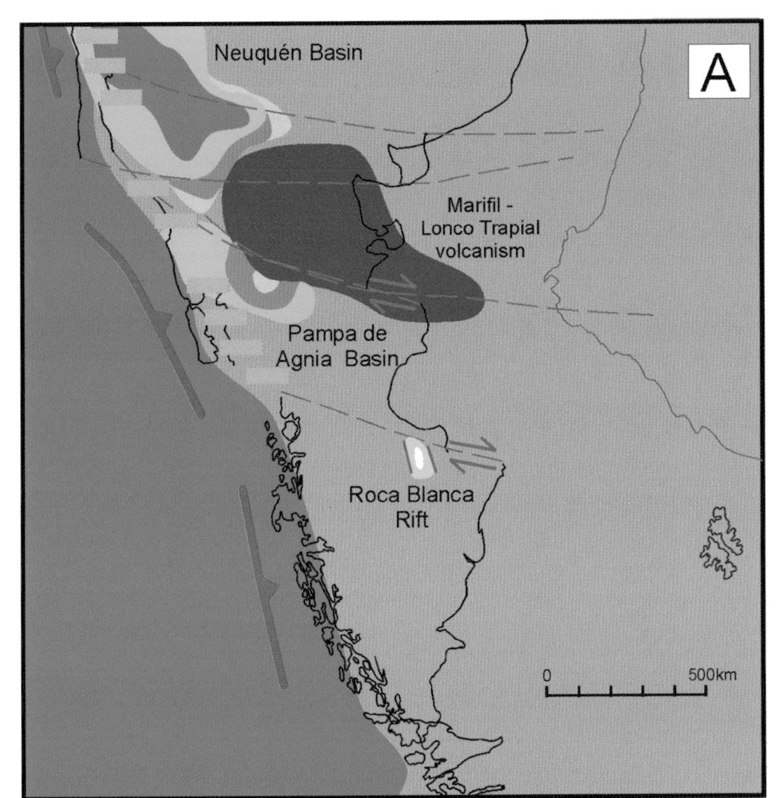

Plate 11. (A) Reconstruction of Patagonia at 180 Ma (Toarcian-Aalenian Early-Middle Jurassic).

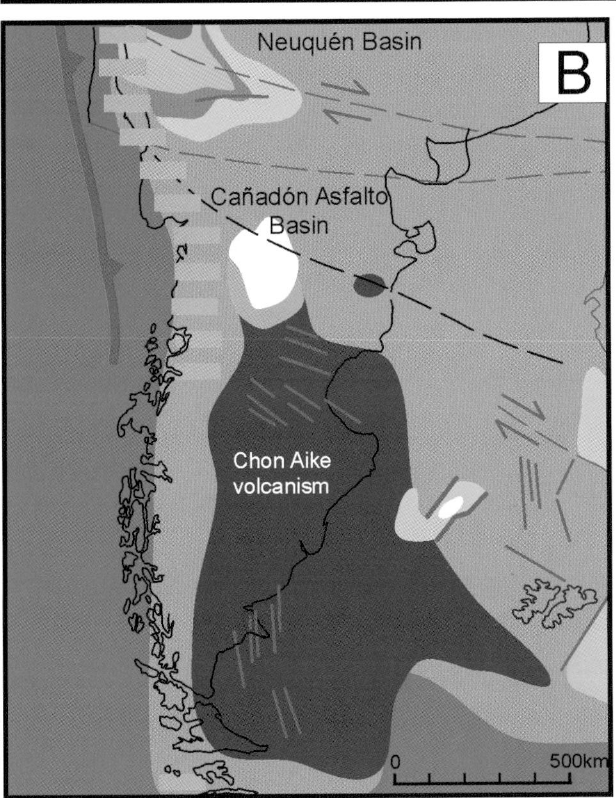

(B) Reconstruction of Patagonia at 165 Ma (Bathonian-Callovian Middle Jurassic).

Plate 12. (A) Reconstruction of Patagonia at 150 Ma (Kimmerigian-Thitonian, Late Jurassic).

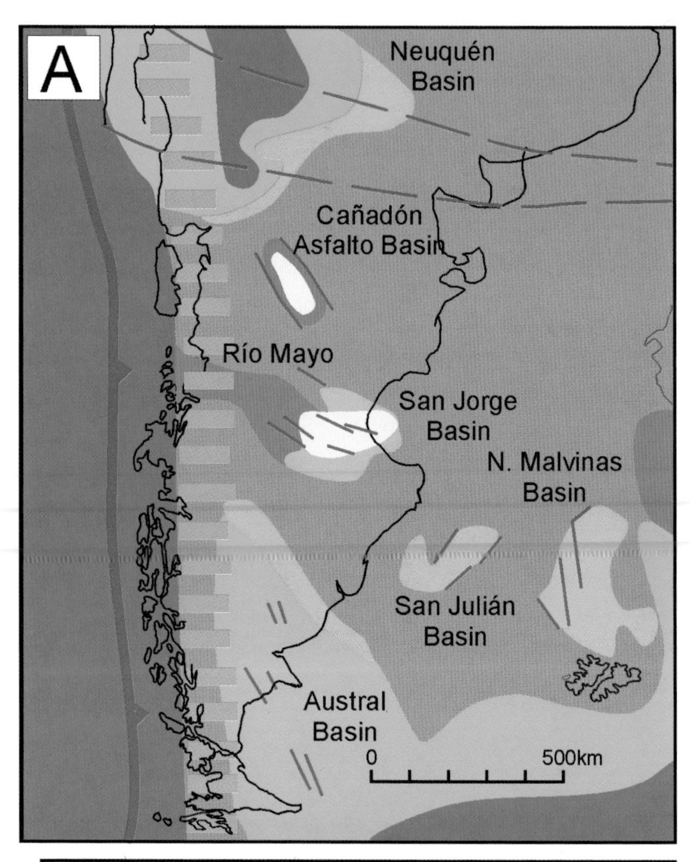

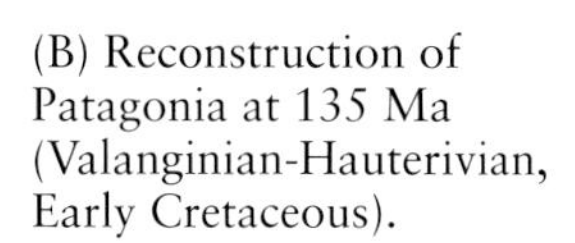
(B) Reconstruction of Patagonia at 135 Ma (Valanginian-Hauterivian, Early Cretaceous).

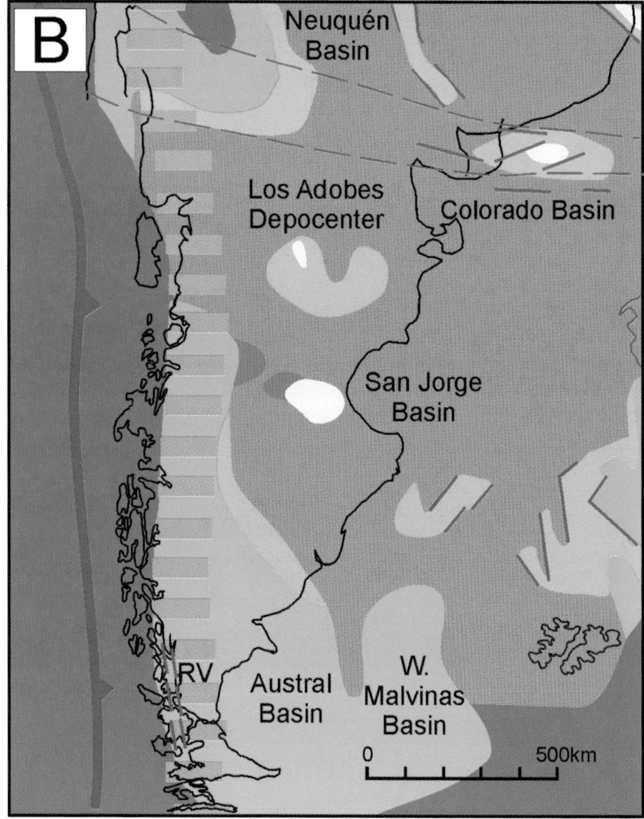

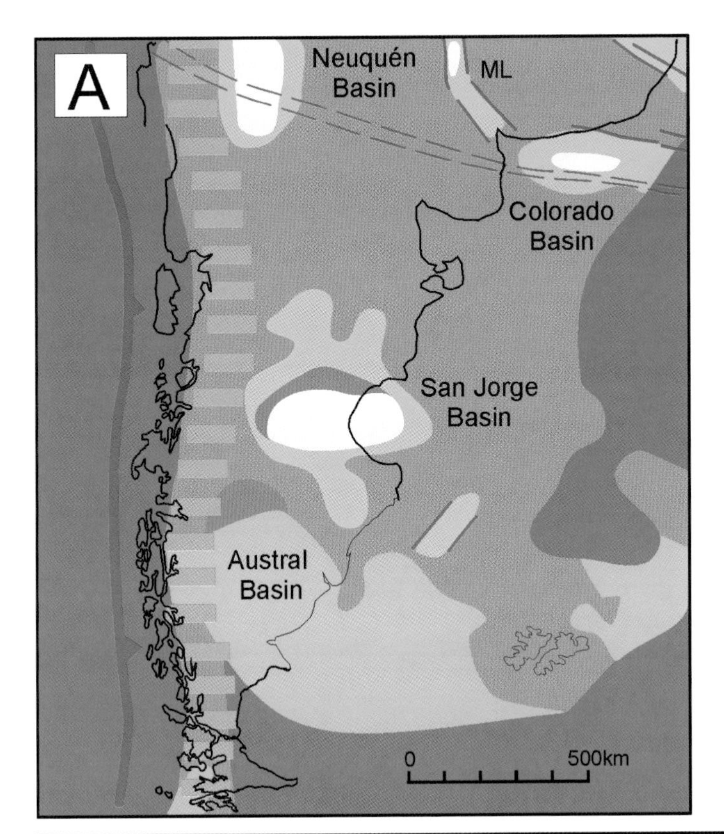

Plate 13. (A) Reconstruction of Patagonia at 120 Ma (Aptian, Early Cretaceous).

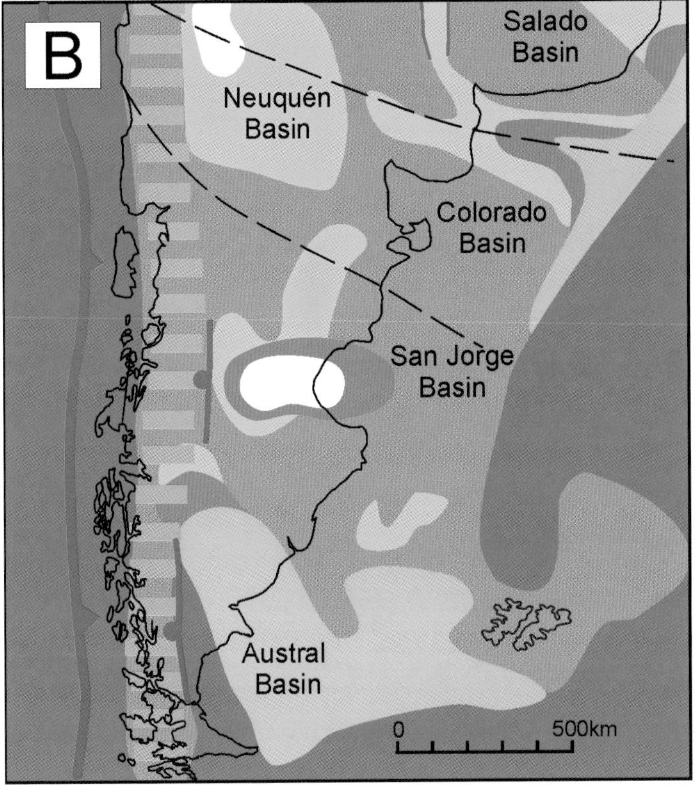

(B) Reconstruction of Patagonia at 105 Ma (Albian, Early Cretaceous). ML = Macachín-Laboulaye rifts.°

Plate 14. (A) Reconstruction of Patagonia at 90 Ma (Cenomanian-Turonian, Late Cretaceous). SFS = Shackleton Fault System.

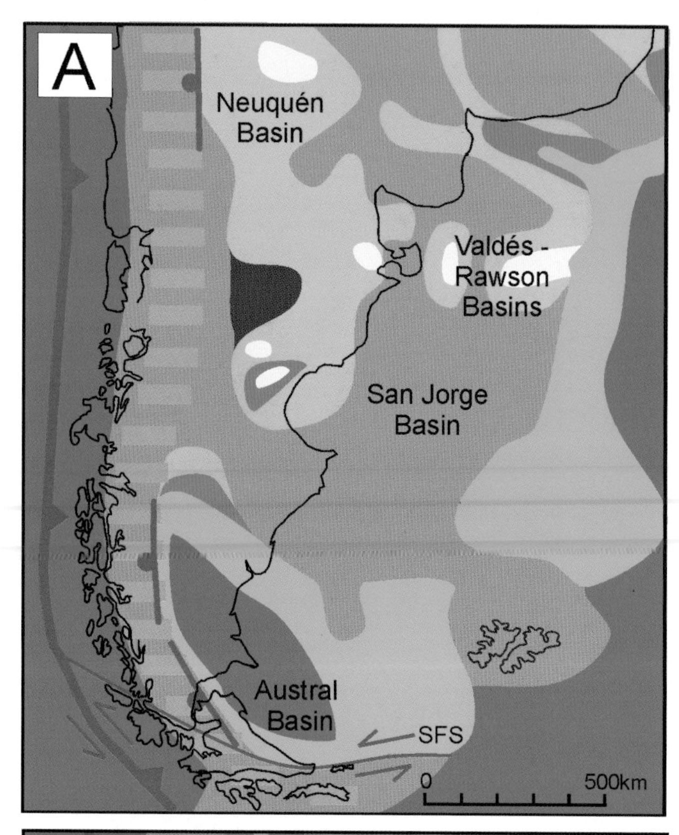

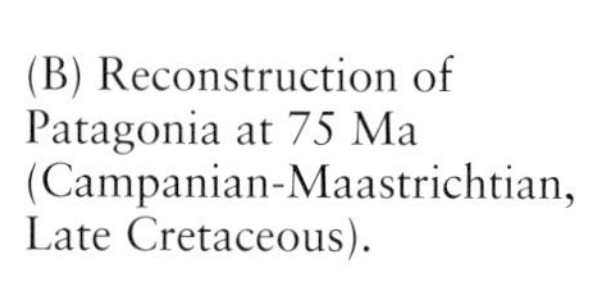

(B) Reconstruction of Patagonia at 75 Ma (Campanian-Maastrichtian, Late Cretaceous).

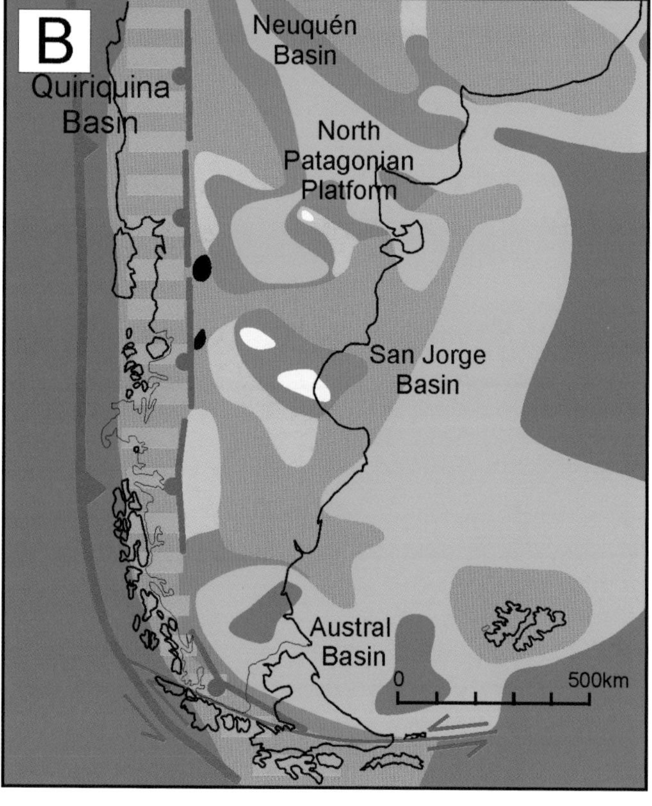

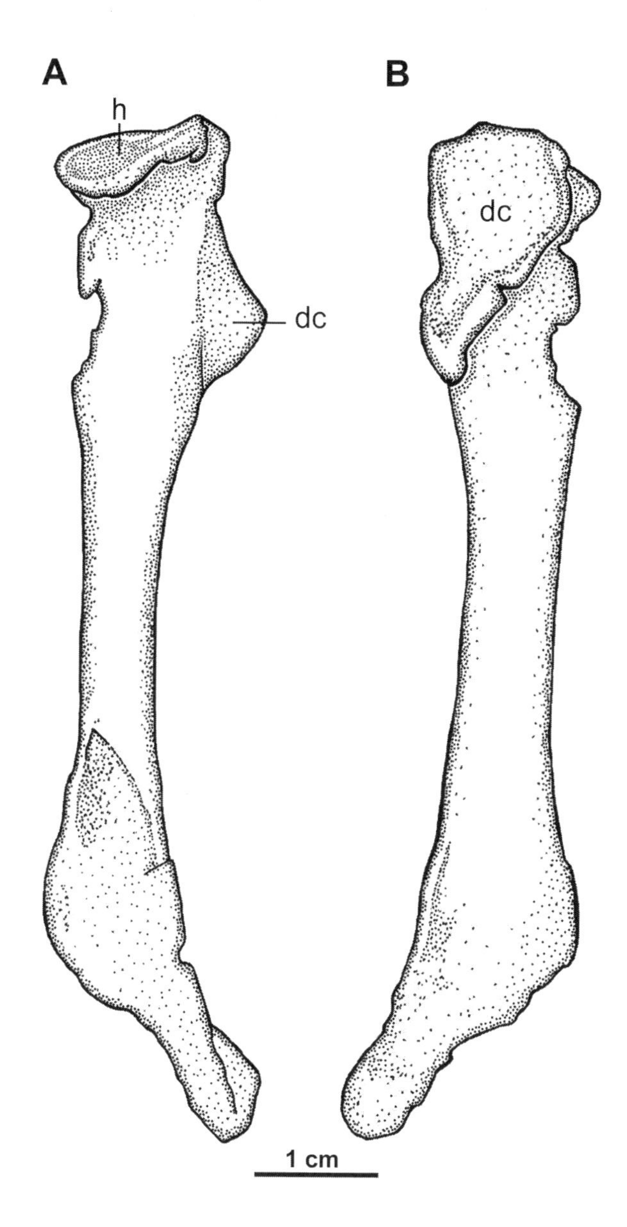

Figure 6.3. Holotype of Pterodaustro guinazui, *right humerus (PVL 2571). (A) Dorsal view. (B) Ventral view. dc = deltopectoral crest displaced ventrally; h = humeral head.*

(i.e., humeri, ulnae) would indicate a size of at least 3 m in wingspan for some individuals of *Pterodaustro guinazui,* a size much larger than the ones previously known for this species (Wellnhofer 1991; Chiappe et al. 1998a). Among the articulated specimens, two small ones (MHIN-UNSL-GEO-V 241 and MMP 1168) represent early life-history stages that have given us clues about the allometric changes that occurred during postnatal development (Codorniú 2002; Codorniú and Chiappe 2004).

The discovery of well-preserved caudal series in unpublished specimens has added new information on the evolution of the caudal morphology of pterosaurs. The high number of caudal vertebrae (more than 22 elements) and the morphological variation

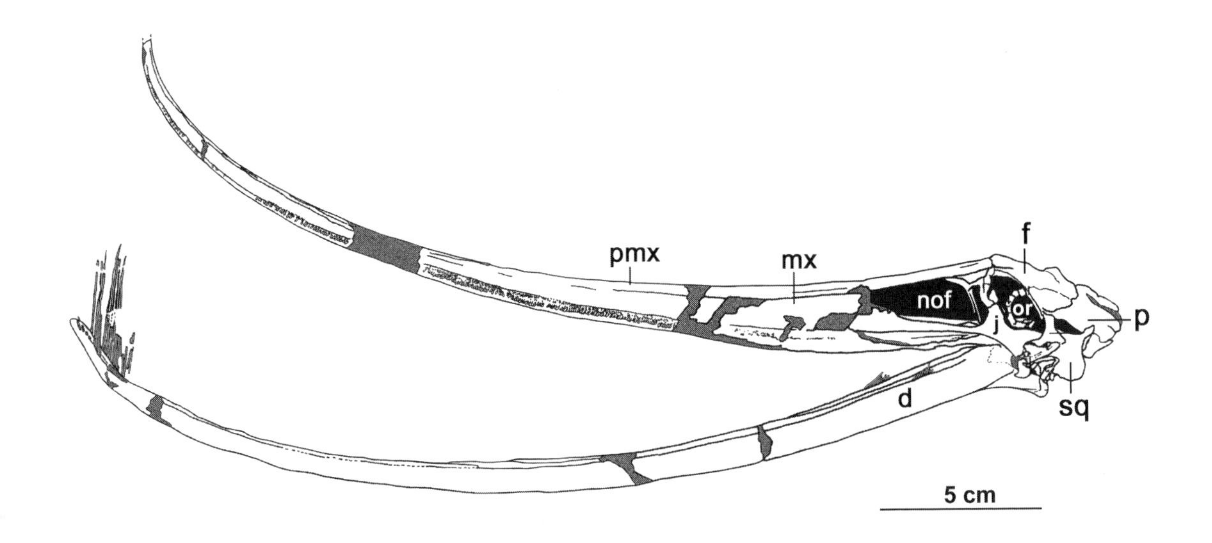

Figure 6.4. Skull and jaws of Pterodaustro guinazui *(MHIN-UNSL-GEO-V 57) (modified from Chiappe et al. 2000). d = dentary; f = frontal; j = jugal; mx = maxilla; nof = nasoantorbital fenestra; p = parietal; pmx = premaxilla; sq = squamosal.*

within the tail of *Pterodaustro* constitute a potential autapomorphy of this pterosaur (Codorniú 2005).

Among the highlights of the *Pterodaustro* collection is a very small specimen interpreted as an embryo. This specimen, MHIN-UNSL-GEO-V 246, is almost complete and has most bones in articulation. This skeleton is inside a small oval surface of approximately 12 cm^2 (Fig. 6.5), and its wingspan approaches 27 cm (Chiappe et al. 2004; Codorniú et al. 2004). The proportions and general anatomy of this baby are similar to those of other juvenile specimens (wingspan 28.9 and 30 cm) of the filter-feeding pterosaur *Pterodaustro guinazui,* collected from the same stratigraphic levels. The correspondence between the wing proportion of the MHIN-UNSL-GEO-V 246 and those of previously published juveniles of *Pterodaustro* (Codorniú and Chiappe 2004) indicates a near-hatching stage of the former and a hatchling state of the latter. Over the bones of MHIN-UNSL-GEO V 246 is a carbonatic material that shows features of eggshell, thus providing important data on the morphology of the pterosaurian egg (Chiappe et al. 2004).

Puntanipterus globosus (PVL 3869), from the La Cruz Formation (upper Lower Cretaceous), was regarded as a new taxon by Bonaparte and Sánchez (1975). The most significant diagnostic features for this determination were the presence of a wide globose articulation, formed by the proximal tarsal bones (astragalus and calcaneum) fused to the tibia, as well as the presence of spiny processes on the medial and lateral sides of the distal end of the tibia. The tibia-fibula is three-dimensionally preserved, unlike most of the remains of *Pterodaustro,* which are two-dimensionally preserved.

In this chapter, new tibia-fibulae of *Pterodaustro guinazui* with less postmortem compression (MHIN-UNSL-GEO-V 169, MHIN-UNSL-GEO-V 90) are compared with that of *Puntanipterus.* The

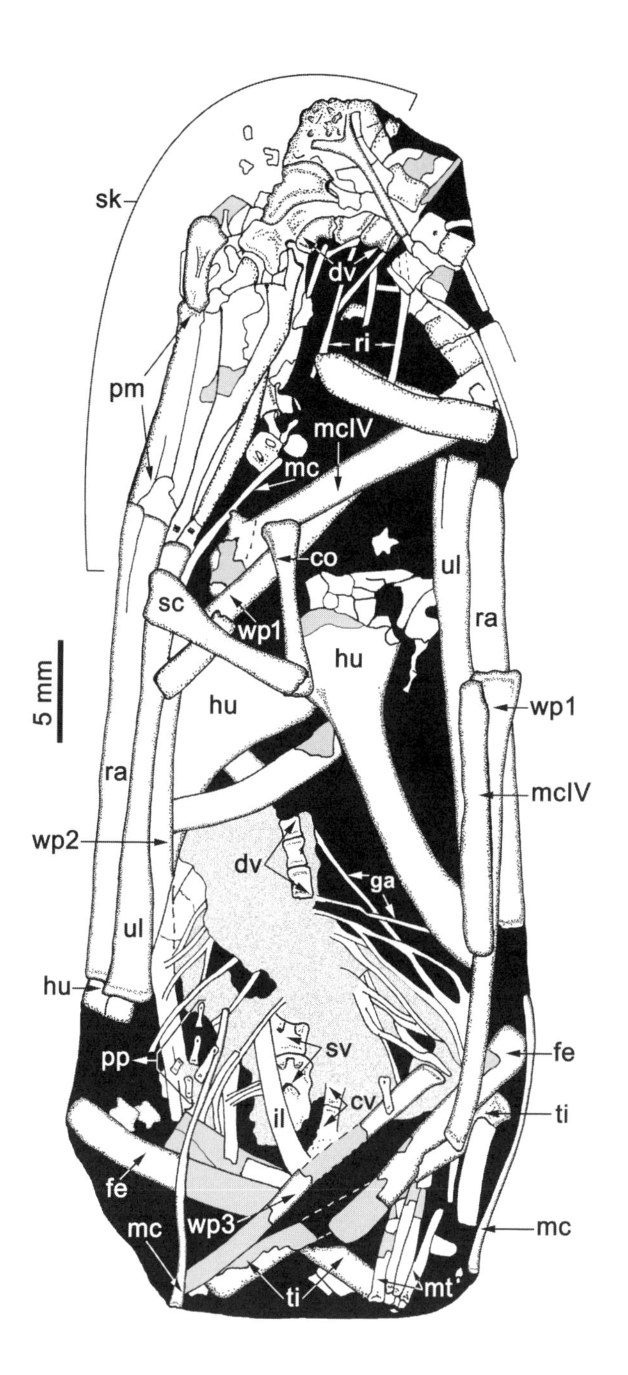

Figure 6.5. Composite drawing of embryo (MHIN-UNSL-GEO-V 246). This specimen is preserved in a slab and a counterslab (modified from Chiappe et al. 2004). co = coracoid; dv = dorsal vertebrae; fe = femur; ga = gastralia; hu = humerus; il = ilium; mc = metacarpal I–III; mcIV = metacarpal IV; mt = metatarsals I–IV; pm = premaxilla; pp = pedal phalanges; ra = radius; ri = rib; sc = scapula; sk = skull; sv = sacral vertebrae; t = tarsals; ti = tibia; ul = ulna; wp 1–3 = first to third phalanx of the wing finger.

great development of the wide globose articulation is also present in *Pterodaustro;* hence, this character is common to both taxa (Fig. 6.6). The morphological differences reported by other authors represent taphonomic factors that deformed some specimens more than others. However, the presence of a spiny process in the lateral condyle of the tibia of *Puntanipterus* has not been observed in any other specimen. This feature is here reported not as diagnostic of *Puntanipterus* but probably as ossified cartilage for the insertion of

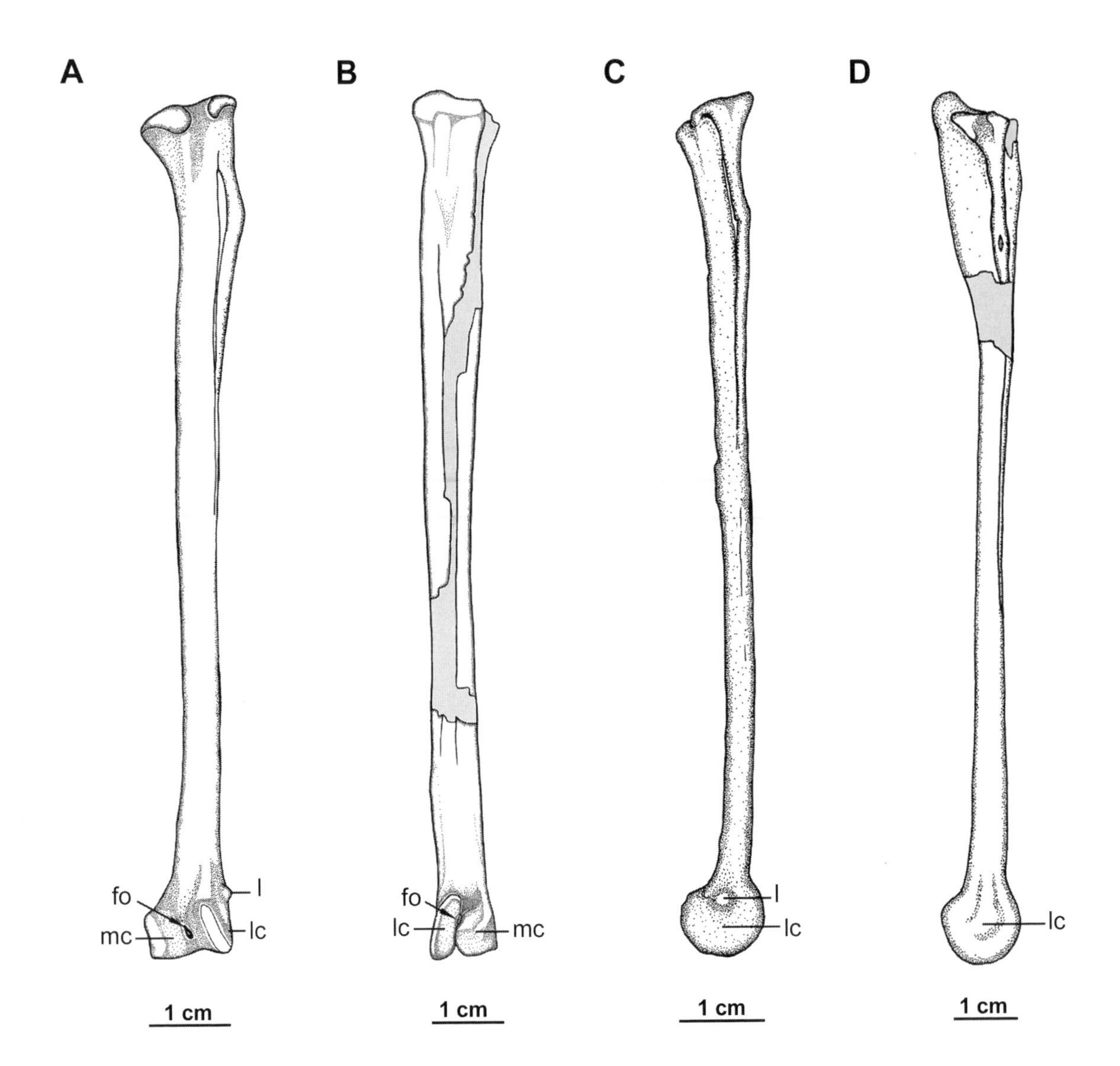

Figure 6.6. (A) Left tibia in anterior view of Puntanipterus *(PVL 3869). (B) Right tibia of* Pterodaustro guinazui *(MHIN-UNSL-GEO-V 169). (C) Tibia in lateral view of* Puntanipterus *(PVL 3869). (D)* Pterodaustro *(MHIN-UNSL-GEO-V 90). fo = foramen; l = lateral ligamentous prominence; lc = lateral condyle; mc = medial condyle.*

a tendon. A similar condition has already been observed in other pterosaurs (Padian 1983a) and was attributed to lateral and medial ligamentous prominences.

The proximal ends of the tibia-fibulae and their general morphology in both taxa (*Pterodaustro* and *Puntanipterus*) have no morphological differences, as already observed (Bonaparte and Sánchez 1975). This confirms the synonymy of *Puntanipterus globosus* with *Pterodaustro guinazui.*

All of the specimens referred to *Pterodaustro guinazui* were found in deposits of a perennial and extensive shallow lake (Chiappe et al. 1995). Paleoenviromental reconstructions, combined with the remarkably low diversity of Loma del Pterodaustro (fishes are the only vertebrates other than the hundreds of *Pterodaustro* fossils), suggest that this pterosaur inhabited an environment unsuitable for most other tetrapods, in agreement with reconstructions of this filter-feeding pterosaur as an ecological analog of the flamingo (Chiappe et al. 2004). The specimens referred to *Puntanipterus globosus* from the La Cruz Formation were found in greenish lacustrine levels intercalated with brown-reddish conglomerates. The paleoenvironment of the La Cruz Formation suggests low transport and chaotic deposition, probably related to high density flows of the alluvial type (Rivarola 1999).

Pterodactyloidea incertae sedis
Herbstosaurus pigmaeus Casamiquela 1975
Fig. 6.7

Holotype. CTES-PZ 1709 (Fig. 6.7), postcranial remains and impressions of a single specimen. Most bones and impressions belong to the pelvic girdle and proximal area of the hind limbs. The bones include a sacrum in ventral view; an impression of a right ilium with very few fragments of bone; an incomplete prepubis; an impression and partial filling of a complete right femur nearly in articulation with the acetabulum of the pelvis; and the impression of the incomplete left femur, which lacks the head and is somewhat displaced from its original position.

Locality and age. Southern sector of the Arroyo Picún Leufú anticlinal, where it is crossed by National Route 40, Neuquén Province (Fig. 6.1, locality 1). Originally, the holotype was cited as from the Middle Jurassic (Callovian) levels of the Lotena Formation (Casamiquela 1975). This age has been so far accepted and cited by other authors (Wellnhofer 1991; Unwin 1996). However, in the same nodule with the pterosaur was found the ammonite *Berriasiella* sp. (A. Riccardi personal communication 2002), which undoubtedly belongs to the upper levels of the Tithonian, in the Vaca Muerta Formation (Leanza et al. 1978), which is also exposed in this area.

Comments. Herbstosaurus pigmaeus was originally assigned to a theropod dinosaur (Casamiquela 1975). Ostrom (1978) suggested for the first time that these remains belonged to a pterosaur and not to a dinosaur. Later, authors placed this taxon within different lineages: Bonaparte (1978) assigned it to the Pterodactylidae, whereas Wellnhofer (1991) compared it to the long-tailed pterosaurs (nonpterodactyloids) on the basis of the pelvis and femur morphology. In other papers, it was considered a pterodactyloid (Unwin 1996) close to Dsungaripteroidea (Unwin 1995; Unwin and Heinrich 1999), a hypothesis sustained by Unwin (2003) in his phylogenetic proposal of the Pterosauria.

In this synthesis, we agree with the assignment of *Herbstosaurus pigmaeus* to the clade Pterodactyloidea because of the presence of the only acceptable character previously observed by other authors: sacrum composed of 5–6 vertebrae (Wellnhofer 1978; Unwin 1996). However, the placement of *Herbstosaurus pigmaeus* with the dsungaripteroids has to be revised. The diagnosis of

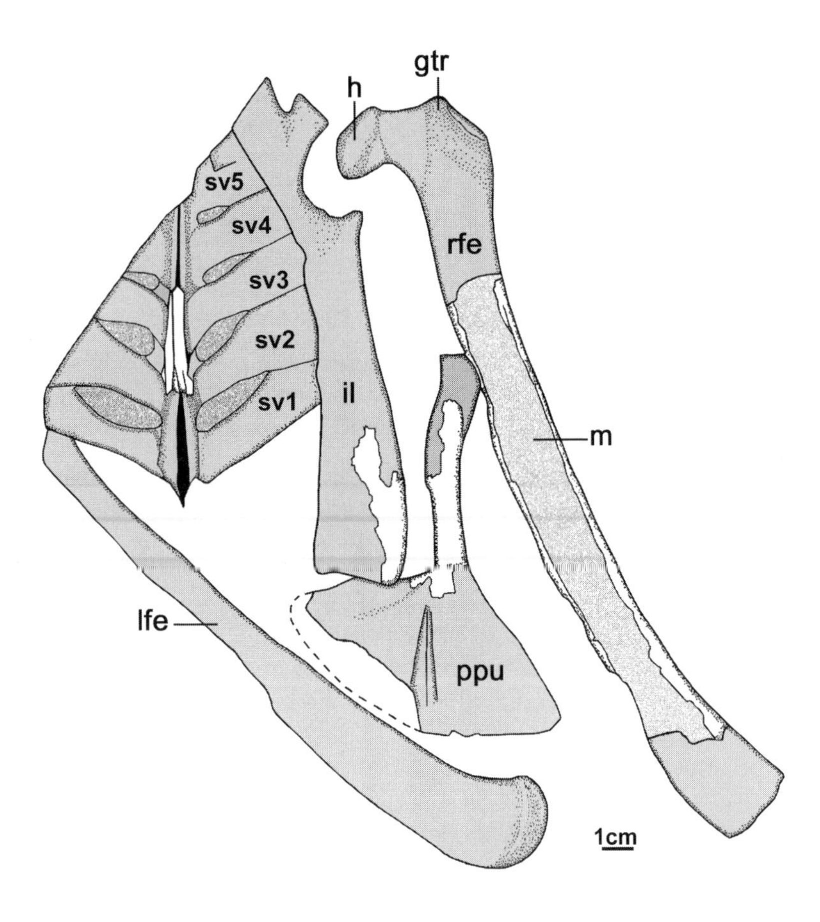

Figure 6.7. Herbstosaurus pigmaeus. *gtr = greater trochanter; h = femoral head; il; ilium; sv1–5 = sacral vertebrae 1 to 5; m = filling matrix; lfe = left femur; ppu = prepubis; rfe = right femur. White represents bone remains; gray represents impressions.*

the clade Dsungaripteroidea (sensu Unwin 2003) is based mainly on characters of the dentition, skull, and mandible, although it is also diagnosed by two postcranial characters (e.g., limb bones with relatively thick bone walls and a strongly bowed femur). The holotype of *Herbstosaurus* has no skull; hence, the comparison is limited to the postcranium.

The walls of the long bones of dsungaripteroids are relatively thick, and consequently the central lumen is narrow (Unwin 2003). The bony walls of the holotype of *Herbstosaurus* are well exposed and could be measured precisely in the right femur (Fig. 6.7); they are only 1 mm thick, and the central lumen is not narrow. Consequently, *Herbstosaurus* does not share this alleged character with the dsungaripteroids. On the contrary, it is similar to the rest of the pterodactyloids pterosaurs that have a quite thin cortex, rarely exceeding 2 mm (Unwin 2003). The second postcranial character of the dsungaripteroids, in which the femur is strongly curved in two planes (both in lateral and anterior view), could not be observed in *Herbstosaurus pigmaeus* because it is two-dimensionally preserved. However, a strongly curved femur in lateral and anterior view is also reported for *Dimorphodon* (Padian 1983a), and it has been considered as a general feature in all pterosaurs that are adequately preserved (Padian 1983b).

The femur of *Herbstosaurus pigmaeus* is 9.15 cm long. The fusion of the sacral vertebrae with the neural spines to form a very solid and differentiated supraneural plate (observed in the cast, in dorsal view) suggests that this specimen of *Herbstosaurus* represents a subadult or perhaps adult animal, which is a small- to medium-sized individual.

The record of the Pterodactyloidea begins in the Tithonian (Wellnhofer 1991; Kellner 2003; Unwin 2003), and with the new assessment of the age of the bearing rocks, *Herbstosaurus* is the first record of this clade in the Southern Hemisphere. The sediments that bear *Herbstosaurus pigmaeus* were deposited in a nearshore marine environment (Leanza et al. 1978).

Pterodactyloidea Pleininger 1901
Archaeopterodactyloidea Kellner 2003
Gen. et sp. indeterminate

Material. The specimen MOZ 3625P (Fig. 6.8) was collected as split slab and counterslab. It lacks the skull but constitutes a nearly complete postcranial skeleton. MOZ 3625P includes some cervical and dorsal vertebrae; a few thoracic ribs; both pectoral girdles exposed in anterior view; left pelvic girdle; a proximal right wing (humerus, ulna, and radius) and right metacarpal IV; a more complete left wing that only lacks of the wing phalanx four; and hind limb bones (both femora, both tibiae, and pes).

Locality and age. El Ministerio Quarry, Los Catutos, 13 km northwest of Zapala, Neuquén Province, Argentina (Fig. 6.1, locality 2). Los Catutos Member limestones of the Vaca Muerta Formation, upper-middle Tithonian (Leanza and Zeiss 1992; Scasso and Concheyro 1999).

Comments. The postcranial skeleton is incomplete and partially preserved. Some bones are articulated, but most of them have been displaced from their original position (Codorniú et al. 2006).

Epiphyses of the long bones (humerus, radio, ulna, femur, tibia) are well ossified; the series of the proximal and distal carpals are fused and constitute proximal and distal syncarpus, respectively; the extensor tendon process is fused to the shaft of wing phalange one; the scapula fused to coracoid without suture; pelvic elements are coosified without suture, suggesting that the individual was osteologically mature. However, the fact that the right pelvic plate appears isolated suggests that it was unfused to the sacrum when this animal died. The latter could be a feature of immaturity and thus could indicate that MOZ 3625P was an individual relatively mature osteologically and consequently a subadult at time of death.

Wingspan was calculated on the basis of an adult specimen of *Pterodactylus antiquus* (BSP 1968 I 95, whose wingspan is

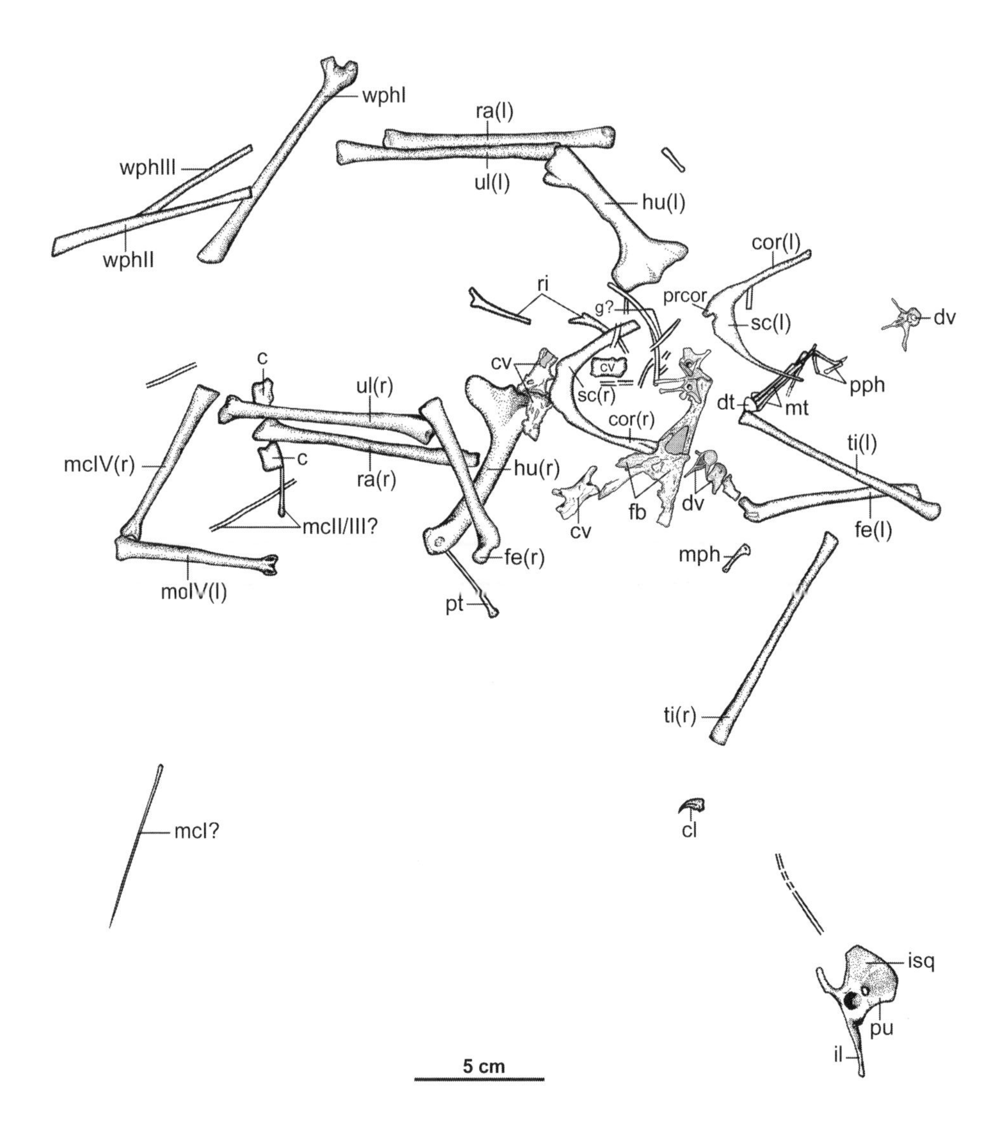

Figure 6.8. Composite drawing of MOZ 3625P. c = carpals; cl = claw; cor = coracoid; cv = cervical vertebrae; dt = distal tarsals; dv = dorsal vertebrae; fe = femur; fb = very fragmentary bones; hu = humerus; il = ilium; isq = isquion; mcI? = wing metacarpal I?; mc II/III? = wing metacarpal II or III; mcIV = wing metacarpal IV; mph = manual phalanges; mt = metatarsals; pph = pedal phalanges; prcor = procoracoid; pu = pubis; pt = pteroid; ra = radius; ri = rib; sc = scapula; ti = tibia; ul = ulna; wph I–III = wing phalanges I–III; (r) = right; (l) = left.

510 mm). Proportions have been taken from Wellnhofer (1970). This results in a wingspan of 109.74 mm for MOZ 3625P. The same procedure was applied to the proportions of a relatively mature individual of *Pterodaustro guinazui* (MHIN-UNSL-GEO-V 243, whose wingspan is 1590 mm), a medium-sized, filter-feeding archaeopterodactyloid (see proportions in Codorniú and Chiappe 2004). The results were very similar, resulting in a 105.75-mm wingspan for MOZ 3625P. These data indicate that MOZ 3625P was a subadult whose wingspan approaches 1.10 m.

The absence of cervical ribs on midcervical vertebrae, length of the wing metacarpal (more than 80% of the humerus length), ulna less than twice the length of metacarpal IV, and femur about the same length or less than metacarpal IV support the assignment of MOZ 3625P to the Pterodactyloidea (Kellner 1996, 2003). Features like the deltopectoral of the humerus proximally placed and curved ventrally; midcervical vertebrae elongated, but not to the same degree as azhdarchids; low neural spines in midcervical that are shaped like a blade suggest a close relationship of MOZ 3625P with basal pterodactyloids, the Archaeopterodactyloidea (Kellner 2003), within Ctenochasmatoidea (sensu Unwin 2003). However, the most precise systematic position between basal pterodactyloids (*Pterodactylus, Ctenochasma, Pterodaustro, Cycnorhamphus, Germanodactylus,* and *Gallodactylus canjuersensis*) is still uncertain. This is because most synapomorphies are based on the skull, rostrum, and dentition and on other portions of the skeleton that are not preserved in MOZ 3625P.

In summary, MOZ 3625P is assigned to the Pterodactyloidea and is probably closely related to the Archaeopterodactyloidea (Kellner 1996, 2003). According to Unwin's (2003) phylogeny, MOZ 3625P is closely related to the Euctenochasmia on the basis of the presence of certain characters, such as the depressed neural arch of the midseries cervicals, with a low neural spine and elongate midseries cervicals.

This specimen is the most complete Jurassic pterosaur known to date from South America. Sedimentological and structural analyses suggest that Los Catutos limestones were deposited in a shallow sea (10–30 m) no more than 100 km from the eastern coast of the Neuquén Basin (Leanza and Zeiss 1990; Scasso et al. 2002). Other marine reptiles, such as ophthalmosaurians, metriorhynchids, and chelonians, were also found in the area of Los Catutos (Gasparini et al. 1995).

Pteranodontoidea sensu Kellner 2003
Gen. et sp. indeterminate

Material. MACN-SC 3617, a right incomplete ulna (Fig. 6.9) and a wing metacarpal? (Fig. 6.10), probably not from the same individual (Kellner et al. 2003).

Locality and age. From 5 km south of Estancia Río Noble, and a few kilometers south of Lake Belgrano (48°S) in Santa Cruz Province (Aguirre-Urreta and Ramos 1981; Kellner et al. 2003) (Fig. 6.1, locality 3). Lower section of the Río Belgrano Formation (Ramos 1979). On the basis of the presence of *Hatchericeras patagonense* assemblage zone, it can be assigned to the lower-middle Barremian (Riccardi 1983).

Comments. The ulna was previously mentioned by Aguirre-Urreta and Ramos (1981), Montanelli (1987), and Bonaparte (1996). It lacks the proximal end, and the preserved length is 19.9 cm. It has been described by Kellner et al. (2003), confirming the assignment to the clade Pterodactyloidea, mainly because of the large size. Kellner et al. (2003) estimated that the total length of the ulna was 28.2 cm, and on the basis of the well-developed ventral ridge, they suggested affinities with the Anhangueridae. Bearing in mind this length estimate, and comparing it with the body sizes of the specimens of *Anhanguera,* Kellner et al. (2003) calculated a wingspan of approximately 3.6 m for MACN-SC 3617.

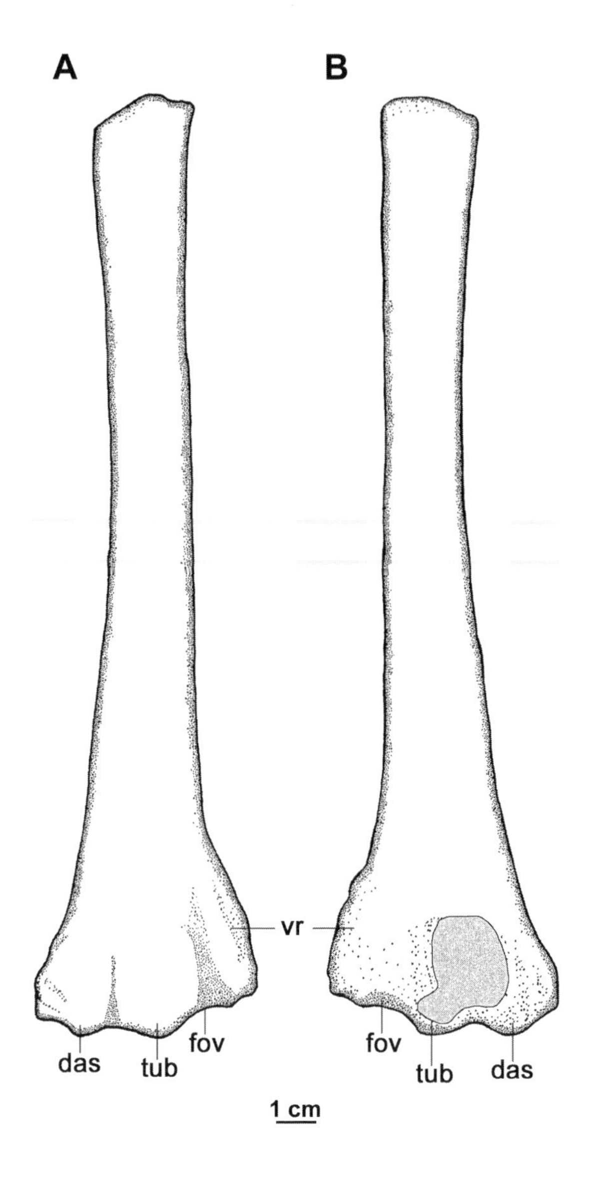

Figure 6.9. Right ulna, MACN 3617. (A) Anterior view. (B) Posterior view. das = dorsal articular surface; fov = ventral fovea; tub = tuberculum; vr = ventral ridge.

The assignment of the material from Santa Cruz to the Anhangueridae further extends the anhanguerid record to the Barremian (Kellner et al. 2003), which was previously limited to the Aptian/Albian (Kellner and Tomida 2000). However, according to Unwin (2003), Anhangueridae is a junior synonym of Ornithocheiridae Seeley; hence, following this criterion, the range of the latter taxon is extended to the Valanginian-Cenomanian (Unwin 2003, 178), and consequently, MACN-SC 3617 (Barremian) is not the oldest record of the clade.

The Patagonian specimen is the southernmost record of pterosaurs in South America. The fossil-bearing rocks of MACN-SC 3617 were deposited in a low-energy marine platform environment (nearshore) (Kellner et al. 2003).

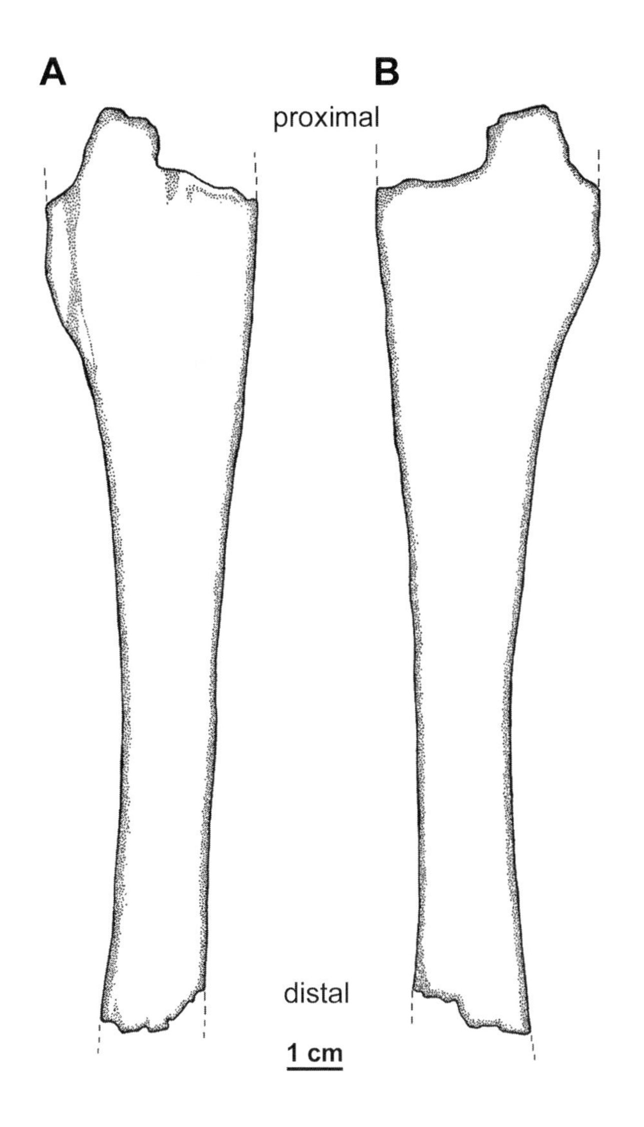

Figure 6.10. Incomplete ?wing metacarpal, MACN 3617. (A) Anterior view. (B) Posterior view.

Azhdarchidae Nesov 1984
Gen. et sp. indeterminate

Material. MUCPv 358, epiphysis of an ulna (Kellner et al. 2004; Porfiri and Calvo 2004).

Locality and age. Loma de la Lata, northern coast of Barreales Lake, Neuquén Province, Argentina (Fig. 6.1, locality 4). Portezuelo Formation (sensu Ramos 1981), upper Turonian–lower Coniacian (Calvo and Grill 2003).

Comments. Features of the wing (epiphysis of ulna) suggest a relationship with Azhdarchidae (Nesov 1984) (this family includes the largest known pterosaur, *Quetzalcoatlus,* with a 12-m wingspan, from the Javelina Formation, Texas). The wingspan estimate of the pterosaur from Barreales Lake is 6 m; if confirmed, this would be the largest flying reptile from South America (Kellner et al.

2004). This material was quite recently discovered and is currently under study.

The depositional environment of the Portezuelo Formation is characteristic of high-energy rivers. In addition to pterosaur remains, bivalves; fish; turtles; crocodiles; dinosaur eggs; ornithopod, sauropod, and theropod dinosaurs; and a large number of gymnosperms and angiosperms have been recovered from this locality (Kellner et al. 2004; Porfiri and Calvo 2004).

Pterodactyloidea gen. et sp. indeterminate

Material. MOZ 2280P, right tibia, lacking the distal end.

Locality and age. As for MOZ 3625P (Fig. 6.1, locality 2).

Comments. Gasparini et al. (1987) referred an isolated right tibia to Pterodactyloidea, which lacks the distal end, us larger than the tibia of the archaeopterodactyloid MOZ 3625P, and has more deformation. The degree of preservation of MOZ 2280P and the absence of a tibia in *Herbstosaurus* constrain a direct comparison to other Patagonian pterosaurs. Given these constraints, the specimen is considered to be Pterodactyloidea gen. et sp. indeterminate.

A better understanding of these pterosaurs depends largely on finding new material, and on a more complete knowledge of these taxa.

Pterodactyloidea gen. et sp. indeterminate

Material. MACN-N 02 (Fig. 6.11), incomplete right femur (Montanelli 1987; Bonaparte 1996).

Locality and age. At 2.5 km southeast from the junction of National Route 40 with Arroyo La Amarga, Neuquén Province, Argentina (Fig. 6.1, locality 5). Lower section of the La Amarga Formation (Parker 1965), referred to the Hauterivian-Barremian (Montanelli 1987), Hauterivian (Bonaparte 1996), or Barremian–Early Aptian (Leanza et al. 2004).

Comments. The material is preserved in three dimensions. The proximal end and diaphysis of the femur can be seen; it lacks the distal end. The length preserved is 11.6 cm. In posteromedial view, the neck is well marked and elongate; the head of the femur has a popliteal groove in the middle region, and the diaphysis is slightly curved in two planes. The major trochanter is well developed. Montanelli (1987) referred this material to the Pterodactyloidea because of the conspicuous head of the femur, and pointed out its possible relationship to *Herbstosaurus*. However, she also observed a difference in the inclination of the neck, which, at an angle of 135° to the diaphysis of the femur, is indeed more marked in MACN-N 02 than in *Herbstosaurus*. This character has been recently interpreted as the plesiomorphic state for pterodactyloids that separates them from the ornithocheiroids by a derived condition in which the neck is relatively robust and the caput is directed upward at 160° to the shaft (Unwin 2003). Taking these criteria into account, MACN-N 02 is interpreted as a nonornithocheiroid pterodactyloid, but it is too incomplete and poorly preserved to as-

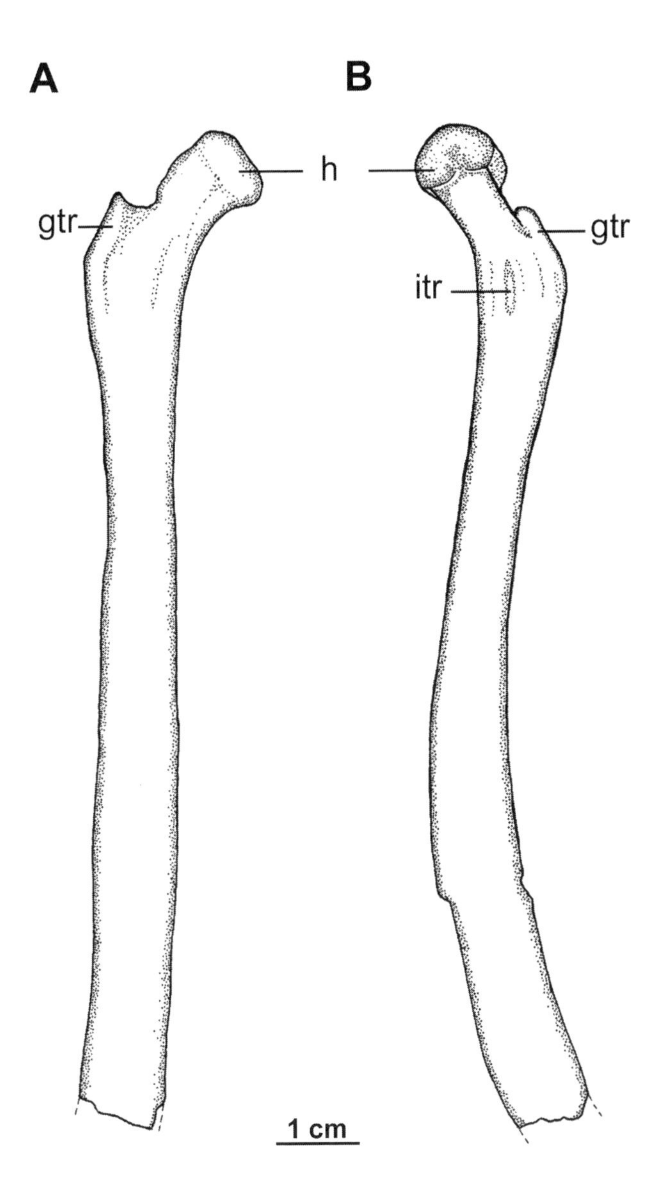

Figure 6.11. Right femur, MACN-N 02. (A) Anterior view. (B) Posteromedial view. gtr = greater trochanter; h = femoral head; itr = internal trochanter.

sign it to a particular genus or species. MACN-N 02 was found in rocks deposited in a continental nearshore environment, and is associated with large theropod teeth and sauropod and mammal remains (Montanelli 1987).

Pterosauria gen. et sp. indeterminate

Material. Uncatalogued specimens consisting of several mandibles, a braincase, shoulder girdle, and disarticulated forelimb bones (two humeri and several wing-finger phalanges) (Unwin et al. 2004).

Locality and age. Cerro Cóndor, Las Chacritas, central Chubut Province, on the middle course of Chubut River (Fig. 6.1, locality 6). Lower section of the Cañadón Asfalto Formation. This sequence is currently dated as Middle Jurassic (Callovian) in age on the basis of conchostracans, palynology, and geological relationships (Unwin et al. 2004).

Comments. These bones are well preserved and uncrushed. Most of the material studied appears to represent a "rhamphorhynchoid" pterosaur approximately 1.5–2 m in wingspan (Unwin et al. 2004). Unwin et al. (2004) attributed a pair of mandibles to Scaphognathinae, a relatively derived clade of nonpterodactyloids that includes *Scaphognathus* and *Sordes* (Unwin 2003; but see Kellner 2003). In addition, the pectoral girdle and the humerus also compare well with the corresponding bones of *Scaphognathus*. Because of this, the Cañadón Asfalto scaphognathinae probably represents a new taxon. Another lower jaw could correspond to Rhamphorhynchinae (Unwin et al. 2004).

These pterosaur remains are the oldest in South America; they were found in a continental Jurassic formation. The Cañadón Asfalto Formation is dominated by lacustrine deposits with strong volcanic influence and has yielded an important vertebrate fauna, including freshwater fish and turtles, a large-sized anuran, squamates, dinosaurs, and mammals (Rauhut and Puerta 2001; Rauhut et al. 2001).

Patagonian Pterosaurs

Although most Patagonian pterosaurs are fragmentary, their remains are distributed over a large area and extensive time span (Fig. 6.12). Together with Loma del Pterodaustro from Central Argentina, the six localities from Patagonia constitute pterosaur sites from South America. The potential for new discoveries of pterosaurs from Patagonia is very high, and it is likely that many more finds will become available as searches concentrate on the already known localities. A recent finding in Chubut indicates that forms inhabiting continental environments were present at least since the Callovian. This occurrence is the oldest record of Pterosauria in South America. Moreover, because one of the most important gaps in the history of Pterosauria belongs to the Middle Jurassic (Aalenian-Callovian) (Unwin 1996), the pterosaurs from Chubut are especially significant.

Likewise, specimens found in Tithonian rocks of the Neuquén Province suggest that small- to middle-sized pterodactyloids flew over the seas of the Neuquén Basin toward the end of the Jurassic, a time especially abundant in records of marine fish and reptiles in the area (Gasparini and Fernández 1997). One of them was referred to *Herbstosaurus pigmaeus*, and the other is a basal pterodactyloid (archaeopterodactyloid, sensu Kellner [2003]; and a ctenochasmatid, sensu Unwin [2003]), the most complete Jurassic pterosaur from South America. The understanding of the phylogenetic relationships of these fossils depends on discovery of more complete material with greater phylogenetic signal.

In the Lower Cretaceous, between the Barremian and Early

Period	Epoch	Stage	Record	Formation
Cretaceous	Upper	Coniacian Turonian	(4) MUCPv 358, epiphysis of an ulna	Portezuelo Fm. high energy rivers
		Cenomanian	(7) ?Pterosaur tracks	Candeleros Fm. fluvial environment
	Lower	Albian Aptian Barremian	(3) MACN-SC 3617, incomplete ulna and an incomplete wing metacarpal?	Río Belgrano Fm. marine nearshore environment
		Hauterivian Valanginian Berriasian	(5) MACN-N 02, incomplete right femur	La Amarga Fm. continental nearshore environment
Jurassic	Upper	Tithonian Kimmeridgian Oxfordian	(2) MOZ 3625P, nearly complete postcranial skeleton MOZ 2280P, incomplete tibia (1) CTES-PZ 1709, *Herbstosaurus pigmaeus*, pelvic girdle, proximal himbdimbs, sacrum	Vaca Muerta Fm. marine environment
	Middle	Callovian	(6) several mandibles, a brancaise, shoulder girdle, two humeri, several wing finger phalanges	Cañadón Asfalto Fm. lacustrine deposits

Figure 6.12. Distribution of Patagonian pterosaurs in Argentina, from Callovian to Coniacian. The numbers correspond to the locality number in Figure 6.1. Fm. = Formation.

Aptian, two pterosaurs inhabited Patagonia. These differ greatly in size and probably represent two different clades. One is from northwestern Patagonia (Neuquén Province, Argentina) and belongs to a middle-sized nonornithocheiroid pterodactyloid. The other is a larger pterodactyloid pteranodontoid, approximately 3.6 m in wingspan, that flew over coastal environments in the southern tip of South America. The latter is the southernmost pterosaur record of South America. This pteranodontoid, a pterosaur from the Upper Cretaceous (Campanian-Maastrichtian) of New Zealand (Wiffen and Molnar 1988), and fragments that represent at least one pterosaur allied to the well-known *Pteranodon* from the Lower Cretaceous (Albian) of Australia (Molnar and Thulborn 1980) are the three southernmost records of Gondwana.

The youngest record of Pterosauria in Patagonia is a large pterosaur, not yet studied in detail (Kellner et al. 2004), from the Upper Turonian–Lower Coniacian of Northwestern Patagonia. This occurrence comes from rocks containing a large diversity of sauropod and theropod dinosaurs, crocodiles, and freshwater turtles (Leanza et al. 2004; Porfiri and Calvo 2004).

Acknowledgments. We thank Dr. Kevin Padian (Department of Integrative Biology and Museum of Paleontology University of California, Berkeley), Dr. Alexander Kellner (Setor de Paleovertebrados, Departamento de Geologia e Paleontologia, Museu Nacional/UFRJ, Rio de Janeiro), and Dr. Luis M. Chiappe (Natural History Museum of Los Angeles County) for help during revision. We thank the institutions mentioned in the text for access to study and review the pterosaurs from their collections, particularly the Departamento de Paleontología, Universidad del Nordeste, for authorization to review *Herbstosurus pigmaeus.* We thank Dr. Cecilia Deschamps (Museo de La Plata) for help with translation. Research was partially supported by Agencia (PICT 8439) and Proyecto de Incentivos, Universidad Nacional de La Plata, to Z. G., and Fundación Antorchas, and Consejo Nacional de Investigaciones Científicas y Técnicas (CONICET) to L. C. This chapter is part of project 340103 from Ciencia y Técnica, Universidad Nacional de San Luis (L. C.).

References Cited

Aguirre-Urreta, M. B., and V. A. Ramos. 1981. Estratigrafía y paleontología de la alta cuenca del río Roble, cordillera patagónica. *Actas del VIII Congreso Geológico Argentino,* 101–138.

Bell, C. M., and K. Padian. 1995. Pterosaur fossils the Cretaceous of Chile: evidence for a pterosaur colony on an inland desert plain. *Geology Magazine* 132: 31–33.

Bennett, S. C. 1989. A Pteranodontid pterosaur from the Early Cretaceous of Perú, with comments on the relationships of Cretaceous pterosaurs. *Journal of Vertebrate Paleontology* 63 (5): 669–677.

———. 1994. Taxonomy and systematics of the Late Cretaceous pterosaur *Pteranodon* (Pterosauria, Pterodactyloidea). *University of Kansas Occasional Papers* 169: 1–70.

Bonaparte, J. F. 1970. *Pterodaustro guiñazui* gen. et sp. nov. Pterosaurio de la Formación Lagarcito, Provincia de San Luis, Argentina y su significado en la geología regional (Pterodactylidae). *Acta Geológica Lilloana* 10: 207–226.

———. 1971. Descripción del cráneo y mandíbulas de *Pterodaustro guiñazui,* (Pterodactyloidea), Pterodaustriidae nov.) de la Formación Lagarcito, San Luis, Argentina. *Publicaciones del Museo Municipal de Ciencias Naturales de Mar del Plata* 1 (9): 263–272.

———. 1978. *El Mesozoico de América del Sur y sus Tetrápodos.* Tucumán, Argentina: Ministerio de Cultura y Educación, Fundación Miguel Lillo.

———. 1996. Cretaceous tetrapods of Argentina. In F. Pfeil and G. Arratia (eds.), *Contributions of Southern South America to Vertebrate Pale-*

ontology. Münchner Geowissenchaftliche Abhandlungen (A) 30: 73–130.
Bonaparte, J. F., and T. Sánchez. 1975. Restos de un Pterosaurio, *Puntanipterus globosus* de la Formación La Cruz, Provincia de San Luis, Argentina. *Actas del I° Congreso Argentino de Paleontología y Bioestratigrafía* 2: 105–113.
Calvo, J. O., and D. Grill. 2003. Titanosaurid sauropod teeth from Futalognko quarry, Barreales lake, Neuquén, Patagonia, Argentina. *Ameghiniana* 40 (4): 52–53.
Calvo, J. O., and M. Lockley. 2001. The first pterosaur tracks from Gondwana. *Cretaceous Research* 22: 585–590.
Casamiquela, R. M. 1975. *Herbstosaurus pigmaeus* (Coeluria, Compsognathidae) n. gen. n. sp. del Jurásico del Neuquén (Patagonia septentrional). Uno de los más pequeños dinosaurios conocidos. *Actas del I° Congreso Argentino de Paleontología y Bioestratigrafía* 2: 87–103.
Casamiquela, R., and G. Chong Díaz. 1980. La presencia de *Pterodaustro* Bonaparte (Pterodactyloidea) del Neojurásico (?) de la Argentina, en los Andes del Norte de Chile. *Actas del II° Congreso Argentino de Paleontología y Bioestratigrafía y I° Congreso Latinoamericano de Paleontología* 1: 201–213.
Chiappe, L. M., D. Rivarola, A. Cione, M. Fregenal, A. Buscalioni, H. Sozzi, L. Buatois, O. Gallego, E. Romero, A. Lopez, S. McGehee, C. Marsicano, S. Adamonis, O. Laza, P. Ortega, and O. Di Iorio. 1995. Inland biota from a Lower Cretaceous lagerstätte of Central Argentina. *Second International Symposium on Lithographic Limestones,* 57–69. Spain: Lleida-Cuenca.
Chiappe, L. M., D. Rivarola, E. Romero, S. Dávila, and L. S. Codorniú. 1998a. Recent advances in the paleontology of the Lower Cretaceous Lagarcito Formation (Parque Nacional Sierra de Las Quijadas, San Luis, Argentina). *New Mexico Museum of Natural History and Science Bulletin* 14: 187–192.
Chiappe, L. M., D. Rivarola, L. Cione, M. Fregenal, H. Sozzi, L. Buatois, O. Gallego, J. H. Laza, E. Romero, A. Lopez-Arbarello, A. Buscalioni, C. Marsicano, S. Adamonis, P. Ortega, S. McGehee, and O. Di Iorio. 1998b. Biotic association and paleoenvironmental reconstruction of the "Loma del *Pterodaustro*" fossil site (Lagarcito Formation, Early Cretaceous, San Luis, Argentina). *Geobios* 31: 349–369.
Chiappe, L. M., A. W. A. Kellner, D. Rivarola, S. Dávila, and M. Fox. 2000. Cranial morphology of *Pterodaustro guinazui* (Pterosauria: Pterodactyloidea) from the Lower Cretaceous of Argentina. *Contributions in Science* 483: 1–19.
Chiappe, L. M., L. Codorniú, G. Grellet-Tinner, and D. Rivarola. 2004. Argentinian unhatched pterosaur fossil. *Nature* 432: 571–572.
Codorniú, L. 2002. Juvenile pterosaurs from the Early Cretaceous of Argentina. *Journal of Vertebrate Paleontology* 22 (Suppl. to 3): 45A.
———. 2005. Morfología caudal de *Pterodaustro guinazui* (Pterosauria: Ctenochasmatidae) del Cretácico de Argentina. *Ameghiniana* 42 (2): 505–509.
Codorniú, L., and L. M. Chiappe. 2004. Early juvenile pterosaurs (Pterodactyloidea: *Pterodaustro guinazui*) from the Lower Cretaceous of Central Argentina. *Canadian Journal of Earth Sciences* 41: 9–18.
Codorniú, L., L. M. Chiappe, and D. Rivarola. 2004. Primer reporte de un embrión de pterosaurio (Cretácico Inferior, San Luis, Argentina). *Ameghiniana* 41: 40R.

Codorniú, L., Z. Gasparini, and A. A. Paulina-Carabajal. 2006. A late pterosaur (Reptilia, Pterodactyloidea) from Northwestern Patagonia, Argentina. *Journal of South American Earth Sciences* 20: 383–389.

Dávila, S, L. M. Chiappe, and D. Rivarola. 1999. Anatomía pélvica y miembros posteriores de pterosaurios del Cretácico de la provincia de San Luis. *Ameghiniana* 36: 98R.

Díaz, H. 1946. *Reconocimiento Geológico de la Región Comprendida entre Marayes, en la Prov. de San Juan y la Sierra de Las Quijadas, en la Prov. de San Luis, incluyendo Sierra del Gigantillo, Sierra Guayaguas y Sierra del Cantantal.* Buenos Aires: YPF Informe Inédito.

Flores, M. A. 1969. El bolsón de Las Salinas en la Prov. de San Luis. *Actas de las IV Jornadas Geológicas Argentinas* 1: 311–327.

Gasparini, Z., and M. Fernández. 1997. Tithonian marine reptiles of the eastern Pacific. In J. Callaway and E. Nicholls (eds.), *Ancient Marine Reptiles,* 435–440. San Diego: Academic Press.

Gasparini, Z., H. Leanza, and J. G. Zubillaga. 1987. Un pterosaurio de la calizas litográficas Titonianas del área de los Catutos, Neuquén, Argentina. *Ameghiniana* 24: 141–143.

Gasparini, Z., M. de la Fuente, and M. Fernández. 1995. Sea reptiles from the lithographic limestones of the Neuquén Basin, Argentina. *II International Symposium on Lithographic Limestones,* Universidad Autónoma de Madrid, 81–84.

Howse, S. C. B. 1986. On the cervical vertebrae of the Pterodactyloidea (Reptilia: Archosauria). *Zoological Journal of the Linnean Society* 88: 307–328.

Kaup, J. J. 1834. Versuch einer eintheilung der saugethiere in 6 stämme und der amphibien in 6 ordnungen. *Isis* 3: 311–315.

Kellner, A. W. A. 1996. Description of New Material of Tapejaridae and Anhangueridae (Pterosauria, Pterodactyloidea) and Discussion of Pterosaur Phylogeny. Ph.D. thesis, Columbia University.

———. 2001. A review of the pterosaur record from Gondwana. In *Two Hundred Years of Pterosaurs: A Symposium on the Anatomy, Evolution, Palaeobiology and Environments of Mesozoic Flying Reptiles. Strata,* ser. 1, 11: 51–53.

———. 2003. Pterosaur phylogeny and comments on the evolutionary history of the group. In E. Buffetaut and J.-M. Mazin (eds.), *Evolution and Paleobiology of Pterosaurs,* 105–137. Special Publications of the Geological Society 217.

Kellner, A. W. A., and J. M. Moody. 2003. Pterosaur (Pteranodontoidea, Pterodactyloidea) scapulocoracoid from the Early Cretaceous of Venezuela. In E. Buffetaut and J.-M. Mazin (eds.), *Evolution and Palaeobiology of Pterosaurs,* 73–77. Special Publications of the Geological Society 217.

Kellner, A. W. A., and Y. Tomida. 2000. *Description of a New Species of Anhangueridae (Pterodactyloidea) with Comments on the Pterosaur Fauna from the Santana Formation (Aptian-Albian) Northeastern Brazil.* National Science Museum Monograph 17. Tokyo: National Science Museum.

Kellner, A. W. A., M. B. Aguirre-Urreta, and V. A. Ramos. 2003. On the pterosaur remains from the Río Belgrano Formation (Barremian), Patagonian Andes of Argentina. *Anais de Academia Brasileira de Ciências* 75 (4): 487–495.

Kellner, A. W. A., J. O. Calvo, J. M. Sayão, and J. D. Porfiri. 2004. First pterosaur from the Portezuelo Formation, Neuquén Group, Patagonia, Argentina. In *Resumos IV Simpósio Brasileiro de Paleontología de Vertebrados*, 29–30.

Leanza, H., and A. Zeiss. 1990. Upper Jurassic lithographic limestones from Argentina (Neuquén Basin): stratigraphy and fossils. *Facies* 22: 169–186.

———. 1992. On the ammonite fauna of the lithographic limestones from the Zapala region (Neuquén Province, Argentine), with the description of a new genus. *Zentralbatt für und Paläontologie* 1: 1841–1850.

Leanza, H. A., H. G. Marchese, and J. C. Riggi. 1978. Estratigrafía del grupo Mendoza con especial referencia a la Formación Vaca Muerta, entre los paralelos 35° y 40° l.s., Cuenca Neuquina-Mendocina. *Revista de la Asociación Geológica Argentina* 32 (3): 190–208.

Leanza, H., S. Apesteguía, F. Novas, and M. de La Fuente. 2004. Cretaceous terrestrial beds from the Neuquén Basin (Argentina) and their tetrapod assemblages. *Cretaceous Research* 25: 61–87.

Martill, D. M., E. Frey, G. Chong Díaz, and C. M. Bell. 2000. Reinterpretation of a Chilean pterosaur and the occurrence of Dsungaripteridae in South America. *Geological Magazine* 137: 19–25.

Molnar, R. E., and R. A. Thulborn. 1980. First pterosaur from Australia. *Nature* 288 (5789): 361–363.

Montanelli, S. B. 1987. Presencia de Pterosauria (Reptilia) en la Formación La Amarga (Hauteriviano-Barremiano) Neuquén, Argentina. *Ameghiniana* 24: 109–113.

Nesov, L. A. 1984. Upper Cretaceous pterosaurs and birds from Central Asia [in Russian]. *Paleontologicheskii Zhurnal* 1: 47–57.

Ostrom, J. H. 1978. The osteology of *Compsognathus longiceps* Wagner. *Zitteliana* 4: 73–118.

Padian, K. 1983a. Osteology and functional morphology of *Dimorphodon macronyx* (Buckland) (Pterosauria: Rhamphorynchoidea) based on new material in the Yale Peabody Museum. *Postilla* 189: 1–44.

———. 1983b. A functional analysis of flying and walking in pterosaurs. *Paleobiology* 9: 218–239.

———. 2003. Pterosaur stance and gait and the interpretation of trackways. *Ichnos* 10: 115–126.

Parker, G. 1965. *Relevamiento Geológico a Escala 1:25.000 entre el Arroyo Picún Leufú y Catán Lil, a Ambos Lados de la Ruta Nacional N° 40*. Buenos Aires: YPF Informe Inédito.

Porfiri, J., and J. O. Calvo. 2004. El Centro paleontológico Lago Barreales. *Ciencia Hoy* 14 (79): 10–21.

Ramos, V. A. 1979. *Descripción Geológica de la Hoja 53a, Lago Belgrano, Provincia de Santa Cruz*. Buenos Aires: Servicio Geológico Nacional, Informe Inédito.

———. 1981. Descripción geológica de la Hoja 33c, Los Chihuidos norte, Provincia del Neuquén. *Boletín del Servicio Geológico Nacional* 182: 1–103.

Rauhut, O., and P. Puerta. 2001. New vertebrate fossils from the Middle-Late Jurassic Cañadón Asfalto Formation of Chubut, Argentina. *Ameghiniana* 38 (4): 16.

Rauhut, O., A. López Arbarello, P. Puerta, and T. Martin. 2001. Jurassic vertebrates from Patagonia. *Journal of Vertebrate Paleontology* 21 (Suppl. to 3): 91A.

Riccardi, A. C. 1983. The Jurassic of Argentina and Chile. In M. Moullade and A. E. M. Nairn (eds.), *The Phanerozoic Geology of the World II: The Mesozoic* B, 201–263. Amsterdam: Elsevier.

Rivarola, D. L. 1999. Estratigrafía y Sedimentología de Secuencias Cretácicas, Parque Nacional Sierra de Las Quijadas, San Luis, Argentina. Ph.D. thesis, Universidad Nacional de San Luis.

Rubilar, D., A. Vargas, and A. W. Kellner. 2002. Vértebras cervicales de Pterodactyloidea (Archosauria: Pterosauria) de la Formación Quebrada Monardes (Cretácico Inferior), norte de Chile. *Ameghiniana* 39: 16–17.

Sánchez, T. M. 1973. Redescripción del cráneo y mandíbula de *Pterodaustro guinazui* Bonaparte (Pterodactyloidea, Pterodaustriidae). *Ameghiniana* 10: 313–325.

Scasso, R., and A. Concheyro. 1999. Nanofósiles calcáreos, duración y origen de ciclos caliza-marga (Jurásico tardío de la Cuenca Neuquina). *Revista de la Asociación Geológica Argentina* 54: 290–297.

Scasso, R., M. Alonso, S. Lanés, H. Villar, and H. Lippai. 2002. Petrología y geoquímica de una micrita marga-caliza del Hemisferio Austral: el Miembro de Los Catutos (Formación Vaca Muerta) Titoniano medio de la Cuenca Neuquina. *Revista de la Asociación Geológica Argentina* 57: 143–159.

Unwin, D. M. 1995. Preliminary results of a phylogenetic analysis of the Pterosauria (Diapsida: Archosauria). In A. Sun and Y. Wang (eds.), *Sixth Symposium on Mesozoic Terrestrial Ecosystems and Biota,* 69–72. Beijing: China Ocean Press.

———. 1996. The fossil record of Middle Jurassic pterosaurs. In M. Morales (ed.), *The Continental Jurassic,* 291–304. Museum of Northern Arizona Bulletin 60.

———. 2003. On the phylogeny and evolutionary history of pterosaurs. In E. Buffetaut and J.-M. Mazin (eds.), *Evolution and Paleobiology of Pterosaurs,* 139–190. Special Publications of the Geological Society 217.

Unwin, D. M., and W.-D. Heinrich. 1999. On a pterosaur jaw from the Upper Jurassic of Tendaguru (Tanzania). *Mitteilungen aus dem Museum für Naturkunde Berlin, Geowissenschaftlichen Reihe* 2: 121–134.

Unwin, D. M., O. W. M. Rauhut, and A. Haluza. 2004. The first "Rhamphorynchoid" from South America and the early history of pterosaurs. *74th Annual Meeting of the Paläontologische Gesellschaft,* 235–237A.

Wellnhofer, P. 1970. Die Pterodactyloidea (Pterosauria) der Oberjura Plattenkalke. Süddeutschlands. *Abhandlungen der Bayerischen Akademie der Wissenschaften zu München, Mathematisch-Naturwisenschaftlichen Klasse* 141: 1–133.

———. 1978. Pterosauria. In P. Wellnhofer (ed.), *Handbuch der Paläoherpetologie. Encyclopedia of Paleoherpetology.* Stuttgart: Gustav Fischer.

———. 1991. *The Illustrated Encyclopedia of Pterosaurs.* London: Salamander Books.

Wiffen, J., and R. E. Molnar. 1988. First pterosaur from New Zealand. *Alcheringa* 12: 53–59.

Yrigoyen, M. R. 1975. La edad Cretácica del Grupo del Gigante (San Luis), su relación con cuencas circunvecinas. *Actas del I° Congreso Geológico Argentino de Paleontología y Bioestratigrafía* 2: 9–56.

7. Ornithischia

Rodolfo A. Coria and
Andrea V. Cambiaso

Introduction

The Patagonian ornithischian dinosaur record, unlike that from the Laurasian landmasses, is significantly less abundant and diverse than that of the saurischian dinosaurs. The Patagonian ornithischians (and to some extent, that of South America as a whole, because of the scarcity of records outside Patagonian) are restricted to a single Triassic sample, with the remaining forms coming from the Late Cretaceous (Fig. 7.1). The poor diversity is probably the result of some biases of the fossil record rather than a real reduction in biological diversity.

At present, ornithopods are the richest ornithischian fauna from Patagonia (Casamiquela 1964; Bonaparte et al. 1984; Coria and Salgado 1996a,b; Martínez 1998; Coria and Calvo 2002; Novas et al. 2004). Isolated specimens assigned to other ornithischian clades are of uncertain affinities (Tapia 1918; Bonaparte 1996). Nevertheless, we will present all these controversial forms as they were originally described, and we will include some comments related to current ideas about the evolution of the ornithischian dinosaurs in Patagonia (Coria 1999).

Institutional abbreviations. CPBA-V, Cátedra de Paleontología de la Facultad de Ciencias Exactas de la Universidad de Buenos Aires; DNM, ex Dirección Nacional de Minería, Buenos Aires, Argentina; FMNH-P, Field Museum of Natural History, Paleontology;

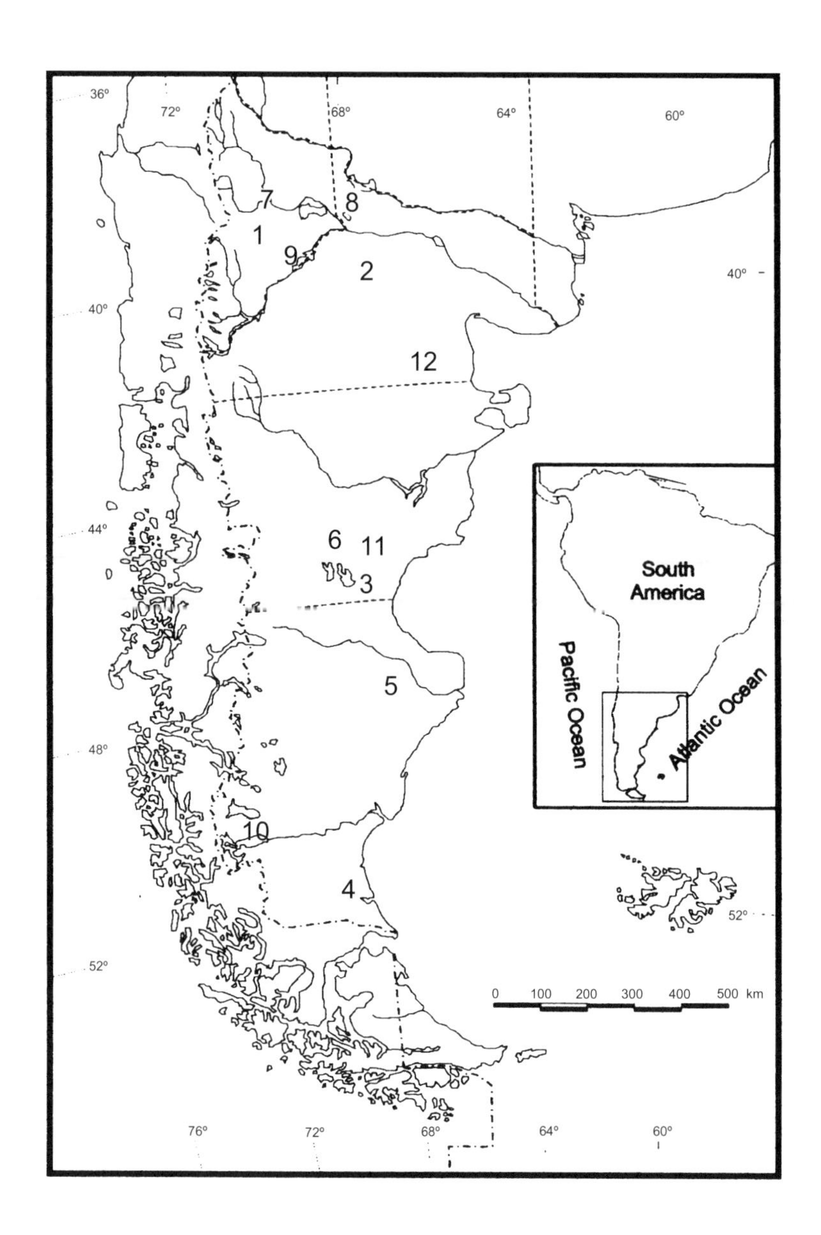

Figure 7.1. Map of Patagonia showing the localities with ornithischian dinosaur remains mentioned in this chapter; 1, La Amarga; 2, Salitral Moreno; 3, Lago Colhue Huapi; 4, Pari Aike; 5, Laguna Colorada; 6, Buen Pasto; 7, Cerro Bandera; 8, Cinco Saltos; 9, Cerro Bayo; 10, Lago Argentino; 11, Río Chico; 12, Los Alamitos.

MACN, Museo Argentino de Ciencias Naturales, Buenos Aires, Argentina; MCF-PVPH, Museo "Carmen Funes," Paleontología de Vertebrados, Plaza Huincul, Neuquén, Argentina; MCS, Museo de Cinco Saltos, Río Negro, Argentina; MPCA-SM, Museo Provincial "Carlos Ameghino," Colección Salitral Moreno, Cipolletti, Río Negro, Argentina; MUCPv, Museo de la Universidad Nacional del Comahue, Vertebrate Paleontology; UNPSJB-Pv, Universidad Nacional de la Patagonia "San Juan Bosco," Vertebrate Paleontology.

Systematic Paleontology

Ornithischia Seeley 1888
Genasauria Sereno 1986
Thyreophora Nopcsa 1915
Eurypoda Sereno 1986
Stegosauria Marsh 1877
Stegosauridae Marsh 1880
Gen. et sp. indeterminate (Bonaparte 1996)

Material. MACN-N-43, an incomplete posterior cervical vertebra, possibly the seventh cervical, without neural spine and prezygapophysial area; incomplete posterior cervical centra, proximal midcaudal centra with the base of the transverse processes (Fig. 7.2A, B).

Locality and age. At 2.5 km southeast of the La Amarga creek, on Road No. 40, Catan Lil Department, Neuquén Province, Argentina. Antigual Member of the La Amarga Formation, Barremian (Leanza et al. 2004).

Comments. Bonaparte identified this specimen as stegosaurian on the basis of the broad neural canal of the cervical vertebrae. In this particular fossil, the width of the neural canal does not reach the apomorphic condition of stegosaurs mentioned by Galton and Upchurch (2004) in having a neural canal in anterior dorsal vertebrae with a diameter greater than half of the centrum. Nevertheless, as mentioned by Bonaparte (1996), the similarities of these Patagonian remains with that of other stegosaurs are remarkable. The vertebral material was found associated with two elements identified as osteoderms; on this basis, Bonaparte inferred that these specimens belong to Stegosauria. These osteoderms are particularly small, as are the vertebrae that clearly correspond to an adult individual. Therefore, if the fossils indeed represent a Lower Cretaceous Patagonian stegosaur, then it would also be one of the smallest stegosaurs ever recorded.

Ankylosauria Osborn 1923
Nodosauridae Marsh 1890
Gen. et sp. indeterminate (Salgado and Coria 1996)

Material. MPCA-SM-1, right femur; MPCA-Pv-68/69/70, three posterior dorsal vertebrae; MPVA-Pv-71, caudal vertebrae; MPCA-Pv-72/73, two caudal centra; MPCA-Pv-41/42/43/74/75/76, six conical dermal plates; MPCA-Pv-78, two fused dermal plates; MPCA-Pv-77, one tooth (Fig. 7.2C–F).

Locality and age. Salitral Moreno, 40 km south of General Roca, Río Negro Province, Argentina. Allen Formation, Campanian-Maastrichtian (Ballent 1980; Powell 2003).

Comments. The ankylosaur material known from Patagonia was collected from the uppermost levels of the Cretaceous. Coria and Salgado (2001) recognized several ankylosaurian remains from the Allen Formation, so far the only Patagonian unit yielding ankylosaurs. This is based on the presence of the following derived fea-

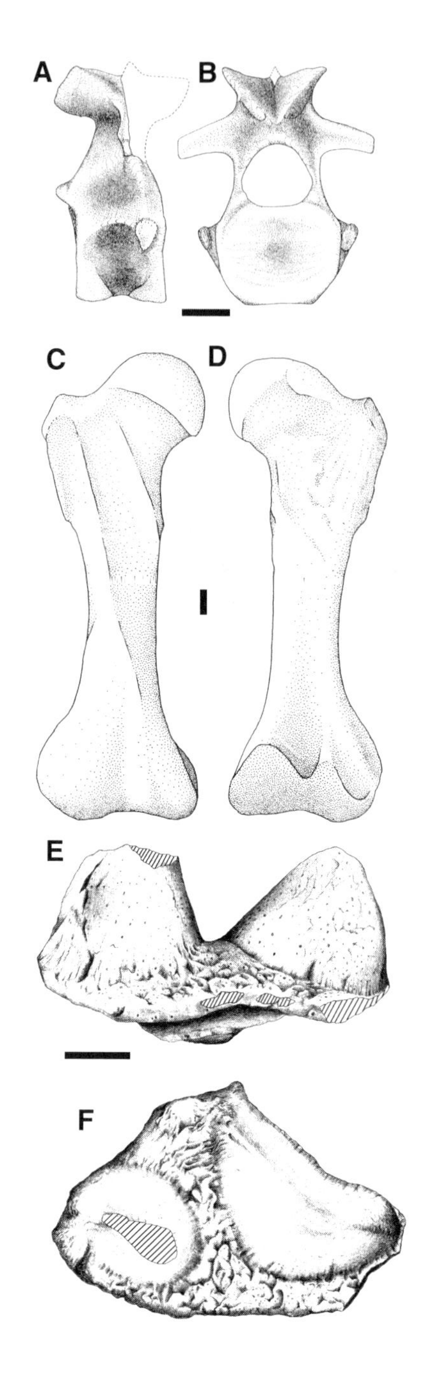

Figure 7.2. Patagonian Thyreophora remains. Stegosaur cervical vertebra from the Hauterivian La Amarga Formation in (A) left lateral and (B) posterior views. Ankylosaur femur from the Campanian-Maastrichtian Salitral Moreno locality in (C) anterior and (D) posterior views. Ankylosaur scutes from the Campanian-Maastrichtian Salitral Moreno locality in (E) side and (F) dorsal views. Scale bar = 1 cm. (A, B) Modified from Bonaparte (1996).

tures: leaf-shaped, laterally compressed tooth crown with a cingulum, U-shaped dorsal prezygapophyses, upwardly directed dorsal transverse processes, fusion of greater and anterior trochanters of the femur, hemispherical femoral head (suggesting a completely closed acetabulum), and dermal scutes with sharp keels and concave ventral surfaces.

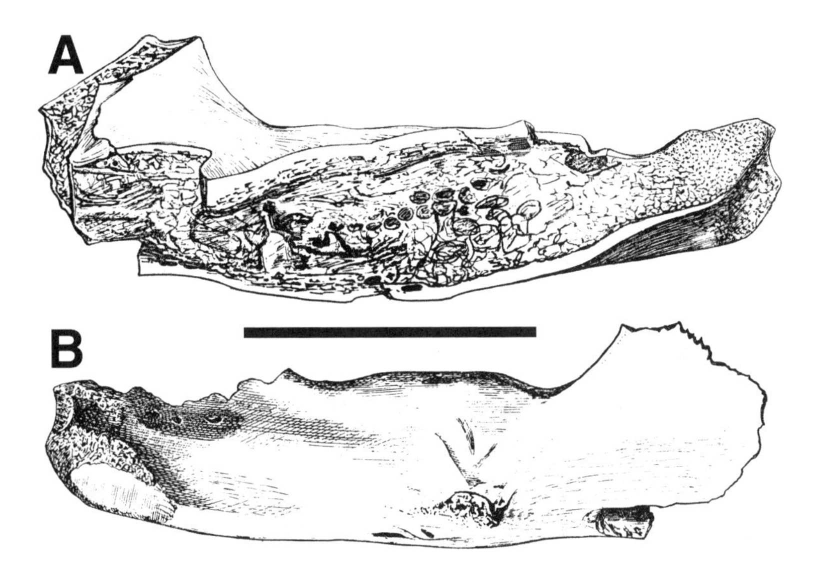

Figure 7.3. Right dentary of the supposed Patagonian ceratopsian Notoceratops *in (A) lateral and (B) medial views (modified from Huene 1929). Scale bar = 10 cm.*

Neornithischia Cooper 1985
?Marginocephalia Sereno 1986
?Ceratopsia Marsh 1890
***Notoceratops* Tapia 1918**
***Notoceratops bonarelli* Tapia 1918**
Fig. 7.3

Holotype. DNM uncataloged, an incomplete left lower jaw (Fig. 7.3).

Locality and age. Uncertain locality of the Chubut Province, Argentina. Laguna Palacios Formation, Maastrichtian (Bonaparte 1996).

Comments. The presence of Cerapoda ceratopsians as represented by the single bone of *Notoceratops* is debatable. The specimen is lost, and only the drawings in the original paper exist. However, the general profile of the lower jaw assigned to *Notoceratops*, as well as the morphology of the tooth row battery, suggest that the specimen is unlikely to be a ceratopsian.

Ornithopoda Marsh 1871
Gen. et sp. indeterminate (Coria and Salgado 1996)
***Loncosaurus argentinus* Ameghino 1899**

Material. MACN-1629, proximal half of a left femur (Fig. 7.4A, B).

Locality and age. Pari-Aike, Río Sehuen, southeast of Santa Cruz Province, Argentina. Mata Amarilla Formation, Santonian–early Campanian (Novas 1997), or probably Coniacian (Goin et al. 2002).

Comments. Loncosaurus argentinus was erected on the basis of a partial femur and one tooth by Ameghino (1899). Formerly assigned to the Theropoda on the basis of the form of the tooth, the

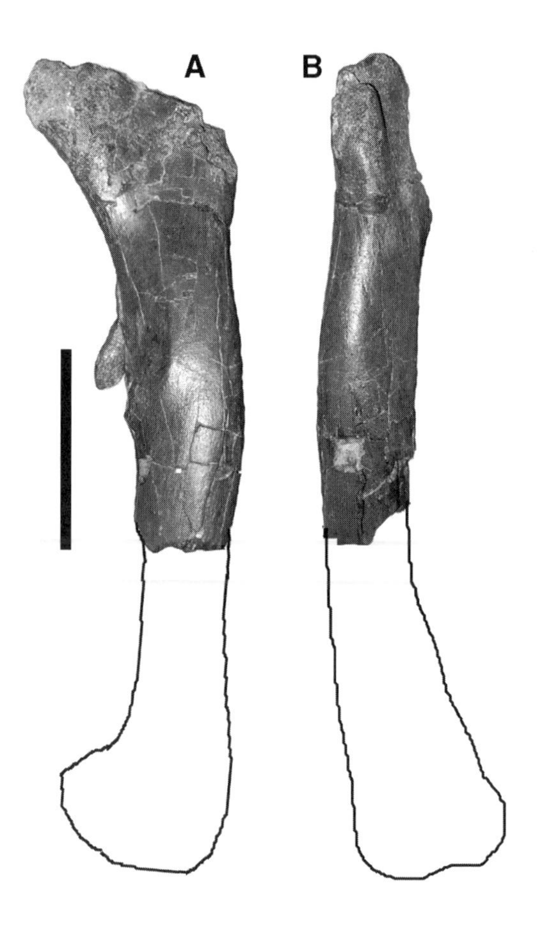

Figure 7.4. Loncosaurus argentinus *(holotype MACN-11629), proximal end of left femur in (A) anteromedial and (B) posterolateral views. Scale bar = 10 cm.*

femur was identified as that of an ornithopod by Molnar (1980). More recently, its taxonomic status was questioned because of the lack of autopomorphic features, being regarded as a nomen vanum by Coria and Salgado (1996a). However, the femur has a distinctive fourth trochanter, likely pendant, and a trochanteric fossa, features that incline us to include it within the Ornithopoda.

Heterodontosauridae Kuhn 1966
***Heterodontosaurus* Crompton and Charig 1962**
***Heterodontosaurus* sp. (Báez and Marsicano 2001)**
Fig. 7.5

Material. CPBA-V-14091, posterior left maxillary fragment with teeth and natural external mold; CPBA-V-14092, a concretion containing a caniniform tooth with serrated cutting edges (Fig. 7.5).

Locality and age. Laguna Colorada, stratotype locality of the Laguna Colorada Formation center of the Santa Cruz Province, Argentina. Laguna Colorada Formation, Late Triassic, ?Norian (Báez and Marsicano 2001).

Comments. The specimen is very fragmentary, consisting of a

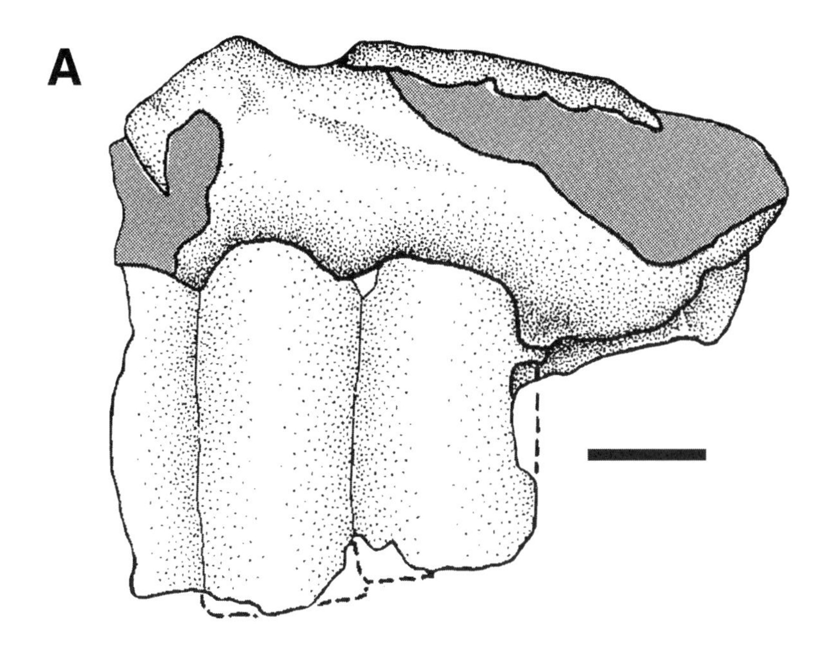

Figure 7.5. Left maxilla fragment assigned as *cf.* Heterodontosaurus from the Triassic El Tranquilo Formation (modified from Báez and Marsicano 2001) in labial view. Scale bar = 2 mm.

weathered portion of the left maxilla with three teeth in situ. The columnar, closely packeted teeth bear flat wear facets and high crowns. These last features, together with the presence of a broad contact between the mesial and distal surfaces of the crown and the lack of a cingulum, suggest relationships particularly with the Lower Jurassic South African ornithopod *Heterodontosaurus* among the South African heterodontosaurids (see Norman et al. 2004). The presence of heterodontosaurids in South America has already been established with the referral of *Pisanosaurus* (Casamiquela 1967; Bonaparte 1976) within Heterodontosauria (Sereno 1991). The discovery of this specimen from the Upper Triassic outcrops of Santa Cruz Province has been suggested as having implications for an earlier origin of heterodontosaurids than indicated by the South African forms (Báez and Marsicano 2001).

Euornithopoda Sereno 1986
Hypsilophodontidae Dollo 1882
***Notohypsilophodon* Martínez 1998**
***Notohypsilophodon comodorensis* Martínez 1998**

Holotype. UNPSJB-Pv-942, partial skeleton that includes four cervical, seven dorsal, four sacral, and six caudal vertebrae; fragments of four ribs; incomplete left scapula; right coracoid; right humerus; both ulnae; incomplete left femur; right tibia; incomplete left tibia; left fibula; incomplete right fibula; right astragalus; left calcaneum; and 13 pedal phalanges.

Locality and age. At 28 km northeast Buen Pasto town, central-south Chubut Province, Argentina. Bajo Barreal Formation, upper part of its lower member (J. C. Sciutto oral communication 2005), possibly Cenomanian (Martínez 1998).

Diagnosis. Humerus with greatly reduced deltopectoral crest; anteromedial bulge on the proximal extremity of the tibia; pronounced narrowing of the fibular shaft; astragalus with the proximal surface disposed in two levels; calcaneum with a pronounced posterodistal projection and ungual pedal phalanges with a flat ventral surface.

Comments. Martínez (1998) based the assignation of *Notohypsilophodon* to the Hypsilophodontidae on the presence of humerus with posterior flexure at the level of the deltopectoral region and a femur with the following characters: an ischiatic groove, anterior trochanter well below the greater trochanter, a shallow intertrochanteric cleft, and the lack of a distal extensor groove (this last character was proposed as an autopomorphy of *Gasparinisaura* [Coria and Salgado 1996]). *Notohypsilophodon* shares with *Gasparinisaura* (Coria and Salgado 1996) the presence of a ventral groove on the centra of proximal caudal vertebrae and an oval-shaped coracoid (also present in *Hypsilophodon* [Galton 1974]). The monophyly of Hypsilophodontidae has recently been challenged (Norman et al. 2004), and as a result, the phyletic positions of all components of the former Hypsilophodontidae have been rearranged. However, the latest analysis did not include *Notohypsilophodon* (Norman et al. 2004), so the phylogenetic relationships of this Patagonian euornithopod are still to be resolved.

Iguanodontia Dollo 1888
Gen. et sp. indeterminate (Coria 1999)

Material. MCF-PVPH-167, proximal half of a tibia (Fig. 7.6).

Locality and age. Minas de Bentonita, Yacimiento Cerro Bandera, west Cerro Portezuelo, 20 km west of Plaza Huincul, Neuquén Province, Argentina. Portezuelo Formation (Turonian?), Río Neuquén Subgroup, Neuquén Group (Leanza et al. 2004).

Comments. This specimen is relevant because it represents a large-sized ornithopod dinosaur that probably indicates the presence of a new form. Although the information given by this single bone is not enough to identify a new species, it is evident that it does not belong to any iguanodontian taxon so far recorded from South America. The size of the tibia matches those of *Iguanodon* and other large, graviportal ornithopods. No comparable form has been yet recognized for Turonian levels from any South America locality.

Euiguanodontia Coria and Salgado 1996
***Gasparinisaura* Coria and Salgado 1996**
***Gasparinisaura cincosaltensis* Coria and Salgado 1996**
Fig. 7.7

Holotype. MUCPv-208, almost complete skull with atlas and axis in articulation; sacrum; pectoral and pelvic girdles; humeri without distal ends; and almost complete hind limbs.

Figure 7.6. Probable iguanodontian left tibia from the Turonian Portezuelo Formation (MCF-PVPH-167) in (A) proximal and (B) lateral views. Scale bar = 10 cm.

Hypodigm. MUCPv-212, nearly complete series of articulated tail; distal ends of tibiae and proximal ends of both metatarsals; MUCPv-213, dentary; posterior dorsal vertebra, last dorsal, dorsosacral, and the first sacral vertebra in articulation, isolated sacral vertebra, incomplete humeri articulated with both radii and ulnae, incomplete right ilium, incomplete left pubis, incomplete right femur, distal portion of tibia and fibula articulated with astragalus and calcaneum, proximal ends of right metatarsals articulated with the distal tarsals, and several fragmentary bones that cannot be determined; MUCPv-214, five incomplete dorsal vertebrae, three cau-

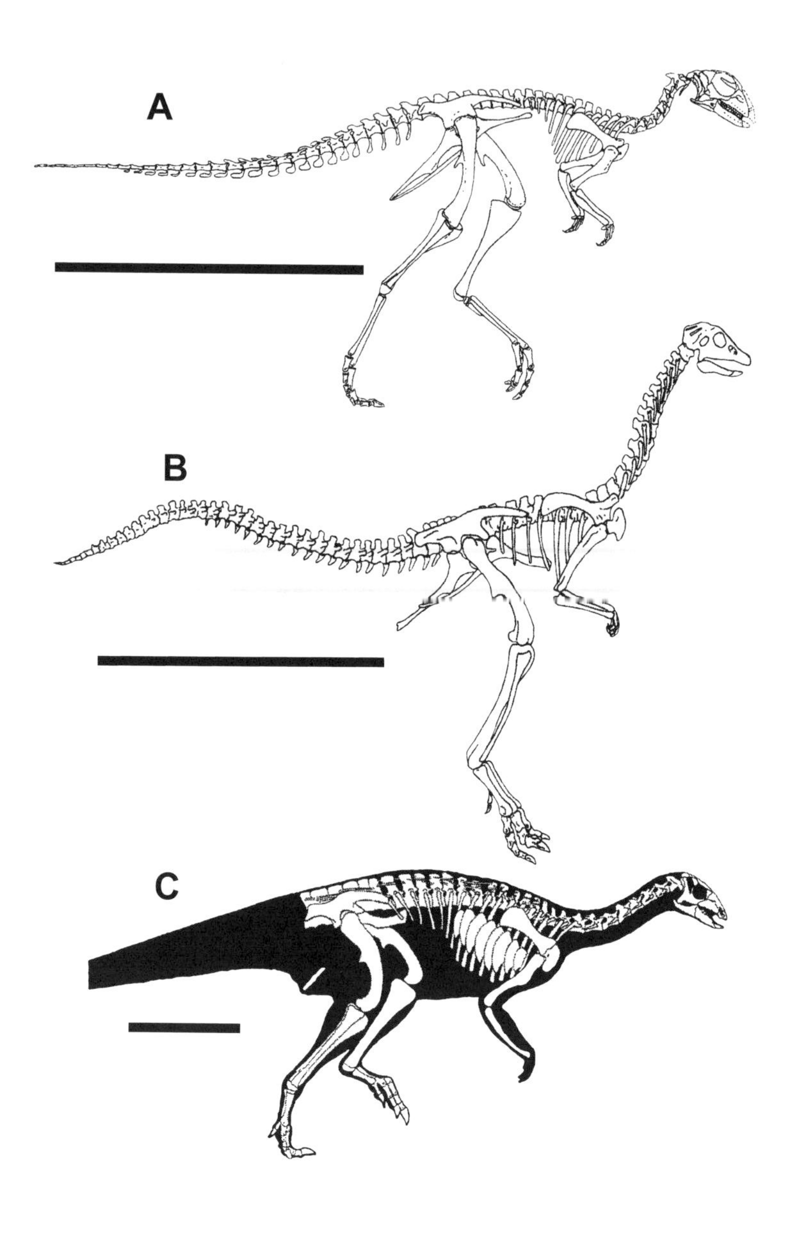

Figure 7.7. Skeletal restorations of basal iguanodotian dinosaurs from the Cretaceous of Patagonia. (A) Gasparinisaura cincosaltensis *(modified from Coria and Salgado 1996). Scale bar = 25 cm. (B)* Anabisetia saldiviai. *Scale bar = 50 cm. (C)* Talenkauen santacrucensis *(modified from Novas et al. 2004). Scale bar = 50 cm.*

dal centra, proximal extremity of left femur, distal extremity of right femur, proximal region of right tibia, distal extremity of left tibia, distal end of right tibia articulated to distal end of fibula, right astragalus and calcaneum, proximal end of right metatarsals articulated to distal tarsals 1 and 2, and appendicular fragments that cannot be determined; MUCPv-215, two dorsal vertebrae, fragmentary sacral vertebra, four caudal centra, distal extremity of radius, distal end of right femur, proximal end of right tibia, distal end of right tibia, distal extremity of phalanx, and fragmentary appendicular bones; MUCPv-216, proximal region of a right tibia; MUCPv-217, proximal region of a left tibia; MUCPv-218, fragmentary vertebral centra as well as fragments that cannot be determined, proximal region of metatarsal, distal end of tibia, phalanges; MUCP-v 226, distal region of right femur and proximal end of both tibiae; MUCPv-227, two caudal vertebrae sheathed in ossified tendons; MUCPv-225, distal end of a left femur; MCS-1, nearly complete tail and part of the pelvis; MCS-2, right tibia and fibula with complete pes and part of left pes; MCS-3, a complete sacrum two caudals, both ilia, partial right ischium, partial left pubis, partial femora and right tibia, right astragalus and fibula, and proximal portion of left metatarsus.

Locality and age. At 2 km southeast of Cinco Saltos, northwestern Río Negro Province, Argentina. Anacleto Formation, Campanian, Río Colorado Subgroup, Neuquén Group (Dingus et al. 2000).

Diagnosis. Anterior process of the jugal wedged between maxilla and lacrimal; anteroposteriorly wide ascending process of jugal; ascending process of jugal posteriorly contacted by the ventral process of postorbital; infratemporal fenestra narrowed ventrally into a ventroposteriorly directed slot bordered by quadratojugal; apex of arched dorsal margin of infratemporal fenestra positioned posterior to mandibular articulation; first and second sacral ribs not fused to each other; subtriangular, anteroposteriorly expanded chevrons in midcaudal region; long pubic peduncle of ilium; fully fused greater and anterior trochanters of femur; absence of anterior intercondylar groove in femur; condyloid of femur laterally placed; tibia with inner and outer malleoli approximately level ventrally; metatarsal II transversely compressed in anterior view (less than 15% of transverse width of three metatarsals) and anteroposteriorly developed.

Comments. Gasparinisaura was the first unquestionable basal ornithopod with iguanodontian affinities to be recognized for South America (Coria and Salgado 1996b) (Fig.7.7A). The first analysis run for this form indicated *Gasparinisaura* as a sister group of Dryomorpha (Coria and Salgado 1996b; Salgado et al. 1997). Recently, *Gasparinisaura* has been regarded instead as more closely related with more basal representatives among ornithopods, forming a clade together with *Thescelosaurus* and *Parksosaurus* (Norman et al. 2004). Indeed, several features present in *Gasparinisaura* are also recognized in European forms (i.e., *Hypsilophodon*), such as the absence of an extensor groove in distal femur and presence of a rhomboidal coracoid. Significantly, the clade formed by *Gasparinisaura* plus *Thescelosaurus* and *Parksosaurus* is linked by the presence of a reduced quadratojugal that does not participate in the mandibular articulation, and a sigmoid dorsal edge to the ilium, two remarkable features present in all iguanodontians. On the other hand, the presence of *Gasparinisaura* in a phyletic arrangement constituted only of Laurasian Cretaceous ornithopods is at least intriguing. Further analysis of the additional information gathered from new *Gasparinisaura* specimens will likely help to clarify its phylogenetic relationships among within Euornithopoda.

Anabisetia Coria and Calvo 2002
Anabisetia saldiviai Coria and Calvo 2002
Fig. 7.7B

Holotype. MCF-PVPH-74, partial skeleton comprising portion of left maxilla with one tooth, both dentaries, ventral portion of braincase, four cervical vertebrae, four dorsal neural arches, four sacral centra, two complete caudal vertebrae, 10 caudal centra, complete left scapulocoracoid, radius, ulna, and manus, nearly complete right scapula and coracoid, proximal end of right humerus; nearly complete left hind limb, nearly complete left pes with metatarsal V but lacking metatarsal I, proximal right metatarsal III and right metatarsal V.

Hypodigm. MCF-PVPH-75, two cervical neural arches, one cervical centrum, two dorsal neural arches, two dorsal centra, one sacral centrum, two caudal neural arches, six caudal centra, both scapulae, distal right humerus, proximal right pubis, almost complete right ischium, portion of left ischium, complete left femur, right tibia, fibula, and proximal tarsals, complete right pes, complete left metatarsals II and III; MCF-PVPH-76, three sacral centra, six caudal centra, four fragments of caudal neural arches, complete right scapula, complete left pelvis, fragment of right ilium, both femora, right tibia and fibula, proximal metatarsal II, one phalanx.

Referred specimens. MCF-PVPH-77, most of an articulated tail, portion of left scapula, humerus and tarsus.

Locality and age. Cerro Bayo Mesa, 30 km south of Plaza Huincul, Neuquén Province, Argentina. Cerro Lisandro Formation (late Cenomanian–early Turonian), Río Limay Subgroup, Neuquén Group (Leanza et al. 2004).

Diagnosis. Cranioventrally oriented occipital condyle; flattened fifth metacarpal; scapula with strong acromial process; ilium with preacetabular process longer than 50% of the total ilium length; when articulated, the preacetabular process reaches the same anterior extent as the prepubic process; ischial shaft triangular proximally and quadrangular distally in cross section; tarsus with fibular-astragalar contact.

Comments. Anabisetia was regarded as a basal iguanodontian, closely related to *Gasparinisaura,* although with unsolved internal relationships (Coria and Calvo 2002) (Fig. 7.7B). The skull material is very limited, which prevents obtaining better-supported results. The postcranial information for this species is abundant, and the preservation of the specimens is excellent. *Anabisetia* is nested among iguanodontians more derived than *Tenontosaurus* by having a primary lateral ridge on the maxillary teeth, a broad brevis shelf on the ilium, and a reduced metatarsal I. Also, the presence of a transversely flattened prepubic process of the pubis and an anteroventrally oriented ischial foot links *Anabisetia* closer to *Dryosaurus* than to *Gasparinisaura.*

Talenkauen Novas, Cambiaso, and Ambrosio 2004
Talenkauen santacrucensis Novas, Cambiaso, and Ambrosio 2004

Fig. 7.7C

Holotype. MPM-100001, partially articulated specimen preserving rostrum, jaws, and teeth, precaudal vertebral column and ribs, pectoral and pelvic girdles, and fore- and hind limb bones.

Locality and age. Cerro Los Hornos, southern shore of Viedma Lake, Santa Cruz Province, Argentina. Pari Aike Formation, Maastrichtian (Novas et al. 2004).

Diagnosis. Well-developed epipophysis on cervical 3 and platelike uncinate processes on the ribcage. The following reversals emerge from the cladistic analysis: lacrimal and premaxilla not in contact, and dentaries convergent rostrally.

Comments. Talenkauen is a remarkable specimen composed by a fairly complete skeleton that was articulated when discovered (Fig. 7.7C). Without question, the most unusual aspects of this specimen are the platelike structures on the ribcage that are reminiscent of uncinate processes. The homologies of these processes are not yet understood. The presence of similar structures has been recorded in a *Thescelosaurus* specimen from North America (Fischer et al. 2000). *Talenkauen* has been regarded as part of an endemic assemblage of basal iguanodontians, together with *Gasparinisaura* and *Anabisetia* (Novas et al. 2004). However, so far the search for clear anatomical features unifying a South American basal euornithopod taxon entity has been unsuccessful.

Hadrosauridae Cope 1869
Hadrosaurinae Lambe 1918
Secernosaurus Brett-Surman 1979
Secernosaurus koerneri Brett-Surman 1979

Fig. 7.8A

Holotype. FMNH-P-13423, a partial braincase, caudal vertebrae, scapula, two ilia, prepubic process, and fibula (Fig 7.8A).

Locality and age. Rio Chico, Chubut Province, Argentina. ?Laguna Palacios Formation (Maastrichtian), San Jorge Basin (Brett-Surman 1979; Bonaparte 1996).

Diagnosis. Postacetabular process of ilium greatly deflected dorsomedially and elongate, unlike any iguanodontian; preacetabular process of ilium deflected ventrally at an angle greater than in *Hadrosaurus;* antitrochanter relatively smaller than in any hadrosaurid of the same size.

Comments. Besides the fragmentary condition of the specimen of *Secernosaurus,* several unknowns regarding the geographical provenance conspire against inferring the biogeographic significance of this form. In his original description, Brett-Surman (1979) pointed to the San Jorge Formation as the fossil-bearing unit, mentioning also Rio Chico as the locality of provenance. In more recent contri-

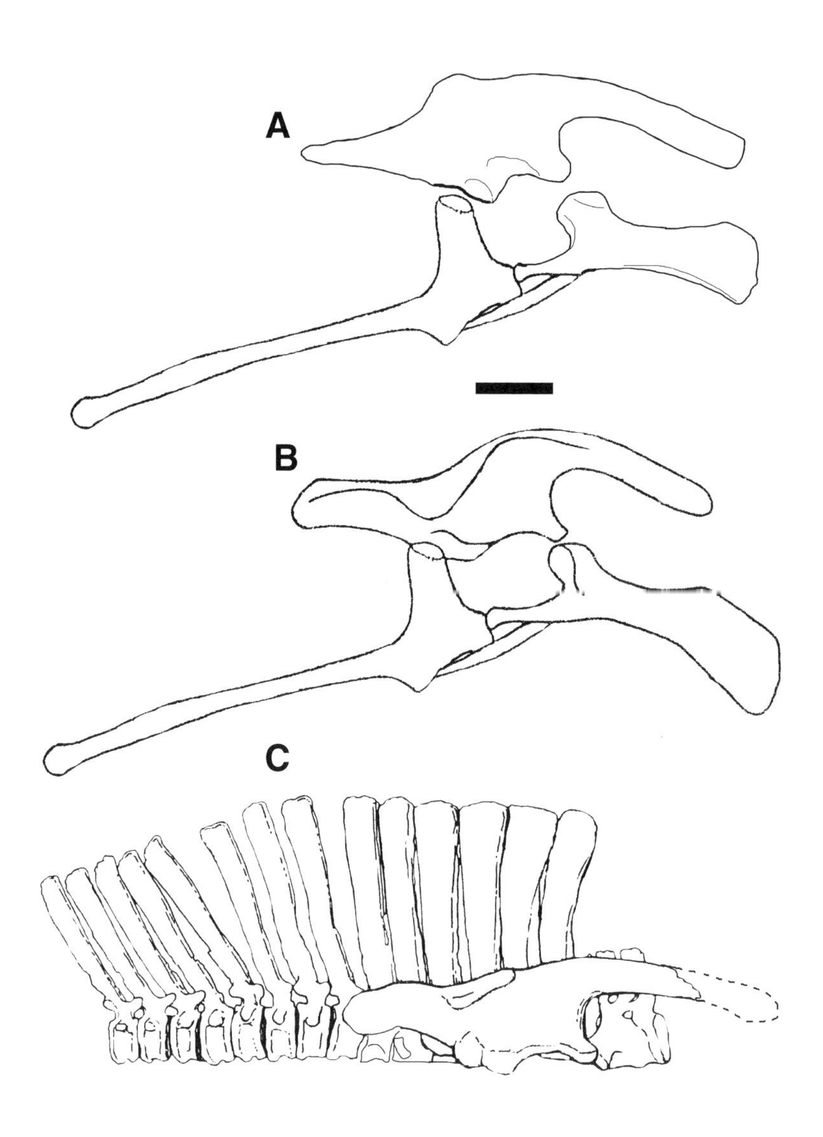

Figure 7.8. Right lateral views of partial pelves of Patagonian hadrosaurs. (A) Secernosaurus koerneri *(FMNH-P-13423). (B)* Kritosaurus australis *(MACN-RN-2). (C) Lambeosaurinae indet (modified from Powell 1987). Scale bar = 10 cm.*

butions, the geological bed of this hadrosaurid was switched to Bajo Barreal Formation (Weishampel and Horner 1990) or, tentatively, to the Laguna Palacios Formation (Bonaparte 1996; Novas 1997). Regardless of the formation, apparently there is agreement that the fossil was found in Late Cretaceous beds. The recent announcement of hadrosaurid remains from the upper levels of Bajo Barreal Formation (Luna et al. 2003) confirms that *Secernosaurus* could also come from a comparable level. The fossils from Bajo Barreal come from deposits near the trangressional beds of the marine Salamanca Formation, currently equivalent to the San Jorge Formation mentioned by Brett-Surman (1979). Therefore, the presence of hadrosaurid dinosaurs in the southern Chubut Province could be supported by more than one specimen, and also they could be considered as the South American highest latitudinal reference. Some primitive features shown by *Secernosaurus,* such as a depressed postacetabular process of the ilium, suggest the vicariant origin an alternative hypothesis for at least some South American hadrosaurids.

Kritosaurus Brown 1910
Kritosaurus australis Bonaparte, Franchi, Powell, and Sepúlveda 1984
Fig. 7.8B

Holotype. MACN-RN-2, an incomplete skeleton with posterior region of the skull; a postorbital; some isolated teeth; cervical, dorsal, sacral, and caudal vertebrae; several ribs; an incomplete scapula; a sternum; both ilia; pubes and ischia; and an incomplete femur (Bonaparte and Rougier 1987).

Hypodigm. MACN-RN-142, incomplete skull and jaw, several vertebrae and scapula; MACN-RN 143, most of the braincase and sphenoid region, some vertebrae and ribs.

Referred specimens. MACN-RN-143, most of the braincase and sphenoid region, incomplete lower jaws, several vertebrae and ribs; MACN-RN-144, an incomplete skull with the frontal region and its ventral projection rather well preserved, and incomplete temporal and occipital areas; MACN-RN 145, fragment of predentary, 11 dorsal and 6 caudal vertebrae, ribs, radius, ulna, distal portion of the femur, two metatarsi; MACN-RN 146, several vertebrae and ribs, a scapula, an almost complete ilium and a portion of the other, fragment of ischium (Fig. 7.8B), and a few more individuals represented by fragmentary material.

Locality and age. Estancia Los Alamitos, southeastern Río Negro Province, Argentina. Middle sector of Los Alamitos Formation (Campanian–early Maastrichtian) (Bonaparte et al. 1984).

Diagnosis. The characteristics observable in the skull, teeth, and both girdles coincide with the different species of *Kritosaurus*, except for the more prominent denticles of the predentary and a slight difference in the dorsal border of the ilium, which is convex in the anterior and concave in its posterior half, whereas in *Kritosaurus* species the joining of both sectors, convex and concave, occurs, in general, more posteriorly.

Comments. Kritosaurus australis was erected by Bonaparte et al. (1984) on the basis of on the similarities between the Patagonian form and the North American *Kritosaurus* (Brown 1910) that was lately synonimized with *Hadrosaurus* (Baird and Horner 1977). Nevertheless, the validity of *Kritosaurus australis* has remained unquestioned, basically because the South American hadrosaurids still await proper study. Some distinguishing features present in the many specimens of *Kritosaurus australis* show unique characters that encourage predicting the recognition of an endemic taxon on the basis of phylogenetic analysis.

Lambeosaurinae Parks 1923
Gen. et sp. indeterminate (Powell 1987)

Material. MPCA, uncataloged specimen, articulated remains of an individual including four incomplete dorsal, eight sacral, and 25 caudal vertebrae, several hemapophyses and rib fragments, fragments of ossified ligaments in original position, incomplete left scapula, both femora, tibiae and fibulae, both calcanea and distal tarsals, and left astragalus and pes (Fig. 7.8C).

Locality and age. Southeast Salitral Moreno, 40 km south General Roca, Río Negro Province, Argentina. Allen Formation (early Maastrichtian), Malargüe Group (Ballent 1980).

Comments. This specimen has a sacrum composed of eight ventrally keeled vertebrae, high neural spines in the caudals, laterally projected antitrochanter process, ventrally inclined distally postacetabular process, and ischium without distal expansion. Those features that indicate lambeosaurine affinities include the height of the caudal neural spines and the ventral keellike ridge of ventral sacrals. However, other features, such as the absence of an ischial boot, as well as the presence of other lambeosaurine apomorphies that apparently are also present in the more primitive *Ouranosaurus* (high neural spines) and *Camptosaurus* (footed ischium), may alter such lambeosaurine assignment (Coria 1999).

Discussion

As mentioned above, Patagonian ornithischian dinosaurs are far less abundant and diverse than the saurischians. Although the most diversified clade of extinct dinosaur worldwide, only two clades of Ornithischia are unquestionably represented so far from Patagonian localities: Thyreophora and Ornithopoda.

The ceratopsian record, consisting solely of *Notoceratops bonarelli,* is indeed controversial. This form has been regarded by several twentieth-century authors as a ceratopsid, along with other bones assigned to ceratopsians, or to ceratopsian-related thyreophoran dinosaurs (Huene 1929). The single bone known from *Notoceratops* is a partial left dentary, now lost from the collection of La Plata Museum. Nonetheless, the specimen was profusely illustrated by Huene (1929). From these drawings, it is not possible to identify any ceratopsid or even ceratopsian features. The tooth row does not seem to extend beyond the coronoid process, and the teeth do not show double roots (a convergent feature for hadrosaurs according to Horner et al. [2004]). In fact, the fossil could correspond instead to a more basal iguanodontian dinosaur given the apparently single tooth row and the anteroposteriorly straight dentary with parallel dorsal and ventral margins. There are also some biogeographic concerns leading one to conclude this. So far, Ceratopsidae has been exclusively recorded from the Cretaceous of Laurasia, and from a very restricted time span. Up to now, no ceratopsid dinosaurs have been recognized anywhere but in North America. The Patagonian record of these forms is nonexistent. Although they could have been part of the North American–South American faunal interchange that occurred in the Late Cretaceous, together with hadrosaurids and ankylosaurs (Casamiquela 1964; Bonaparte and Kielan-Jawarowska 1987; Salgado and Coria

1996; Coria 1999), the evident absence of ceratopsians from the South American fossil record—even though this line of thought is based on negative evidence—must be taken into consideration. Therefore, we think that the presence of ceratopsians in South America, and particularly in Patagonia, has yet to be supported by fossil evidence.

Ornithopoda is a clade of ornithischian dinosaurs that has a worldwide distribution. Their remains are present on every continent, from the Late Triassic up to the end of the Cretaceous. The South America record, and especially that from Patagonia, involves discoveries that match with that time span. The heterodontosaurid from the El Tranquilo Formation (Báez and Marsicano 2001), likely related to *Pisanosaurus* from the Ischigualasto Formation, indicates that this group occupied a wide latitudinal range.

Although no Jurassic ornithopods have yet been recorded for Patagonia, the rich fossil-bearing beds of the Cañadon Asfalto Formation (Middle Jurassic, Rauhut et al. 2001) also present a great potential to also yield ornithopod remains.

Basal iguanodontian ornithopods (sensu Norman 2004), or to some extent, more basal ornithopods, have been found often over the last 10 years, and our knowledge of their evolutionary history is beginning to be understood (Coria et al. in press).

The phylogenetic relationships of some Patagonian euornithopods have become a controversial issue. Several contributions regard *Gasparinisaura* and *Anabisetia* (Coria and Salgado 1996b; Salgado et al. 1997; Coria 1999; Novas et al. 2004) as having a phylogenetic position within Iguanodontia. On the other hand, more recent phylogenetic analyses depict *Gasparinisaura* and *Anabisetia* as having a more basal position. According to Norman et al. (2004), both *Anabisetia* and *Gasparinisaura* are noniguanodontian euornithopods. These authors explain this switch of the phyletic relationships of the Patagonian ornithopods in the paraphyly of Hypsilophodontidae, which has reorganized the basal part of the Ornithopoda clade (Norman et al. 2004). However, the cladogram presented lacks the more derived iguanodontians as terminal taxa. We think that although the new arrangement of *Hypsilophodon*-related forms has had an important role in the results of the analysis, the inclusion of more derived ornithopods like *Iguanodon, Camptosaurus,* and even hadrosaurids would produce significant changes in the trees obtained. Finally, despite the cladograms currently known, it is to be expected that both *Gasparinisaura* and *Anabisetia* (but also including *Talenkauen* and *Notohypsilophodon*) will show closer relationships than suspected. Nevertheless, the scarcity of well-preserved skeletons prevents fine-tuning their phylogenetic relationships with the iguandontian faunas from other continents. Close relationships with the African forms like *Dryosaurus lettowvorbecki* (Galton 1981) are predictable, although yet to be confirmed by character analysis.

The presence of hadrosaurids in South America has been known since the mid-twentieth century (Casamiquela 1964; Salgado this volume) and confirmed by further discoveries in different Patagonian localities (Bonaparte et al. 1984; Powell 1987; Luna et al. 2003). So far, the phylogenetic relationships of the many forms within Hadrosauridae are yet to be resolved. Only a single species has been recognized, *Kritosaurus australis* (Bonaparte et al. 1984), apparently on the basis of many similarities with the North American genus *Kritosaurus,* although recently this has been challenged (Horner et al. 2004). The denticulated predentary and differences in the ilium suggest probable derived features unique to this Patagonian species. Another possible local form is a well-preserved, articulated partial skeleton of a lambeosaurine hadrosaurid. This specimen was found in beds that are correlated with the *Kritosaurus australis* level. The long neural spines on presacral and caudal vertebrae are very different from those of *Kritosaurus australis,* which makes it clear that they represent different forms. The third hadrosaurid species recorded from Patagonia is *Secernosaurus koerneri,* only known from an ilium and the proximal part of a left pubis. The specimen is too fragmentary to form any solid phylogenetic hypothesis. However, certain features, such as the low, dorsoventrally depressed postacetabular blade, make it difficult to link this specimen with other hadrosaurids. Such a feature has been regarded as primitive among hadrosaurids.

Acknowledgments. We thank Drs. Peter Galton and Dave Weishampel for reviewing the manuscript and for their insightful comments. This work was supported by the "Museo Carmen Funes," Plaza Huincul, Argentina.

References Cited

Ameghino, F. 1899. Nota preliminar sobre el *Loncosaurus argentinus,* un representante de la familia Megalosauridae de la República Argentina. *Anales de la Sociedad Científica Argentina* 49: 61–62.

Báez, A. M., and C. A. Marsicano. 2001. A heterodontosaurid ornithischian dinosaur from the Upper Triassic of Patagonia. *Ameghiniana* 38 (3): 271–279.

Baird, D., and J. R. Horner. 1977. A fresh look at the dinosaurs of New Jersey and Delaware. *Bulletin of New Jersey Academy of Sciences* 22: 50.

Ballent, S. C. 1980. Ostrácodos de ambiente salobre de la Formación Allen (Cretácico Superior) en la Provincia de Río Negro (República Argentina). *Ameghiniana* 17: 67–82.

Bonaparte, J. F. 1976. *Pisanosaurus mertii* Casamiquela and the origin of the Ornithischia. *Journal of Vertebrate Paleontology* 50: 808–820.

———. 1996. Cretaceous tetrapods of Argentina. *Münchner Geowissenschaftliche Abhandlungen* 30: 73–130.

Bonaparte, J. F., and G. Rougier. 1987. The Late Cretaceous fauna of Los Alamitos, Patagonia, Argentina. Part VII. The hadrosaurs. *Revista del Museo Argentino de Ciencias Naturales "Bernardino Rivadavia," Paleontología* 3: 155–161.

Bonaparte, J. F., M. R. Franchi, J. E. Powell, and E. C. Sepúlveda. 1984. La Formación Los Alamitos (Campaniano-Maastrichtiano) del sudeste de Río Negro, con descripción de *Kritosaurus australis* nov. sp.

(Hadrosauridae). Significación paleobiogeográfica de los vertebrados. *Revista de la Asociación Geológica Argentina* 39: 284–299.

Bonaparte, J. F., and Z. Kielan-Jawarowska. 1987. Late Cretaceous dinosaurs and mammals faunas of Laurasia and Gondwana. In P. J. Currie and E. H. Koster (eds.), *Fourth Symposium on Mesozoic Terrestrial Ecosystems,* Short Papers, 24–29. Drumheller, Alberta, Canada: Royal Tyrrell Museum of Palaeontology.

Brett-Surman, M. K. 1979. Phylogeny and palaeobiogeography of hadrosaurian dinosaurs. *Nature* 277: 560–562.

Brown, B. 1910. The Cretaceous Ojo Alamo beds of New Mexico with description of the new dinosaur genus *Kritosaurus. Bulletin of the American Museum of Natural. History* 28: 267–274.

Casamiquela, R. M. 1964. Sobre un dinosaurio hadrosaurio de la Argentina. *Ameghiniana* 3: 285–308.

———. 1967. Un nuevo dinosaurio ornithisquio triásico (*Pisanosaurus mertii;* Ornithopoda) de la Formación Ischigualasto, Argentina. *Ameghiniana* 5: 47–64.

Coria, R. A. 1999. Ornithopod dinosaurs from the Neuquén Group, Patagonia, Argentina: phylogeny and biostratigraphy. In: Y. Tomida, T. Rich, and P. Vickers-Rich (eds.), *Proceedings of the Second Gondwanan Dinosaur Symposium,* 47–60. Tokyo: National Science Museum Monographs.

Coria, R. A., and J. O. Calvo. 2002. A new iguanodontian ornithopod from Neuquén Basin, Patagonia, Argentina. *Journal of Vertebrate Paleontology* 22: 503–509.

Coria, R. A., and L. Salgado. 1996a. "*Loncosaurus argentinus*" Ameghino, 1899 (Ornithischia, Ornithopoda): a revised description with comments on its phylogenetic relationships. *Ameghiniana* 33 (4): 373–376.

———. 1996b. A basal iguanodontian (Ornithischia: Ornithopoda) from the Late Cretaceous of South America. *Journal of Vertebrate Paleontology* 16: 445–457.

———. 2001. South American ankylosaurs. In K. Carpenter (ed.), *The Armored Dinosaurs,* 159–168. Bloomington: Indiana University Press.

Coria, R. A., A. V. Cambiaso, and L. Salgado. In press. New records of basal ornithopod dinosaurs in the Cretaceous of North Patagonia. *Ameghiniana.*

Dingus, L., J. Clarke, G. R. Scott, C. C. Swisher, L. M. Chiappe, and R. A. Coria. 2000. Stratigraphy and magnetostratigraphic/faunal constraints for the age of sauropod embryo-bearing rocks in the Neuquén Group (Late Cretaceous, Neuquén Province, Argentina). *American Museum Novitates* 3290: 1–11.

Fischer, P. E., D. Russell, M. Stoskopf, R. Barrick, M. Hammer, and A. Kusmitz. 2000. Cardiovascular evidence for an intermediate of higher metabolic rate in an ornithischian dinosaur. *Science* 288: 503–505.

Galton, P. M. 1974. The ornithischian dinosaur *Hypsilophodon foxii* from the Wealden of the Isle of Wight. *Bulletin of the British Museum (Natural History), Geology* 25: 1–152.

———. 1981. *Dryosaurus,* a hypsilophodontid dinosaur from the Upper Jurassic of North America and Africa: postcranial skeleton. *Paläontologische Zeitschrift* 55: 271–312.

Galton, P. M., and P. Upchurch. 2004. Stegosauria. In D. Weishampel, P. Dodson, and H. Osmólska (eds.), *The Dinosauria,* 2nd ed., 343–362. Berkeley: California University Press.

Goin, F. J., D. G. Poiré, M. S. de la Fuente, A. L. Cione, F. E. Novas, E. S. Bellosi, A. Ambrosio, O. Ferrer, N. D. Canessa, A. Carlini, J. Ferigolo, A. M. Ribeiro, M. S. Sales Viana, M. A. Reguero, M. G. Vucetich, S. Marenssi, M. F. de Lima Filho, and S. Agostinho. 2002. Paleontología y geología de los sedimentos del Cretácico Superior aflorantes al sur del río Shehuen (Mata Amarilla, Provincia de Santa Cruz, Argentina). *Actas del XV Congreso Geológico Argentino,* El Calafate, 1: 603–608.

Horner, J. R., D. B. Weishampel, and C. A. Forster. 2004. In D. Weishampel, P. Dodson, and H. Osmólska (eds.), *The Dinosauria,* 2nd ed., 438–463. Berkeley: California University Press.

Huene, F. von. 1929. Los Saurisquios y Ornitisquios del Cretácico Argentino. *Anales del Museo de La Plata* (2) 3: 147–154.

Leanza, H. A., S. Apesteguía, F. E. Novas, and M. S. de la Fuente. 2004. Cretaceous terrestrial beds from the Neuquén Basin (Argentina) and their tetrapod assemblages. *Cretaceous Research* 25: 61–87.

Luna, M., G. Casal, R. D. Martínez, M. Lamanna, L. Ibiricu, and E. Ivany. 2003. La presencia de un Ornithopoda (Dinosauria: Ornithishia) en el Miembro Superior de la Formación Bajo Barreal (Campaniano-Maastrichtiano?) del Sur del Chubut. *Ameghiniana* 40: 61R.

Martínez, R. D. 1998. *Notohypsilophodon comodorensis* gen. et. sp. nov., un Hypsilophodontidae (Ornithischia: Ornithopoda) del Cretácico Superior de Chubut, Patagonia central, Argentina. *Acta Geológica Leopoldensia* 21: 119–135.

Molnar, R. E. 1980. Australian late Mesozoic terrestrial tetrapods: some implications. *Mémoirs du Societe Geologique du France* 139: 131–143.

Norman, D. B. 2004. Basal Iguanodontia. In D. Weishampel, P. Dodson, and H. Osmólska (eds.), *The Dinosauria,* 2nd ed., 413–437. Berkeley: California University Press.

Norman, D. B., H.-D. Sues, L. M. Witmer, and R. A. Coria. 2004. Basal Ornithopoda. In D. Weishampel, P. Dodson, and H. Osmólska (eds.), *The Dinosauria,* 2nd ed., 393–392. Berkeley: California University Press.

Novas, F. E. 1997. South American dinosaurs. In K. Padian and P. Currie (eds.), *Encyclopedia of Dinosaur,* 678–689. San Diego: Academic Press.

Novas, F. E., A. V. Cambiaso, and A. Ambrosio. 2004. A new basal iguanodontian (Dinosauria, Ornithischia) from the Upper Cretaceous of Patagonia. *Ameghiniana* 41 (1): 75–82.

Powell, J. E. 1987. Hallazgo de un dinosaurio hadrosáurido (Ornithischia, Ornithopoda) en la Formación Allen (Cretácico Superior) de Salitral Moreno, Provincia de Río Negro, Argentina. *Actas del X Congreso Geológico Argentino* 3: 149–152. Tucumán. Argentina.

———. 2003. Revision of South American titanosaurid dinosaurs: palaeobiological, palaeobiogeographical and phylogenetic aspects. *Records of the Queen Victoria Museum* 111: 1–173.

Rauhut, O. W. M., A. López Albarello, P. Puerta, and T. Martin. 2001. Jurassic vertebrates from Patagonia. *Journal of Vertebrate Paleontology* 21: 91A.

Salgado, L., and R. A. Coria. 1996. First evidence of an armoured ornithischian dinosaur in the late Cretaceous of North Patagonia, Argentina. *Ameghiniana* 33: 367–371.

Salgado, L., R. A. Coria, and S. E. Heredia. 1997. New materials of *Gas-*

parinisaura cincosaltensis (Ornithischia, Ornithopoda) from the Upper Cretaceous of Argentina. *Journal of Paleontology* 71: 933–940.
Sereno, P. C. 1991. *Lesothosaurus*, "Fabrosaurids" and the early evolution of Ornithischia. *Journal of Vertebrate Paleontology* 11: 168–197.
Tapia, A. 1918. Una mandíbula de dinosaurio procedente de Patagonia. *Physis* 4: 369–370.
Weishampel, D. B., and J. Horner. 1990. Hadrosauridae. In D. B. Weishampel, P. Dodson, and H. Osmólska (eds.), *The Dinosauria*, 534–561. Berkeley: University of California Press.

8. Sauropodomorpha

Leonardo Salgado and
José F. Bonaparte

Introduction

Sauropodomorphs were small bipedal to large quadrupedal herbivorous dinosaurs. The name Sauropodomorpha was introduced by Huene (1932) to designate the group that includes prosauropods and sauropods, which are now thought to be sister taxa (Sereno 1998). The oldest members of the clade are recorded in the Upper Triassic (Flynn et al. 1999).

The record of Patagonian sauropodomorphs is abundant, especially that of sauropods. Soldiers, settlers, and farmers excavated the first reported Patagonian sauropodomorph bones during the late nineteenth century (Coria and Salgado 2000; Salgado this volume). To date, there are at least 24 valid species of Patagonian sauropods. In contrast, prosauropods are still poorly known, and their record is significantly less abundant than that of sauropods. Only one species of Patagonian prosauropod (*Mussaurus patagonicus*) has been identified. Casamiquela (1964) first described bones now referred to this species.

In this chapter, we summarize current knowledge of Patagonian sauropodomorphs, supplying basic data on well-established species, including stratigraphic provenance, diagnosis, and phylogenetic relationships, and briefly consider some aspects of Patagonian sauropodomorph evolution.

Institutional abbreviations. IANIGLA-PV, Instituto Argentino de Nivología, Glaciología y Ciencias Ambientales (CONICET), Paleontología de Vertebrados, Mendoza, Argentina; MACN, Museo Argentino de Ciencias Naturales "Bernardino Rivadavia" (CONICET), Buenos Aires, Argentina; MCF-PVPH, Museo "Carmen

Funes," Paleontología de Vertebrados, Plaza Huincul, Neuquén, Argentina; MCS, Museo de Cinco Saltos, Río Negro, Argentina; MJG, Museo "Jorge Gerhold," Ingeniero Jacobacci, Río Negro, Argentina; MLP, Museo de La Plata, Universidad Nacional de La Plata, Buenos Aires, Argentina; MPCA, Museo Provincial "Carlos Ameghino," Cipolletti, Río Negro, Argentina; MPEF-Pv, Museo Paleontológico "Egidio Feruglio," Paleontología de Vertebrados, Trelew, Chubut, Argentina; MRS-Pv, Museo de Rincón de Los Sauces, Paleontología de Vertebrados, Neuquén, Argentina; MUCPv, Museo de la Universidad Nacional del Comahue, Paleontología de Vertebrados, Neuquén, Argentina; PVL, Instituto "Miguel Lillo," Universidad Nacional de Tucumán, Paleontología de Vertebrados, Tucumán, Argentina; Pv-MOZ, Museo "Profesor Dr. Juan Olsacher," Dirección Provincial de Minería, Paleontología de Vertebrados, Zapala, Neuquén; UNPSJB-PV, Universidad Nacional de la Patagonia "San Juan Bosco," Paleontología de Vertebrados, Comodoro Rivadavia, Chubut, Argentina.

Systematic Palaeontology

Sauropodomorpha Huene 1932

Prosauropoda Huene 1920

Plateosauridae Marsh 1895

***Mussaurus* Bonaparte and Vince 1979**

***Mussaurus patagonicus* Bonaparte and Vince 1979**

Holotype. PVL 4068, a juvenile partial skeleton.

Referred specimens. Apart from the holotype, Bonaparte and Vince (1979) referred six juvenile specimens (all of which were found all together, presumably within a single nest) to this species: PVL 4208, skull and mandibles, cervical and dorsal vertebrae, scapula, forelimbs, pelvis, hind limbs; PVL 4209, partial skull and mandibles, vertebrae and other disarticulated postcranial elements; PVL 4210, skull and mandibles, cervical and dorsal vertebrae, scapular girdle and articulated forelimbs, pelvis and articulated hind limbs; PVL 4211, incomplete skull and mandibles and most of the postcranial skeleton; PVL 4212, partially articulated postcranial skeleton; PVL 4213, postcranial elements; PVL 4214 and PVL 4215, two incomplete eggs.

Many adult skeletons from the same locality and horizon partially described by Casamiquela (1964, 1980) may also be referable to *Mussaurus patagonicus* (Galton 1990; Galton and Upchurch 2004; Pol and Powell 2005): MLP 61-III-20-22, skeleton preserving the last four dorsal vertebrae, three sacral vertebrae, most caudal vertebrae, right ilium, both femora, left tibia and fíbula, portion of right tibia and fibula, almost complete left and right pedes, and many unidentified fragments; MLP 61-III-20-23, partial skeleton consisting of 14 dorsal vertebrae (three of them articulated), sacrum, three caudal vertebrae, most of the right manus, both ilia, fragments of both pubes, both ischia and incomplete femora, complete right tibia and fragments of the left tibia, rib fragments, and

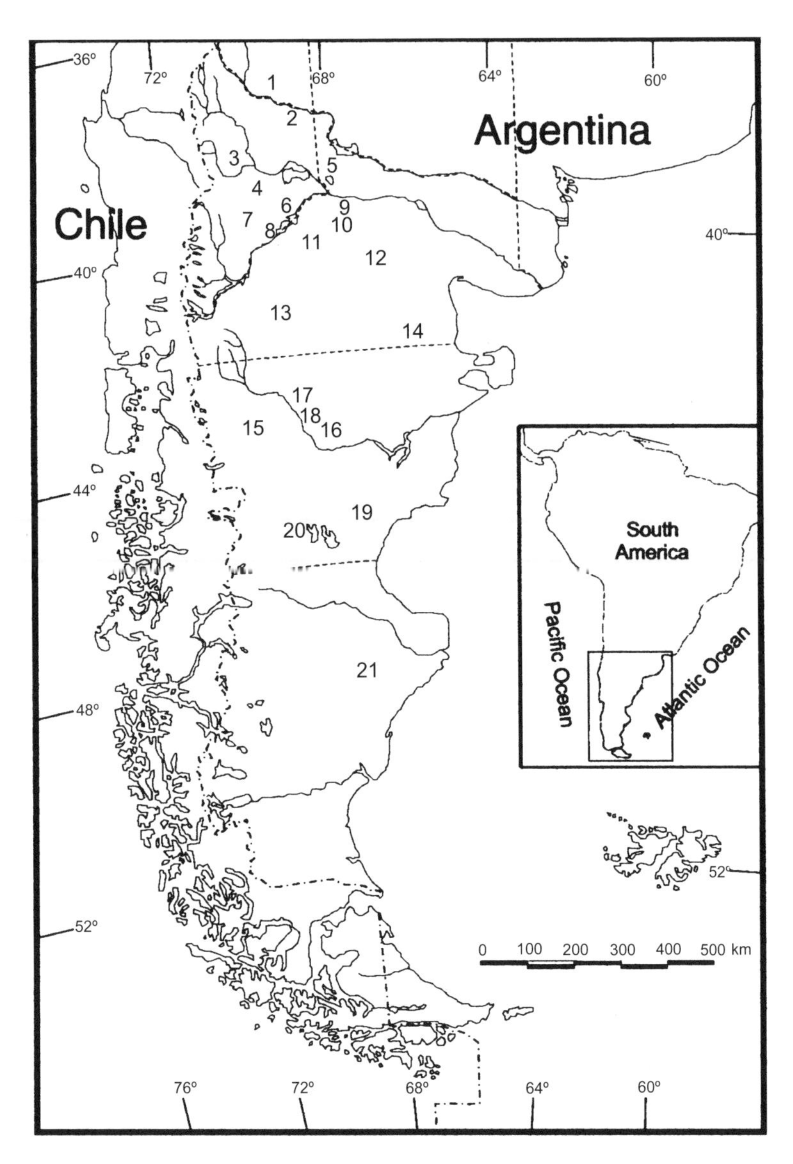

Figure 8.1. Principal Patagonian localities that have yielded sauropodomorph remains. 1, Cerro Guillermo; 2, Rincón de los Sauces; 3, Agrio del Medio; 4, Plaza Huincul; 5, Cinco Saltos–Lago Pellegrini; 6, Villa El Chocón; 7, La Amarga; 8, Cerro Leones; 9, Paso Córdoba; 10, Salitral Moreno; 11, La Bonita; 12, Salitral de Santa Rosa; 13, Casa de Piedra; 14, Los Alamitos; 15, Cañadón Puelman; 16, Paso de Indios; 17, Estancia Fernández; 18, Cerro Cóndor; 19, Pampa Pelada; 20, Estancia Ocho Hermanos; 21, Laguna Colorada.

many unidentified fragments; MLP 68-II-27-1, specimen represented by eight dorsal vertebrae, sacrum, 32 caudal vertebrae, scapula and part of the right coracoid, both humeri, ulnae, radii, manus, ilia, incomplete pubes, ischia, femora, tibiae, fibulae, and pedes; MLP 68-II-28-1, poorly preserved skull and mandibles, which may belong to MLP 68-II-27-1 (Casamiquela 1980).

Locality and age. Laguna La Colorada, Estancia Cañadón Largo, north-central Santa Cruz Province (Fig. 8.1, locality 21), Laguna Colorada Formation, Norian (Jalfin and Herbst 1995).

Diagnosis. Mussaurus patagonicus was diagnosed by Bonaparte and Vince (1979) on the basis of the following characters: short and high skull, with elongate frontals and parietals (1), very short beak (2), short and high anteorbital opening, placed near the orbit (3), long teeth (4), lower jaw with thick symphysis (5), short and high cervical vertebrae (6). In turn, Pol and Powell (2005) listed the following diagnostic characters: anterior margin of the premaxilla directed posterodorsally, forming an angle of 45° with the alveolar margin; thin ridge along the entire ventral ramus of the lacrimal; dorsally projected peglike process on the anterior tip of the dentary; high anterior end of mandibular symphysis; premaxillary teeth and anterior maxillary teeth with sigmoid posterior edge lacking serrations on their margins.

Comments. Because the prosauropod material from El Tranquilo has still not been studied as a whole, we cannot ensure that the holotype of *Mussaurus patagonicus* and material assigned to it, as well as the material described by Casamiquela (1964, originally assigned to *Plateosaurus* sp.), all belongs to a single species. In fact, many of the characters originally proposed as diagnostic of the species (e.g., characters 1–3) are most probably due to the juvenile condition of the specimens, upon which the original description was based (Bonaparte and Vince 1979). Moreover, the assignment of *Mussaurus* to the Plateosauridae made by Galton (1990) is also tentative and remains to be corroborated. In this regard, Pol and Powell (2005) note that *Mussaurus* is phylogenetically closer to "melanorosaurids" than to more basal forms like *Plateosaurus.*

Sauropoda Marsh 1878
Eusauropoda Upchurch 1995
***Amygdalodon* Cabrera 1947**
***Amygdalodon patagonicus* Cabrera 1947**

Lectotype. MLP 46-VIII-21-1/2, posterior dorsal vertebra (Rauhut 2003a).

Referred specimens. MLP 46-VIII-21-1/1, MLP 36-XI-10-3/1, posterior dorsal vertebrae, with an attached head of a dorsal rib (Rauhut 2003a).

Locality and age. Cañadón Puelman, Pampa de Agnia, Chubut Province (Fig. 8.1, locality 15). Cerro Carnerero Formation, Pampa de Agnia Basin, Toarcian to Bajocian (Riccardi and Damborenea 1993).

Diagnosis. Rauhut's (2003a) diagnosis of *A. patagonicus* includes lateral walls of the neural canal and centropostzygapophyseal laminae flared laterally posteriorly, and neural canal strongly flexed anteroposteriorly within the dorsal neural arches.

Comments. Amygdalodon is the oldest indisputable South American sauropod. Cabrera (1947) and Casamiquela (1963) regarded *Amygdalodon* as a member of the "Cetiosauridae" (sensu Lydekker 1888). Wilson (2002) placed this dinosaur within the clade *Patagosaurus* + (Omeisauridae + [*Jobaria* + Neosauropoda]) because it possesses cervical ribs positioned ventrolateral to the centrum (his character 139). Finally, Upchurch et al. (2004) regarded *Amygdalodon patagonicus* as Eusauropoda incertae sedis.

Volkheimeria Bonaparte 1979
Volkheimeria chubutensis Bonaparte 1979

Holotype. PVL 4077, incomplete cervical vertebra, two posterior dorsal vertebrae, an incomplete dorsal neural arch, two incomplete sacral vertebrae, incomplete ilia, pubis, ischium, left femur and tibia.

Locality and age. Cerro Cóndor, Chubut Province (Fig. 8.1, locality 18). Lower section of the Cañadón Asfalto Formation, Callovian (Bonaparte 1986; Page et al. 1999).

Diagnosis. Diagnostic characters of *Volkheimeria chubutensis* include posterior dorsal vertebrae with the neural arch lower than in *Patagosaurus;* neural spines laterally compressed, with a rectangular basal cross section, without the diverging laminae present in *Patagosaurus;* anteriorly twisted iliac blade; ischium with subcircular shaft; pubis more gracile than in *Patagosaurus* (Bonaparte 1986).

Comments. Although *V. chubutensis* has been included within the "Cetiosauridae" by Bonaparte (1979, 1986), according to this author, this species exhibits a series of primitive features that are absent in other "cetiosaurids" found at the same locality. In particular, the morphology of the dorsal neural spines of *Volkheimeria* is, according to Bonaparte (1999a), an evolutionary precursor of the "X cross-section stage" typical of more advanced "cetiosaurids."

Cetiosauridae Lydekker 1888 (sensu Upchurch et al. 2004)
Patagosaurus Bonaparte 1979
Patagosaurus fariasi Bonaparte 1979
Fig. 8.2

Holotype. PVL 4170, adult specimen preserving four anterior cervical vertebrae, three posterior cervical vertebrae, three anterior dorsal vertebrae (articulated to the last cervical vertebra), five mid- and posterior dorsal vertebrae, two dorsal centra, complete sacrum, six anterior caudal vertebrae, 12 mid- and posterior caudals, incomplete hemapophyses, proximal part of the right scapula associated with the right coracoid, proximal portion of the right humerus, right ilium, right pubis, two partially fused ischia, and right femur.

Hypodigm. PVL 4615; PVL 4616; PVL 4171; PVL 4172; MACN-CH935, four middorsal neural arches, three dorsal centra, four sacral vertebrae, one sacral centrum, two sacral neural arches, one sacral neural arch preserving the neural spine, one anterior caudal centrum, six mid- and posterior caudal vertebrae, two incomplete neural spines, three hemapophyses, one ilium fragment, right pubis and two ischia; MACN-CH932, two cervical centra, one anterior dorsal centrum, six dorsal neural arches, eight dorsal centra, one sacral neural arch, sacral ribs, right scapula, coracoid, humerus, radius and ulna, pubis, metatarsals, two pedal phalanges, and one ungual phalanx; MACN-CH 933, incomplete juvenile left mandible, two cervical centra, one anterior dorsal centrum, four

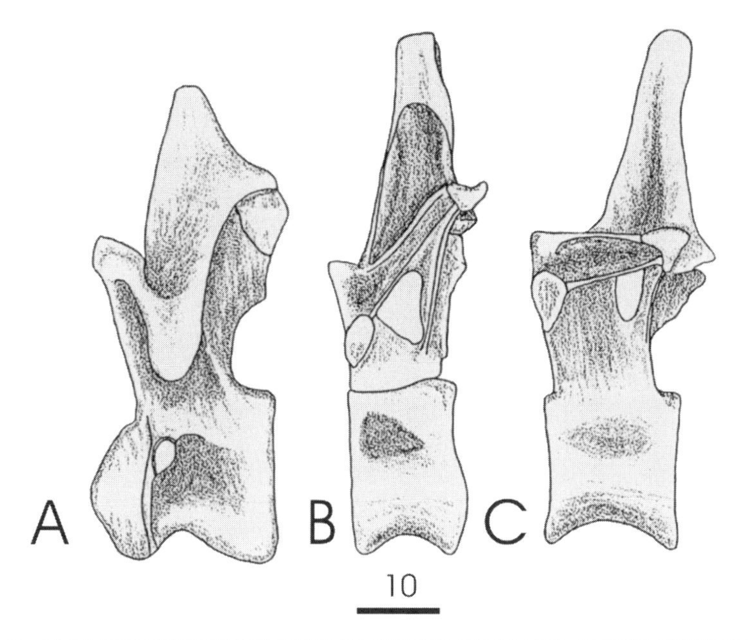

Figure 8.2. Patagosaurus fariasi *(PVL 4170). Anterior (A), midanterior (B), and posterior (C) dorsal vertebrae in left lateral view. Scale in centimeters. After Bonaparte (1999a). Below, reconstruction of the skeleton mounted in the Museo Argentino de Ciencias Naturales "Bernardino Rivadavia" of Buenos Aires, Argentina.*

dorsal neural arches, three dorsal centra, one sacral neural arch, one ilium fragment, one pubis, right femur and tibia.

Referred specimens. PVL 4076, left premaxilla, five midcervical vertebrae, two incomplete posterior cervical vertebrae, four dorsal centra, one anterior dorsal arch, one dorsal centrum preserving part of the neural arch, many dorsal ribs, 17 midanterior and midcaudal vertebrae, incomplete right pubis, right femur and tibia; PVL 4075, three dorsal centra, left scapula and humerus, one ischium fragment, left femur; PVL 4617, six incomplete caudal vertebrae, one cervical rib, one scapula, one coracoid, and one ungual phalanx.

Locality and age. North Cerro Cóndor, Chubut Province (Fig. 8.1, locality 18). Lower third of the Cañadón Asfalto Formation, Callovian (Bonaparte 1986; Page et al. 1999).

Diagnosis. The diagnostic characters of *Patagosaurus fariasi* proposed by Bonaparte (1986) are these: posterior dorsal vertebrae with high neural arches and well-developed neural spines formed by four diverging vertical laminae; cavities of the neural spines dorsoventrally oriented; sacrum with high neural spines and with an expanded neural canal, which forms a neural cavity; pubis with proximolateral expansions; sublaminar, transversely compressed ischium with a ventromedial crest and a distal expansion.

Wilson (2002, 271) provided the following autapomorphies of *Patagosaurus fariasi:* cervical vertebrae with elongate centroprezygapophyseal laminae and "hooded" infraprezygapophyseal coels, anterior dorsal vertebrae with elongate centropostzygapophyseal and postzygodiapophyseal laminae, middle and posterior dorsal neural arches with "infradiapophyseal" pneumatopore opening into the neural canal, transversely narrow third sacral vertebra, proximal humerus with median ridge on posterior aspect, humeral distal condyles exposed on anterior aspect of shaft, tibial cnemial crest projects anteriorly.

Comments. According to Rauhut (2002), the sauropod species from Cerro Cóndor (*Volkheimeria chubutensis* and *Patagosaurus fariasi*) are basal eusauropods. Wilson (2002) considers *Patagosaurus* as more derived than *Barapasaurus* (from the Middle Jurassic of India) because the former possesses presacral pneumatopores (his character 83), and five sacral vertebrae (his character 108), among other characters.

Marginal denticles are present in an in situ tooth in the dentary MACN-CH 933, described by Bonaparte (1986), and in a recently described dentary (MPEF-Pv 1670) referable to *Patagosaurus* (Rauhut 2003b). Denticles are absent on teeth in the maxillae of MACN-CH-934, but this specimen may belong to a new neosauropod species (Rauhut 2003b). The proximal end of the tibia of *Patagosaurus* is transversely compressed, as in other non-neosauropods (Wilson and Sereno 1998, 48) except *Jobaria* (character 203 of Wilson 2002). Moreover, the tibial cnemial crest of *Patagosaurus fariasi* projects anteriorly, as in basal sauropodomorphs. However, Wilson (2002) considered this condition as an autapomorphic reversal of this species. According to Upchurch et al. (2004), *Patagosaurus* is a member of the Cetiosauridae, a stem-based taxon defined as all sauropods more related to *Cetiosaurus* than to *Saltasaurus*.

Specimen MACN-CH 934, consisting of two maxillae, five dorsal neural arches, an ilium, pubis, ischium, and unidentified appendicular elements, was referred to *Patagosaurus fariasi* by Bonaparte (1986). Nevertheless, Rauhut (2003b) noted characters in this material that could justify its taxonomic differentiation from *P.*

fariasi. Among these is the complete absence of marginal denticles on the maxillary teeth. Furthermore, in the postcranium, MACN-CH 934 differs from *Patagosaurus* in the morphology of the neural canal, the inclination of the zygapophyses, the possession of a short spinodiapophyseal lamina, and the proportions of the ilium. The absence of marginal denticles suggests that this form pertains to Neosauropoda (Wilson 2002, character 72). If so, the specimen represents the oldest neosauropod found in Patagonia, and one of the most ancient recorded in the world.

Neosauropoda Bonaparte 1986
Macronaria Wilson and Sereno 1998
***Tehuelchesaurus* Rich, Vickers-Rich, Giménez, Cúneo, Puerta, and Vacca 1999**
***Tehuelchesaurus benitezi* Rich, Vickers-Rich, Giménez, Cúneo, Puerta, and Vacca 1999**

Holotype. MPEF-Pv 1125, 10 partial dorsal vertebrae, two partial sacral vertebrae, one caudal vertebra, dorsal rib fragments, right scapulocoracoid, left humerus, left radius, left ulna, left and right femora, partial ilium, left pubis, partial right pubis, left ischium, distal part of right ischium, skin impressions.

Locality and age. Estancia Fernández, Chubut Province (Fig. 8.1, locality 17). Cañadón Calcáreo Formation, Kimmerdigian-Tithonian (Rich et al. 1999; Rauhut et al. 2005).

Diagnosis. Rich et al. (1999, 63) presented the following diagnosis of *Tehuelchesaurus benitezi:*

> [Sauropod] [d]istinguished from *Omeisaurus* by (1) subequal measurements on the coracoid of (a) the distance from the scapular surface to the anteroventral border and (b) the distance from the glenoid surface perpendicular to it across to the opposite side of the bone rather than (b) being markedly greater than (a); (2) anteroposterior length of distal end of the humerus only slightly less than mediolateral width rather than anteroposteriorly compressed; (3) radius and ulna stouter bones; i.e., comparative widths greater relative to the lengths of these two bones; (4) anterior border of the distal shaft of the pubis deeply concave rather than modestly so; and (5) expansion in acetabular region of the ischium is noticeably deeper dorsoventrally. Distinguished from the cetiosaurs *Patagosaurus* and *Barapasaurus* by all rather than just the anterior dorsal vertebrae having pseudopleurocoels and being opisthocoelous rather than the posterior dorsals being amphiplatyan with only modest lateral depressions in the centra. Pseudopleurocoels (sensu Bonaparte 1986) are "deep lateral depressions . . . in the centra." Unlike true pleurocoels, however, there are no internal chambers within the centrum of the vertebra with the pseudopleurocoelous condition.

Comments. Rich et al. (1999) and Upchurch et al. (2004) related this species to *Omeisaurus tianfuensis,* a primitive sauropod from the Middle Jurassic of China. According to Rauhut (2002), *Tehuelchesaurus benitezi* is the sister taxon of the Titanosauriformes (defined by Wilson and Sereno [1998] as the common ancestor of *Brachiosaurus* and *Saltasaurus* and all of its descendants). Wilson (2002, table 13), in turn, regarded *Tehuelchesaurus* as a member of the clade *Patagosaurus* + more derived sauropods because it possesses presacral pleurocoels. Recently, Alifanov and Averianov (2003) placed *Tehuelchesaurus* among the Euhelopodidae (sensu Upchurch 1995, 1998), outside the Neosauropoda.

The opisthocoelous condition of the posterior dorsal vertebrae (Wilson 2002, character 105) of *Tehuelchesaurus benitezi* confirms its affiliation to the Macronaria, defined by Wilson and Sereno (1998) as neosauropods more closely related to *Saltasaurus* than to *Diplodocus,* and to the Camarasauromorpha (=*Camarasaurus* + Titanosauriformes of Wilson and Sereno [1998], defined by Salgado et al. [1997a] as Camarasauridae, Titanosauriformes, their common ancestor, and all its descendants), whereas the pronounced lateral bulge on the proximal one-third of the femur indicates affinity with the Titanosauriformes. *Tehuelchesaurus* is one of the oldest indisputable neosauropods from Patagonia and all of South America.

Titanosauriformes Salgado, Coria, and Calvo 1997
Titanosauria Bonaparte and Coria 1993
***Chubutisaurus* del Corro 1975**
***Chubutisaurus insignis* del Corro 1975**
Fig. 8.3

Holotype. MACN 18222; includes an incomplete dorsal vertebral centrum; an incomplete dorsal neural arch; a dorsal neural spine; fragmentary dorsal ribs; a caudosacral vertebra; 11 caudal centra; the left humerus, radius, and ulna; four metacarpals and the distal portion of another; the left femur and tibia; and several fragments of appendicular elements.

Locality and age. Estancia Paso de Indios, 43°30'S; 68°20'W, Chubut Province (Fig. 8.1, locality 16). Bayo Overo Member of the Cerro Barcino Formation, early Late Cretaceous (Rauhut et al. 2003).

Diagnosis. The only diagnosis of *Chubutisaurus insignis* is that proposed by del Corro (1975, 231, translated from the Spanish):

> big sauropod of massive construction, with concave-convex cervical vertebrae, with large pneumatic cavities on both halves separated by an osseous bar; big, flat, dorsal vertebrae, probably with big neural spines, with double pleurocoels, and of a cavernous structure; solid caudal vertebrae, with the proximal surface flat and the distal surface concave, with wide transverse apophyses and divided, short, and robust neural spines; big femur, of massive construction, with the fourth trochanter poorly developed; big

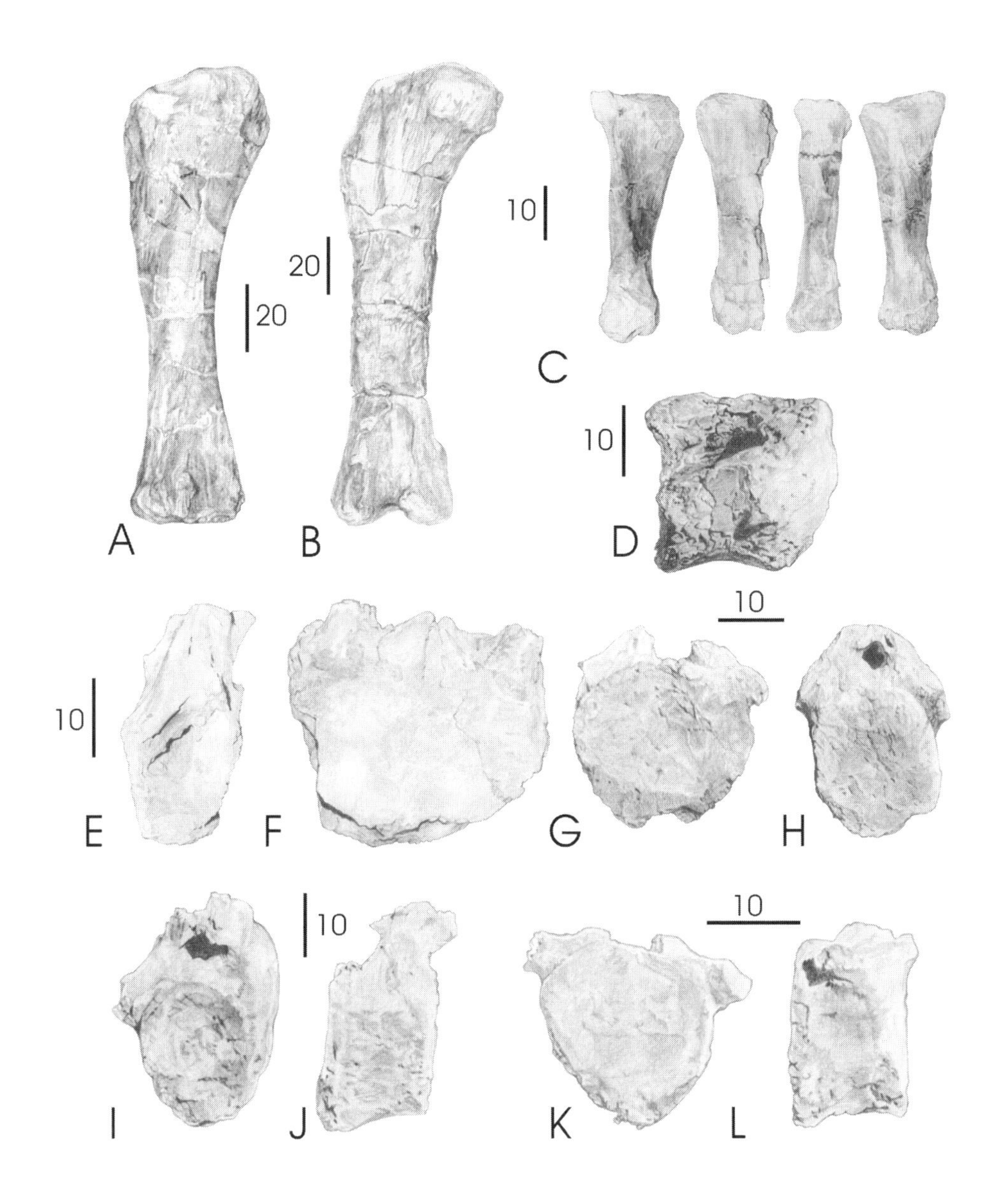

Figure 8.3. Chubutisaurus insignis *(MACN 18222). (A) Left humerus in posterior view. (B) Left femur in posterior view. (C) Metacarpals I to IV in posterior view. (D) Dorsal centrum in right lateral view. Caudosacral centrum in right lateral (E) and anterior (F) views. (G, H) Midcaudal centra in posterior view. Midposterior caudal centra in posterior (I) and right lateral (J) views. Posterior caudal centrum in posterior (K) and right lteral (L) views. Scale in centimeters.*

humerus with prominent deltopectoral crest, similar to that of *Bothriospondylus madagascariensis;* wide ribs, mainly those of the mid part of the body.

The reference to caudal vertebrae with "divided neural spines" is based on an error: these structures are not paired neural spines but rather prezygapophyses (see del Corro 1975, fig. 4). Because del Corro confused (anteriorly projecting) caudal prezygapophyses with (posteriorly inclined) neural spines, he misinterpreted the anterior articular face as the posterior, and vice versa. Actually, the caudal anterior articular surfaces are slightly concave, and the posterior surfaces are flattened.

The only characters observed in *Chubutisaurus insignis* that are probably autapomorphies are two of those listed by Salgado (1993): (1) presence of one or more caudosacral vertebrae having articular faces in different planes, and (2) the complete formation of a prespinal lamina on posterior dorsal vertebrae (convergent in diplodocoids and eutitanosaurians).

Comments. Del Corro (1975) placed *Chubutisaurus* into a new family created for this taxon: "Chubutisauridae." Conversely, Salgado (1993) and Salgado et al. (1997a) considered this genus as a titanosauriform, specifically as the sister taxon to Titanosauria. *Chubutisaurus* shares a transversely expanded distal end of the tibia with Titanosauria (Salgado et al. 1997a, character 7). Wilson (2002, table 13) considered *Chubutisaurus* a titanosaurian because of its procoelous anterior caudals. We also regard *Chubutisaurus* as a titanosaurian (defined as all somphospondylians closer to *Saltasaurus* than to *Euhelopus* [Sereno 1998]) because it possesses slightly procoelous anterior caudals and the aforementioned morphology of the tibial distal end. Within this clade, we cannot determine whether *Chubutisaurus* is placed within the Andesauroidea or the Titanosauroidea (sensu Salgado 2003).

?Titanosauria Bonaparte and Coria 1993
Agustinia Bonaparte 1999
Agustinia ligabuei Bonaparte 1999

Holotype. Specimen MCF-PVPH-110, represented by three incomplete dorsal, six incomplete sacral, and 10 incomplete caudal vertebrae, almost complete right tibia and fibula, five articulated left metatarsals, and nine dermal ossifications.

Locality and age. Northern face of Cerro Leones, 8 km west of Picún Leufú, Neuquén Province (Fig. 8.1, locality 8). Cullín Grande Member of the Lohan Cura Formation, Aptian-Albian (Bonaparte 1999b; Leanza and Hugo 2001; Leanza et al. 2004).

Diagnosis. Bonaparte (1999b, 2) diagnosed this species as follows: sauropod with the apex of the neural spines transversely expanded in the posterior dorsal, all sacral, and the three anteriormost caudal vertebrae; three types of osteoderms articulating with the apices of the neural spines: type a, unpaired leaf-shaped; type b, laminar, transversely wide lateral projections; and type c, elongate, flat, or cylindrical, dorsolaterally projected; internally directed bend bounding the cnemial crest of the tibia; fibula with pronounced posterior projection on its proximal region; and metatarsals of the type present in titanosaurians.

Comments. Bonaparte (1999b) asserted that *Agustinia* was more similar to members of Rebbachisauridae than to "Titanosauridae." However, he did not include this genus within the first clade but rather a new one: "Agustiniidae." Wilson (2002, table 13) and Upchurch et al. (2004, table 13.1) considered *Agustinia* as a titanosaurian because it possesses osteoderms. However, it is uncertain whether these osteoderms are homologous to those present in some titanosaurians. Indeed, as Bonaparte (1999b) noted, *Agustinia* most closely resembles diplodocoids in the morphology of the neural spines.

Andesauroidea Salgado 2003
***Andesaurus* Calvo and Bonaparte 1991**
***Andesaurus delgadoi* Calvo and Bonaparte 1991**
Fig. 8.4

Holotype. MUCPv-132, four articulated posterior dorsal vertebrae, 27 partially articulated caudal vertebrae, dorsal rib fragments, incomplete right humerus, left pubis and almost complete ischia, partial femur and two metacarpals.

Referred specimen. MUCPv-271, partial pelvis and some caudal vertebrae, and 11 rounded gastroliths (Calvo 1999, fig. 5).

Locality and age. At 5 km southwest of Villa El Chocón, Confluencia Department, Neuquén Province (Fig. 8.1, locality 6). Candeleros Formation (lower Cenomanian) (Leanza and Hugo 2001).

Diagnosis. Calvo and Bonaparte (1991, 304, translated from the Spanish) provided the following diagnosis of *Andesaurus delgadoi:*

> Large Titanosauridae, with high posterior dorsal vertebrae, formed by vertebral centra proportionally lower and longer than in *Argyrosaurus,* and by the neural arch proportionally higher than in remaining titanosaurids, with hyposphene-hypantrum articulations, well developed prespinal lamina forked on its lower part; amphiplatyan caudal vertebrae, without evidence of procoely, with broad laminar neural spines, and the neural arch placed more posteriorly than in other titanosaurs; long humerus, proportionally more gracile than in *Argyrosaurus;* pubis with broad proximolateral process and pubic foramen placed far from the lateral border; ischium with small iliac peduncle.

Salgado (2000) proposed mid- and posterior caudal vertebrae with anteroposteriorly wide and slightly expanded neural spines as a probable autapomorphy of *A. delgadoi.* In turn, Powell (2003, 85) proposed a well-developed, branching prespinal lamina on mid- and posterior dorsal vertebrae as a diagnostic character of this taxon (even though this character also occurs in *Epachthosaurus*).

Comments. Calvo and Bonaparte (1991), Salgado et al. (1997a), and González Riga (2003) considered *Andesaurus delgadoi* a titanosaurian on the basis of the presence of a series of characters such as eye-shaped pleurocoels in dorsal vertebrae, posterior dorsal vertebrae with ventrally widened, slightly forked "infradiapophyseal" (=posterior centrodiapophyseal [Wilson 1999]) laminae, presence of centroparapophyseal lamina in posterior dorsal vertebrae, anterior caudals procoelous, and pubis considerably longer than ischium. Wilson (2002) regarded *Andesaurus* as a titanosaurian because it exhibits a platelike ischial blade with no emargination distal to the

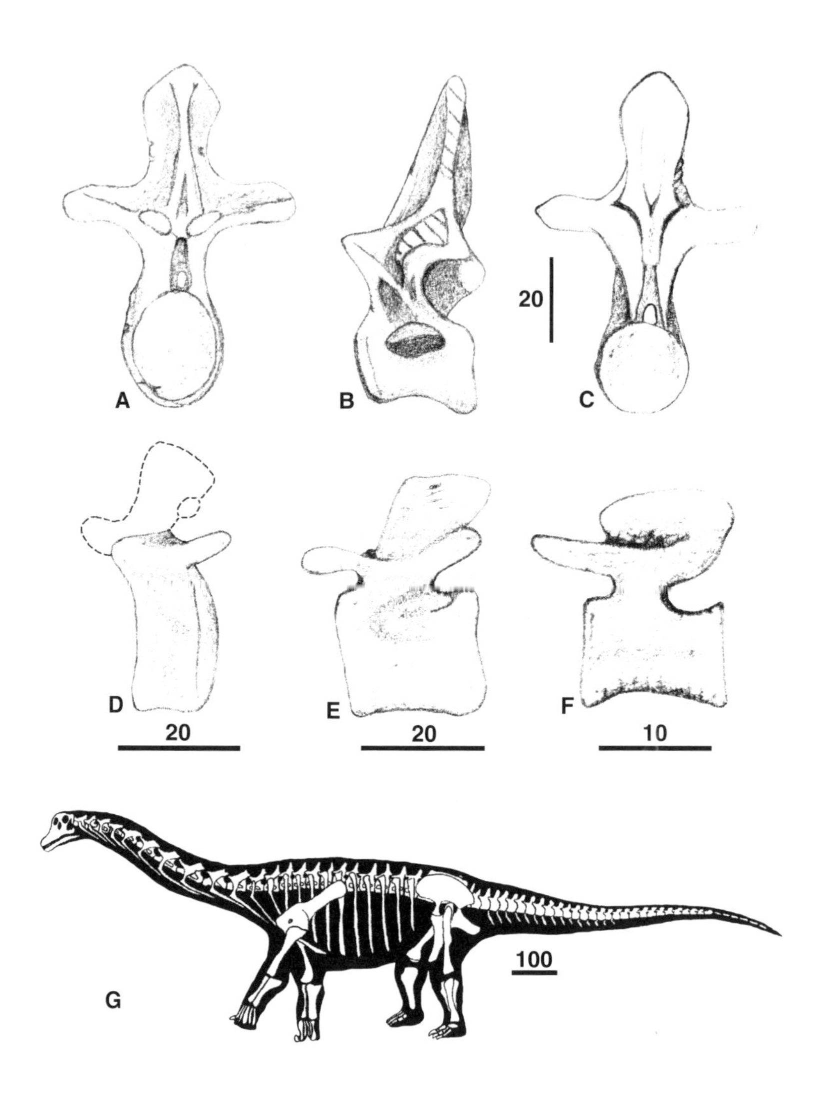

Figure 8.4. Andesaurus delgadoi *(MUCPv-132). Reconstructed posterior dorsal vertebra in anterior (A), left lateral (B), and posterior (C) views; anterior (D), mid- (E), and midposterior (F) caudal vertebrae in left lateral view. (G) Reconstruction of the skeleton. Modified from Wilson and Sereno (1998, foldout). Scale in centimeters.*

pubic peduncle (his character 193, table 13). Wilson (2002, character 192) considered the character "ischial blade shorter than pubic blade," which is evident in *Andesaurus,* as a synapomorphy of Nemegtosauridae + (*Isisaurus colberti* + Saltasauridae). The anterior caudals of *Andesaurus delgadoi* are slightly procoelous (Salgado et al. 1997a; contra Calvo and Bonaparte 1991). Although Wilson (2002, character 132) considered procoelous anterior caudal centra as a titanosaurian synapomorphy, he did not list this condition as present in *Andesaurus delgadoi* (contra Salgado et al. 1997a). Both Sanz et al. (1999) and Powell (2003) placed *Andesaurus* outside of Titanosauria, although their definition of this clade (the most recent common ancestor of *Epachthosaurus, Saltasaurus, Argyrosaurus,* and *Lirainosaurus* plus the Peirópolis titanosaurian and all its descendants) differs from previous definitions given by Salgado et al. (1997a) (the most recent common ancestor of *Andesaurus delgadoi* and Titanosauridae and all of its descendants) and Wilson and

Sereno (1998) (titanosauriforms more closely related to *Saltasaurus* than to *Euhelopus* or *Brachiosaurus*). Importantly, Sanz et al. (1999) regarded the strong bifurcation of the prespinal lamina in the posterior dorsals as an apomorphic condition independently developed in *Andesaurus* and *Epachthosaurus*. The prespinal lamina formed down to the base of the neural spine had been considered by Salgado et al. (1997a) to be a synapomorphy of their unnamed taxon II (here interpreted as equivalent to Eutitanosauria).

Lithostrotia Upchurch, Barret, and Dodson 2004
(=Titanosauridae sensu Salgado, Coria, and Calvo 1997)
***Mendozasaurus* González Riga 2003**
***Mendozasaurus neguyelap* González Riga 2003**

Holotype. IANIGLA-PV 065/1-24, 22 articulated caudal vertebrae and two anterior hemapophyses.

Paratypes. IANIGLA-PV 066, anterior dorsal vertebra; IANIGLA-PV 068, scapula; IANIGLA-PV 067, sternal plate; IANIGLA-PV 069, humerus; IANIGLA-PV 070/1-2, radius and ulna; IANIGLA-PV 071/1-4, four metacarpals; IANIGLA-PV 072, pubis fragment; IANIGLA-PV 073/1-2, femur and tibia; IANIGLA-PV 074/1-3, two tibiae and fibula; IANIGLA-PV 077/1-5, five metatarsals; IANIGLA-PV 078, 079, two ungual phalanges; and IANIGLA-PV 080/1-2, 081/1-2, four osteoderms.

The holotype and paratypes belong to three individuals (González Riga 2003).

Locality and age. South of Cerro Guillermo, Malargüe Department, Mendoza Province (Fig. 8.1, locality 1). Levels provisionally assigned to the upper section of the Río Neuquén Formation (or Río Neuquén Subgroup, according to Leanza and Hugo [2001]), Late Cretaceous, upper Turonian–upper Coniacian (González Riga 2003).

Diagnosis. Large titanosaurian (18–25 m long) characterized by the following autapomorphies: (1) two subtriangular infrapostzygapophyseal fossae in anterior dorsal vertebrae; (2) postzygapostspinal laminae (running from the postzygapophyses up to the base of the postspinal lamina) parallel to the plane of postzygapophyseal facets in anterior dorsal vertebrae; (3) interzygapophyseal cavity dorsoventrally extended and limited by the spinopostzygapophyseal and spinoprezygapophyseal laminae in anterior caudal vertebrae; (4) middle caudal vertebrae slightly procoelous with reduced posterior condyles displaced dorsally; (5) laminar midposterior caudal neural spines with horizontal and straight dorsal border, and anterodorsal corner forming right angle; (6) large subconical to spherical osteoderms lacking cingulum (González Riga 2003).

Comments. Mendozasaurus was originally included within the Titanosauridae (González Riga 2003). According to González Riga's (2003) analysis, which includes *Malawisaurus* but not *Epachthosaurus,* we can only state that *Mendozasaurus* pertains to Lithostrotia (Upchurch et al. 2004), the least inclusive clade containing *Malawisaurus dixeyi* and *Saltasaurus loricatus*. According to González Riga (2003), the sister-group relationship between *Mendozasaurus* and *Malawisaurus* is supported by three synapomorphies: presence of laminar and anteroposteriorly elongate neural spine located on the middle of the centrum in middle caudal vertebrae; relatively long prezygapophyses in middle caudal vertebrae; and semilunar sternal plate with straight posterior border.

Titanosauroidea Upchurch 1995
Epachthosaurinae Salgado 2003
***Epachthosaurus* Powell 1990**
***Epachthosaurus sciuttoi* Powell 1990**

Holotype. MACN-CH 1317, an incomplete posterior dorsal vertebra (Powell 1990, 2003).

Paratype. MACN-CH 18689, cast of six incomplete posterior dorsal vertebrae articulated to the sacrum, and a fragment of the iliac pubic peduncle.

Referred specimens. UNPSJB-PV 920, articulated skeleton lacking the skull, cervical, and anterior dorsal vertebrae (Martínez et al. 2004). Also, Salgado and Coria (2005) tentatively assigned MUCPv-Pv 244, "Titanosauridae" indet. (Calvo 1999, 24, figs. 8–10), from Villa El Chocón (Neuquén Province, Candeleros Formation) to *Epachthosaurus*, although actually it could correspond to a new genus (Simón personal communication 2005).

Locality and age. Estancia Ocho Hermanos, southern Chubut Province (Fig. 8.1, locality 20). Lower Member of the Bajo Barreal Formation, early Late Cretaceous (Cenomanian-Turonian; Martínez et al. 2004).

Diagnosis. Powell (2003, 54) provided this diagnosis for *Epachthosaurus sciuttoi:* "centra of the dorsal vertebrae relatively large, very broad and depressed, pleurocoels very large and deep as in the other titanosaurids, neural spines of the posterior dorsals anteroposteriorly compressed, with prespinal laminae thickened and well developed, forked at the lower part, forming the spinal-prezygapophyseal lamina, accessory articulations are present but distinct from the hyposphene-hypantrum structures, the diapophysis without a flat surface at the dorsal end."

Salgado (2000) proposed the presence of an "interprezygapophyseal shelf" in the dorsal vertebrae as a probable autapomorphy of *Epachthosaurus*. Powell (2003, 85) included the following features that characterize *Epachthosaurus:* U-shaped condylar grooves in the posterior articular surfaces of distal caudal vertebrae; well-developed branching of the prespinal laminae on mid- and posterior dorsal vertebrae (also in *Andesaurus*); presence of a depression ventral to the prezygapophysis in anterior caudal vertebrae (also in *Lirainosaurus*).

A revised diagnosis of *Epachthosaurus sciuttoi* has been recently published by Martínez et al. (2004, 108–109): "middle and caudal dorsal vertebrae possessing accessory articular processes extending ventrolaterally from the hyposphene, a strongly developed intraprezygapophyseal lamina, and aliform processes projecting laterally from the dorsal portion of the spinodiapophyseal lamina; hyposphene-hypantrum articulations in caudals 1–14; and a pedal phalangeal formula of 2-2-3-2-0."

Comments. Salgado et al. (1997a), taking into account all of the material assigned to *E. sciuttoi,* placed this species within their "unnamed taxon II," here interpreted as equivalent to Eutitanosauria (sensu Sanz et al. 1999), as the sister group of other "titanosaurids." Sanz et al. (1999) considered *Epachthosaurus* a titanosaurian on the basis of its possession of the following characters: cancellous osseous tissue in the axial skeleton (their character 16.1); horizontal surface at the end of diapophyses on posterior dorsal vertebrae (their character 5); developed posterior condyles in anterior caudal vertebrae (their character 18); anteroventral ridge in sternal plates (their character 30); and short, laminar ischiadic process of the ischium (their character 39). Elsewhere (p. 252), they stated that the strong ventral bifurcation of the prespinal lamina is an autapomorphy of *Epachthosaurus* (acquired independently by *Andesaurus*).

Wilson (2002), considering only the holotype, placed *Epachthosaurus sciuttoi* within the Titanosauria on the basis of the presence of posterior dorsal neural arches lacking hyposphene-hypantrum articulations (his character 106). However, the holotype does not preserve the part of the neural arch where these articulations, if present, would be evident. As such, Wilson probably inferred the absence of hyposphene-hypantrum articulations from the presence of an "interprezygapophyseal shelf," visible on the holotype, which would prevent the articulation of a true hyposphene. Martínez et al. (2004), in describing the nearly complete skeleton UNPSJB-PV 920, considered *Epachthosaurus* a basal titanosaurian.

According to the phylogenetic analysis of Wilson (2002), *Epachthosaurus* is a member of the clade *Isisaurus colberti* + Saltasauridae, because its iliac blade is oriented perpendicular to its body axis (Wilson 2002, character 187). If the taxonomy proposed by Salgado (2003) is applied to the cladogram presented by Salgado et al. (1997a), *Epachthosaurus* is a basal titanosauroid.

Eutitanosauria Sanz, Powell, Le Loeuff, Martínez, and Pereda-Suberbiola 1999

Argentinosaurus Bonaparte and Coria 1993

Argentinosaurus huinculensis Bonaparte and Coria 1993

Figs. 8.5A, B, 8.6

Holotype. MCF-PVPH-1, three anterior and three posterior dorsal vertebrae, partial sacrum, the proximal end of a dorsal rib, and a left fibula (or an eroded tibia, according to Bonaparte and Coria [1993], fig. 8).

Locality and age. Plaza Huincul, Neuquén Province (Fig. 8.1, locality 4). Huincul Formation, upper Cenomanian (Leanza and Hugo 2001).

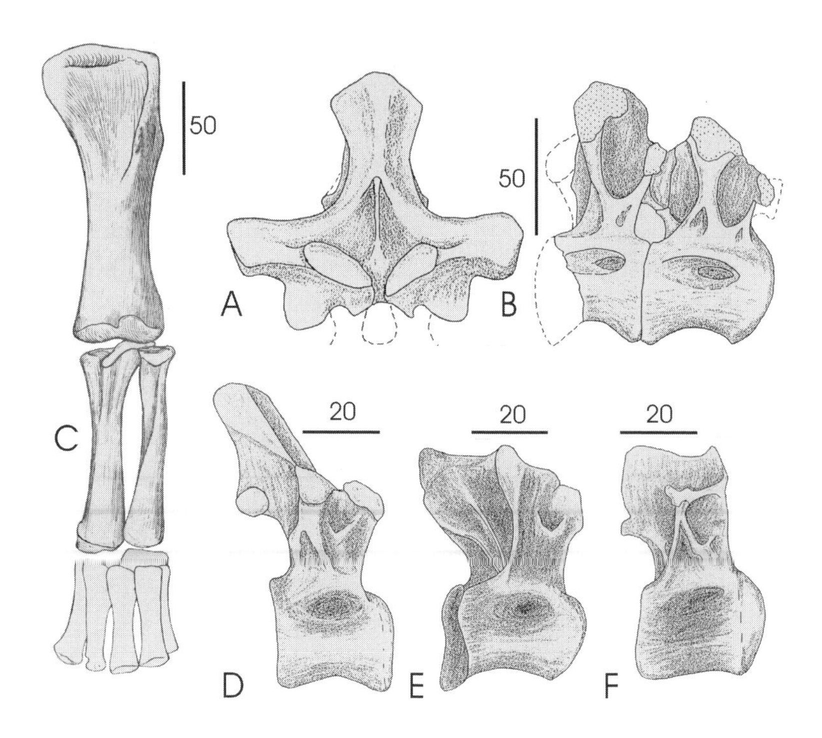

Figure 8.5. Argentinosaurus huinculensis *(MCF-PVPH-1). (A) Anterior dorsal neural arch in anterior view. (B) Posterior dorsal vertebrae in right lateral view.* Argyrosaurus superbus *(MLP 77-V-29-1). (C) Left forelimb in anterior view. Cf.* Argyrosaurus *(PVL 4628), (D) posterior, (E) midanterior, and (F) anterior dorsal vertebrae in right lateral view. Scale in centimeters. After Bonaparte (1999a).*

Diagnosis. Bonaparte and Coria (1993) diagnosed *Argentinosaurus huinculensis* as follows: opisthocoelous dorsal vertebrae with well-developed hyposphene-hypantrum complex and associated accessory articulations (1); anterior dorsals with transversely broad and anteroposteriorly flat neural spines with robust prespinal laminae (2); mid- and posterior dorsals with low and wide centra with flat ventral surfaces and pleurocoels placed on their anterior halves (3); sacral centra two to five very reduced (4); cylindrical and hollow dorsal ribs of tubular structure (5); osseous macrocells in presacral and sacral vertebrae (6); slender tibia with short cnemial crest (7).

Comments. Salgado and Martínez (1993) questioned whether the accessory articular complex observed by Bonaparte and Coria (1993) actually corresponds to the hyposphene-hypantrum. Salgado and Martínez argued that these structures were instead derived from the centropostygapophyseal laminae. Other diagnostic characters listed by Bonaparte and Coria (1993) are doubtful; for instance, it is not certain whether the ribs are actually hollow, or whether they were secondarily hollowed by erosion (character 5) (Salgado 2000). Other characters (1–4, 6) are found in other titanosaurians. Nevertheless, the presence of an unusual hypertrophied hyposphene in the mid- and posterior dorsals probably represents an autapomorphy of this species.

Bonaparte and Coria (1993), on the basis of symplesiomorphies, united *Argentinosaurus huinculensis* and *Andesaurus delgadoi* within the family Andesauridae, as part of the suprafamiliar group Titanosauria. In contrast, Salgado and Martínez (1993) and

Figure 8.6. Argentinosaurus huinculensis. *Reconstruction of the skeleton mounted in the Museo "Carmen Funes" of Plaza Huincul, Neuquén, Argentina.* Photo courtesy of R. Coria.

Salgado et al. (1997a) considered *Argentinosaurus* a titanosaurid more derived than *Epachthosaurus*, because its posterior dorsals exhibit a prespinal lamina formed to the base of the neural spine and lack typical hyposphene-hypantrum articulations.

Nevertheless, Wilson (2002) simply considered *A. huinculensis* as a macronarian on the basis of the presence of opisthocoelous posterior dorsal centra. However, as mentioned above, *Argentinosaurus* lacks typical hyposphene-hypantrum articulations in its posterior dorsal neural arches. Wilson (2002) proposed this condition as a titanosaurian synapomorphy. If Salgado's (2003) taxonomic scheme is applied to the cladogram of Salgado et al. (1997a), *Argentinosaurus* is a representative of Eutitanosauria. According to Salgado et al. (1997a), another character that supports the inclusion of *Argentinosaurus* within the Titanosauria is its possession of an accessory posterior centrodiapophyseal lamina. This condition was formerly stated as "posterior trunk vertebrae with ventrally widened, slightly forked infradiapophyseal laminae" (Salgado et al. 1997a, character 21).

Argyrosaurus Lydekker 1893
Argyrosaurus superbus Lydekker 1893
Fig. 8.5C–F

Holotype. MLP 77-V-29-1, left forelimb, including the humerus, radius, ulna, and five metacarpals (Fig. 8.5C).

Referred specimen. PVL 4628, three dorsal and three caudal vertebrae, left scapula, humerus, radius, left and right ulnae, right pubis, incomplete right femur, incomplete left tibia and various rib fragments (Fig. 8.5D–F).

Locality and age. Left bank of the Río Chico, near Pampa Pelada, northeast of Lago Colhué Huapi, Chubut Province (Fig. 8.1, locality 19). Bajo Barreal Formation of the Chubut Group (Upper Cretaceous) (Powell 2003). New evidence indicates that the holotype of *Argyrosaurus superbus* may be Campanian or Maastrichtian in age, whereas the referred specimen PVL 4628 may be Cenomanian-Turonian (Casal personal communication 2004).

Diagnosis. Powell's (2003, 50) diagnosis of *Argyrosaurus superbus* is as follows: "A huge-sized titanosaurid; stout humerus with broad proximal end that has a straight upper edge (margin) perpendicular to the long axis of the bone; pectoral muscle insertion area far more prominent and projected forward and medially; ulna with extremely robust proximal end showing prominent edges delimiting markedly concave facets; stout metacarpals approximately one-third the length of the humerus."

On the basis of the specimen PVL 4628, Powell (2003, 85) also included the character "slightly developed opisthocoely in posterior dorsal vertebrae" as diagnostic of *Argyrosaurus.*

Comments. It is unclear whether all of the material assigned to *Argyrosaurus superbus* pertains to the same species. Doubts increase if, as is probable, the holotype and referred specimen come from different stratigraphic levels. In view of this, our phylogenetic considerations on *Argyrosaurus superbus* are provisional, pending resolution on the specific assignment of the holotype and referred specimens.

Sanz et al. (1999, 252) considered the monospecific genus *Argyrosaurus* a titanosaurian on the basis of the presence of developed posterior condyles in anterior caudal vertebrae (their character 18) and cancellous osseous tissue in the axial skeleton (their character 16), thus confirming the original assignation of Huene (1929). Sanz et al. (1999) further interpreted *Argyrosaurus* as a member of the Eutitanosauria because of its lack of hyposphene-hypantrum articulations in its posterior dorsal vertebrae (their character 6) and reduced femoral fourth trochanter (their character 41).

Wilson (2002, table 3) regarded *Argyrosaurus* as a member of the clade Nemegtosauridae + (*Isisaurus colberti* + Saltasauridae) because of the presence of procoelous middle and posterior caudal vertebrae (his character 134). Powell (2003) considered this genus a member of his "clade 7" (together with *Isisaurus colberti* and *Ampelosaurus*) because it displays incipient prespinal and postspinal laminae in the anterior caudal vertebrae.

Antarctosaurus Huene 1929
Antarctosaurus wichmannianus Huene 1929

Holotype. MACN 6904, incomplete cranium and mandible; rib fragments; left scapula; incomplete right humerus; two fragments of the distal radius; a fragment of the proximal end of an ulna; seven metacarpals (two of these incomplete); an ilium fragment; a pubis fragment; an almost complete right ischium and part of the left; left femur, tibia, fibula, and metatarsus; and some pedal phalanges.

Locality and age. Paso Córdoba, General Roca, Río Negro Province (Fig. 8.1, locality 9). According to A. Garrido (personal communication 2003), the unit that yielded the holotypic material of *Antarctosaurus wichmannianus* corresponds to the Allen Formation (Campanian-Maastrichtian) instead of the Anacleto Formation (lower Campanian), as initially believed (Powell 1986, 2003).

Comments. In surveying the material collected by Wichmann from the quarry, Huene (1929) assigned a series of caudal vertebrae (currently lost) and a procoelous caudal vertebra to *Laplatasaurus araukanicus* (Huene 1929, 70). He also doubted the taxonomic assignment of an ischium and excluded it from the holotypic specimen of *Antarctosaurus wichmannianus* (Huene 1929, 72). Conversely, he did not doubt the association of the other elements, including all the cranial material, which he considered to belong to a single individual.

Because of the lack of quarry maps or sketches, Powell (1986, 2003) claimed that it was impossible to determine whether or not all of the postcranial material from the type locality of *A. wichmannianus* belongs to one individual. Nevertheless, the recovered cranial bones likely pertain to a single skull, in view of the lack of repeated elements and their general size correspondence. As such, the consideration of mandibular characters separately from those derived from other parts of the skeleton, especially from other parts of the skull, is not justified.

Diagnosis. Powell's (2003, 45) diagnosis of *Antarctosaurus wichmannianus* is as follows: High posterior portion of the skull; large orbits; wide muzzle; short, broad frontal and parietal; upper temporal fenestra reduced but partially open dorsally; wide, short, and triangular basipresphenoid complex; foramen for nerves IX–XI separated from the fenestra ovalis; basipterygoid processes long, bar-shaped, and very divergent; basioccipital tuberosities separated; paraoccipital processes long and recurved as in *Saltasaurus;* straight and broad symphyseal region perpendicular to the mandibular rami; scapula with narrow blade and wide supraglenoid region; axis of the scapular blade almost perpendicular to the long axis of the supraglenoid portion; ischium with narrow, distally expanded blade and well-developed pubic peduncle; tibia with stout articular ends; and fibula with lateral tuberosity formed by two parallel rugosities that are parallel to the long axis of the bone.

Comments. Because we consider the holotype to represent a single individual, our interpretation of the phylogenetic relationships of *Antarctosaurus wichmannianus* differs substantially from that of Wilson (2002, table 13), who, on the basis of mandibular characters only, regarded this species as having affinity with the rebbachisaurid *Nigersaurus.* The first caudal of *Antarctosaurus wichmannianus* is biconvex, which indicates that it could be related to Saltasauridae (Wilson 2002).

Laplatasaurus Huene 1929
Laplatasaurus araukanicus Huene 1929
Fig. 8.7A, B

Titanosaurus araukanicus Powell 1986, 2003

Lectotype. MLP CS 1128, right tibia and MLP CS 1127, right fibula (Fig. 8.7A, B).

Hypodigm. MLP CS 1316, cervical vertebra; MLP CS 1145/1131/1136/1146, dorsal vertebrae; MLP CS 1348/1315/1352; MPCA 1501, caudal vertebrae; MLP 1031, right scapula; MLP CS 1262, coracoids; MLP CS 1322, left sternal plate; MLP CS 1174, incomplete left humerus; MLP CS 1299, left radius; MLP CS 1170/1168/1192/1196, metacarpals; MLP CS 2002/1217, incomplete metacarpals; MLP CS 1059, left pubis; MLP CS 2202/2217, incomplete metatarsals.

Regrettably, it is not possible to demonstrate that the lectotype and hypodigm of *Laplatasaurus araukanicus* all pertain to a single species.

Locality and age. Cinco Saltos-Lago Pellegrini, Río Negro Province (Fig. 8.1, locality 5). Río Colorado Subgroup (top of the Anacleto Formation) (Leanza 1999), lower Campanian (Leanza 1999; Leanza and Hugo 2001).

Diagnosis. Powell (2003) proposed this diagnosis of *Laplatasaurus araukanicus:* Slender tibia (robustness index = 0.40), with transversely wide distal end and cnemial crest longer than in *Antarctosaurus,* separated from the axis of the proximal end by a clear depression on the lateral face; slender fibula with prominent double lateral tuberosities, with a clearly defined depression on the anteroproximal corner of the lateral surface.

Comments. Although, as mentioned above, the material is fragmentary and its assignment to a single species is doubtful, we can assume, on the basis of Wilson (2002), that at least some of the specimens assigned to *Laplatasaurus araukanicus* pertain to the Saltasauridae, because they include short, biconvex distalmost caudal centra (his characters 136 and 137) and a dorsolaterally bevelled distal radius (his character 171).

Nemegtosauridae Upchurch 1995
Bonitasaura Apesteguía 2004
Bonitasaura salgadoi Apesteguía 2004

Holotype. MPCA 300, subadult skeleton preserving left frontal and parietal; right dentary with 15 teeth; two cervical, six dorsal, and 12 caudal vertebrae; several cervical and dorsal ribs; two hemapophyses; humerus, radius, two metacarpals; femur, tibia, and two metatarsals.

Locality and age. La Bonita Hill Quarry, Cerro Policía, Río

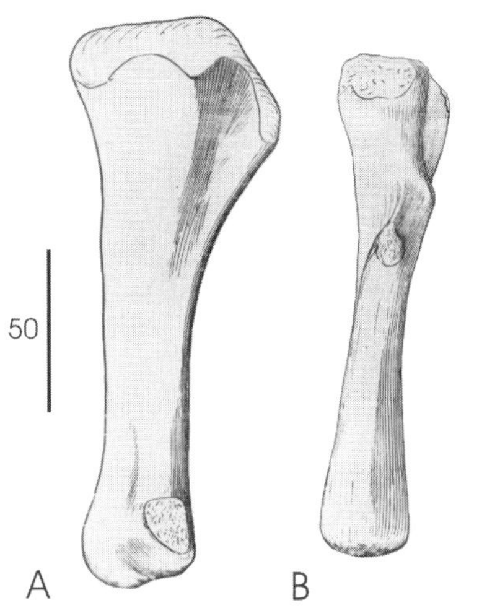

Figure 8.7. Laplatasaurus araukanicus. *(A) MLP CS 1128, right tibia in lateral view. (B) MLP CS 1127, right fibula in lateral view. Scale in centimeters. After Huene (1929). Right, first reconstruction of the skeleton of* Neuquensaurus australis *mounted in the Museo de La Plata, Buenos Aires, Argentina, by technician Antonio Castro, under the supervision of Á. Cabrera.*

Negro Province (Fig. 8.1, locality 11). Uppermost layers of the Bajo de la Carpa Formation. Santonian (Leanza and Hugo 2001).

Diagnosis. Dentary alveoli reduced in number (three in the main ramus, one in the angle, and up to seven in the anterior region); middle and posterior region of the dentary edentulous and forming a sharp dorsal edge, with profusely vascularized lateral side; very robust, diagonal neural arch pillars and bulging neural spine summits on anterior dorsal vertebrae (Apesteguía 2004).

Comments. According to Apesteguía (2004), *Bonitasaura salgadoi* shares with *Rapetosaurus krausei, Nemegtosaurus mongoliensis,* and *Quaesitosaurus orientalis* the following characters: sculptured frontal borders, a dentary symphysis that is almost perpendicular to the mandibular rami, and narrow, chisel-like teeth that are cylindrical in cross section and mostly restricted to the anteriormost portion of the lower jaw.

Saltasauridae Powell 1992
Saltasaurinae Powell 1992 (sensu Sereno 1998)
***Pellegrinisaurus* Salgado 1996**
***Pellegrinisaurus powelli* Salgado 1996**

Holotype. MPCA 1500, four dorsal centra, 26 incomplete caudal vertebrae, and an incomplete right femur.

Referred specimen. MCS-unnumbered, midcaudal vertebra.

Locality and age. Cinco Saltos-Lago Pellegrini, Río Negro Province (Fig. 8.1, locality 5). Río Colorado Subgroup (top of the Anacleto Formation) (Leanza 1999), lower Campanian (Leanza 1999; Leanza and Hugo 2001).

Diagnosis. Large eutitanosaurian characterized by strongly depressed dorsal centra (whose transverse width is approximately twice their maximum dorsoventral depth), and midposterior and posterior caudals with anteroposteriorly elongate and dorsoventrally depressed neural spines whose anterior ends are higher than their posterior counterparts (Salgado 1996).

Comments. Salgado (1996, 363) considered *Pellegrinisaurus powelli* to be the sister group of the Saltasaurinae (sensu Salgado et al. 1997a) because it displays low mid- and posterior caudal centra with dorsoventrally convex lateral surfaces. If Sereno's (1998) phylogenetic definitions are applied to the cladogram of Salgado et al. (1997a), *Pellegrinisaurus powelli* is placed within Saltasaurinae. According to Salgado (1996), characters demonstrating the affinity of *P. powelli* to *Saltasaurus loricatus* include the presence of low mid- and posterior caudal centra with dorsoventrally convex lateral surfaces.

Bonatitan Martinelli and Forasiepi 2004
Bonatitan reigi Martinelli and Forasiepi 2004

Holotype. MACN-Pv RN 821, complete braincase, middle dorsal vertebra, anterior caudal vertebra, middle caudal neural arch, left humerus, fragment of metacarpal, both femora, both tibiae, left fibula, left calcaneous, left metatarsal I, and some fragmentary elements.

Referred specimens. MACN-Pv RN 1061, complete braincase, incomplete anterior cervical vertebra, left radius, left ulna, left femur, left tibia, calcaneous, metatarsal III, a few incomplete hemal arches, and some fragmentary indeterminate elements.

Locality and age. Salitral de Santa Rosa, Río Negro Province. 40°3'32"S; 66°48'5"W (Fig. 8.1, locality 12). Middle Member of the Allen Formation, middle Campanian–lower Maastrichtian (Ballent 1980).

Diagnosis. Small-sized titanosaur characterized by the following association of characters: longitudinal groove located on the suture between parietals that continues posteriorly over the supraoccipital to the foramen magnum; basisphenoid tubera long and narrow (more than twice long as wide); dorsal to middle caudal vertebrae with deep oval to circular pits on both sides of the prespinal lamina; anterior caudal vertebra with spinopostzygapophyseal and spinoprezygapophyseal laminae; neural arch of anterior caudals with deep interzygapophyseal fossae with numerous pits; anterior caudal vertebra with an accessory subhorizontal lamina extending from the anteroventral portion of the postzygapophysis to the midportion of the spinoprezygapophyseal lamina; anterior caudal vertebra with a prominent axial crest on the ventral surface of the centrum (Martinelli and Forasiepi 2004).

Comments. Martinelli and Forasiepi (2004) considered *Bonatitan reigi* a Saltasaurinae because it exhibits the following diagnostic features of the clade: anterodorsal edge of the neural spine at the posterior level of the postzygapophyses in the middle caudal neural arch; distal femoral condyles anteriorly exposed; cancellous osseous tissue in the presacral and caudal vertebrae.

Aeolosaurini Costa Franco-Rosas, Salgado, Rosas, and Carvalho 2004
Aeolosaurus Powell 1991
Aeolosaurus rionegrinus Powell 1991

Holotype. MJG-R 1, seven anterior caudal vertebrae; three hemapophyses; incomplete left and right scapulae; both humeri; right radius and ulna; right pubis; both ischia; five metacarpals; right tibia, fibula, and astragalus; and indeterminate fragments.

Referred specimens. Salgado and Coria (1993) and Salgado et al. (1997b) reported *Aeolosaurus* in the Allen (MPCA 27174) and Los Alamitos (MPCA 27100) Formations (at Los Alamitos locality, Fig. 8.1, locality 14) (Campanian-Maastrichtian), in Río Negro Province (Fig. 8.8). Powell (1987) referred a partial caudal series from the Los Alamitos Formation to *Aeolosaurus* (MACN-RN 107), but Salgado and Coria (1993) challenged this taxonomic assignment.

Locality and age. Casa de Piedra, Estancia Maquinchao, Río Negro Province (Fig. 8.1, locality 13). Angostura Colorada Formation, Upper Cretaceous (Powell 2003).

Recently, Casal et al. (2002) linked a caudal series from the ?Campanian-Maastrichtian Upper Member of the Bajo Barreal Formation in Chubut Province to *Aeolosaurus.*

Diagnosis. Salgado et al. (1997b, 45) provided a modified diagnosis of *Aeolosaurus rionegrinus:* "Mid-sized titanosaurid with mid- and posterior caudal vertebrae with the neural arch placed on the anterior border of the vertebral centrum and the neural spine anteriorly inclined."

Powell (2003) diagnosed *Aeolosaurus rionegrinus* as follows: caudal vertebrae with compressed centra, with high lateral walls and narrow ventral face, from the third to fourth caudal; prezygapophysis longer than in any other known titanosaurid, projecting forward and upward on the anterior caudals, and slightly forward on the fourth caudal; neural arch inclined somewhat forward; neurapophysis slightly inclined forward, situated on the anterior half of the centrum; facets of the postzygapophyses more inclined than in *Saltasaurus,* almost paralleling the sagittal plane in anterior caudals, and located on the anterior half of the centrum; hemapophyses with separated articular ends; articular facets of hemapophyses divided into two angled surfaces; wide scapular blade, distally expanded; prominence for muscular attachment upon the internal face close to the upper margin of the scapula, as in *Saltasaurus loricatus;* humerus robust with a prominent apex on the deltoid crest for the insertion of the pectoral muscle; metacarpals relatively short and stout as in *Saltasaurus loricatus;* pubis with wide distal end of pubic blade.

Comments. According to Salgado et al. (1997a), *Aeolosaurus rionegrinus* is the probable sister group of *Alamosaurus* + Saltasaurinae (sensu Powell 1986), in view of the presence of (1) a biconvex first caudal (a character considered by Curry Rogers and Forster [2001] as a synapomorphy of their Saltasaurinae, and by Wilson [2002] to be a synapomorphy of his Saltasauridae [Saltasaurinae + Opisthocoelicaudiinae]), and (2) the presence of a dorsal prominence on the medial face of the scapula. Powell (2003) placed *Aeolosaurus* as the sister group of *Saltasaurus* + (*Argyrosaurus* + [*Isisaurus colberti* + *Ampelosaurus*]) because of the absence of an anteroventral ridge on the sternal plate. When Sereno's (1998) definition of Saltasaurinae is applied to the cladogram of Salgado et al. (1997a), *Aeolosaurus* is a saltasaurine, because the presence of a dorsal prominence on the medial face of the scapula and the absence of an anteroventral ridge on the sternal plate indicate that this genus is closer to *Saltasaurus* than to *Opisthocoelicaudia*.

Rinconsaurus Calvo and González Riga 2003
Rinconsaurus caudamirus Calvo and González Riga 2003

Holotype. MRS-Pv 26, 13 articulated anterior-middle and middle-posterior caudal vertebrae and two ilia.

Paratypes. The following bones associated with the holotype are included: MRS-Pv 117, 263, teeth; MRS-Pv 102, prefrontal; MRS-Pv 112, angular and surangular; MRS-Pv 2, 3, 8, 4, 21, cervical vertebrae; MRS-Pv 5, 6, 9, 11, 13, 16–19, dorsal vertebrae; MRS-Pv 22–25, 27, anterior caudal vertebrae; MRS-Pv 27, 28, 31, middle caudal vertebrae; MRS-Pv 29, 30, 32–40, posterior caudal vertebrae; MRS-Pv 20, 42, 93, 99, 109, 113, hemapophyses; MRS-Pv 43, scapula and coracoid; MRS-Pv 46, 103, 104, sternal plates; MRS-Pv 47, humerus; MRS-Pv 98, metacarpals; MRS-Pv 96, ilia; MRS-Pv 97, 100, pubes; MRS-Pv 94, 101, ischia; MRS-Pv 49, 92, femora; and MRS-Pv 111, metatarsals.

Both the holotype and paratypes of *Rinconsaurus caudamirus* belong to three individuals (Calvo and González Riga 2003). Seemingly, specimen MRS-Pv 5 and the holotype are part of a single individual (Calvo and González Riga 2003).

Locality and age. Cañadón Río Seco, 2 km north of Rincón de los Sauces, Neuquén Province (Fig. 8.1, locality 2). Río Neuquén Subgroup, upper Turonian-Coniacian (Leanza and Hugo 2001; Calvo and González Riga 2003).

Diagnosis. Calvo and González Riga (2003, 335) diagnosed this early representative of Aeolosaurini as follows: slender titanosaurid characterized by the following association of autapomorphies: (1) neural spines in midanterior dorsal vertebrae inclined posteriorly more than 60° with respect to the vertical; (2) middle caudal vertebrae with bony processes that support the articular facets of postzygapophyses; (3) procoelous posterior caudal centra with the eventual intercalation of a series of amphicoelous-biconvex or amphicoelous-opisthocoelous-biconvex centra.

Comments. Calvo and González Riga (2003) indicated that *Rinconsaurus* is the sister group of *Aeolosaurus* because it has relatively long prezygapophyses. We used the definition of Costa Franco-Rosas et al. (2004) for Aeolosaurini (the most inclusive group containing *Aeolosaurus* and *Gondwanatitan*, but not

Saltasaurus or *Opisthocoelicaudia*); *Rinconsaurus* is a member of this clade. Indeed, Costa Franco-Rosas et al. (2004) proposed the character "relatively long prezygapophyses" as a synapomorphy of Aeolosaurini. Combining the phylogenetic analysis of Salgado et al. (1997a) and the phylogenetic taxonomy of Sereno (1998) and Costa Franco-Rosas et al. (2004), Aeolosaurini is a subclade of Saltasaurinae. However, Calvo and González Riga (2003) placed Aeolosaurini outside Saltasauridae.

The existence of biconvex distal caudal centra, which, according to Calvo and González Riga (2003) is an autapomorphy of *Rinconsaurus caudamirus,* is interpreted by Wilson et al. (1999) and Wilson (2002) as diagnostic of Saltasauridae (=Opisthocoelicaudiinae + Saltasaurinae). The latter interpretation is in better agreement with the phylogenetic position of the Aeolosaurini proposed by Salgado et al. (1997a).

Saltasaurini nov.

Definition. The less inclusive clade containing *Saltasaurus loricatus* and *Neuquensaurus australis* (=Saltasaurinae sensu Salgado et al. 1997a; Saltasaurinae sensu Powell 2003).

***Neuquensaurus* Powell 1992**
***Neuquensaurus australis* (Lydekker 1893)**
Figs. 8.7 right, 8.8D, E

Holotype. MLP Ly 1/2/3/4/5/6, caudal vertebrae.

Referred specimens. MLP Ly 1-3 and MLP CS 1-4, seven series of bones housed in the Museo de La Plata (see Powell 2003, 39); MCS-5, an articulated partial skeleton preserving one posterior cervical vertebra, six mid- and posterior dorsal vertebrae, the sacrum articulated to both ilia, 15 caudal vertebrae, many rib fragments, three hemapophyses, the left ischium, both femora, right tibia, fibula, and astragalus, and osteoderms; MCS-6, right tibia; MCS-8, left humerus; MCS-9, right femur; MCS-10, metatarsal (Salgado et al. 2005).

Locality and age. Cinco Saltos-Lago Pellegrini, Río Negro Province (Fig. 8.1, locality 5). Río Colorado Subgroup (top of the Anacleto Formation) (Leanza 1999), lower Campanian (Leanza 1999; Leanza and Hugo 2001).

Diagnosis. Salgado et al. (2005) provided this diagnosis of *Neuquensaurus australis:* Anteriormost dorsal vertebrae lacking centroprezygapophyseal and centropostzygapophyseal laminae, seven sacral vertebrae, sacral centra 3 to 5 narrowed, mid- and posterior caudal vertebrae bearing a transversely wide, unkeeled ventral depression delimited by lateral rounded ridges that culminate in the articular facets of the hemapophyses, lateral walls of the caudal vertebral centra little exposed in ventral view, fibula with strong lateral tuberosity and bent shaft.

Comments. It is clear that *Neuquensaurus australis* is a member of Saltasaurinae, closely related to *Saltasaurus loricatus* and *Rocasaurus muniozi.* Salgado et al. (1997a) defined the Saltasaurinae as

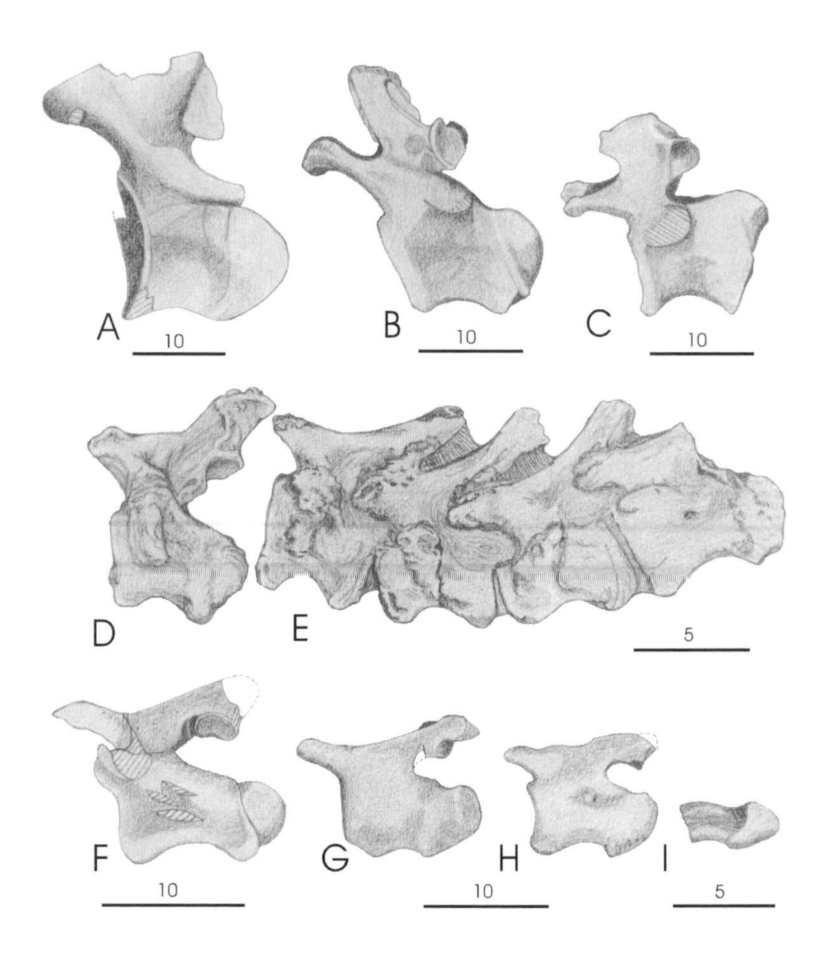

Figure 8.8. Aeolosaurus *sp. (MPCA 27174): (A–C) Anterior, mid-, and midposterior caudal vertebrae in left lateral view. Redrawn from Salgado and Coria (1993).* Neuquensaurus australis: *(D) MLP CS 1394, third caudal vertebra in left lateral view. Redrawn from Huene (1929). (E) MLP CS 1403, fourth to senventh caudal vertebrae in left lateral view. (F) Cf.* Rocasaurus *sp. (MPCA-Pv 58), midcaudal vertebra in left lateral view. (G–I)* Rocasaurus muniozi *(MPCA-Pv 46), midposterior and posterior caudal vertebrae in left lateral view. Redrawn from Salgado and Azpilicueta (2000). Scale in centimeters.*

the clade including the most recent common ancestor of *Neuquensaurus australis* and *Saltasaurus loricatus* and all of its descendants. Sereno (1998) substituted this definition for another: all saltasaurids closer to *Saltasaurus* than to *Opisthocoelicaudia*. Under this new definition, many taxa that are not included within Powell's (2003) subfamily Saltasaurinae—such as *Aeolosaurus, Pellegrinisaurus,* or even *Alamosaurus*—would be part of this clade. Consequently, Saltasaurini is here proposed to refer to all saltasaurines included within the less inclusive clade comprising *Neuquensaurus* and *Saltasaurus* (Saltasaurinae sensu Salgado et al. 1997a). Synapomorphies of the new clade are those proposed by Salgado et al. (1997a) for their Saltasaurinae: depressed anterior caudal centra, anterodorsal edge of the neural spine placed posteriorly with respect to the anterior root of the midcaudal postzygapophyses, and spongy (=camellate) caudal bone texture (Powell 2003).

Regarding its phylogenetic position within Saltasaurini, *Neuquensaurus* is the sister group of *Rocasaurus* + *Saltasaurus* because it lacks a ventral longitudinal septum in the anterior and midcaudals and possesses a conspicuous ischiadic peduncle of the ilium.

Rocasaurus Salgado and Azpilicueta 2000
Rocasaurus muniozi Salgado and Azpilicueta 2000
Fig. 8.8F–I

Holotype. MPCA-Pv 46, a juvenile individual represented by a cervical centrum, a cervical neural arch, two dorsal centra, three dorsal neural arches, two sacral neural arches, a midcaudal vertebra, a midposterior caudal vertebra, the left ilium, a fragment of the right ilium, the left pubis, both ischia, and the left femur.

Referred specimens. MPCA-Pv 47, 48, 60, three anterior caudal centra; MPCA-Pv 49, a midcaudal centrum; MPCA-Pv 50, a posterior caudal centrum; MPCA-Pv 51-56, six posterior caudal vertebrae.

Locality and age. Salitral Moreno, 25 km south of General Roca, Río Negro Province (Fig. 8.1, locality 10). Middle Member of the Allen Formation, middle Campanian–lower Maastrichtian (Ballent 1980).

Diagnosis. Salgado and Azpilicueta (2000) characterized *Rocasaurus muniozi* by the presence of caudal vertebrae with a deep ventral depression divided by a longitudinal septum, and the posterior articulation notably compressed and ventrally extended; a distal expansion on the lateral border of the pubis, and an ischium with a broad blade.

Comments. *Rocasaurus muniozi* is, following Salgado and Azpilicueta (2000), the sister group of *Saltasaurus loricatus* on the bases of the presence of a longitudinal septum dividing a ventral depression in the anterior and midcaudals and a poorly developed ischiadic peduncle of the ilium.

Diplodocoidea Upchurch 1995
Rebbachisauridae Bonaparte 1997
Limaysaurus Salgado, Garrido, Cocca, and Cocca 2004
Limaysaurus tessonei (Calvo and Salgado 1995)
Fig. 8.9

Holotype. MUCPv-205, an articulated, well-preserved skeleton, including the basicranium, teeth, eight disarticulated cervical vertebrae, 12 dorsal vertebrae (six of them articulated), 40 articulated caudal vertebrae, complete pectoral and pelvic girdles, nearly complete fore- and hind limbs lacking the manus, and gastroliths.

Referred specimens. MUCPv-206, a disarticulated skeleton composed of two anterior and two posterior cervical vertebrae, one posterior dorsal vertebra, ribs, a sternal plate, four metacarpals, and gastroliths; MUCPv-153, an articulated partial skeleton composed of two sacrals, the first six caudals, pubis and ischium; MUCPv-272, four articulated dorsal vertebrae, a pubis, ischium, and six gastroliths; MUCPv-273, two anterior, two middle, and one posterior caudal vertebrae (Calvo 1999, fig. 3), unnumbered dorsal vertebra described by Nopcsa (1902), now reposited in Switzerland (Salgado this volume).

Locality and age. Approximately 5 km southwest of Villa El

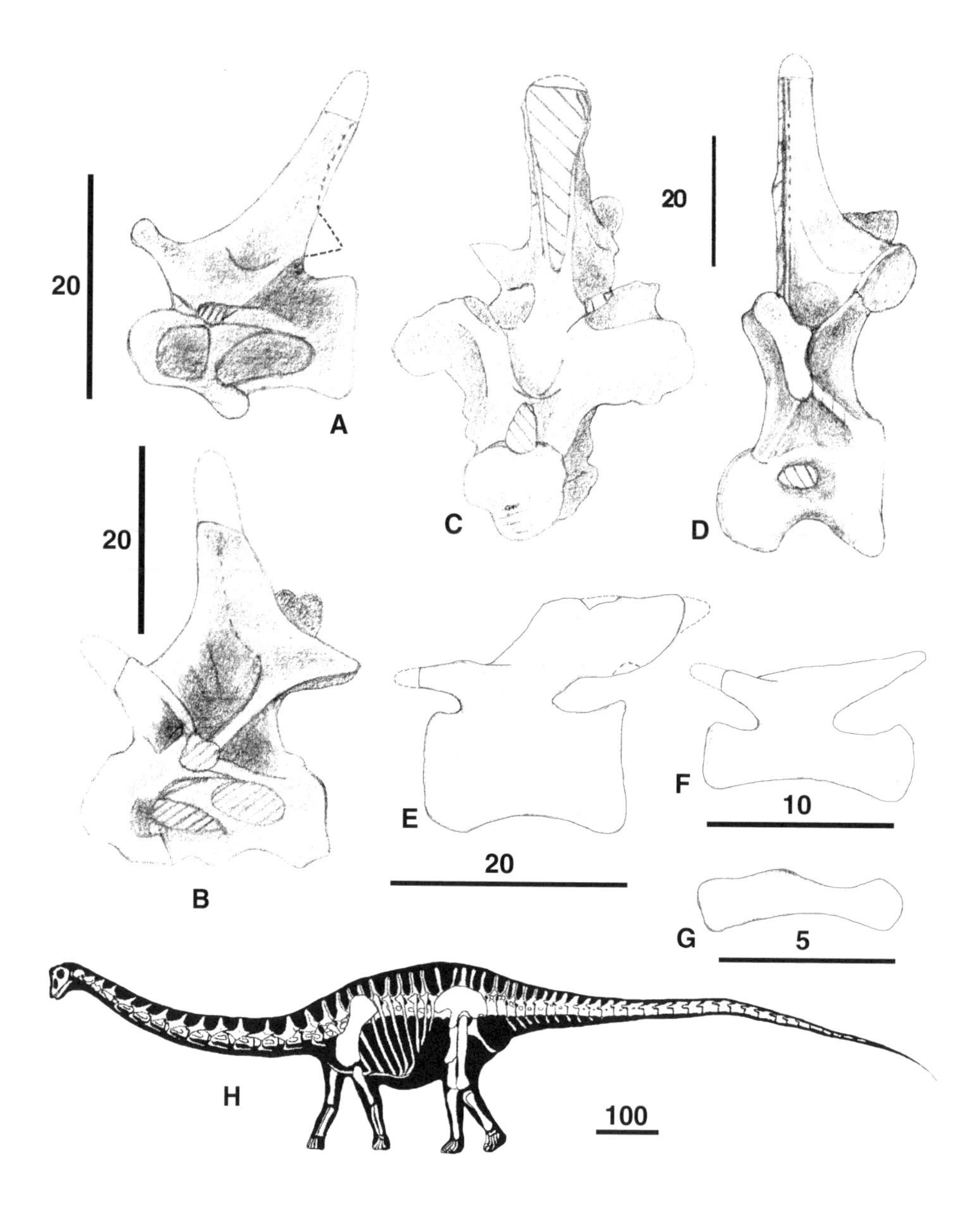

Figure 8.9. Limaysaurus tessonei *(MUCPv-205): (A) Anterior and (B) posterior cervical vertebrae in left lateral view. Anterior dorsal in (C) anterior and (D) left lateral views. (E) Mid-, (F) midposterior, and (G) posterior caudal vertebrae in left lateral views. (H) Reconstruction of the skeleton, redrawn from Calvo and Salgado (1995). Scale in centimeters.*

Chocón, Neuquén Province (Fig. 8.1, locality 6). Top of the Candeleros Formation (MUCPv-205 and 206) and base of the Huincul Formation (MUCPv-153). Lower Cenomanian (Leanza and Hugo 2001; Leanza et al. 2004).

Diagnosis. Calvo and Salgado's (1995) diagnosis of *Limaysaurus tessonei* includes the following characters: basipterygoid processes very thin and short (1); posterior process of the postorbital absent (2); anteroposteriorly elongate articular condyle of the quadrate (3); basal tubera very reduced (4); paraoccipital processes not distally expanded (5); neural spine in posterior cervical and anterior dorsal vertebrae with an accessory lamina connecting the postzygodiapophyseal and spinoprezygapophyseal laminae (6); anterior dorsals with both spinoprezygapophyseal laminae contacting at the apex of the spine (7); transverse process in anterior caudal vertebrae formed by dorsally directed dorsal and ventral bars, differing from *R. garasbae* that has a true winglike transverse process (8); shaft of the pubis oval in cross section (9).

Comments. Character 6 (or its reformulation as "cervical neural arches with accessory lamina extending from the postzygodiapophyseal lamina anterodorsally") and character 8 were proposed by Wilson (2002, 274) as autapomorphies of *Rayososaurus* (this author considered *Limaysaurus* [="*Rebbachisaurus*"] *tessonei* referable to the genus *Rayososaurus*). More recently, Salgado et al. (2004) proposed characters 8 and 9 as autapomorphies of *Limaysaurus.*

Limaysaurus is a member of the Rebbachisauridae. Although the first rebbachisaurid remains were assigned to disparate, previously erected sauropod clades (Romer 1956; McIntosh 1990), they were later recognized as members of a new family (Bonaparte 1997). Calvo and Salgado (1995) were the first to provide cladistic evidence supporting a relationship between *Limaysaurus* and diplodocids. Rebbachisaurids are currently regarded as the sister group of Flagellicaudata (=Diplodocidae + Dicraeosauridae; Harris and Dodson 2004). Wilson (2002) lists 14 characters supporting this relationship.

Rayososaurus Bonaparte 1996
Rayososaurus agrioensis Bonaparte 1996

Holotype. MACN-N 41, a complete left scapula, distal half of the right scapular blade, an incomplete left femur, proximal half of the left fibula, and indeterminate fragments.

Locality and age. At 3 km south of Agrio del Medio, Departamento Picunches, Neuquén Province (Fig. 8.1, locality 3). Upper section of the Rayoso Formation, Aptian (Bonaparte 1996).

Diagnosis. Bonaparte (1996, 99) diagnosed *Rayososaurus agrioensis* as follows: "Sauropod provided with a prominent spinous acromial process on scapula, internally flat, but externally with a rounded ridge running obliquely to scapular long axis, along the middle of the acromial process. Deep anterior (dorsal) border of scapular blade. The latter well expanded distally. Acromial depression dorsoventrally elongated."

Comments. Bonaparte (1996, 99) noted "certain similarities with the scapula of *Rebbachisaurus garasbae.*" *Rayososaurus* and *Rebbachisaurus* are rebbachisaurid diplodocoids.

Dicraeosauridae Huene 1927
Brachytrachelopan Rauhut, Remes, Fechner, Cladera, and Puerta 2005
Brachytrachelopan mesai Rauhut, Remes, Fechner, Cladera, and Puerta 2005

Holotype. MPEF-PV 1716, an articulated partial skeleton, including eight cervical, 12 dorsal, and three sacral vertebrae; the proximal parts of the posterior cervical and all dorsal ribs, the right ilium, distal part of the left femur, and proximal end of the left tibia.

Locality and age. At 25 km north-northeast of Cerro Cóndor, Chubut Province. Cañadón Calcáreo Formation, Tithonian (Upper Jurassic) (Rauhut et al. 2005) (Fig. 8.1, locality 18).

Diagnosis. This dicraeosaurid differs from all other sauropods in its very short neck, with individual cervical vertebrae being as long as, or shorter in anteroposterior length than, high posteriorly; pronounced, pillarlike centropostzygapophyseal lamina in the cervical vertebrae; a pronounced anterior inclination of the midcervical neural spines, with the tip of the spine extending beyond the anterior end of the centrum, and anterior dorsal neural spines 1 to 6 with vertical bases and anteriorly flexed tips (Rauhut et al. 2005).

Comments. Brachytrachelopan is closely related to *Dicraeosaurus* from the Upper Jurassic Tendaguru Formation of East Africa. Rauhut et al (2005) listed three derived characters shared by these two genera: oblique lamina dividing cervical pleurocoels absent in anterior and midcervicals; midcervical neural spines inclined anteriorly; and posterior cervical neural spines inclined anteriorly.

Virtually all authors accept that the dicraeosaurids and diplodocids are sister groups (Upchurch 1998; Wilson 2002; Alifanov and Averianov 2003; Harris and Dodson 2004). Derived characters shared by these groups include presacral neural spines bifid, sacral neural spines approximately four times the length of centrum; anterior caudal vertebrae with "winglike" transverse processes and spinoprezygapophyseal laminae on lateral aspect of neural spines. Conversely, Bonaparte (1999a, 158, translated from the Spanish) interpreted the phylogenetic position of dicraeosaurids as seeming "to correspond to the level of *Patagosaurus* or an even more primitive one. To justify a different interpretation, we cite several concomitant, correlated reversals (such as the reduced number of cervicals, the lack of elaboration in the lateral depressions in the cervical vertebral centra, lack of pleurocoels in most of the dorsal vertebrae, modest size of known individuals). I don't believe that all of these characters together can reasonably be interpreted as reversals."

Amargasaurus Salgado and Bonaparte 1991
Amargasaurus cazaui Salgado and Bonaparte 1991
Fig. 8.10

Holotype. MACN-N 15, most of a skeleton preserving the braincase and temporal portion of the skull; 22 articulated presacral vertebrae; one proximal end of cervical rib and many incomplete dorsal ribs; sacrum consisting of five fused vertebrae; three midanterior caudal vertebrae; one posterior caudal and numerous caudal centra; three hemapophyses; right scapula and coracoid; left humerus, radius, ulna, ilium, femur, tibia, fibula, and astragalus; and two metatarsals.

Referred specimen. PV-6126-MOZ, a posterior cervical vertebra.

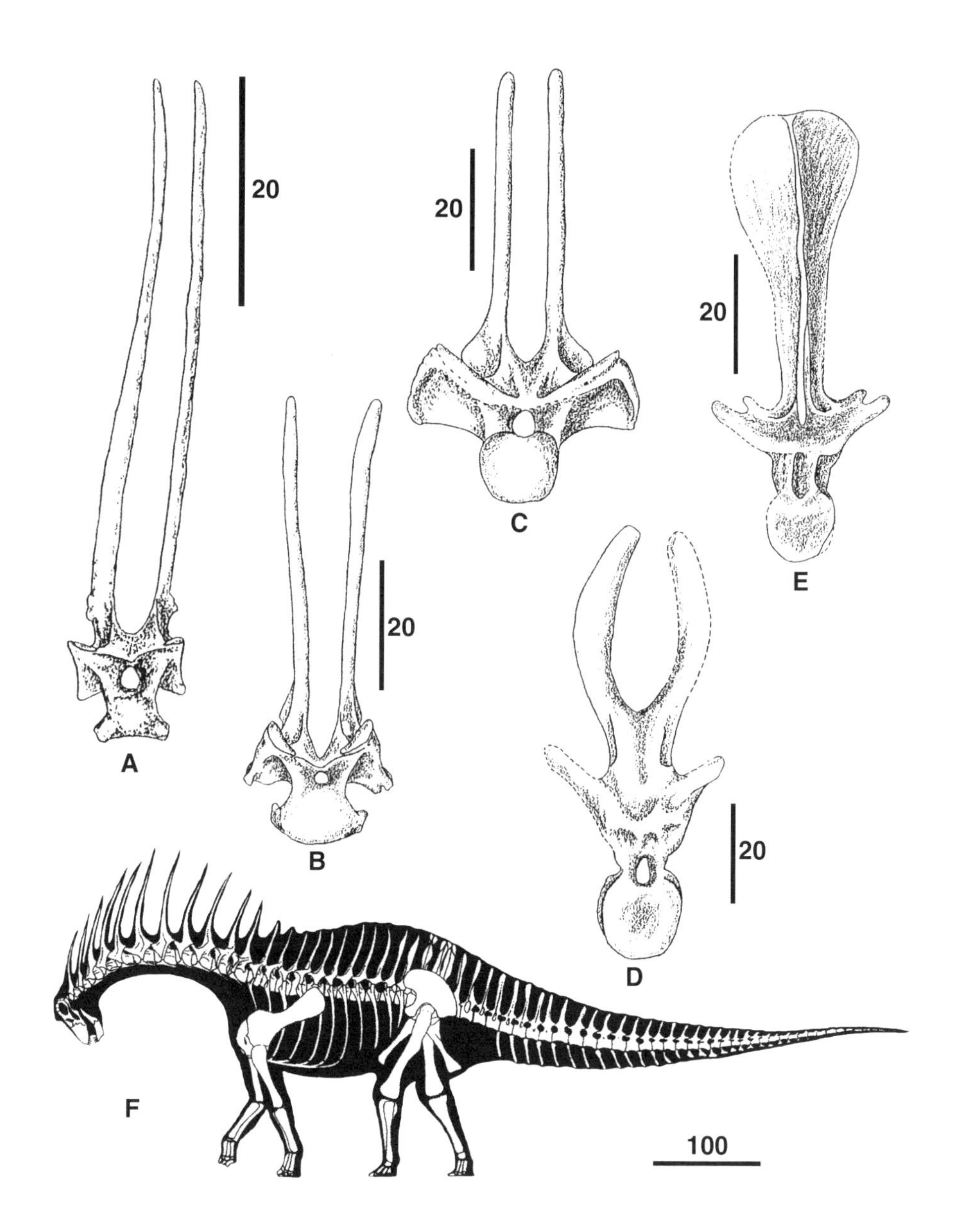

Figure 8.10. Amargasaurus cazaui *(MACN-N 15), (A) sixth and (B) eighth cervical vertebrae in anterior view. (C) First, (D) eighth, and (E) ninth dorsal vertebrae in anterior view, after Bonaparte (1999a). (F) Reconstruction of the skeleton. Scale in centimeters.*

Locality and age. At 2.5 km southeast of La Amarga creek, on Road 40, Catan Lil Department, Neuquén Province (Fig. 8.1, locality 7). Antigual Member of the La Amarga Formation, Barremian (Leanza et al. 2004).

Diagnosis. Salgado and Bonaparte (1991) diagnosed *Amargasaurus cazaui* as follows: Dicraeosauridae comparable in size to *Dicraeosaurus hansemanni;* temporal portion of the skull with parietal and postparietal fenestrae, as in *Dicraeosaurus;* position and shape of the nares comparable to *Dicraeosaurus hansemanni;* fused basioccipital tubera; robust orbitosphenoid with large opening for the olfactory nerve, bordered by an osseous process; presacral vertebrae taller than in *Dicraeosaurus;* bifurcation of the neural spines more accentuated than in that genus, extending to the penultimate dorsal vertebra; cervical neural spines longer than in *Dicraeosaurus,* and with a subcylindrical cross section; pleurocoels secondarily obliterated, as in *Dicraeosaurus;* five fused sacrals; appendicular bones comparable to those of *Dicraeosaurus.*

Wilson (2002, 273) listed a series of potential autapomorphies of *Amargasaurus cazaui:* multiple foramina leading into endocranial cavity in a depression located between the supraoccipital and exoccipital; basioccipital depression between foramen magnum and basal tubera; marked ventral excavation of paraoccipital processes; basal tubera fused together; presacral pneumatopores absent; extremely elongate cervical neural spines; 10 or fewer dorsal vertebrae; first dorsal vertebra with fused diapophysis and parapophysis; ulnar proximal condylar processes subequal in length.

Comments. Amargasaurus is, according to Rauhut et al. (2005), the sister taxon of all other dicraeosaurids.

Evolutionary Considerations

The Middle Jurassic of Patagonia is characterized by the presence of basal eusauropods (*Patagosaurus* and *Volkheimeria*) and the first record of neosauropods (specimen MACN-CH 934). In the Upper Jurassic, basal eusauropods (MCF-PVPH-379, García et al. 2003) and neosauropods (*Brachytrachelopan mesai* and *Tehuelchesaurus benitezi*) are still insufficiently represented, although it is possible that neosauropods were more diverse than in the Middle Jurassic.

On the basis of its sauropod associations, the Patagonian Cretaceous can be divided into three time intervals: (1) pre-Turonian Cretaceous, (2) Turonian-Santonian, and (3) Campanian-Maastrichtian (="Senonian"). This scheme summarizes our current knowledge of Patagonian sauropod evolution well.

Pre-Turonian Cretaceous

The record of *Amargasaurus cazaui* in this interval reveals the persistence, at least in northern Patagonia, of the dicraeosaurids, a lineage whose first members are recorded in the Upper Jurassic of Africa (Salgado and Bonaparte 1991) and Patagonia (Rauhut et al. 2005). Seemingly, the sister group of the dicraeosaurids (the diplodocids) did not survive into the Cretaceous.

Basal diplodocoids (rebbachisaurids and their relatives) are an important component of Patagonian sauropod faunas during the Early and mid-Cretaceous. In the Neuquén Basin, the genera *Limaysaurus* (Aptian-Albian Lohan Cura Formation, lower Cenomanian Candeleros Formation, and upper Cenomanian Huincul Formation) and *Rayososaurus* (Aptian Rayoso Formation) are recorded.

Macronarians are also present during this interval. In northern Patagonia, the oldest indubitable macronarians come from the Lohan Cura Formation (Leanza et al. 2004). Central Patagonia has produced *Chubutisaurus insignis,* a basal titanosaurian from levels of similar age (Salgado 1993), and indeterminate remains from the Aptian Matasiete and Cenomanian-Turonian Lower Bajo Barreal formations (Martínez et al. 1989; Powell et al. 1989). Basal titanosaurians survived until the end of this interval, represented by *Andesaurus delgadoi* from the Candeleros Formation (Calvo and Bonaparte 1991) and a new form from the base of the Huincul Formation (Calvo 1999).

Titanosauroids (=Titanosauridae sensu Salgado et al. 1997a), identifiable by their strongly procoelous caudal vertebrae, are known in the Neuquén Basin (Candeleros Formation) and in the San Jorge Basin (Lower Member of the Bajo Barreal Formation) beginning from the last stages of the pre-Turonian Cretaceous interval (Powell 1990; Calvo and Salgado 1998; Martínez et al. 2004). The first definitive eutitanosaur (*Argentinosaurus*) is recorded toward the end of this interval (Huincul Formation).

Most pre-Turonian titanosaurians such as "*Campylodoniscus*" and MUCPv-Pv 244 ("Titanosauridae" indet.) had broad teeth. Occasionally, isolated slender teeth have been recorded, but they may instead be referable to diplodocoids (Salgado et al. 2004).

Unlike the Patagonian record, the pre-Turonian Cretaceous period in North America and Australia is characterized by the complete absence of diplodocoids. In Europe, titanosauroids (*Iuticosaurus valdensis,* from the Barremian) and basal diplodocoids (Pereda-Suberbiola et al. 2003) are recorded earlier than in Patagonia, whereas dicraeosaurids (and diplodocids) are absent. *Jobaria tiguidensis,* the sister group of the Neosauropoda (Sereno et al. 1999), is present in the Neocomian of Africa. This lineage of basal eusauropods has not been recorded from the Patagonian Cretaceous.

Turonian-Santonian

Patagonian sauropod faunas of this period can be characterized by the following:

1. Persistence of a relictual group of basal titanosaurians, with amphiplatyan (MUCPv-204, Salgado and Calvo 1993) to slightly procoelous (*Mendozasaurus,* González Riga 2003) mid- and posterior caudal vertebrae.
2. Probable diversification of the Eutitanosauria. Eutitanosaurians from this period are poorly known, possibly represented by a specimen referred to *Argyrosaurus* (PVL 4628; Powell 2003) and a new form from the Portezuelo Formation (Turonian) of Lago Los Barreales in Neuquén Province (Calvo et al. 2001).
3. Complete absence of diplodocoids (both basal forms and dicraeosaurids). There is an isolated record of a possible post-Cenomanian diplodocoid (UNPSJB-PV 581, possibly a rebbachisaurid) from the uppermost levels of the Lower Member of the Bajo Barreal Formation (Sciutto and Martínez 1994).

Dinosaurs from this interval are little known in North America and Europe. This period roughly corresponds to the postulated "sauropod hiatus" recorded in these continents (Lucas and Hunt 1989; Le Loeuff 1993).

Campanian-Maastrichtian

At least in northern Patagonia, the sauropod record is most complete in the "Senonian." Sauropod faunas of this period are characterized by the following:

1. Complete absence of basal titanosaurians.
2. Diversification of titanosaurians with an opisthocoelous sacrum (Saltasauridae according to Wilson [2002]; "Opisthocoelisacralia" according to Salgado [2000]).
3. Diversification of Saltasaurinae.
4. First record and diversification of Saltasaurini.
5. Diversification of Aeolosaurini (Costa Franco-Rosas et al. 2004), the proposed sister group of the Saltasaurini.

Most sauropod clades characteristic of this interval are recorded in other continents, except for the Aeolosaurini (which are only recorded in northern and central Patagonia, and Brazil) and the Saltasaurini (which are exclusive to northern Patagonia and northwestern Argentina).

Faunal interchange between North and South America that apparently took place by the end of the Cretaceous (Leanza et al. 2004) seemingly had a minimal effect on Patagonian sauropod faunas. The same major sauropod groups (saltasaurids of uncertain phylogenetic position, basal saltasaurines, and Aeolosaurini and Saltasaurini) are present in the lower Campanian Río Colorado Subgroup (represented by the genera *Laplatasaurus, Pellegrinisaurus,* and *Neuquensaurus*) and the Campanian-Maastrichtian Allen Formation (*Antarctosaurus, Aeolosaurus, Rocasaurus, Neuquensaurus,* and *Bonatitan*). The immigration of large herbivorous ornithischian dinosaurs from North America (e.g., ankylosaurs, hadrosaurs) did not have obvious consequences for Patagonian sauropod evolution.

Acknowledgments. We thank Prebiterio Pacheco for providing us with most of the illustrations of sauropodomorph specimens. We are also thankful to Jeff Wilson, Matt Lamanna, and Oliver Rauhut for their reviews and suggestions that improved this work. This work was partially supported by a grant (PIP 6455) from the CONICET.

References Cited

Alifanov, V. R., and A. Averianov. 2003. *Ferganasaurus verzilini,* gen. et sp. nov., a new neosauropod (Dinosauria, Saurischia, Sauropoda) from the Middle Jurassic of Fergana Valley, Kirghizia. *Journal of Vertebrate Paleontology* 23: 358–372.

Apesteguía, S. 2004. *Bonitasaura salgadoi:* a beaked sauropod in the Late Cretaceous of Gondwana. *Naturwissenschaften* 91 (10): 493–497.

Ballent, S. C. 1980. Ostrácodos de ambiente salobre de la Formación Allen (Cretácico Superior) en la Provincia de Río Negro (República Argentina). *Ameghiniana* 17 (1): 67–82.

Bonaparte, J. F. 1979. Dinosaurs: a Jurassic assemblage from Patagonia. *Science* 205: 1377–1378.

———. 1986. Les dinosaures (Carnosaures, Allosauridés, Sauropodes, Cétiosauridés) du Jurassique Moyen de Cerro Cóndor (Chubut, Argentina). *Annales de Paléontologie* 72: 325–386.

———. 1996. Cretaceous tetrapods of Argentina. In F. Pfeil and G. Arratia (eds.), *Contributions of Southern South America to Vertebrate Paleontology,* 73–130. Münchner Geowissenchaftliche Abhandlungen (A) 30.

———. 1997. *Rayososaurus agrioensis* Bonaparte, 1995. *Ameghiniana* 34: 116.

———. 1999a. Evolución de las vértebras presacras en Sauropodomorpha. *Ameghiniana* 36 (2): 115–187.

———. 1999b. An armoured sauropod from the Aptian of northern Patagonia, Argentina. In Y. Tomida, T. Rich, and P. Vickers-Rich (eds.), *Proceedings of the Second Gondwanan Dinosaur Symposium,* 1–12. National Science Museum Monographs 15.

Bonaparte, J. F., and R. A. Coria. 1993. Un nuevo y gigantesco saurópodo titanosaurio de la Formación Río Limay (Albiano-Cenomaniano) de la Provincia del Neuquén, Argentina. *Ameghiniana* 30: 271–282.

Bonaparte, J. F., and M. Vince. 1979. El hallazgo del primer nido de dinosaurios triásicos (Saurischia, Prosauropoda), Triásico Superior de Patagonia, Argentina. *Ameghiniana* 16: 173–182.

Cabrera, Á. 1947. Un saurópodo nuevo del Jurásico de Patagonia. *Notas del Museo de La Plata 12 Paleontología* 95: 1–17.

Calvo, J. O. 1999. Dinosaurs and other vertebrates of the Lake Ezequiel Ramos Mexía area, Neuquén-Patagonia, Argentina. In Y. Tomida, T. H. Rich, and P. Vickers-Rich (eds.), *Proceedings of the Second Gondwanan Dinosaur Symposium,* 13–45. National Science Museum Monographs 15.

Calvo, J. O., and J. F. Bonaparte. 1991. *Andesaurus delgadoi* gen. et sp. nov. (Saurischia-Sauropoda), dinosaurio Titanosauridae de la Formación Río Limay (Albiano-Cenomaniano), Neuquén, Argentina. *Ameghiniana* 28: 303–310.

Calvo, J. O., and B. J. González Riga. 2003. *Rinconsaurus caudamirus* gen. et sp. nov., a new titanosaurid (Dinosauria, Sauropoda) from the Late Cretaceous of Patagonia, Argentina. *Revista Geológica de Chile* 30: 333–353.

Calvo, J. O., and L. Salgado. 1995. *Rebbachisaurus tessonei* sp. nov.: a new Sauropoda from the Albian-Cenomanian of Argentina—new evidence on the origin of Diplodocidae. *Gaia* 11: 13–33.

———. 1998. Nuevos restos de Titanosauridae (Sauropoda) en el Cretácico Inferior de Neuquén, Argentina. In *Resúmenes VII Congreso Argentino de Paleontología y Bioestratigrafía,* 59. Bahía Blanca, Buenos Aires: Asociación Paleontológica Argentina.

Calvo, J. O., J. Porfiri, C. Veralli, and F. Poblete. 2001. A giant titanosaurid sauropod from the Upper Cretaceous of Neuquén, Patagonia, Argentina. *Ameghiniana* 38: 5R.

Casal, G., M. Luna, L. Ibiricu, E. Ivany, R. D. Martínez, M. Lamanna, and

A. Koprowsky. 2002. Hallazgo de una serie caudal articulada de Sauropoda de la Formación Bajo Barreal, Cretácico Superior del sur de Chubut. *Ameghiniana* 39 (Suppl.): 8R.

Casamiquela, R. M. 1963. Consideraciones acerca de *Amygdalodon* Cabrera (Sauropoda, Cetiosauridae) del Jurásico medio de la Patagonia. *Ameghiniana* 3: 79–95.

———. 1964. Sobre el hallazgo de dinosaurios triásicos en la Provincia de Santa Cruz, Argentina. *Austral* 35: 10–11.

———. 1980. La presencia del género *Plateosaurus* (Prosauropoda) en el Triásico superior de la Formación El Tranquilo, Patagonia. *Actas del II Congreso Argentino de Paleontología y Bioestratigrafía y I Congreso Latinoamericano de Paleontología* 1: 143–158.

Coria, R. A., and L. Salgado. 2000. Los dinosaurios de Ameghino. In S. Vizcaíno (ed.), *Obra de los Hermanos Ameghino,* 43–49. Publicación Especial, Universidad Nacional de Luján.

Costa Franco-Rosas, A., L. Salgado, I. S. Carvalho, and C. F. Rosas. 2004. Nuevos materiales de titanosaurios (Sauropoda) en el Cretácico Superior de Mato Grosso, Brazil. *Revista Brasileira de Paleontología* 7 (3): 329–336.

Curry Rogers, K., and C. A. Forster. 2001. The last of the dinosaur titans: a new sauropod from Madagascar. *Nature* 412: 530–534.

Del Corro, G. 1975. Un nuevo saurópodo del Cretácico Superior, *Chubutisaurus insignis* gen. et sp. nov. (Saurischia, Chubutisauridae, nov.) del Cretácico Superior (Chubutiano), Chubut, Argentina. *Actas del I° Congreso Argentino de Paleontología y Bioestratigrafía* 2: 229–240. Tucumán.

Flynn, J. J., J. M. Parrish, B. Rakotosamimanana, W. F. Simpson, R. L. Whatley, and A. R. Wyss. 1999. A Triassic fauna from Madagascar, including early dinosaurs. *Science* 285: 763–765.

Galton, P. M. 1990. Basal Sauropodomorpha. In D. B. Weishampel, P. Dodson, and H. Osmólska (eds.), *The Dinosauria,* 320–344. Berkeley: University of California Press.

Galton, P. M., and P. Uphurch. 2004. Prosauropoda. In D. B. Weishampel, P. Dodson, and H. Osmólska (eds.), *The Dinosauria,* 2nd ed., 232–258. Berkeley: University of California Press.

García, R. A., L. Salgado, and R. A. Coria. 2003. Primeros restos de dinosaurios saurópodos en el Jurásico de la Cuenca Neuquina, Patagonia, Argentina. *Ameghiniana* 40: 123–126.

González Riga, B. J. 2003. A new titanosaur (Dinosauria, Sauropoda) from the Upper Cretaceous of Mendoza Province, Argentina. *Ameghiniana* 40: 155–172.

Harris, J. D., and P. Dodson. 2004. A new diplodocoid sauropod dinosaur from the Upper Jurassic Morrison Formation of Montana, USA. *Acta Palaeontologica Polonica* 49 (2): 197–210.

Huene, F. 1932. *Die fossile Reptile-Ordnung Saurischia, ihre entwicklung und geschichte.* Monographien zur Geologie und Palaeontologie, ser. 1, vol. 4.

———. 1929. Los saurisquios y ornitisquios del Cretáceo Argentino. *Anales del Museo de La Plata* 3: 1–194.

Jalfin, J. A., and R. Herbst. 1995. La flora triásica del Grupo El Tranquilo, Provincia de Santa Cruz (Patagonia). Estratigrafía. *Ameghiniana* 32: 211–229.

Leanza, H. A. 1999. *The Jurassic and Cretaceous Terrestrial Beds from*

Southern Neuquén Basin, Argentina. Miscellaneous Publication 4. Instituto Superior de Correlación Geológica (Insugeo).

Leanza, H. A., and C. A. Hugo. 2001. Cretaceous red beds from southern Neuquén Basin (Argentina): age, distribution and stratigraphic discontinuities. In *VII International Symposium on Mesozoic Terrestrial Ecosystems,* 117–122. Special Paper 7. Asociación Paleontológica Argentina.

Leanza, H. A., S. Apesteguía, F. E. Novas, and M. S. de la Fuente. 2004. Cretaceous terrestrial beds from the Neuquén Basin (Argentina) and their tetrapod assemblages. *Cretaceous Research* 25: 61–87.

Le Loeuff, J. 1993. European titanosaurids. *Revue de Paléobiologie* 7: 105–117.

Lucas, S. G., and A. P. Hunt. 1989. *Alamosaurus* and the sauropod hiatus in the Cretaceous of the North American Western Interior. In J. O. Farlow (ed.), *Paleobiology of the Dinosaurs,* 75–85. Special Paper 238. Boulder, Colo.: Geological Society of America.

Lydekker, R. 1888. *Catalogue of the Fossil Reptilia and Amphibia in the British Museum. Pt. I. Containing the orders Ornithosauria, Crocodilia, Dinosauria, Squamata, Rhynchocephalia, and Proterosauria.* London: British Museum of Natural History.

Martinelli, A. G., and A. M. Forasiepi. 2004. Late Cretaceous vertebrates from Bajo de Santa Rosa (Allen Formation), Río Negro province, Argentina, with the description of a new sauropod dinosaur (Titanosauridae). *Revista del Museo Argentino de Ciencias Naturales* 6 (2): 257–305.

Martínez, R. D., O. Giménez, J. Rodríguez, and M. Luna. 1989. Hallazgo de restos de saurópodos en Cañadón Las Horquetas, Formación Matasiete (Aptiano), Chubut. *Resúmenes de las VI Jornadas Argentinas de Paleontología de Vertebrados,* 49–51.

Martínez, R. D., O. Giménez, J. Rodríguez, M. Luna, and M. C. Lamanna. 2004. An articulated specimen of the basal titanosaurian (Dinosauria, Sauropoda) *Epachthosaurus sciuttoi* from the early Late Cretaceous Bajo Barreal Formation of Chubut Province, Argentina. *Journal of Vertebrate Paleontology* 24: 107–120.

McIntosh, J. S. 1990. Sauropoda. In D. Weishampel, P. Dodson, and H. Osmólska (eds.), *The Dinosauria,* 345–401. Berkeley: University of California Press.

Nopcsa, F. 1902. Notizen über cretacische Dinosaurier. *Sitzungsberichte der Kaiserlichen Akademie der Wissenschaften in Wien* 9: 1–20.

Page, R., A. Ardolino, R. E. Barrio, M. de Franchi, A. Lizuain, S. Page, and D. S. Nieto. 1999. Estratigrafía del Jurásico y Cretácico del Macizo de Somún Curá, Provincias de Río Negro y Chubut. In R. Caminos (ed.), *Geología Argentina,* 460–488. Buenos Aires: Subsecretaría de Minería de la Nación.

Pereda-Suberbiola, X., F. Torcida Fernandez Balbor, L. A. Izquierdo, P. Huerta, D. Montero, and G. Perez. 2003. First rebbachisaurid dinosaur (Sauropoda, Diplodocoidea) from the Early Cretaceous of Spain: palaeobiogeographical implications. *Bulletin de la Société Géologique de France* 174 (5): 471–479.

Pol, D., and J. E. Powell. 2005. Anatomy and phylogenetic relationships of *Mussaurus patagonicus* (Dinosauria, Sauropodomorpha) from the Late Triassic of Patagonia. *Boletim de Resumos/II Congresso Latino-Americano de Paleontología de Vertebrados,* 208.

Powell, J. E. 1986. Revisión de los Titanosáuridos de América del Sur. Ph.D. diss., Universidad Nacional de Tucumán.
———. 1987. The titanosaurids. In J. F. Bonaparte (ed.), *The Late Cretaceous Fauna of Los Alamitos, Patagonia, Argentina,* 147–153. Revista del Museo Argentino de Ciencias Naturales "Bernardino Rivadavia" 3 (3).
———. 1990. *Epachthosaurus sciuttoi* (gen. et sp. nov.), un dinosaurio saurópodo del Cretácico de Patagonia (Provincia de Chubut, Argentina). *Actas del V Congreso Argentino de Paleontologia y Bioestratigrafia* 1: 123–128.
———. 2003. Revision of South American titanosaurid dinosaurs: palaeobiological, palaeobiogeographical and phylogenetic aspects. *Records of the Queen Victoria Museum* 111: 1–173.
Powell, J. E., O. Giménez, R. D. Martínez, and J. Rodríguez. 1989. Hallazgo de saurópodos en la Formación Bajo Barreal de Ocho Hermanos, Sierra de San Bernardo, Provincia de Chubut (Argentina) y su significado cronológico. *Anais do XI Congresso Brasileiro de Paleontologia,* Curitiba, 165–176.
Rauhut, O. W. 2002. Los dinosaurios de la Formación Cañadón Asfalto. *Ameghiniana* 39 (4): 15R–16R.
———. 2003a. Revision of *Amygdalodon patagonicus* Cabrera, 1947 (Dinosauria, Sauropoda). *Mitteilungen aus dem Museum für Naurkunde in Berlin, Geowissenschaftliche Reihe* 6: 173–181.
———. 2003b. A dentary of *Patagosaurus* (Sauropoda) from the Middle Jurassic of Patagonia. *Ameghiniana* 40: 425–432.
Rauhut, O. W., G. Cladera, P. Vickers-Rich, and T. H. Rich. 2003. Dinosaur remains from the Lower Cretaceous of the Chubut Group, Argentina. *Cretaceous Research* 24: 487–497.
Rauhut, O. W., K. Remes, R. Fechner, G. Cladera, and P. Puerta. 2005. Discovery of a short-necked sauropod dinosaur fom the Late Jurassic period of Patagonia. *Nature* 435: 670–672.
Riccardi, A. C., and S. E. Damborenea (eds.). 1993. *Léxico Estratigráfico de la Argentina.* Vol. 9, *Jurásico*. Buenos Aires: Asociación Geológica Argentina.
Rich, T. H., P. Vickers-Rich, O. Giménez, R. Cúneo, P. Puerta, and R. Vacca. 1999. A new sauropod dinosaur from Chubut Province, Argentina. In Y. Tomida, T. H. Rich, and P. Vickers-Rich (eds.), *Proceedings of the Second Gondwanan Dinosaur Symposium,* 61–84. National Science Museum Monograph 15.
Romer, A. S. 1956. *Osteology of the Reptiles.* Chicago: University of Chicago Press.
Salgado, L. 1993. Comments on *Chubutisaurus insignis* Del Corro (Saurischia; Sauropoda). *Ameghiniana* 30: 265–270.
———. 1996. *Pellegrinisaurus powelli* nov. gen. et sp. (Sauropoda, Titanosauridae) from the Upper Cretaceous of Lago Pellegrini, northwestern Patagonia, Argentina. *Ameghiniana* 33: 355–365.
———. 2000. Evolución y Paleobiología de los Saurópodos Titanosauridae. Ph.D. diss., Universidad Nacional de La Plata.
———. 2003. Should we abandon the name Titanosauridae? Some comments on the taxonomy of titanosaurian sauropods (Dinosauria). *Revista Española de Paleontología* 18: 15–21.
Salgado, L., and C. Azpilicueta. 2000. Un nuevo saltasaurino (Sauropoda, Titanosauridae) de la Provincia de Río Negro (Formación Allen,

Cretácico Superior), Patagonia, Argentina. *Ameghiniana* 37: 259–264.
Salgado, L., and J. F. Bonaparte. 1991. Un nuevo saurópodo Dicraeosauridae, *Amargasaurus cazaui* gen. et sp. nov. de la Formación La Amarga, Neocomiano de la Provincia del Neuquén, Argentina. *Ameghiniana* 28: 333–346.
Salgado, L., and J. O. Calvo. 1993. Report of a sauropod with amphiplatyan mid-caudal vertebrae from the Late Cretaceous of Neuquén Province (Argentina). *Ameghiniana* 30: 215–218.
Salgado, L., and R. A. Coria. 1993. El género *Aeolosaurus* (Sauropoda, Titanosauridae) en la Formación Allen (Campaniano-Maastrichtiano) de la Provincia de Río Negro. *Ameghiniana* 30: 119–128.
———. 2005. Sauropods of Patagonia: systematic update and notes on sauropod global evolution. In V. Tidwell and K. Carpenter (eds.), *Thunder-Lizards: The Sauropodomorph Dinosaurs,* 430–453. Bloomington: Indiana University Press.
Salgado, L., and R. A. García. 2002. Variación morfológica en la secuencia de vértebras caudales de algunos saurópodos titanosaurios. *Revista Española de Paleontología* 17: 211–216.
Salgado, L., and R. D. Martínez. 1993. Relaciones filogenéticas de los titanosáuridos basales *Andesaurus delgadoi* y *Epachthosaurus* sp. *Ameghiniana* 30: 339.
Salgado, L., R. A. Coria, and J. O. Calvo. 1997a. Evolution of titanosaurid sauropods I: phylogenetic analysis based on the postcranial evidence. *Ameghiniana* 34: 3–32.
———. 1997b. Presencia del género *Aeolosaurus* (Sauropoda, Titanosauridae) en la Formación Los Alamitos, Cretácico Superior de la Provincia de Río Negro, Argentina. *Revista Universidade Guarulhos, Geociências* II (6): 44–49.
Salgado, L., A. Garrido, S. Cocca, and J. R. Cocca. 2004. Lower Cretaceous rebbachisaurid sauropods from Cerro Aguada del León (Lohan Cura Formation), Neuquén Province, northwestern Patagonia, Argentina. *Journal of Vertebrate Paleontology* 24 (4): 903–912.
Salgado, L., S. Apesteguía, and S. E. Heredia. 2005. A new specimen of *Neuquensaurus australis,* a Late Cretaceous Saltasaurinae titanosaur from north Patagonia. *Journal of Vertebrate Paleontology* 25 (3): 623–634.
Sanz, J. L., J. E. Powell, J. Le Loeuff, R. D. Martínez, and X. Pereda-Suberbiola. 1999. Sauropod remains from the Upper Cretaceous of Laño (northcentral Spain): titanosaur phylogenetic relationships. *Estudios del Museo de Ciencias Naturales de Álava.* 14 (Special No. 1): 235–255.
Sciutto, J. C., and R. D. Martínez. 1994. Un nuevo yacimiento fosilífero de la Formación Bajo Barreal (Cretácico Tardío) y su fauna de saurópodos. *Naturalia Patagónica, Ciencias de la Tierra* 2: 27–47.
Sereno, P. C. 1998. A rationale for phylogenetic definitions, with applications to the higher-level taxonomy of Dinosauria. *Neues Jahrbuch für Geologie und Paläontologie Abhandlungen* 210: 41–83.
Sereno, P. C., A. L. Beck, D. B. Dutheil, H. C. E. Larsson, G. H. Lyon, B. Moussa, R. W. Sadleir, C. A. Sidor, D. J. Varrichio, G. P. Wilson, and J. A. Wilson. 1999. Cretaceous sauropods from the Sahara and the uneven rate of skeletal evolution among dinosaurs. *Science* 286: 1342–1347.

Upchurch, P. 1995. The evolutionary history of sauropod dinosaurs. *Philosophical Transactions of the Royal Society of London* 349: 365–390.
———. 1998. The phylogenetic relationships of sauropod dinosaurs. *Zoological Journal of the Linnean Society* 124: 43–103.
Upchurch, P., P. M. Barrett, and P. Dodson. 2004. Sauropoda. In D. B. Weishampel, P., Dodson, and H. Osmólska (eds.), *The Dinosauria*, 2nd ed., 259–322. Berkeley: University of California Press.
Wilson, J. A. 1999. A nomenclature for vertebrae laminae in sauropods and other saurischian dinosaurs. *Journal of Vertebrate Paleontology* 19: 639–653.
———. 2002. Sauropod dinosaur phylogeny: critique and cladistic analysis. *Zoological Journal of the Linnean Society* 136: 217–276.
Wilson, J. A., and P. C. Sereno. 1998. Early evolution and higher-level phylogeny of sauropod dinosaurs. Society of Vertebrate Paleontology Memoir 5. *Journal of Vertebrate Paleontology* 18 (Suppl. to 2): 1–68.
Wilson, J. A., R. N. Martínez, and O. Alcober. 1999. Distal tail segment of a titanosaur (Dinosauria: Sauropoda) from the Upper Cretaceous of Mendoza, Argentina. *Journal of Vertebrate Paleontology* 19: 591–594.

9. Nonavian Theropods

RODOLFO A. CORIA

Introduction

The Theropoda comprise all dinosaurs closer to birds than to sauropodomorphs. The most plesiomorphic taxon of this group is *Eoraptor* and includes ceratosaurs, coelurosaurs, birds, and all descendants of their common ancestor. For organizing purposes, in this chapter, all avian theropods (*Archaeopteryx,* modern birds, and all descendants) will be excluded from consideration because they are analyzed elsewhere (Chiappe this volume).

The Patagonian nonavian theropod record includes exclusively Neotheropoda (=*Coelophysis,* Neornithes, their most recent common ancestor and all descendants; Sereno 1998) taxa. No Triassic Patagonian meat-eating dinosaurs are known, and the Jurassic record is quite scarce, besides some Late Jurassic theropod diversity represented only by footprints from Santa Cruz Province (Casamiquela 1964; Calvo this volume). In this regard, two Jurassic theropod dinosaurs have been properly described and named, *Piatnitzkysaurus floresi* (Bonaparte 1979), and *Condorraptor currumili* (Rauhut 2005), both probably related with basal Tetanurae (sensu Gauthier 1986).

By far, the richest and most diverse sample of meat-eating Patagonian dinosaurs comes from Cretaceous beds. Their remains have been collected from most of the Patagonian provinces (Fig. 9.1), and the specimens range from almost complete and articulated skeletons, up to fragmentary, sometimes single bone–based species. They will be summarized in light of the most recent reports, taking under consideration most published species and certain indeterminate specimens that are worth mention in the discussion.

Institutional abbreviations. MACN, Museo Argentino de

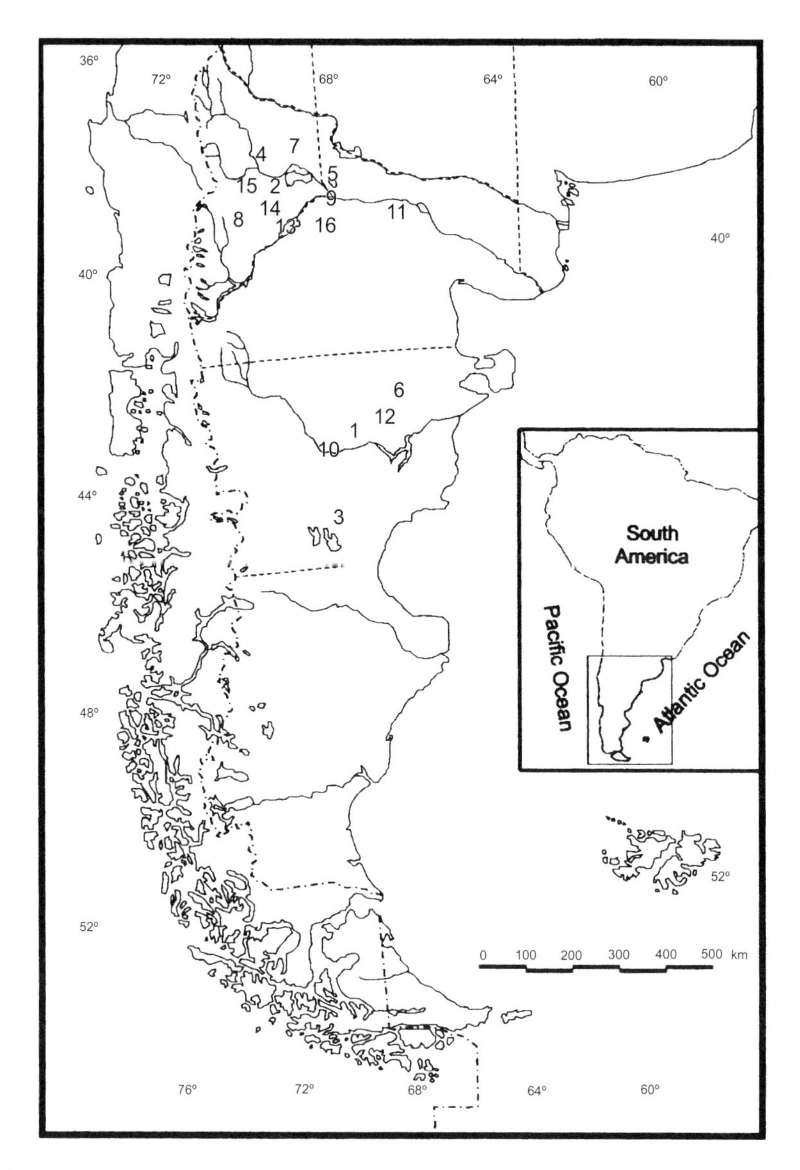

Figure 9.1. Map depicting the provenance locations of most significant specimens of theropod dinosaurs recorded in Patagonia. Localities numbered as they are mentioned throughout the text. 1, Genyodectes serus, *Cañadón Grande, Department of Paso de Indios, Chubut Province; 2,* Ilokelesia aguadagrandensis, *Aguada Grande, Department of Confluencia, Neuquén Province; 3,* Xenotarsosaurus bonapartei, *6 km north of Estancia Ocho Hermanos, Department of Sarmiento, Chubut Province; 4,* Ekrixinatosaurus novasi, *approximately 34 km northwest of Añelo, Neuquén Province; 5,* Abelisaurus comahuensis, *sand quarries of Pellegrini Lake, Department of General Roca, Río Negro Province; 6,* Carnotaurus sastrei, *Estancia Pocho Sastre, near Bajada Moreno, Department of Telsen, Chubut Province; 7,* Aucasaurus garridoi, *Auca Mahuevo (Chiappe et al. 1998), near Mina La Escondida, in the northeastern corner of the Neuquén Province; 8,* Ligabueino andesi, *5 km to the southeast of La Amarga Creek, Catan Lil Department, Neuquén Province; 9,* Velocisaurus unicus *and* Alvarezsaurus calvoi, *campus of University of Comahue, 200 m east of the main building, Neuquén City, Neuquén Province; 10,* Condorraptor currumili *and* Piatnitzkysaurus floresi *Bonaparte, Cerro Cóndor, Chubut Province; 11,* Quilmesaurus curriei, *Salitral Ojo de Agua, 40 km south of Gral. Roca, Río Negro Province; 12,* Tyrannotitan chubutensis, *Estancia La Juanita, 28 km northeast of Paso de Indios, Chubut Province; 13,* Giganotosaurus carolinii, *15 km south of Villa El Chocón, Neuquén Province; 14,* Mapusaurus roseae, *Cañadón del Gato in the Cortaderas area 20 km southwest of Plaza Huincul, Neuquén Province; 15,* Megaraptor namunhuaiquii, Neuquenraptor argentinus, Unenlagia comahuensis, *and* Patagonykus puertai, *Sierra del Portezuelo, 22 km west of Plaza Huincul, Neuquén Province; 16,* Buitreraptor gonzalezorun, *La Buitrera, 40 km southeast of El Chocón, Río Negro Province.*

Ciencias Naturales, Buenos Aires, Argentina; MCF-PVPH, Museo "Carmen Funes," Paleontología de Vertebrados, Plaza Huincul, Neuquén, Argentina; MLP, Museo de La Plata, Buenos Aires, Argentina; MPCA, Museo Provincial "Carlos Ameghino," Cipolletti, Río Negro, Argentina; MPEF-Pv, Museo Paleontológico "Egidio Feruglio," Paleontología de Vertebrados, Trelew, Chubut, Argentina; PVL, Instituto "Miguel Lillo," Paleontología de Vertebrados, Tucumán, Argentina; MUCPv, Museo de la Universidad Nacional del Comahue, Paleontología de Vertebrados, Neuquén, Argentina; UNPSJB-Pv, Universidad Nacional de la Patagonia "San Juan Bosco," Paleontología de Vertebrados, Comodoro Rivadavia, Chubut, Argentina.

Systematic Paleontology
Theropoda Marsh 1881
Neotheropoda Bakker 1986
Ceratosauria Marsh 1884
Ceratosauria incertae sedis
***Genyodectes* Woodward 1901**
***Genyodectes serus* Woodward 1901**
Fig. 9.2

Locality and age. Cañadón Grande, Department of Paso de Indios, Chubut Province, ?Cerro Castaño Member, Cerro Barcino Formation, Aptian-Albian, Lower Cretaceous.

Holotype. MLP-26–39, incomplete premaxillae united to the anteroventral sections of the maxillae, and the anterior half of both dentaries, bearing upper and lower teeth, parts of both supradentaries, and fragments of the left splenial (Woodward 1901).

Diagnosis. Rauhut (2004) proposed the following diagnostic features for *Genyodectes serus:* premaxillary teeth arranged in an overlapping en-echelon pattern and longest maxillary tooth crowns longer apicobasally than the minimal dorsoventral depth of the mandible. In his diagnosis, Rauhut recognizes that those features are probably also present in *Ceratosaurus,* which differs from *Genyodectes* in having three premaxillary teeth.

Comments. Originally, *Genyodectes* was tentatively related with tyrannosaurids on the bases of the large size of the specimen and some features connected with the shape of the snout (Huene 1926). During the last two decades, the affinities with the tyrannosaurids have been dismissed; researchers now agree that most of the similarities are either primitive features or homoplasies (Molnar 1990; Carpenter 1997). Therefore, the definition of its phylogenetic relationships of *Genyodectes* with other theropods remains uncertain. The presence of some features, such as the ornamented surface of maxillae and premaxillae and the shape of the teeth, strongly suggest that *Genyodectes* could be some kind of abelisaurid theropod. In several aspects, the *Genyodectes* snout evokes the morphology exhibited by *Carnotaurus* and *Aucasaurus.*

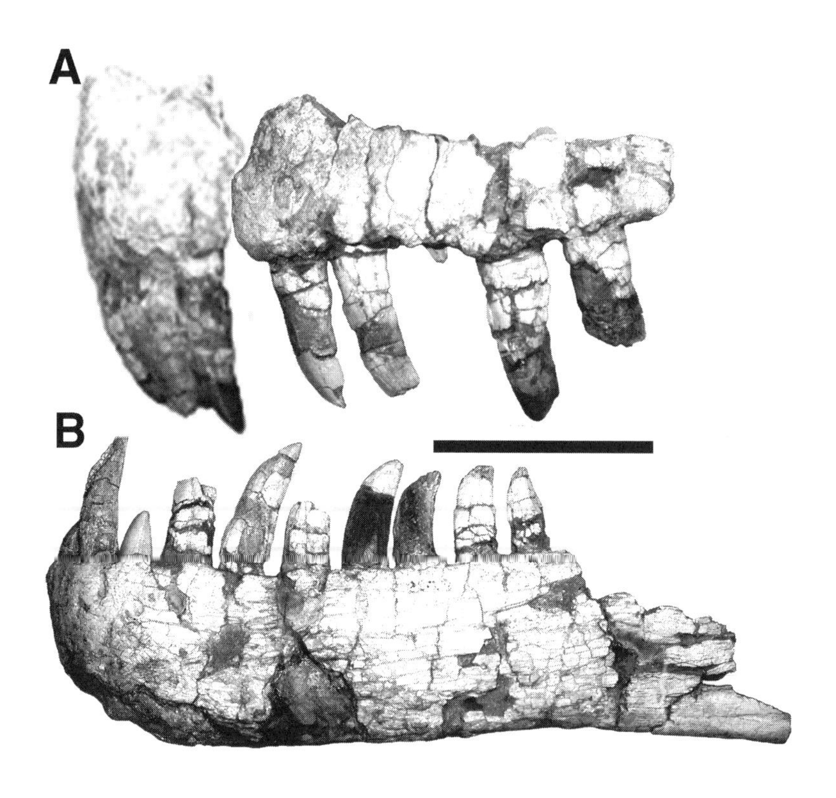

Figure 9.2. Genyodectes serus *(type specimen MLP-26-39) in left lateral view. (A) Maxilla. (B) Dentary. Scale bar = 10 cm.*

Neoceratosauria Novas 1992
Abelisauria Novas 1992
Abelisauridae Bonaparte and Novas 1985
Abelisaurinae Sereno 1998
***Abelisaurus* Bonaparte and Novas 1985**
***Abelisaurus comahuensis* Bonaparte and Novas 1985**

Locality and age. Sand quarries of Pellegrini Lake, Department Gral. Roca, Río Negro Province. Anacleto Formation (lower Campanian), Río Colorado Subgroup, Neuquén Group (Heredia and Salgado 1997).

Holotype. MPCA-11098, most of a skull that lacks right maxilla, jugal, quadrate-jugal, squamosal, and most of palate.

Diagnosis. Abelisaurus was diagnosed by Bonaparte and Novas (1985) as a large-sized theropod with a high, elongate, and deeply excavated skull, with prominent nasal rugosities, preorbital fenestrae proportionally larger than Tyrannosauridae and other Jurassic and Cretaceous carnosaurs, small accessory preorbital opening placed on the anterior border of the preorbital fenestra, interorbital region broader than other carnosaurs, with lateral lacrimal-postorbital contact that forms a supraorbital shelf, orbital fenestra very high, framed by postorbital and lacrimal, slightly wider ventrally, with low jugal, remaining the condition in *Tyrannosaurus rex;* almost horizontal, backwardly projected squamosal with its quadrato-jugal process oriented ventrally unlike Tyrannosauridae, but similar to *Ceratosaurus* and *Allosaurus;* quadrate longer than Tyrannosauridae, comparable to *Ceratosaurus;* extensive infratemporal fenestra as in *Ceratosaurus* and longer than in Tyrannosauridae; supratemporal fenestrae very short anteroposteriorly, extensive horizontal ramus of maxilla, with laterally compressed teeth; braincase similar to *Piatnitzkysaurus* in having large winglike laterosphenoid processes and the very transversely compressed basisphenoid.

Comments. Several anatomical features present in the skull of *Abelisaurus* leave no doubt about its close affinities with other abelisaurids like *Carnotaurus, Majungatholus, Aucasaurus,* and *Rugops* (Bonaparte 1985; Sampson et al. 1998; Coria et al. 2002; Sereno et al. 2004), such as ornamented premaxillae and maxillae, supraorbital shelf formed by an ossified palpebral (convergent with Carcharodontosauridae; Coria and Currie 2006), and very tall quadrate. *Abelisaurus* has no derived structures on the skull such as horns or swells, like the Nigerian *Rugops* (Sereno et al. 2004) and unlike *Carnotaurus, Majungatholus,* and *Aucasaurus* (Bonaparte 1985; Sampson et al. 1998; Coria et al. 2002). This fact, together with the belief that the skull of *Abelisaurus* is longer than it is tall, has led researchers to regard this abelisaurid as a more primitive form among currently known abelisaurids. The skull was collected by unskilled hands and assembled following the Laurasian stereotype of large theropods with long and low skulls, like *Allosaurus* and *Tyrannosaurus,* some time before the discovery of *Carnotaurus* and *Majungatholus* (Bonaparte 1985; Sampson et al. 1998). Therefore, the length of the *Abelisaurus* skull is perhaps the result of a preparation artifact, bearing instead the typical stout, short-faced skull of the Abelisauridae. Finally, all claimed *Abelisaurus comahuensis* autopomorphies are currently considered simplesiomorphies at different levels of Theropoda. Restudying this skull is necessary in order to resolve the phylogenetic diversity within the most derived abelisaurid taxa.

Carnotaurinae Sereno 1998
Carnotaurini Coria, Chiappe, and Dingus 2002
***Carnotaurus* Bonaparte 1985**
***Carnotaurus sastrei* Bonaparte 1985**
Fig. 9.3A

Locality and age. Estancia Pocho Sastre, near Bajada Moreno, Department of Telsen, Chubut Province. Lower Section of La Colonia Formation, Campanian-Maastrichtian.

Holotype. MACN-CH-894, an almost complete, articulated skeleton with skin impressions, lacking most of the tail, both tibiae, and feet.

Diagnosis. Bonaparte et al. (1990) diagnosed *Carnotaurus* as follows: Abelisaurid with skull shorter and higher than in *Abelisaurus* and other theropods, and with deep snout and prominent frontal horns; orbits divided into two parts: an upper, rounded

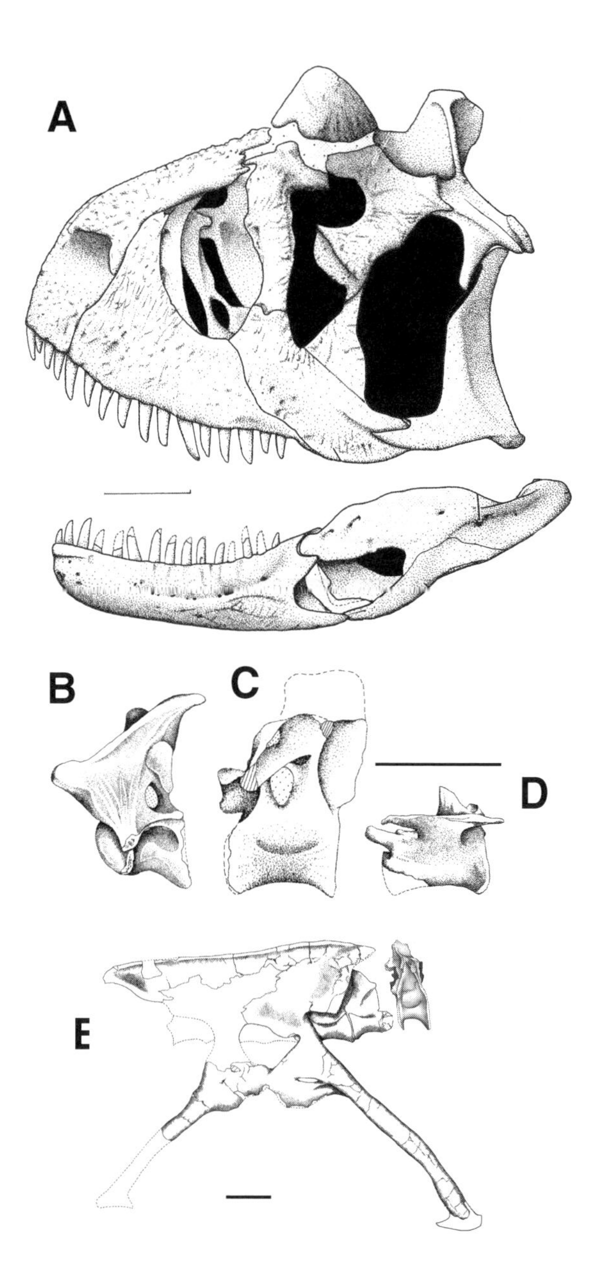

Figure 9.3. Patagonian abelisaurs. (A) Skull of Carnotaurus sastrei. *Cervical (B), dorsal (C), and caudal (D) vertebra of* Ilokelesia aguadagrandensis. *(E) Last dorsal vertebrae and pelvis of MCF-PVPH-237. All illustrations in lateral view. Scale bar = 10 cm.* (A) Modified from Bonaparte et al. (1990). (B, C, D) Partially taken and modified from Coria and Salgado (1998). (E) Modified from Paulina Carabajal et al. (2003).

section anterolaterally projected for the eyes, and a lower, dorsoventrally elongated section; small supratemporal opening, with parietal and squamosal forming a high posterior wall and having a low, lateral border; infratemporal and preorbital openings smaller than in *Abelisaurus;* quadrate very high, and squamosal having a short, rodlike ventral projection; loose contact between dentary and postdentary bones, forming a large mandibular fenestra; cervical vertebrae behind the axis with reduced neural spines and high, well-developed epipophyses; sacrum with seven fused vertebrae; forelimbs reduced, with extremely short and stout radius and ulna; both provided with large, convex distal ends; ilia long and square shaped; pubes, ischia, and femora long and slender.

Comments. The holotype specimen of *Carnotaurus* was long regarded as the most complete abelisaurid known. For that reason, many of the unique features shown by *Carnotaurus* remained as specific autopomorphies for many years. Also, the extreme derived condition of several characters throughout the entire postcranial skeleton conspired the comparison with another abelisaurid taxa from which only cranial information was provided (Bonaparte and Novas 1985; Sampson et al. 1998). Within the last 10 years, the discoveries of several abelisaurid-related taxa have allowed researchers to examine which features are unique for *Carnotaurus* and which ones have a wider distribution within the clade. The extreme reduction of the forearm, especially the radius and ulna, seems to be a linking feature for at least two taxa, *Carnotaurus* and *Aucasaurus* (Coria et al. 2002). Also, the peculiar vertebral morphology described for *Carnotaurus* is one more abelisaurid characteristic found in all the clade members whose vertebrae are known (*Aucasaurus, Majungatholus, Ekrixinatosaurus;* Sampson et al. 1998; Coria et al. 2002; Calvo et al. 2004b). *Carnotaurus sastrei* is nonetheless a valid taxon based on conspicuous characters such as the paired, robust frontal horns above the orbits; the anteroposterior epipophysial projections on cervical vertebrae, the stout, cylindrically shaped humerus, and the large, elongated metacarpal I.

Aucasaurus Coria, Chiappe, and Dingus 2002
Aucasaurus garridoi Coria, Chiappe, and Dingus 2002

Locality and age. Auca Mahuevo (Chiappe et al. 1998), near Mina La Escondida, in the northeastern corner of the Province of Neuquén. Anacleto Formation (lower Campanian), Río Colorado Subgroup, Neuquén Group (Dingus et al. 2000).

Holotype. MCF-PVPH-236, mid- to large-sized skeleton of an adult specimen that preserved most of the skull, fragments of lower jaws, posterior presacral vertebrae, sacrum, and almost two-thirds of the tail, complete pectoral girdle, forearms, pelvic girdle, and hind limbs.

Diagnosis. Aucasaurus was diagnosed by Coria et al. (2002) as a carnotaur theropod whose skull differs from *Carnotaurus sastrei* by having a longer and lower rostrum and external antorbital fenestra, a horizontal ventral margin of the latter, complete lateral exposure of the maxillary fenestra, frontal swells instead of horns, a sigmoid outline of the dentigerous margin of the maxilla, and chevrons with dorsally open hemal canals.

Comments. Aucasaurus exhibits many features that have allowed researchers to improve the number of synapomorphies that link the Abelisauridae. Before the discovery of *Aucasaurus,* this family was based on cranial features, mainly because postcranial information for *Abelisaurus* was lacking. The finding of a very well preserved, partially articulated *Aucasaurus* specimen that preserved part of the skull, posterior presacrals, sacrum, most of the tail, and complete fore- and hind limbs allowed it to be related to the diversified but quite unknown abelisaurid theropods (Coria et al. 2002).

Noasauridae Bonaparte and Powell 1980
Ligabueino Bonaparte 1996
Ligabueino andesi Bonaparte 1996

Locality and age. Three kilometers to South East of the La Amarga creek, on Road No. 40, Catan Lil Department, Neuquén Province. Antigual Member of the La Amarga Formation, lower Barremian (Leanza and Hugo 2001).

Holotype. MACN-N-42, most of a cervical neural arch, a dorsal centrum, two posterior dorsal neural arches, a complete caudal vertebra, most of a right ilium, both incomplete pubes, a complete left femur, two phalanges, indeterminate fragments.

Diagnosis. Ligabueino was diagnosed by Bonaparte (1996) on the basis of the following features: small abelisaur with (1) very reduced neural spines in cervical and posterior vertebrae; (2) a rather flat dorsal surface and a triangular, well-marked axial depression in front of the neural spine base of anterior cervicals; (3) long prezygapophyseal process with a ridge defining lateral and dorsal planes in cervical vertebrae; (4) elongated posterior dorsal vertebrae bearing winglike transverse processes and reduced neural spines; (5) femur with trochanteric shelf and a modest but well-defined lesser trochanter on anterior dorsal half of trochanteric shelf; (6) very low and thin fourth trochanter; (7) distal section of femur with a defined but modest anteromedial crest; (8) ilium low, elongate, with postacetabular section of its blade larger than preacetabular section.

Comments. Although fragmentary, *Ligabueino* is a very interesting specimen because it represents the oldest South American abelisaur record and the only Patagonian noasaurid known. Several features present in the holotype specimen of *Ligabueino* permit no doubt about its abelisaur affinities (characters 1, 2, and 8). Other features are better indications of its theropod condition. However, some of the diagnostic features proposed in the original description may be considered as unique characteristics of *Ligabueino* (characters 3 and 4). Noasaurids are a clade of small-sized theropods with an increasing incipient record (Carrano et al. 2002). Recent phylogenetic approaches to abelisaur internal relationships face the problem of the incompleteness of data as a result of the fragmentary condition of the specimens (Carrano et al. 2002). Thus, the phylogenetic acceptance of a clade Noasauridae awaits more data.

Velocisaurus Bonaparte 1991
Velocisaurus unicus Bonaparte 1991

Locality and age. Campus of University of El Comahue, 200 m east of the main building, Neuquén City, Neuquén Province. Bajo de la Carpa Formation (Santonian), Río Colorado Subgroup, Neuquén Group.

Holotype. MUCPv-41, right tibia, astragalus, metatarsals II, III, and IV, and digits II to IV almost complete.

Diagnosis. Small theropod with unfused metatarsals, metatarsal III straight, two to three times broader than lateral metatarsals; metatarsals II and IV very thin, astragalus unfused to tibia, which it contacts all along its distal end and anterior side.

Comments. Velocisaurus was primarily assigned by Bonaparte (1991) to a different and unusual family with uncertain abelisaur affinities. The scarcity of the available material, which was limited to the distal half of a hind limb, impeded ascertainment of its precise phylogenetic relationships. The unusual metatarsals arrangement, which shows the opposite of the artometatarsalian condition, having the metatarsal III proximally larger than the lateral ones, was proposed by Bonaparte as indicating a completely new clade of theropod dinosaurs. The recently described noasaurid from Malagasy, *Masiakasaurus* (Sampson et al. 2001), presents a similar condition of reduction of metatarsal II, supporting the abelisaur affinities of *Velocisaurus,* as supposed by Bonaparte in his original description.

Abelisauria incertae sedis
Xenotarsosaurus Martínez, Giménez, Rodríguez, and Bochatey 1986
Xenotarsosaurus bonapartei Martínez, Giménez, Rodríguez, and Bochatey 1986

Locality and age. Six kilometers north of Estancia Ocho Hermanos, Department of Sarmiento, Chubut Province. Lower Member of the Bajo Barreal Formation (Cenomanian-Turonian).

Holotype. UNPSJB-Pv-184 and -612, two incomplete cervicodorsal vertebrae; right femur, tibia, fibula, and calcaneum-astragalus complex.

Diagnosis. Midsized carnosaur, with anterior dorsal vertebrae with centrum wider than high; anterior side concave, not flat, as in *Carnotaurus;* higher neural arch with lateral pits over the neural canal deeper than in the latter; femur similar to *Carnotaurus;* tibia with slender proximal end, with pits at the base of the cnemial crest; tibia and fibula tight against each other with indication of the presence of ligament insertions; astragalus and calcaneum fused; contact between tibia and astragalus-calcaneum not visible in posterior view; fibula with very compressed shaft; femur-tibia ratio of 1/0.94.

Comments. Martínez et al. (1986) identified *Xenotarsosaurus* within the family Abelisauridae on the basis of number of similarities between the anterior dorsal vertebrae and the hind limbs of this species and *Carnotaurus sastrei.* The two available dorsal vertebrae of *Xenotarsosaurus* are badly weathered, making it hard to actually see features that link it more closely to *Carnotaurus.* Novas (1989) agreed with the familiar assignation of *Xenotarsosaurus bonapartei* and diagnosed the species on the basis of a single autapomorphy: presence of a concave articular surface in the dorsal centra, a character also present in *Ilokelesia* (Coria and Salgado 1998). Perhaps the most relevant anatomical peculiarity of the Chubut species is the preservation of a complete left hind limb, in which the full coosification of the tibia with the proximal tarsals is the most remarkable feature. This condition, which parallels the avian tibiotarsus, was proposed as an abelisaurid synapomorphy by Bonaparte (1991).

Abelisaurid affinities of *Xenotarsosaurus* were challenged by Coria and Rodríguez (1993), supporting earlier suggestions that the derived feature shared by *Xenotarsosaurus* and *Carnotaurus*—i.e., presence of deep pre- and postspinal basins—is a synapomorphy that links the clade formed by *Ilokelesia* + Abelisauroidea (Novas 1989; Coria and Rodríguez 1993; Coria and Salgado 1998). However, the recent finding of *Aucasaurus garridoi*, an abelisaurid closely related to *Carnotaurus* with hind limbs virtually indistinguishable from that of *Xenotarsosaurus* (including a fused tibiotarsus), suggests that this character would have a broader distribution within a more inclusive assemblage of ceratosauroids (Rowe and Gauthier 1990), reversed in Noasauridae, rather than an exclusive synapomorphy for Abelisauridae (Bonaparte 1991).

Ilokelesia Coria and Salgado 1998
Ilokelesia aguadagrandensis Coria and Salgado 1998
Fig. 9.3C, D

Locality and age. Aguada Grande, 15 km south of Plaza Huincul, Neuquén Province. Huincul Formation (upper Cenomanian), Río Limay Subgroup, Neuquén Group.

Holotype. MCF-PVPH-35, right postorbital and quadrate, occipital condyle and unidentified fragments of the braincase, partial third? cervical vertebra, fourth cervical vertebra, partial fifth cervical vertebra, partial centum of sixth cervical vertebra, incomplete posterior dorsal, five articulated midcaudal vertebrae with chevrons, three fragmentary cervical ribs, nine preungual phalanges, two ungual phalanges.

Diagnosis. Medium-sized theropod distinguished by having a quadrate with lateral condyle very reduced and posterior border of the ventral articular surface formed completely by the medial condyle; cervical vertebrae with poorly defined diapopostzygapophyseal laminae; posterior dorsal vertebrae with infraparapophyseal laminae ventrally concave and parapophyses ventrally oriented; posterior dorsal vertebrae lacking pleurocoels; caudal vertebrae in central third of the tail with distally expanded transverse processes bearing cranially and caudally projecting processes; distal edge of caudal transverse processes slightly concave in the mid part.

Comments. Although fragmentary, *Ilokelesia* is the best available evidence of the early evolutionary stages of the Abelisauria (Coria and Salgado 1998; Carrano et al. 2002). Some recent contributions have suggested that *Ilokelesia* could be nested within the Abelisauridae (Coria et al. 2002; Sereno et al. 2004). Nevertheless, the fragmentary condition of the holotype specimen—currently the only one known for this species—does not permit a more confident affiliation among the abelisaurids. On the other hand, some features, such as absence of anterior projection of cervical epipophyses, show a primitive condition absent in both Noasauridae and Abelisauridae.

The presence of *Ilokelesia* in rocks of the Huincul Formation, from which a *Giganotosaurus*-related form was also recovered, indicates that the former might have been one of the last abelisaurs associated with carcharodontosaurid theropods before their Turonian extinction (Coria and Salgado 2005).

Ekrixinatosaurus Calvo, Rubilar-Rogers, and Moreno 2004
Ekrixinatosaurus novasi Calvo, Rubilar-Rogers, and Moreno 2004

Locality and age. Approximately 34 km northwest of Añelo, Neuquén Province, northwestern Patagonia, Argentina, Candeleros Formation (lower Cenomanian), Río Limay Subgroup, Neuquén Group.

Holotype. MUCPv-294, a well-preserved disarticulated skeleton with elements including left and partial right maxillae; basicranium; both dentaries; teeth; cervical, a dorsal, sacral, and caudal vertebrae; hemal arches; ribs; ilia; pubis; and proximal ischia; left and distal end of right femur; left tibia; left astragalus and calcaneum; proximal end of left fibula and right tibia; metatarsals; phalanges; and a pedal ungual.

Diagnosis. Ekrixinatosaurus was diagnosed by Calvo et al. (2004b) as a large abelisaur theropod distinguished by having a fenestra between the postorbital and the anterior border of the frontal, a protuberance directed backward on the contact between the parietals with the paraoccipital, cervical vertebrae craniocaudally compressed, cervical neural spines as tall as the epipophyses, midposterior cervical centrum with ventral side flattened, two wide foramina in the midposterior cervicals, small prespinal depression with a pneumatic excavation connected to the neural canal in the middle-posterior cervicals, small prespinal lamina in midcervicals, tibia with a swelling at midshaft.

Comments. The identification of *Ekrixinatosaurus* as an abelisauroid theropod is unquestionable on the basis of the presence of stout snout inferred from a short maxilla, the fused interdental plates in the maxilla, the convex ventral edge of the dentary, the presence of a prezygoepipophyseal lamina, and the unique anteroposterior distal expansions of the caudal vertebrae. Unfortunately, none of the autapomorphies that diagnose this form was properly illustrated in the original description. Therefore, it is not possible from the original paper to evaluate their phylogenetic significance. It is worthy of remark that the lateral, robust projection that produces a fenestra between the postorbital and the frontal (Calvo et al. 2004b) is also present in some carcharodontosaurids (i.e., *Giganotosaurus* and *Mapusaurus;* Coria and Salgado 1995;

Coria and Currie 2006) and inferred in *Ilokelesia aguadagrandensis* (Coria and Salgado 1998). This element has been identified as the palpebral that bridged between lacrimal and postorbital (Coria and Currie 2006). Indeed, the inclusion of *Ekrixinatosaurus* within Abelisauridae is preliminary because of the rather modest 13-character data matrix almost completely composed of abelisaurid synapomorphies. It is likely that adding additional scoring information from the incomplete but quite informative holotype skeleton of *Ekrixinatosaurus* will allow researchers to develop a more realistic hypothesis about the phylogenetic relationships of this interesting specimen.

Tetanurae Gauthier 1986
Spinosauroidea Stromer 1931
Megalosauridae Huxley 1869
***Piatnitzkysaurus* Bonaparte 1979**
***Piatnitzkysaurus floresi* Bonaparte 1979**

Locality and age. Cerro Cóndor, 1 km West Estancia Farías, Chubut Province. Lower Section of the Cañadón Asfalto Formation, Callovian (Bonaparte 1979).

Holotype. PVL 4073, most of the skeleton of a medium-sized adult specimen.

Hypodigm. MACN-CH-895, one individual represented by part of the right maxilla; right humerus; proximal half of pubes; one ischium; left tibia; left metatarsals II, II, and IV; two posterior dorsal vertebrae; two dorsal centra; and four sacral vertebrae.

Diagnosis. *Piatnitzkysaurus floresi* was diagnosed by Bonaparte (1979) on the basis of the following characters: presence of profound depressions between the basioccipital condyle and the opisthotic apophyses; braincase similar to *Eustreptospondylus* although with more developed lateral depressions of basisphenoid and laterosphenoid wings; scapula shorter than *Allosaurus;* subcircular coracoid; pubis with obturator foramen completely enclosed by bone; ischium ventrally longer than *Allosaurus;* ulna, tibia, and femur more slender than *Allosaurus* but more robust than *Dilophosaurus.*

Comments. Bonaparte (1979) nested *Piatnitzkysaurus* among Allosauridae. Other authors, however, have considered it as a basal Tetanurae with uncertain affinities (Novas 1997a; Sereno 1997). Indeed, many, if not all, of the features proposed at the original diagnosis are plesiomorphies widely distributed among tetanurans. Nevertheless, later assumptions regarding the phylogenetic relationships of *Piatnitzkysaurus* are basically expressions of conservative thought rather than being based on character analysis. Retention of primitive features, such as the presence of obturator foramen or subcircular coracoid, distinguish it from *Allosaurus* and other Jurassic forms. Therefore, its placement as a basal taxon within Tetanurae seems reasonable, although its precise phylogenetic relationships remain uncertain.

Avetheropoda Paul 1988
Carnosauria Huene 1920
Carcharodontosauridae Stromer 1931
Tyrannotitan chubutensis Novas, de Valais, Vickers-Rich, and Rich 2005

Locality and age. Estancia La Juanita, 28 km northeast of Paso de Indios, Chubut Province, Argentina. Possibly Cerro Castaño Member, Cerro Barcino Formation, Aptian (Musacchio and Chebli 1975; Codignotto et al. 1978; Rich et al. 1998), Chubut Group, San Jorge Basin.

Holotype. MPEF-PV 1156, partial dentaries, isolated teeth; dorsals 3–8 and 11–14; proximal caudal vertebra; isolated ribs and hemal arches; incomplete left scapulocoracoid and right humerus and ulna; pubes, ischia, and fragments of left ilium; almost complete left femora, fibula, and metatarsal II.

Paratype. MPEF-PV 1157, jugals; right dentary; isolated teeth; atlas; cervical 9?; dorsals 7?, 10, and 13; partially preserved fused centra of sacrals 1–5; isolated distal caudals; ribs; right femur; incomplete left metatarsal II; pedal phalanges 2.I, 2.II, and 3.III. Paratype specimen is approximately 7% larger than that of the holotype.

Diagnosis. Teeth with bilobate denticles on rostral carina, deep mental groove on dentary, posterior dorsal vertebrae with strongly developed ligament scars on neural spines.

Comments. Currently, we know that the specimens are somewhat more complete: several additional bones have been identified and prepared. A preliminary description and phylogenetic analysis of these specimens has been recently published (Novas et al. 2005). As far we know, these Chubutian forms comprise the oldest South American carcharodontosaurid record, and therefore they likely represent a new species. Several features, like the broad lacrimal process of the jugal and deeply excavated dorsal neural spines, suggest close relationships with the carcharodontosaurid *Mapusaurus roseae* from the Huincul Formation (Coria and Currie 1997, 2006).

Giganotosaurinae Coria and Currie 2006
Giganotosaurus Coria and Salgado 1995
Giganotosaurus carolinii Coria and Salgado 1995

Fig. 9.4A

Locality and age. Fifteen kilometers south of Villa El Chocón, Neuquén Province. Candeleros Formation (lower Cenomanian), Río Limay Subgroup, Neuquén Group.

Holotype. MUCPv-CH-1, a disarticulated skeleton represented by a partial skull, most of the vertebral column, complete pectoral and pelvic girdles, both femora, left tibia, and fibula.

Diagnosis. Dorsoventrally wide main body of the maxilla, with subparallel dorsal and ventral edges; eavelike supraorbital lacrimal-postorbital contact, two pneumatic foramina in the inter-

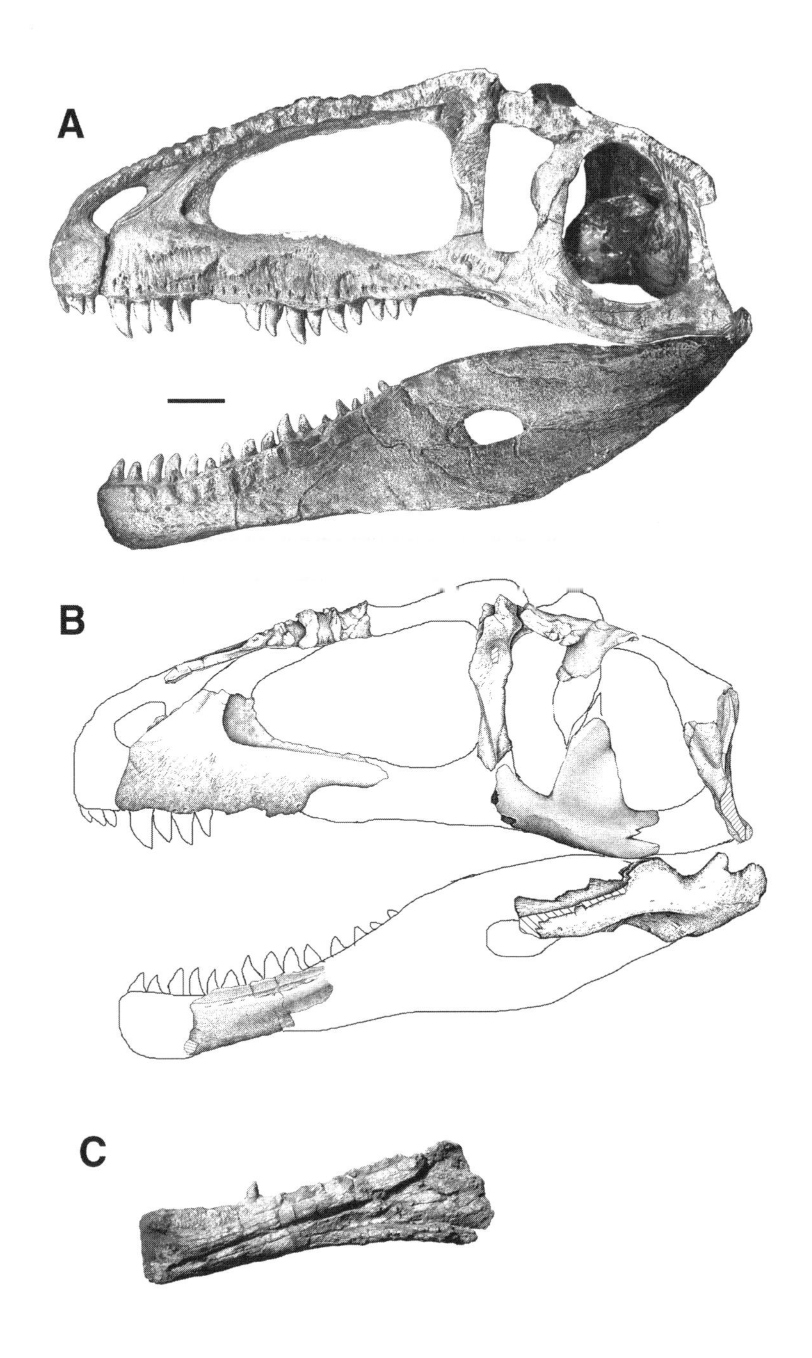

Figure 9.4. Patagonian carcharodontosaurs. (A) Reconstructed skull of Giganotosaurus carolinii. *(B) Skull restoration of* Mapusaurus roseae *based on cranial bones pertaining to several individuals (MCF-PVPH-108.1/169/183/177/168/102/2/15;* Tyrannotitan chubutensis *MPEF-PV 1157). Scale bar = 10 cm. (A) Modeled by M. C. Gravino.*

nal side of the quadrate; dorsal expansion of supraoccipital almost three times width of foramen magnum; supraoccipital with superficial ventral extension on either side of foramen magnum contacting dorsal surface of occipital condyle; occipital condyle much broader than high; ventral articular surface of condyle extending anteriorly on both sides of ventral midline depression; symphyseal end of the dentary dorsoventrally expanded, bearing a ventral process; proximal end of the scapula forwardly projected above the coracoid; tubercle-like insertion for triceps in the ventral border of the scapula; lobule-shaped obturator process of the ischium; dorsally projected femoral head, and posterior intercondylar groove in the proximal end of the tibia.

Comments. Giganotosaurus is characterized by having a long, massive, and low skull with conspicuous ornamentations on the premaxillae, maxillae, and nasals, very reduced shoulder bones, robust hand (based on a single metacarpal recovered), dorsal vertebrae with long and pneumatic neural spines, and robust hind limbs. Moreover, the presence of enamel wrinkles in maxillary and dentary teeth, a large infratemporal fenestra, a large pubic boot and an upwardly directed femoral head place *Giganotosaurus* within the Carcharodontosauridae.

Mapusaurus Coria and Currie 2006
Mapusaurus roseae Coria and Currie 2006

Fig. 9.4B

Locality and age. Cañadón del Gato in the Cortaderas area 20 km southwest of Plaza Huincul, Neuquén Province, Argentina. Huincul Formation, Río Limay Group (upper Cenomanian), of the Neuquén Group.

Holotype. MCF-PVPH-108.1, right nasal.

Paratypes. MCF-PVPH-108.5, left lacrimal/prefrontal; MCF-PVPH-108.45, right humerus; MCF-PVPH-108.83, axis; MCF-PVPH-108.90, cervical neural arch; MCF-PVPH-108.115, right maxilla; MCF-PVPH-108.125, left dentary; MCF-PVPH-108.128, left ilium; MCF-PVPH-108.165, left ischium; MCF-PVPH-108.167, jugal; MCF-PVPH-108.177, right postorbital-palpebral; MCF-PVPH-108.179, right splenial; MCF-PVPH-108.202, right fibula.

Comments. These specimens constitute a new form currently described elsewhere (Coria and Currie 1997, 2006). Several features link *Mapusaurus* with *Giganotosaurus* such as facial bones (maxilla, nasal) heavily sculptured to edge of antorbital fenestra, present of postorbital-lacrimal contact, presence of a large lacrimal pneumatic recess, maxillary and dentary teeth bladelike with wrinkles in the enamel next to the serrations, cervical vertebrae with only a pleurocoel fossa and femur with perpendicular angle between the head and the shaft in anterior or posterior view. Two apomorphic features shared by *Giganotosaurus* and *Mapusaurus* have been recognized: femur with a weak fourth trochanter and a shallow and broad extensor groove. Despite some carcharodontosaurid teeth reported from Late Cretaceous beds (Alcober et al. 1998), *Mapusaurus* is currently the younger record of a carcharodontosaurid theropod known before the saurischian fauna turnover that occurred in the Patagonian Turonian (Coria and Salgado 2005).

Coelurosauria Huene 1914
Maniraptoriformes Holtz 1996
Maniraptora Gauthier 1986
Dromaeosauridae Matthew and Brown 1922
***Buitreraraptor* Makovicky, Apesteguía, and Agnolin 2005**
***Buitreraraptor gonzalezorum* Makovicky, Apesteguía, and Agnolin 2005**

Locality and age. La Buitrera, 80 km southwest of Cipolletti, Río Negro Province. Candeleros Formation (lower Cenomanian).

Holotype. MPCA-245, near-complete, articulated, adult skeleton.

Referred specimen. MPCA-238, articulated partial skeleton comprising the hips, right hind limb, and sacrum.

Diagnosis. Makovicky et al. (2005) characterized this species by having a long skull, which exceeds femoral length by 25%; teeth small, unserrated, without root-crown constriction; quadrate with large lateral flange and pneumatic foramen; posterior cervical centra with ventrolateral ridge; furcula pneumatic; brevis shelf expanded and lobate, projects laterally from caudal end of ilium.

Comments. Buitreraraptor has been regarded as the earlier representative of a South American dromaeosaurid theropod (Makovicky et al. 2005). Several dromaeosaurid features support this assignation, such as lateral process of the quadrate, unconstricted teeth, stalked trunk parapophyses, bifid chevrons, and a ginglymoid distal articulation on metatarsal II.

***Neuquenraptor argentinus* Novas and Pol 2005**

Locality and age. Sierra del Portezuelo, 20 km west Plaza Huincul, Neuquén Province. Portezuelo Formation (Turonian–Early Coniacian), Neuquén Group.

Holotype. MCF-PVPH-77, fragments of cervical vertebra, dorsal ribs, hemal arches, left proximal radius, right femur and distal tibia, proximal tarsals, and most of the left foot.

Diagnosis. Novas and Pol (2005) diagnosed *Neuquenraptor argentinus* as a probable dromaeosaurid with the following combined features: metatarsal II with lateral expansion over the caudal surface of metatarsal III (autapomorphic), metatarsal III proximally pinched, extensor sulcus on proximal half of metatarsus, distal end of metatarsal III is incipiently ginglymoid (to a lesser degree than other dromaeosaurids), pedal digit II with phalanges 1 and 2 subequal in length, and bearing a trenchant ungual phalanx.

Comments. Neuquenraptor is a fragmentary specimen found associated with sauropod remains in the same levels that yielded *Unenlagia comahuensis* and *Patagonykus puertai.* The presence of a short and robust femur in the holotype specimen of *Neuquenraptor* caused the authors to consider it to be a juvenile specimen of *Unenlagia.* Despite the lack of overlapping anatomical information, the similarities between the remains (in both general proportions and dimensions), the clear close phylogenetic relationships, the scarce anatomical information provided by the fragmentary *Unenlagia* specimen, and the same provenance horizon of both forms cast suspicious about the taxonomic validity of *Neuquenraptor,* which might be instead considered to be another specimen of *Unenlagia* (see also comments in Makovicky et al. [2005]).

Unenlagia Novas and Puerta 1997
Unenlagia comahuensis Novas and Puerta 1997
Fig. 9.5

Locality and age. Sierra del Portezuelo, 22 km west of Plaza Huincul, Neuquén Province. Portezuelo Formation (Turonian–early Coniacian), Río Neuquén Subgroup, Neuquén Group.

Holotype. MCF-PVPH-78, three presacral vertebrae (presumably correspond to eighth, 10th and 13th dorsals, the latter articulated to the sacrum), sacrum, dorsal ribs, and two proximal chevrons, left scapula and incomplete humerus, ilia, pubes, right ischium, right femur, and left tibia.

Referred specimen. MUCPv-349, left humerus and left pubis.

Diagnosis. Small-sized theropod that possesses tall neural spines in posterior dorsals and anterior sacral vertebrae; being nearly twice the height of the centrum; deep lateral pits in the base of the neural spines of these vertebrae, twisted scapular shaft, inflected dorsal margin of postacetabular iliac blade.

Comments. Clark et al. (2002) recognized that several features of *Unenlagia* are indeed only present in Avialae, such as the presence of a hypopubic cup on the caudal surface of the distal end of the pubis, the presence of a pubic apron, a laterally oriented glenoid fossa on the scapulocoracoid, and esspecially the orientation of the glenoid from the primitive position on the caudal surface of the scapula. However, the presence of this feature in *Velociraptor,* together with stalked parapophyses on the dorsal vertebrae and the mediolateral expansion of the tip of the neural spine in the caudal dorsal vertebrae, suggest that this Patagonian form is likely related to dromaeosaurids (Norell and Makovicky 1999).

Recently, the new species *Unenlagia paynemili* was communicated (Calvo et al. 2004a). The extremely fragmentary condition of the specimen MUCPv-349 conspires against accurate comparison with *Unenlagia comahuensis.* Apparently, from the methodological point of view, both forms of *Unenlagia* has been treated as different genera, because the original paper of *Unenlagia paynemili* lacks of a proper diagnosis of the genus *Unenlagia,* plus a modified diagnosis of *Unenlagia comahuensis.* Therefore, it is not possible to recognize a new species of *Unenlagia* on the bases of the slight and superficial features of this fragmentary specimen, which casts doubt on the assignation to *Unenlagia* itself. The authors recognize, furthermore, that the inflected dorsal margin of the postacetabular blade of the ilium is present in both specimens, as well as the lateral pits at the base of the neural arch of one of the dorsal vertebrae (Calvo et al. 2004a). Curiously, the original paper raises some questions regarding the assignation to the genus *Unenlagia,* sug-

Figure 9.5. Unenlagia comahuensis, *based on holotype specimen MCF-PVPH-78. Scale bar = 1 m.*

gesting it could represent another genus, but in the same paragraph, the authors mention that several anatomical differences lead them to propose a new species (Calvo et al. 2004a, 562). In sum, the specimen MUCPv-349 represents a probable dromeosaurid theropod, with some similarities with *Unenlagia,* but with not enough information to establish their phylogenetic relationships. Therefore, *Unenlagia paynemili* should be regarded as nomen vanum.

Metornithes Perle, Norell, Chiappe, and Clark 1993
Alvarezsauridae Bonaparte 1991
***Alvarezsaurus* Bonaparte 1991**
***Alvarezsaurus calvoi* Bonaparte 1991**

Locality and age. Campus University of El Comahue, badlands placed 500 m northeast of the main buildings, Neuquén City, Neuquén Province. Bajo de La Carpa Formation (Santonian), Río Colorado Subgroup, Neuquén Group.

Holotype. MUCPv-54, five cervical neural arches and three anterior dorsal neural arches articulated, one cervical centrum, two incomplete dorsal neural arches; three sacral centra articulated with the right ilium; one sacral centrum; 13 almost complete proximal and midcaudal articulated with some chevrons; most of right scapula and coracoid; most of right ilium and fragment of the left one; proximal halves of both ilia; distal half of right tibia articulated with astragalus and calcaneum; fragment of left tibia; incomplete right metatarsals; metatarsal IV articulated to the complete digit IV; three distal phalanges of digit II; incomplete left metatarsals articulated with four phalanges of digit IV; two phalanges of digit III; one isolated phalange, two incomplete phalanges.

Diagnosis. Bonaparte (1991) diagnosed *Alvarezsaurus* as follows: Small theropod with cervical vertebrae with subcircular expansions in the postzygapophyses; vestigial neural spines and amphicoelous centra with pleurocoels; anterior dorsal vertebrae with vestigial neural spines; anterior sacral vertebrae with shallow axial depression on the ventral border of the centra; small, reduced scapula lacking acromial expansion; low and long ilium with postacetabular expansion broader than preacetabular; metatarsals II, III, and IV subequal in width, with no indication of fusion, with metatarsal II being slightly thinner in posterior view and metatarsal IV broader in proximal view; unfused astragalus with wide external and internal condyles. Later, as a result of discovery of other alvarezsaurids in the Cretaceous of Mongolia, Chiappe et al. (2002) were able to add the following characters: abrupt transition between cervicals and thoracic vertebrae; synsacrum less compressed than in *Patagonykus, Mononykus,* and *Shuvuuia;* spinous processes of proximal caudals absent or weakly developed; distal caudal vertebrae twice as long as proximal ones; scapula significantly smaller than in *Mononykus* and *Shuvuuia;* scapular blade curved in dorsal view; postacetabular wing not depressed as in *Shuvuuia* and *Parvicursor;* metatarsal IV longer than metatarsal II; ungual phalanx of digit I with a short, ventral keel.

Comments. Alvarezsaurus occupies the most basal position among the monophyletic Alvarezsauridae. This clade of highly derived nonavian theropod shows a worldwide distribution during the Cretaceous, with representatives in South America, North America, and Asia (Chiappe et al. 2002).

Patagonykus Novas 1997
Patagonykus puertai Novas 1997

Locality and age. Sierra del Portezuelo, 22 km west of Plaza Huincul, Neuquén Province. Portezuelo Formation (Turonian–early Coniacian), Río Neuquén Subgroup, Neuquén Group.

Holotype. MCF-PVPH-37, two incomplete dorsal vertebrae, incomplete sacrum, two proximal and two distal caudal vertebrae, incomplete left and right coracoids, proximal and distal ends of both humeri; right proximal portions of ulna and radius, and distal portion of left ulna, articulated carpometacarpus and first phalanx of digit I of the right manus; incomplete ungual phalanx probably corresponding to digit I, portions of ilia, proximal ends of ischia, and portions of pubes, proximal and distal portions of right femur, and distal end of the left one, proximal and distal ends of both tibiae, fused with proximal tarsals, metatarsals II and III fused to distal tarsal III, several pedal phalanges.

Diagnosis. Novas (1997b) diagnosed *Patagonykus puertai* as an alvarezsaurid avialian theropod that has the following apomorphic features: postzygapophyses in dorsal vertebrae with ventrally curved, tongue-shaped lateral margin; dorsal, sacral, and caudal vertebrae with a bulge on the caudal base of the neural arch; humeral articular facet of coracoid transversely narrow; internal tuberosity of humerus subcylindrical, wider at its extremity rather than at its base; humeral entepicondyle conical shaped and strongly projecting medially; first phalanx of manual digit I with proximomedial hooklike processes; ectocondylar tuber of femur rectangular shaped in distal view.

Comments. The presence of several features such as an ulna with a hypertrophied olecranon process and a single cotyla, large and proximally expanded radiocarpal facet of radius, hypertrophied and strongly depressed metacarpal I, and robust digit I with claw bearing two proximomedial foramina, allow *Patagonykus* to be included within the clade formed with the Mongolians *Mononykus olecranus, Parvicursor remotus,* and *Shuvuuia deserti* (Chiappe et al. 2002).

Tetanurae incertae sedis
Megaraptor Novas 1998
Megaraptor namunhuaiquii Novas 1998

Locality and age. Sierra del Portezuelo, 22 km west of Plaza Huincul, Neuquén Province. Portezuelo Formation (Turonian-early Coniacian), Río Neuquén Subgroup, Neuquén Group.

Holotype. MCF-PVPH-79. right ulna; left manual phalanx 1.I; distal half of right metatarsal III; ungual phalanx of digit II of the right pes.

Referred specimen. MUCPv-341, cervical vertebra, two anterior caudal vertebrae with a fused hemal arch, two isolated hemal arches, left scapulocoracoid, right ulna, radius, and complete manus, right pubis, and right metatarsal IV.

Diagnosis. Large but slender theropod with enlarged pedal ungual on digit II, which exhibits the following autapomorphies: bladelike olecranon process on proximal ulna; distal end of ulna stout and triangular in distal aspect; manual phalanx 1.I subquadrangular in proximal view, with the dorsal portion transversely wider than the ventral one; metatarsal III with deep and wide extensor ligament pit.

Comments. Novas (1998) proposed coelurosaur affinities of *Megaraptor,* likely with some dromeosaurs, on the basis of the shape of an enormous claw that this author identified as belonging to a pedal toe. Recent findings (Calvo et al. 2004c) prove that the claw identifies by Novas (1998) as a pedal ungual of digit II is actually a manual claw. The new findings allow the identification of *Megaraptor* as a large-sized theropod with large and long hands, armed with big claws. Several aspects of *Megaraptor* are reminiscent of the morphology present in large ornithomimids like the Mongolian *Deynokeirus,* although as a result of the incompleteness of the specimens from both continents, a phylogenetic hypothesis cannot be proposed with any certainty.

Condorraptor Rauhut 2005
Condorraptor currumili Rauhut 2005

Locality and age. Las Chacritas locality, 28 km west of Cerro Cóndor, Chubut Province. Cañadón Asfalto Formation, Callovian.

Holotype. MPEF-PV-1672, an incomplete left tibia.

Hypodigm. MPEF-PV-1694/1695, teeth; MPEF-PV-1673/1675, cervical vertebrae; MPEF-PV-1676/1680,1697,1700,1705, dorsal vertebrae and vertebral fragment; MPEF-PV-1681,1701, sacral vertebrae; MPEF-PV-1682/1683,1702, caudal vertebrae; MPEF-PV-1684/1685, 1703, rib fragments and chevron; MPEF-PV-1686/1689, 1696, 1704, partial ilium, pubis and ischium; MPEF-PV-1690/1693, partial femora, metatarsal IV, and pedal ungual.

Diagnosis. Condorraptor currumili was diagnosed by Rauhut (2005) by having posterior incision between fibular condyle and medial part of proximal tibia absent; and large, shallow depression laterally on the base of the cnemial crest. Other apomorphic characters found in the referred material are these: pleurocoel in anterior cervical vertebrae placed behind the posteroventral corner of the parapophyses; large nutrient foramina on the lateral side of the ischial peduncle of the ilium; metatarsal IV with a distinct step dorsally between shaft and distal articular facet.

Comments. Skeletal remains of Jurassic theropods are extremely rare in South America. *Condorraptor* is, together with *Piatnitzkysaurus floresi* (Bonaparte 1979), the only Theropoda record from that age in that continent. As mentioned in its original description, general morphology of the bones of *Condorraptor* reminds that present in *Piatnitzkysaurus* (Rauhut 2005). Nonetheles, some theropod diversity is expectable for these levels.

Quilmesaurus Coria 2001
Quilmesaurus curriei Coria 2001

Locality and age. Salitral Ojo de Agua, 40 km south of Gral. Roca, Río Negro Province. Allen Formation (Campanian-Maastrichtian), Malargüe Group.

Holotype. MPCA-PV-100, distal right femur, complete right tibia.

Diagnosis. Medium-sized theropod, femur with strong, well-developed mediodistal crest; tibia with hook-shaped cnemial crest; lateral malleolus twice the size of medial, asymmetrical distal end.

Comments. Quilmesaurus curriei was erected on the basis of the unique condition of the development presented in the knee area, formed by the combination of two strong bony struts, the femoral mediodistal crest and the tibial cnemial crests (Coria 2001). These two elements, arranged in an unusual bone system placed in the knee articulation, show a unique and unrecorded anatomical feature, with evident functional implicances. The singularity of these features strongly supports the validity of the taxon. The Tetanurae condition of *Quilmesaurus* was challenged, suggesting some shared characters with the sibling taxon of Ceratosauria. It has been proposed that one typical feature of Ceratosauria is the presence of a tibiotarsus formed by the fused tibia, astragalus, and calcaneum (Kellner and Campos 2002). No evidences of such a feature is present in the tibia of *Quilmesaurus.*

Evolutionary Considerations

Patagonian nonavian theropods include all members of Neotheropoda (the node-based taxon that includes *Ceratosaurus,*

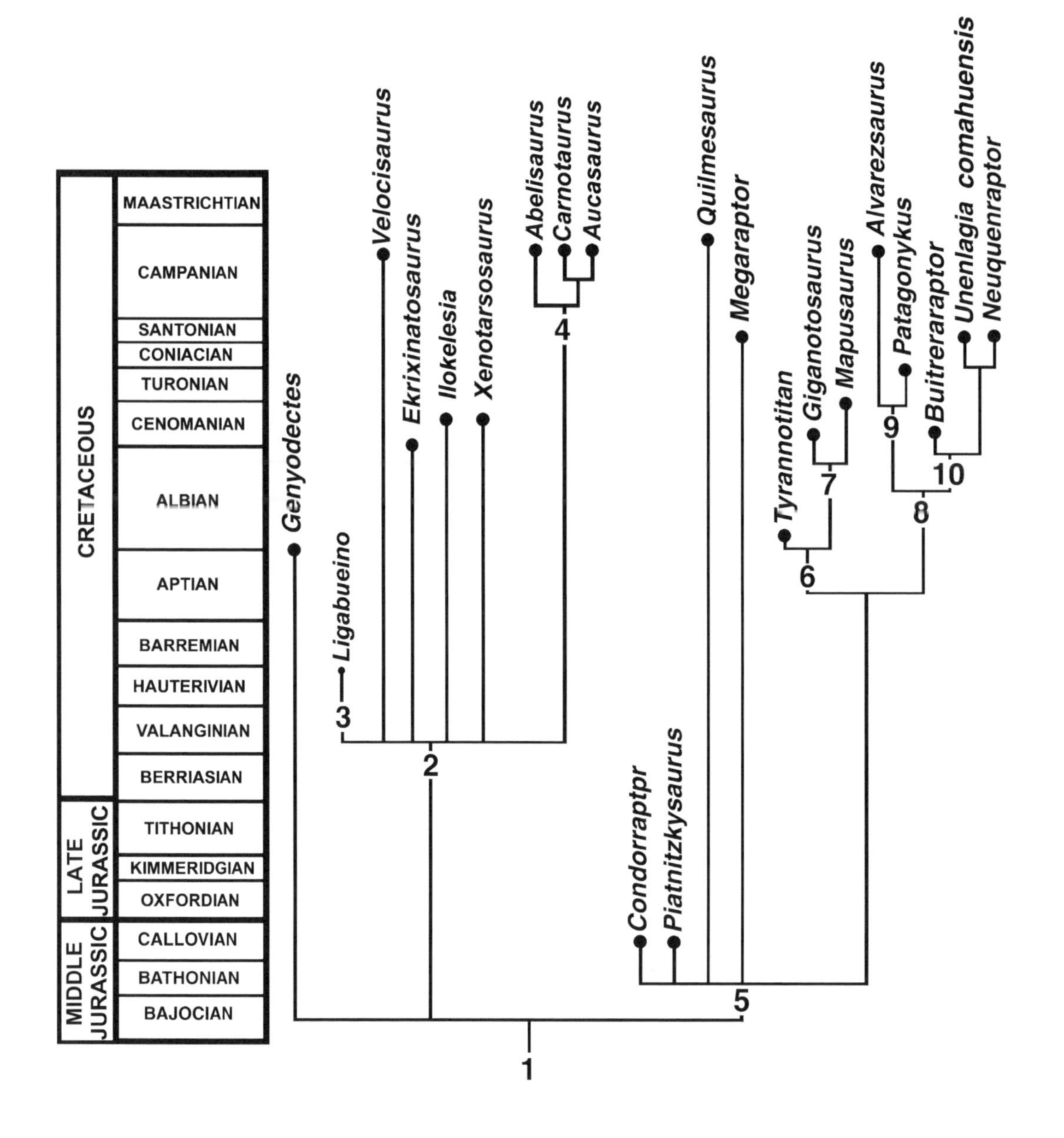

Figure 9.6. Calibrated estratigraphic cladogram of the phylogenetic relationships of nonavian Patagonian theropods. 1, Neotheropoda; 2, Abelisauroidea; 3, Noasauridae; 4, Abelisauridae; 5, Tetanurae; 6, Carcharodontosauridae; 7, Giganotosaurinae; 8, Maniraptora; 9, Alvarezsauridae; 10, Dromaeosauridae.

Allosaurus, birds, their common ancestor, and all descendants) (Fig. 9.6). Nevertheless, the current lack of Triassic records of theropods may hide a wider diversity.

So far, the oldest theropod record from Patagonia corresponds to the poorly known and fragmentary basal Tetanurae *Piatnitzkysaurus* and *Condorraptor* (Bonaparte 1979; Rauhut 2005) from the Jurassic rocks of Chubut Province. Other Jurassic theropods are only known from footprints that represent an interesting ichnofauna of small-sized theropods (Casamiquela 1964; Coria and Paulina 2004; Calvo this volume).

In the Cretaceous, Aptian-Cenomanian beds bear an incipient diversity of carcharodontosaurids. The recently described *Tyrannotitan* is the oldest member of this South American assemblage of Carcharodontosauridae (Novas et al. 2005). Together with the slightly younger *Giganotosaurus* and *Mapusaurus* (Coria and Salgado 1995; Coria and Currie 2006), they indicate a strong biological presence of the clade in Patagonia, which is, up to now, the highest diversity of carcharodontosaurids recorded. Remains assigned to Carcharodontosauridae have been reported from several Patagonian localities (Alcober et al. 1998; Veralli and Calvo 2003), but the specimens are either still undescribed (Alcober et al. 1998) or represented by isolated teeth that are insufficient to appropriate character analysis. Patagonian carcharodontosaurid extinction during the Turonian has been proposed as an explanation of the absence of the record of this family beyond that time boundary, which also correlated with the disappearance of diplodocoid sauropods in Late Cretaceous beds (Coria and Salgado 2005).

Giganotosaurinae comprises all carcharodontosaurids closer to *Giganotosaurus* than to *Carcharodontosaurus* (sensu Coria and Currie 2006), and includes *Mapusaurus* and probably *Tyrannotitan,* likely forming a South American assemblage eventually biogeographically differentiated from the African set, whose main representative is *Carcharodontosaurus.*

Late Cretaceous Patagonian theropods are represented by two clearly identifiable groups: the abelisaurids and the highly derived dromeosaur-like theropods and alvarezsaurids. The abelisauroids are recorded from the Lower Cretaceous up to the Upper Cretaceous. *Ligabueino* is a small abelisauroid probably allied with noasaurids (Bonaparte 1996) and is the only Patagonian member recorded for this family. The remaining genera so far recognized of the Abelisauroidea clade correspond to a wealth of forms distributed along different phyletic levels of abelisaurs. Recently, a probable new abelisauroid form was communicated. The specimen cataloged as MCF-PVPH-237 comes from Cerro Bayo Mesa, 30 km south of Plaza Huincul (Neuquén Province, Argentina), Cerro Lisandro Formation (Turonian?), Río Limay Subgroup, Neuquén Group, Neuquén Basin (Fig. 3C). It consists of the last presacral vertebrae, fragments of the sacrum, an incomplete right ilium, pubes lacking distal ends, and fragments of the proximal region of the right ischia. It exhibits several characters described for other abelisaurs, such as pendant process cranial to the prezigapophyses, very transversely compresed sacral centra 2 and 3, well-developed longitudinal crest located dorsally on the lateral side of the ilium, and obturador foramen, and it is characterized by having narrow, anteroposteriorly compressed bladelike pubic shafts that are crescent shaped in cross section. This specimen is so far the only theropod described for the Cerro Lisandro Formation (Paulina Carabajal et al. 2003; Coria et al. in press). It is very likely that basal forms like this and *Ilokelesia, Xenotarsosaurus,* and *Ekrixinatosaurus* will be found to be abelisaurids once the discovery of more complete specimens allows us to fine-tune their phylogenetic relationships.

The Campanian abelisaurids recovered from Patagonia are *Abelisaurus, Carnotaurus,* and *Aucasaurus* (Bonaparte 1985; Bonaparte and Novas 1985; Coria et al. 2002). These South American forms, together with other Gondwanaland forms (i.e., *Majungatholus;* Sampson et al. 1998) have allowed us to build up a consistent monophyletic assemblage of the Abelisauridae. Abelisaurids seem to have spread throughout the Late Cretaceous, increasing their diversity perhaps because of the carcharodontosaurid extinction mentioned above.

Alvarezsaurids have their most plesiomorphic representative in Patagonia, *Alvarezsaurus calvoi* (Bonaparte 1991), from the Santonian Bajo de la Carpa Formation of Neuquén Basin; curiously, the more derived mononykid-related *Patagonykus* has as its provenance the Turonian beds of the same basin. A biogeographical explanation of the presence of such closely related forms of flightless theropods in both South America and Asia (Chiappe et al. 2002) has not yet been proposed.

South American forms like *Unenlagia* and *Neuquenraptor* bear on their fragmentary skeletons conspicuous morphologies that are reminiscent of the North American deinonychosaurians (Novas and Puerta 1997; Novas and Pol 2005). Even though *Neuquenraptor* is likely synonymous with *Unenlagia* (although the lack of overlapping bones allows us to provisionally maintain the generic differentiation), the similarities with *Deinonychus* and *Dromeosaurus* are noteworthy. However, the incompleteness of the specimens is important, making it possible that the resemblance with the deinonychosaurians is based on homoplasies. The unusual regional diversity of Patagonian dromaeosaurids, currently represented by *Unenlagia comahuensis* (Novas and Puerta 1997), *U. paynemilli* (Calvo et al. 2004a), and *Neuquenraptor argentinus,* is, at least, surprising. All the taxa were collected from the same unit, and even from the same locality, a few meters away from each other. As was mentioned above, there is no support to distinguish between the two species of *Unenlagia.* On other hand, at present, it is not possible to compare analogous bones from both *Unenlagia comahuensis* and *Neuquenraptor argentinus.* Indeed, the two specimens (MCF-PVPH-77 and -78) are very fragmentary, and it is possible to recognize autopomorphic features in both of them. Provisionally, however, *Neuquenraptor argentinus* is sustained as a valid taxon, although further discoveries may change this assessment.

Finally, there are many specimens, some of them preliminarily described but most still unpublished, that suggest the evolutionary history of South American nonavian theropods is far more complex than suspected (Novas 1997a; Calvo et al. 2004c).

Acknowledgments. I thank A. Kramarz (MACN), J. C. Muñoz (MPCA), M. Reguero, M. S. Bargo (MLP), and E. Ruigomez (MEF) for access to the collections under their care. A. Paulina Carabajal edited preliminary versions of the manuscript. Dr. M.

Carrano reviewed the manuscript. Thanks are extended to the Secretaría de Estado de Cultura from Neuquén Province and the Municipality of Plaza Huincul for permanent support and assistance.

References Cited

Alcober, O., P. C. Sereno, H. C. F. Larsson, R. Martínez, and D. J. Varricchio. 1998. A Late Cretaceous carcharodontosaurid (Theropoda: Allosauroidea) from Argentina. *Journal of Vertebrate Paleontology* 18 (Suppl. to 3): 23A.

Bonaparte, J. F. 1979. Dinosaurs: a Jurassic assemblage from Patagonia. *Science* 205: 1377–1379.

———. 1985. A horned Cretaceous carnosaur from Patagonia. *National Geographic Research* 1: 149–151.

———. 1991. Los vertebrados fósiles de la Formación Río Colorado, de la ciudad de Neuquén y cercanías, Cretácico Superior, Argentina. *Revista del Museo Argentino de Ciencias Naturales "Bernardino Rivadavia," Paleontología* 4: 17–123.

———. 1996. Cretaceous tetrapods of Argentina. *Münchner Geowissenschftliche Abhandlungen* A 30: 73–130.

Bonaparte, J. E., and F. E. Novas. 1985. *Abelisaurus comahuensis,* n.g. et n.sp., Carnosauria del Cretácico Tardío de Patagonia. *Ameghiniana* 21: 259–265.

Bonaparte, J. F., F. E. Novas, and R. A. Coria. 1990. *Carnotaurus sastrei* Bonaparte, the horned, lightly built carnosaur from the Middle Cretaceous of Patagonia. *Contributions in Science* 416: 1–41.

Calvo, J. O., J. D. Porfiri, and A. W. A. Kellner. 2004a. On a new maniraptoran dinosaur (Theropoda) from the Upper Cretaceous of Neuquén, Patagonia, Argentina. *Arquivos do Museu Nacional,* Rio de Janeiro 62 (4): 549–566.

Calvo, J. O., D. Rubilar-Rogers, and K. Moreno. 2004b. A new Abelisauridae (Dinosauria: Theropoda) from northwest Patagonia. *Ameghiniana* 41 (4): 555–563.

Calvo, J. O., J. D. Porfiri, C. Veralli, F. Novas, and F. Poblete. 2004c. Phylogenetic status of *Megaraptor namunhuaiquii* Novas based on a new specimen from Neuquén, Patagonia, Argentina. *Ameghiniana* 41 (4): 565–575.

Carpenter, K. 1997. Tyrannosauridae. In P. J. Currie and K. Padian (eds.), *Encyclopedia of Dinosaurs,* 766–768. San Diego: Academic Press.

Carrano, M. T., S. D. Sampson, and C. A. Forster. 2002. The osteology of *Masiakasaurus knopfleri,* a small abelisauroid (Dinosauria: Theropoda) from the Late Cretaceous of Madagascar. *Journal of Vertebrate Paleontology* 22: 510–534.

Casamiquela, R. M. 1964. *Estudios Icnológicos.* Gobierno de la Provincia de Río Negro, Ministerio de Asuntos Sociales, 1–229. Buenos Aires.

Chiappe, L. M., R. A. Coria, L. Dingus, F. Jackson, A. Chinsamy, and M. Fox. 1998. Sauropod dinosaur embryos from the Late Cretaceous of Patagonia. *Nature* 396: 258–261.

Chiappe, L. M., M. A. Norell, and J. M. Clark. 2002. The Cretaceous, short-armed Alvarezsauridae: *Mononykus* and its kind. In L. M. Chiappe and L. M. Witmer (eds.), *Mesozoic Birds: Above the Heads of Dinosaurs,* 87–120. Berkeley: University of California Press.

Clark, J. M., M. Norell, and P. Makovicky. 2002. Cladistic approaches to the relationships of birds to other theropods. In L. Chiappe and L.

Witmer (eds.), *Mesozoic Birds: Above the Heads of Dinosaurs,* 31–64. San Diego: University of California Press.

Codignotto J., F. Nullo, J. Panza, C. Proserpio. 1978. Estratigrafía del Grupo Chubut entre Paso de Indios y Las Plumas, Provincia del Chubut, Argentina. *Actas del VII Congreso Geológico Argentino,* 471–480.

Coria, R. A., 2001. A new theropod from the Late Cretaceous of Patagonia. In D. H. Tanke and K. Carpenter (eds.), *Mesozoic Vertebrate Life,* 3–9. Bloomington: Indiana University Press.

Coria, R. A., and P. Currie. 1997. A new theropod from the Río Limay Formation. *Journal of Vertebrate Paleontology* Suppl. 3: 40A.

———. 2006. A new carcharodontosaurid (Dinosauria: Theropoda) from the Upper Cretaceous of Argentina. *Geodiversitas* 28: 71–118.

Coria, R. A., and A. Paulina Carabajal. 2004. Nuevas huellas de Theropoda (Dinosauria: Saurischia) del Jurásico de Patagonia, Argentina. *Ameghiniana* 41 (3): 393–398.

Coria, R. A., and J. Rodríguez. 1993. Sobre *Xenotarsosaurus bonapartei* Martinez et al, 1986; un problemático Neoceratosauria (Novas, 1989) del Cretacico del Chubut. *Ameghiniana* 30 (3): 326–327.

Coria, R. A., and L. Salgado. 1995. A new giant carnivorous dinosaur from the Cretaceous of Patagonia. *Nature* 377 (6546): 224–226.

———. 1998. A basal Neoceratosauria Novas, 1989 (Theropoda-Ceratosauria) from the Cretaceous of Patagonia, Argentina. *Gaia* 15: 89–102.

———. 2005. Mid-Cretaceous turnover of saurischian dinosaur communities: evidence from the Neuquén Basin. In G. D. Veiga, L. A. Spalletti, J. A. Howell, and E. Schwarz (eds.), *The Neuquén Basin, Argentina: A Case Study in Sequence Stratigraphy and Basin Dynamics,* 317–327. Special Publication 252. London: Geological Society.

Coria, R. A., L. M. Chiappe, and L. Dingus. 2002. A new close relative of *Carnotaurus sastrei* Bonaparte (Abelisauridae: Theropoda) from the Late Cretaceous of Patagonia. *Journal of Vertebrate Paleontology* 22 (2): 460–465.

Coria, R. A., P. J. Currie, and A. Paulina Carabajal. In press. A new abelisaur theropod from northwestern Patagonia. *Canadian Journal of Earth Sciences.*

Dingus, L., J. Clarke, G. R. Scott, C. C. Swisher, L. M. Chiappe, and R. A. Coria. 2000. Stratigraphy and magnetostratigraphic/faunal constraints for the age of sauropod embryo-bearing rocks in the Neuquén Group (Late Cretaceous, Neuquén Province, Argentina). *American Museum Novitates* 3290: 1–11.

Gauthier, J. 1986. Saurischian monophyly and the origin of birds. *Memoirs of the California Academy of Sciences* 8: 1–55.

Heredia S., and L. Salgado, 1997. Consideraciones sobre la estratigrafía y paleontología de los depósitos cretácicos continentales en la localidad de Lago Pellegrini, Provincia de Río Negro, Argentina. *Ameghiniana* 34: 120.

Huene, F. von 1926. The carnivorous Saurischia in the Jura and Cretaceous formations, principally in Europe. *Revista del Museo de la Plata* 29: 35–167.

Kellner, A. W. A., and D. d. A. Campos. 2002. On a theropod dinosaur (Abelisauria) from the continental Cretaceous of Brazil. *Arquivos do Museu Nacional, Río de Janeiro* 60: 163–170.

Leanza, H. A., and C. A. Hugo. 2001. Cretaceous red beds from southern

Neuquén Basin (Argentina): age, distribution and stratigraphic discontinuities. In *VII International Symposium on Mesozoic Terrestrial Ecosystems,* 117–122. Special Paper 7. Asociación Paleontológica Argentina.

Makovicky, P. J., S. Apesteguía, and F. L. Agnolin. 2005. The earliest dromaeosaurid theropod from South America. *Nature* 437: 1007–1011.

Martínez, R., O. Giménez, J. Rodríguez, and G. Bochatey 1986. *Xenotarsosaurus bonapartei* gen. et sp. nov. (Carnosauria, Abelisauridae), un nuevo Theropoda de la Formación Bajo Barreal, Chubut, Argentina. *Actas IV Congreso Argentino de Paleontología y Bioestratigrafía* 2: 23–31.

Molnar, R. E. 1990. Problematic Theropoda: "Carnosaurs." In D. B. Weishampel, P. Dodson, and M. Osmólska (eds.), *The Dinosauria,* 306–307. Berkeley: University of California Press.

Musacchio E., and W. Chebli. 1975. Ostrácodos no marinos y carófitas del Cretácico inferior de las provincias de Chubut y Neuquén, Argentina. *Ameghiniana* 12: 70–96.

Norell, M. A., and P. J. Makovicky 1999. Important features of the dromaeosaur skeleton. II. Information from newly collected specimens of *Velociraptor mongoliensis. American Museum Novitates* 3282: 1–45.

Novas, F. E. 1989. *Los Dinosaurios Carnívoros de la Argentina.* Ph.D. thesis, Universidad Nacional de La Plata, Facultad de Ciencias Naturales.

———. 1997a. South American dinosaurs. In P. J. Currie and K. Padian (eds.), *Encyclopedia of Dinosaurs,* 678–689. San Diego: Academic Press.

———. 1997b. Anatomy of *Patagonykus puertai* (Theropoda, Avialae, Alvarezsauridae) from the Late Cretaceous of Patagonia. *Journal of Vertebrate Paleontology* 17: 137–166.

———. 1998. *Megaraptor namunhuaiquii* gen. et sp. nov., a large clawed, Late Cretaceous theropod from Patagonia. *Journal of Vertebrate Paleontology* 18: 4–9.

Novas, F. E., and D. Pol 2005. New evidence on deinonychosaurian dinosaurs from the Late Cretaceous of Patagonia. *Nature* 3285: 858–861.

Novas, F. E., and P. Puerta. 1997. New evidence concerning avian origins from the Late Cretaceous of Patagonia. *Nature* 387: 390–392.

Novas, F. E., S. de Valais, P. Vickers-Rich, and T. Rich. 2005. A large Cretaceous theropod from Patagonia, Argentina, and the evolution of carchaodontosaurids. *Naturwissenschaften* 92: 226–230.

Paulina Carabajal, A., R. A. Coria, and P. J. Currie. 2003. Primer hallazgo de Abelisauria en la Formación Lisandro (Cretácico Tardío) Neuquén. *Ameghiniana* 40: 65R.

Perle, A., M. A. Norell, L. M. Chiappe, and J. M. Clark 1993. Flightless bird from the Cretaceous of Mongolia. *Nature* 362: 623–626.

Rauhut, O. W. M. 2004. Provenance and anatomy of *Genyodectes serus,* a large-toothed ceratosaur (Dinosauria: Theropoda) from Patagonia. *Journal of Vertebrate Paleontology* 24 (4): 894–902.

———. 2005. Osteology and relationships of a new theropod dinosaur from the Middle Jurassic of Patagonia. *Palaeontology* 48 (Part 1): 87–110.

Rich, T. H., P. Vickers-Rich, F. Novas, R. Cúneo, P. Puerta, and R. Vacca. 1998. Theropods from the "Middle" Cretaceous Chubut Group of the San Jorge sedimentary basin, Central Patagonia: a preliminary note. *Gaia* 15: 111–115.

Rowe, T., and J. A. Gauthier. 1990. Ceratosauria. In D. Weishampel, P. Dodson, and M. Osmólska (eds.), *The Dinosauria,* 151–168. San Diego: University of California Press.

Sampson, S. D., L. M. Witmer, C. A. Forster, D. W. Krause, P. M. O'Connor, P. Dodson, and F. Ravoary. 1998. Predatory dinosaur remains from Madagascar: implications for the Cretaceous biogeography of Gondwana. *Science* 280: 1048–1051.

Sampson, S. D., M. T. Carrano, and C. A. Forster. 2001. A bizarre predatory dinosaur from the Late Cretaceous of Madagascar. *Nature* 409: 504–506.

Sereno, P. C. 1997. The origin and evolution of dinosaurs. *Annual Review of Earth Planet Sciences* 25: 435–489.

———. 1998. A rationale for phylogenetic definitions, with application to the higher-level taxonomy of Dinosauria. *Neues Jahrbuch für Geologie und Paläontologie Abhandlungen* 210 (1): 41–83.

Sereno P. C., J. A. Wilson, and J. L. Conrad. 2004. New dinosaurs link southern landmasses in the Mid-Cretaceous. *Proceedings of the Royal Society of London* B 271: 1325–1330.

Veralli, C., and J. O. Calvo. 2003. New findings of carcharodontosaurid teeth on Futalognko quarry (Upper Turonian), north Barreales Lake, Neuquén, Argentina. *Ameghiniana* 40: 74R.

Woodward, A. S. 1901. On some extinct reptiles from Patagonia, of the genera *Miolania, Dinilysia* and *Genyodectes. Proceedings of the Zoological Society of London* 1901: 169–184.

10. Aves

Luis M. Chiappe

Introduction

The fossil record of Mesozoic birds from Patagonia is scarce (Fig. 10.1). The first evidence of Mesozoic birds from this part of the world was reported by Lambrecht (1929). It consisted of an isolated tarsometatarsus (the holotype of *Neogaeornis wetzeli*), from the latest Cretaceous marine rocks of Quiriquina Island, Chile. Fifty years passed until additional Mesozoic-aged avian fossils from Patagonia were reported. These and other discoveries of the last 20 years are here summarized, with an update drawn from my previous revisions of the Cretaceous record of Patagonian birds (Chiappe 1991, 1996). As in these previous revisions, the present summary includes osteological, oological, and ichnological evidence. The short-armed alvarezsaurids (Chiappe et al. 2002) are, however, omitted because most recent studies (Chiappe 2001; Clark et al. 2002; Novas and Pol 2002) have placed them outside Aves. Given the problems associated with track-maker taxonomic identification (e.g., Padian and Olsen 1984; Unwin 1989), the ichnological occurrences here listed should be taken with caution. Nonetheless, the small size, slender digital impressions, and wide divarication angle of these footprints strongly supports their identification as those of birds (Lockley et al. 1992; Calvo this volume).

Institutional abbreviations. GPMK, Geologisch-Paläontologisches Institut und Museum, Kiel (Germany); MACN, Museo Argentino de Ciencias Naturales (N, colección Neuquén; RN, colección Río Negro), Buenos Aires (Argentina); MCF-PVPH-SB, Museo "Carmen Funes," Paleontología de Vertebrados Plaza Huincul (colección Sierra Barrosa), Plaza Huincul (Argentina); MUCPv, Museo de Ciencias Naturales, Universidad Nacional del Comahue, Neuquén City,

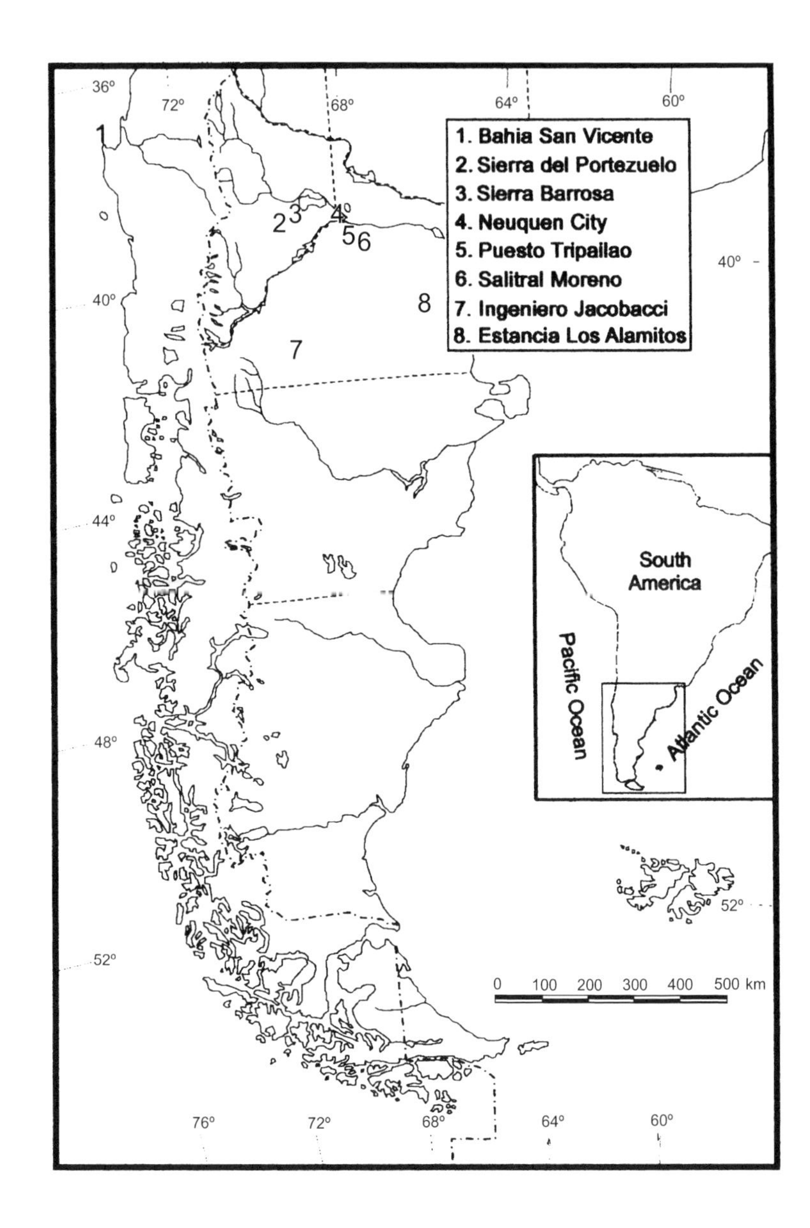

Figure 10.1. Geographical location of Patagonian localities with remains of Mesozoic birds. All occurrences are of Late Cretaceous age. 1, Bahía San Vicente; 2, Sierra del Portezuelo; 3, Sierra Barrosa; 4, Neuquén City; 5, Puesto Tripailao; 6, Salitral Moreno; 7, Ingeniero Jacobacci; 8, Estancia Los Alamitos.

(Argentina); PVL, Paleontología de Vertebrados, Instituto "Miguel Lillo," Universidad Nacional de Tucumán (Argentina).

Systematic Paleontology

The discussion of the following material is organized according to its increasing level of inclusiveness within the cladogram illustrated in Fig. 10.2. Thus, although material that can only be identified as Aves is discussed first, fossils assigned to particular clades of neornithines (modern birds) are discussed last. Because the record of avian footprints (Calvo this volume) is reviewed elsewhere in this volume, these occurrences are minimally discussed.

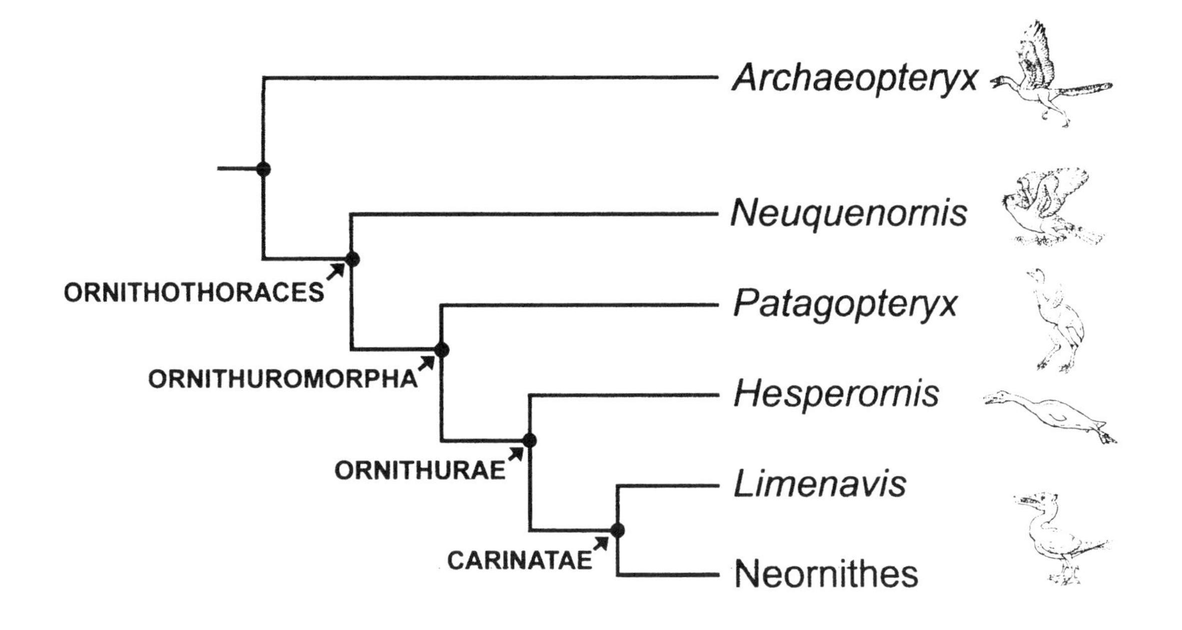

Figure 10.2. Cladogram illustrating the phylogenetic relationships of the taxa here discussed.

Aves Linnaeus 1758
Gen. et sp. indeterminate

Material. An indeterminate number of footprints with no assigned repository (Casamiquela 1987; Calvo this volume).

Locality and age. Vicinity of Ingeniero Jacobacci, Province of Río Negro, Argentina (Fig. 10.1). The age of these footprint-bearing rocks was interpreted as late Maastrichtian (Casamiquela 1987).

Comments. Although never described or illustrated, these footprints were named as "*Patagonichnornis venetiorum*" by Casamiquela (1987), who assigned them to Cimolopterygidae, an extinct group of putative charadriiforms. However, no available evidence can support such a systematic hypothesis (Chiappe 1991).

Aves gen. et sp. indeterminate

Material. An indeterminate number of footprints with no assigned repository (Leonardi 1987; Calvo this volume).

Locality and age. "Montón-I10" (near Ingeniero Jacobacci), Province of Río Negro, Argentina (Fig. 10.1, locality 7). The age of these footprint-bearing rocks was interpreted as late Maastrichtian (Casamiquela 1987).

Comments. A cast of a plaque containing several avian footprints from the latest Cretaceous of Río Negro Province was illustrated by Leonardi (1987). It is unclear whether these footprints are those of "*Patagonichnornis venetiorum.*"

Aves gen. et sp. indeterminate

Material. MCF-PVPH-SB-415, numerous footprints representing at least three ichnotaxa located in two sites separated by less than 100 m (Coria et al. 2002; Calvo this volume).

Locality and age. Sierra Barrosa, 30 km northeast of Plaza Huincul, Province of Neuquén, Argentina (Coria et al. 2002) (Fig. 10.1, locality 4). The footprints are contained in the Anacleto Formation, whose age has been estimated to be early Campanian (Dingus et al. 2000) to late Santonian (Leanza et al. 2004).

Comments. The most common footprints are morphologically comparable to the ichnotaxa *Aquatilavipes* from the Lower Cretaceous of Canada (Currie 1981). Other footprints from Sierra Barrosa are similarly sized to those referred as *Aquatilavipes,* but they display a distinct hallux impression. Coria et al. (2002) identified these footprints as the ichnotaxon *Ignotornis.* A different, much smaller type of footprint was named *Barrosopus slobodai* by these authors. Although these footprints document a diversity of birds inhabiting the Late Cretaceous riverine environments of northwestern Patagonia, it is difficult to relate their pedal morphology to any known lineage of birds.

Neornithes gen. et sp. indeterminate

Material. MACN-RN-976, an isolated cervical vertebra (Chiappe 1992, 1996).

Locality and age. Estancia Los Alamitos, Province of Río Negro, Argentina (Fig. 10.1, locality 8). This specimen comes from the early Maastrichtian beds (Leanza et al. 2004) of the Los Alamitos Formation (Bonaparte 1987).

Comments. Very little can be said about this specimen. The presence of fully heterocoelous articulations (saddle-shaped centra) supports its placement within Ornithothoraces (Fig. 10.2). Rocks of the Los Alamitos Formation were deposited in a shallow, brackish body of water near the shore of the Atlantic Ocean (Bonaparte 1987).

Ornithothoraces gen. et sp. indeterminate

Material. MUCPv-284, 305-306, 350-355, several isolated eggs, some containing embryonic remains (Schweitzer et al. 2002). Other similar eggs are housed at the collections of the Museo Argentino de Ciencias Naturales (Buenos Aires, Argentina) and the "Carmen Funes" Museum (Plaza Huincul, Neuquén Province, Argentina).

Locality and age. Neuquén City, Province of Neuquén, Argentina (Fig. 10.1, locality 4). The eggs are contained in rocks of the Bajo de La Carpa Formation, the age of which is estimated as middle Santonian (Leanza et al. 2004).

Comments. Although some osteological characters from the embryonic material inside these eggs support their assignation within Ornithothoraces, other features suggest a phylogenetic placement outside Ornithuromorpha (Fig. 10.2). The embryonic remains are contained in calcitic and smooth-shelled eggs of asymmetric shape.

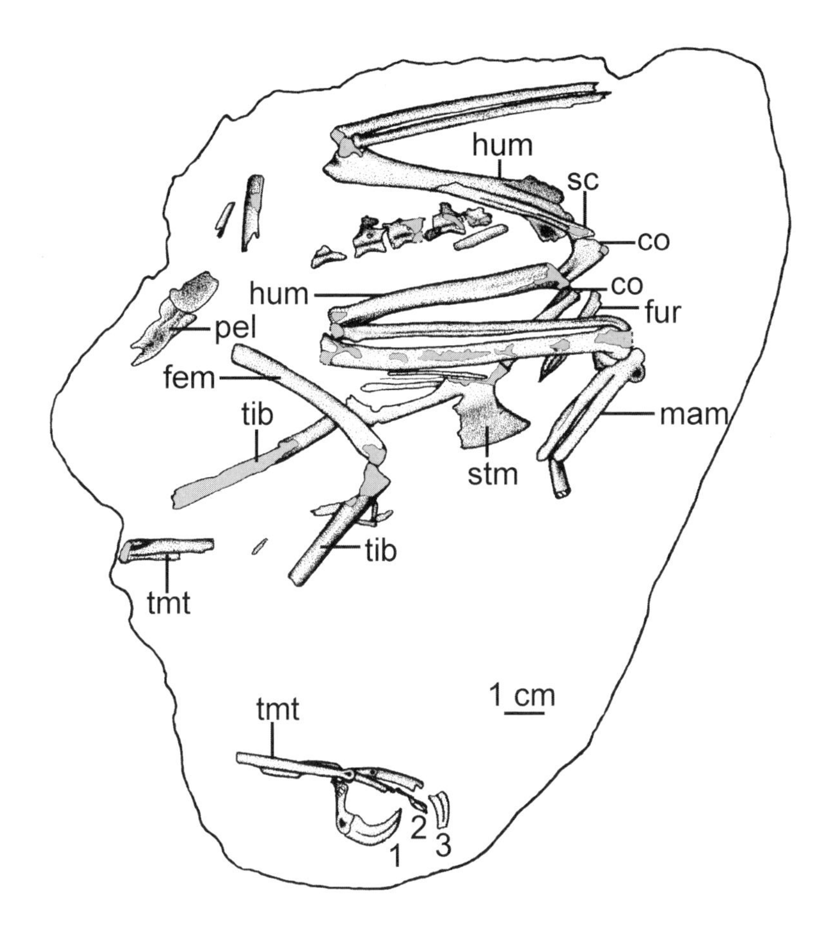

Figure 10.3. Enantiornithine Neuquenornis volans *(Chiappe and Calvo 1994) from the Late Cretaceous of the Bajo de la Carpa Formation (Neuquén City). co = coracoid; fem = femur; fur = furcula; hum = humerus; mam = major metacarpal; pel = pelvis; sc = scapula; stm = sternum; tib = tibia; tmt = tarsometatarsus; 1–3 = pedal digits I–III.*

Enantiornithes Walker 1981
Euenantiornithes Chiappe 2002
Neuquenornis volans Chiappe and Calvo 1994
Fig. 10.3

Holotype. MUCPv-142, partial skeleton including the postorbital region of the skull, thoracic vertebrae, portions of the forelimb and shoulder girdle, and parts of the hind limb (Chiappe and Calvo 1994) (Fig. 10.3).

Locality and age. Same as that of the ornithothoracine eggs from Neuquén City (see above).

Geographical distribution. City of Neuquén and Puesto Tripailao (MACN-RN-977, a distal end of a humerus), Neuquén and Río Negro Provinces, respectively, Argentina (Fig. 10.1, localities 4 and 5).

Diagnosis. Enantiornithine with a gracile tarsometatarsus, a winglike posterior trochanter on the proximal end of the femur, subequal widths of major and minor metacarpals, and a pronounced sternal keel that projects cranially beyond the margin of the sternum.

Comments. The discovery of *Neuquenornis volans* was crucial for the interpretation of the Enantiornithes as a monophyletic group (Chiappe and Calvo 1994; Chiappe and Walker 2002). Before its discovery, the disarticulated nature of the available enantiornithines (primarily those from the latest Cretaceous Lecho Formation of northwestern Argentina [Walker 1981]) had prevented the discovery of characters supporting the common ancestry of this clade. The morphology of the wing, shoulder, and sternum of *Neuquenornis* also highlighted the aerodynamic capabilities of enantiornithines, a group that had previously been considered to have had poor flight abilities (Walker 1981). The raptorial morphology of the foot of *Neuquenornis* suggest this kestrel-sized bird could have used the enlarged claw of its hallux to seize prey (its feet were also suitable for perching).

Ornithuromorpha Chiappe 2002
***Patagopteryx deferrariisi* Alvarenga and Bonaparte 1992**
Fig. 10.4

Holotype. MACN-N-03, partial skeleton including cervical, thoracic, synsacral, and caudal vertebrae; portions of the forelimb and shoulder girdle; portions of the ilium and hind limb (Alvarenga and Bonaparte 1992).

Locality and age. Same as that of the holotype of *Neuquenornis volans* (see above).

Diagnosis. A large number of autapomorphies diagnose this taxon (Chiappe 2002a). The most notable are the presence of a quadrate fused to the pterygoid, the biconvex articular facets of the fifth thoracic vertebra, the procoelous condition and very broad centra of the six vertebrae following this biconvex element (thoracics 6–11), the presence of a minor metacarpal that is more robust than the major metacarpal, the straplike morphology and distal cranioventral curvature of the pubis, the paddlelike shape of the ischium, and the pamprodactyl condition (four toes facing forward) of the feet.

Comments. Known from several specimens (Chiappe 2002a), *Patagopteryx deferrariisi* is the best represented Mesozoic bird of Patagonia (Fig. 10.4). This hen-sized, flightless bird was characterized by having reduced forelimbs (the ratio between humerus + ulna + carpometacarpus and femur + tibiotarsus + tarsometatarsus is approximately 0.47) and robust hind limbs. Original claims that this flightless bird was a member of the living ratites (ostriches and their kin) (Bonaparte 1986; Alvarenga and Bonaparte 1992; Alvarenga 1993) were rejected on the basis of more extensive cladistic analyses (Chiappe 1995, 2002a, 2002b). These hypotheses have firmly placed *Patagopteryx* within basal ornithuromorphs (Fig. 10.2). Indeed, *Patagopteryx* is the basalmost member of this clade to evolve flightlessness.

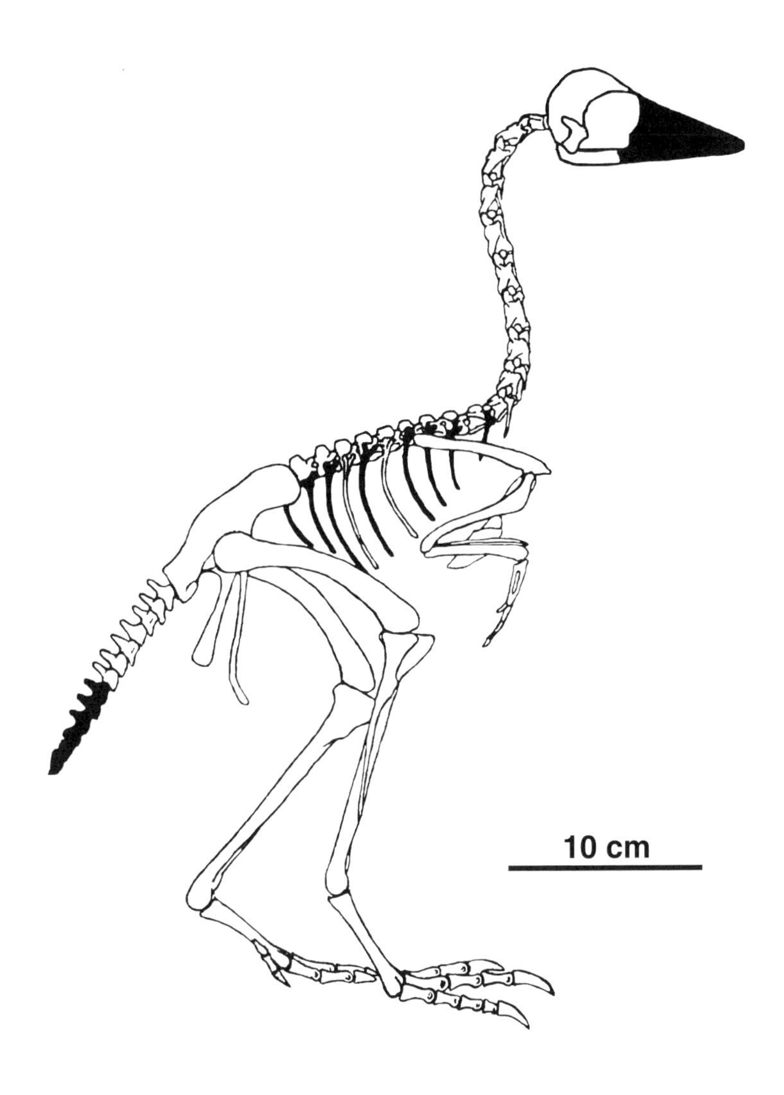

Figure 10.4. Basal ornithuromorph Patagopteryx deferrariisi *(Alvarenga and Bonaparte 1992; Chiappe 2002a) from the Late Cretaceous of the Bajo de la Carpa Formation (Neuquén City).*

Ornithurae Haeckel 1866
Carinatae Merrem 1813
***Limenavis patagonica* Clarke and Chiappe 2001**
Fig. 10.5

Holotype. PVL-4731, fragmentary specimen including associated portions of humerus, ulna, radius, ulnare and radiale, carpometacarpus, the proximal phalanx of manual digit II (major digit), and several indeterminate fragments (Clarke and Chiappe 2001) (Fig. 10.5).

Locality and age. Salitral Moreno, 20 km south of the city of General Roca, Río Negro Province, Argentina (Fig. 10.1, locality 6). The holotype of *Limenavis patagonica* is contained in the early Maastrichtian Allen Formation (Powell 1987; Leanza et al. 2004).

Diagnosis. Carinate bird having several autapomorphies (Clarke and Chiappe 2001) including a carpometacarpus with a deep infratrochlear fossa and a pisiform process proximally leveled

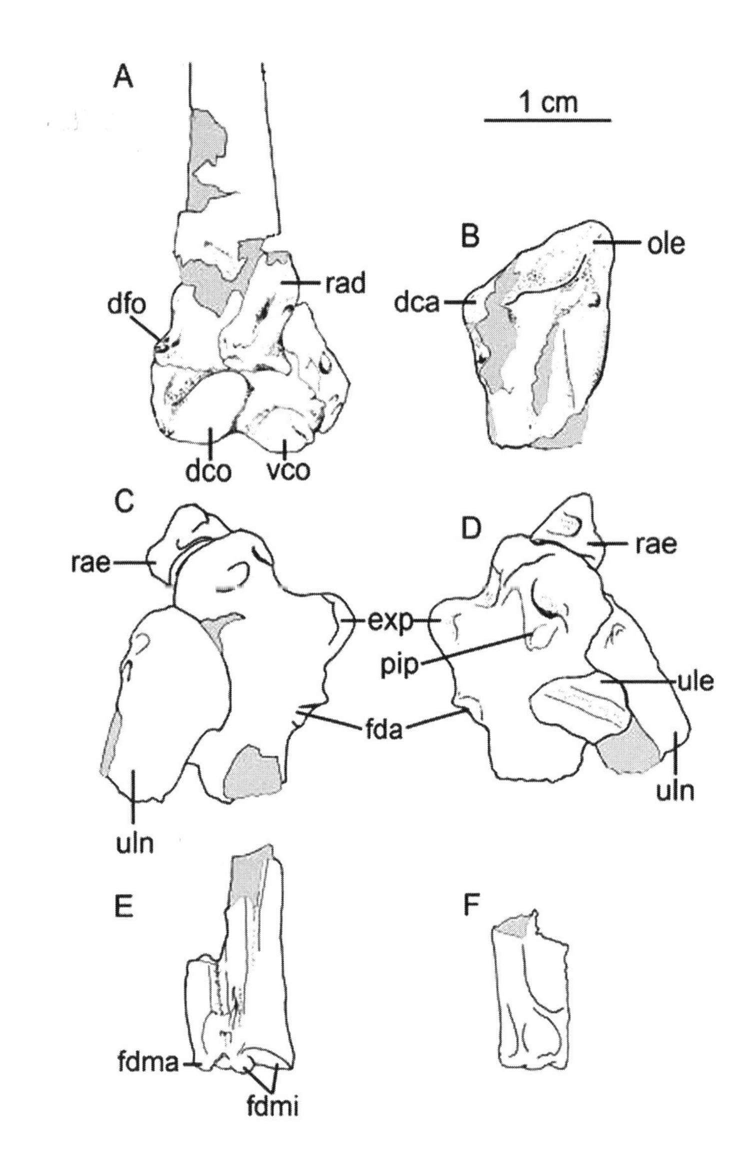

Figure 10.5. Basal carinate Limenavis patagonica *(Clarke and Chiappe 2001) from the Late Cretaceous of the Allen Formation (Salitral Moreno). (A) Distal half of humerus in craneal view. (B) Proximal end of ulna in ventral view. Proximal end of carpometacarpus in dorsal (C) and ventral (D) views. (E) Distal end of carpometacarpus in dorsal view. (F) Proximal phalanx of major digit. dca = dorsal cotyla; dco = distal condyle; dfo = distal fossae of humerus; exp = extensor process; fda = articular facet for alular digit; fdma = articular facet for major digit; fdmi = articular facet for minor digit; ole = olecranon; pip = pisiform process; rad = radius; rae = radiale; ule = ulnare; uln = ulna; vco = ventral condyle.*

with the proximal surface of the alular metacarpal, a well-developed tendinal groove on the ulnare, and three fossae on the proximal surface of the dorsal supracondylar process of the humerus.

Comments. Despite being quite incomplete, the anatomy of *Limenavis patagonica* indicates this bird was able to fly. Although anatomically modern in most respects, some characters of *Limenavis* prevent placing this bird within neornithines (Fig. 10.2). This fossil, however, appears to be phylogenetically closer to neornithines than other basal carinates such as *Ichthyornis,* thus being important for better understanding the early divergences of modern avians. As highlighted in the discussion of the preceding fossil (PVL-4730), *Limenavis* inhabited the shore of the Late Cretaceous south Atlantic.

Ornithurae gen. et sp. indeterminate

Material. PVL-4730, proximal half of tibiotarsus (Chiappe 1996).

Locality and age. Salitral Moreno, 20 km south of the city of General Roca, Province of Río Negro, Argentina (Fig. 10.1, locality 6). This specimen comes from the fluvial-estuarine deposits of the Allen Formation, the age of which is usually regarded as early Maastrichtian (Powell 1987; Leanza et al. 2004).

Comments. Little can be said about this incomplete specimen. The presence of two cnemial crests in its proximal end suggests an ornithurine relationship (Chiappe 1996) (Fig. 10.2). The overall appearance and size of this tibiotarsus resembles that of neornithine burhinids (wading birds such as the living thick-knees and stone-curlews). The sedimentology and the associated fauna of Salitral Moreno indicate that this bird lived in littoral environments near the ancient shore of the Atlantic Ocean.

Neornithes Gadow 1893
Gaviidae Allen 1897
Neogaeornis wetzeli Lambrecht 1929

Holotype. GPMK-123, a right tarsometatarsus (Lambrecht 1929).

Locality and age. West end of Bahía San Vicente, Province of Concepción, Chile (Fig. 10.1, locality 1). The holotype of *Neogaeornis wetzeli* is contained in the nearshore marine rocks of the Quiriquina Formation, the age of which is usually regarded as Maastrichtian (Biró Bagóczky 1982; Stinnesbeck 1986).

Comments. The phylogenetic relationships of the single known bone of *Neogaeornis wetzeli* are not entirely clear. On the one hand, it was regarded as a close relative of modern loons (Gaviiformes) and grebes (Podicipediformes) by Lambrecht (1933) and a member of the latter by Brodkorb (1963). Subsequently, *Neogaeornis* was included within the Hesperornithiformes (a group of Cretaceous foot-propelled divers that includes *Hesperornis regalis* [Fig. 10.2]) by Martin and Tate (1976). A recent study of the reprepared specimen identified it as a gaviid (Olson 1992), the group including all living loons. Details of the proximal end of the tarsometatarsus (e.g., presence of a hypotarsus and of proximal metatarsal foramins) suggest *Neogaeornis* is likely not a hesperornithiform (Olson 1992), thus restricting the record of these foot-propelled divers to Laurasia. Although the overall morphology of this taxon does resemble that of gaviids (Olson 1992), identifications of fragmentary Cretaceous remains as those of living clades need to be taken with caution (Clarke and Chiappe 2001), especially in light that no definitive remains of gaviids have been recorded in Cretaceous rocks (the Cretaceous age of the putative loon *Polarornis gregorii* [Chatterjee 2002] remains inconclusive [Clarke and Chiappe 2001]).

Galliformes Garrod 1873
Gen. et sp. indeterminate

Material. A tiny, isolated coracoid (Agnolin et al. 2003).

Locality and age. Sierra del Portezuelo, Province of Neuquén, Argentina (Fig. 10.1, locality 2). This coracoid comes from the fluvial deposits of the Portezuelo Formation, which Leanza et al. (2004) dated as Coniacian.

Comments. Although anatomically of modern appearance, the fragmentary nature of this fossil warrants caution in classifying it as a neornithine (see Clarke and Chiappe [2001] and Chiappe and Dyke [2002] for concerns regarding identification of fragmentary Mesozoic fossils to modern groups). If so correctly identified, this material would not only be the earliest record of a bird from Patagonia but also the oldest of neornithines for the entire Southern Hemisphere.

Discussion

The Mesozoic record of Patagonian birds, albeit small and restricted to the last 20 million years of the Cretaceous (Coniacian-Maastrichtian), reflects a rich avifauna that is morphologically, phylogenetically, and ecologically very diverse. The fossil record tells us of the existence of flightless land birds and flying enantiornithines that inhabited subhumid environments, anatomically modern birds that lived near the shore of Atlantic Ocean, and foot-propelled divers that swam in the waters of the South Pacific.

The Mesozoic record of Patagonian birds has the potential for testing novel hypotheses about the early divergence of modern avians. Few occurrences that could be interpreted as neornithines are known from the Mesozoic (Chiappe and Dyke 2002; Dyke and van Tuinen 2004), and these Late Cretaceous fossils postdate most molecular estimates for the origin of the group by more than 30 million years (e.g., Hedges et al. 1996; Cooper and Penny 1997; Kumar and Hedges 1998). The absence of any Early Cretaceous fossil that could be defensibly regarded as a neornithine (Chiappe and Dyke 2002) makes it difficult to embrace the explosive Early Cretaceous radiation proposed for these birds by some molecular phylogenies. In the face of this, some researchers have argued that the earliest divergences of neornithines could have taken place in regions that remain paleontologically poorly sampled, such as the ones that once formed the supercontinent of Gondwana (Cracraft 2001; Chatterjee 2002). This southern origin of neornithines receives support from the fact that many primitive lineages (paleognaths, galloanserines, gruiforms, and others) are predominantly distributed across the former Gondwanan landmasses (Cracraft 2001). It has also received some support from the meager Antarctic fossil record (Noriega and Tambussi 1995; Case and Tambussi 1999; Chatterjee 2002; Clarke et al.

2004), which has yielded a few anatomically modern birds from the latest Maastrichtian. This view has also been favored by molecular biologists, who in some instances also see a causal correlation between the fragmentation of Gondwana and the divergence of groups distributed in the Southern Hemisphere (Hedges et al. 1996). The Mesozoic sediments of Patagonia can thus provide critical evidence in favor or against these hypotheses, but the known fossil record of birds from this region is yet too small and too young to tell.

Acknowledgments. I am grateful to the editors of this volume for inviting me to contribute, to R. Urabe for preparation of the figures, and to F. Novas for his review of the article in manuscript.

References Cited

Agnolin, F. L., F. E. Novas, and G. Lio. 2003. Restos de un posible galliforme (Aves, Neornithes) del Cretácico tardío de Patagonia. *Ameghiniana* 40 (4): 49R.

Alvarenga, H. M. F. 1993. A origen das aves seus fósseis. In M. A. de Andrade (ed.), *A Vida das Aves,* 16–26. Belo Horizonte, Brazil: Editora Litera Maciel.

Alvarenga, H. M. F., and J. F. Bonaparte. 1992. A new flightless land bird from the Cretaceous of Patagonia. In K. E. Campbell (ed.), *Papers in Avian Paleontology, Honoring Pierce Brodkorb,* 51–64. Series 36. Los Angeles, Calif.: Natural History Museum of Los Angeles County.

Biró Bagóczky, L. 1982. Revisión y redefinición de los "Estratos de Quiriquina," Campaniano-Maastrichtiano, en su localidad tipo, en la Isla Quiriquina, 36° 37' Lat. Sur, Chile, Sudamérica, con un perfil complementario en Cocholque. *Actas del III Congreso Geológico Chileno* 1: A29–A64.

Bonaparte, J. F. 1986. History of the terrestrial Cretaceous vertebrates of Gondwana. *Actas del IV Congreso Argentino de Paleontología y Bioestratigrafía* 2: 63–95.

———. 1987. The Late Cretaceous fauna of Los Alamitos, Patagonia, Argentina. Parts I to IX. *Revista del Museo Argentino de Ciencias Naturales "Bernardino Rivadavia," Paleontologia* 3 (3): 103–178.

Brodkorb, P. 1963. Catalogue of fossil birds. *Bulletin of the Florida State Museum, Biological Sciences* 7: 179–293.

Casamiquela, R. 1987. Novedades en icnología de vertebrados en la Argentina. *Anais do X Congreso Brasileiro de Paleontologia* 1: 445–456. Rio de Janeiro.

Case, J. A., and C. P. Tambussi. 1999. Maestrichtian record of neornithine birds in Antarctica: comments on a Late Cretaceous radiation of modern birds. *Journal of Vertebrate Paleontology* 19 (3): 37A.

Chatterjee, S. 2002. The morphology and systematics of *Polarornis,* a Cretaceous loon (Aves: Gaviidae) from Antarctica. In Z. Zhou and F. Zhang (eds.), *Proceedings of the 5th Symposium of the Society of Avian Paleontology and Evolution,* 125–155. Beijing: Society of Avian Paleontology and Evolution.

Chiappe, L. M. 1991. Cretaceous birds of Latin-America. *Cretaceous Research* 12: 55–63.

———. 1992. Enantiornithine tarsometatarsi and the avian affinity of the

Late Cretaceous Avisauridae. *Journal of Vertebrate Paleontology* 12 (3): 344–350.
———. 1995. The phylogenetic position of the Cretaceous birds of Argentina: Enantiornithes and *Patagopteryx deferrariisi. Courier Forschungsinstitut-Senckenberg* 181: 55–63.
———. 1996. Early avian evolution in the Southern Hemisphere: fossil record of birds in the Mesozoic of Gondwana. *Memoirs of the Queensland Museum* 39: 533–556.
———. 2001. Phylogenetic relationships among basal birds. In J. Gauthier and L. F. Gall (eds.), *New Perspectives on the Origin and Early Evolution of Birds: Proceedings of the International Symposium in Honor of John H. Ostrom,* 125–139. New Haven, Conn.: Peabody Museum of Natural History.
———. 2002a. Osteology of the flightless *Patagopteryx deferrariisi* from the Late Cretaceous of Patagonia (Argentina). In L. M. Chiappe and L. Witmer (eds.), *Mesozoic Birds: Above the Heads of Dinosaurs,* 281–316. Berkeley: University of California Press.
———. 2002b. Early bird phylogeny: problems and solutions. In L. M. Chiappe and L. Witmer (eds.), *Mesozoic Birds: Above the Heads of Dinosaurs,* 448–472. Berkeley: University of California Press.
Chiappe, L. M., and J. O. Calvo. 1994. *Neuquenornis volans,* a new Upper Cretaceous bird (Enantiornithes: Avisauridae) from Patagonia, Argentina. *Journal of Vertebrate Paleontology* 14 (2): 230–246.
Chiappe, L. M., and G. Dyke. 2002. The Mesozoic radiation of birds. *Annual Review of Ecology and Systematics* 33: 91–124.
Chiappe, L. M., and C. Walker. 2002. Skeletal morphology and systematics of the Cretaceous Enantiornithes. In L. M. Chiappe and L. Witmer (eds.), *Mesozoic Birds: Above the Heads of Dinosaurs,* 240–267. Berkeley: University of California Press.
Chiappe, L. M., M. A. Norell, and J. Clark. 2002. The Cretaceous, short-armed Alvarezsauridae: *Mononykus* and its kind. In L. M. Chiappe and L. Witmer (eds.), *Mesozoic Birds: Above the Heads of Dinosaurs,* 87–120. Berkeley: University of California Press.
Clarke, J. A., and L. M. Chiappe. 2001. A new carinate bird from the Late Cretaceous of Patagonia (Argentina). *American Museum Novitates* 3323: 1–23.
Clark, J., M. A. Norell, and P. J. Makovicky. 2002. Cladistic approaches to the relationships of birds to other theropod dinosaurs. In L. M. Chiappe and L. Witmer (eds.), *Mesozoic Birds: Above the Heads of Dinosaurs,* 31–61. Berkeley: University of California Press.
Clarke, J. A., C. P. Tambussi, J. I. Noriega, G. M. Erickson, and R. A. Ketcham. 2004. Definitive fossil evidence for the extant avian radiation in the Cretaceous. *Nature* 3150: 1–4.
Cooper, A., and D. Penny. 1997. Mass survival of birds across the Cretaceous-Tertiary boundary: molecular evidence. *Science* 275: 1109–1113.
Coria, R. A., P. J. Currie, D. Eberth, and A. Garrido. 2002. Bird footprints from the Anacleto Formation (Late Cretaceous), Neuquén, Argentina. *Ameghiniana* 39 (4): 453–463.
Cracraft, J. 2001. Avian evolution, Gondwana biogeography and the Cretaceous-Tertiary mass extinction event. *Proceedings of the Royal Society of London,* series B, 268: 459–469.
Currie, P. 1981. Bird footprints from the Gething Formation (Aptian,

Lower Cretaceous) of northeastern British Columbia, Canada. *Journal of Vertebrate Paleontology* 22: 460–465.
Dingus, L., J. Clarke, G. R. Scott, C. C. Swisher III, L. M. Chiappe, and R. Coria. 2000. Stratigraphy and magnetostratigraphic/faunal constraints for the age of sauropod embryo-bearing rocks in the Neuquén Group (Late Cretaceous, Neuquén Province, Argentina). *American Museum Novitates* 3290: 1–11.
Dyke, G. J., and M. van Tuinen. 2004. The evolutionary radiation of modern birds (Neornithes): reconciling molecules, morphology and the fósil record. *Zoological Journal of the Linnean Society* 141: 153–177.
Hedges, S. B., P. H. Parker, C. G. Sibley, and S. Kumar. 1996. Continental breakup and the ordinal diversification of birds and mammals. *Nature* 381: 226–229.
Kumar, S., and S. B. Hedges. 1998. A molecular timescale for vertebrate evolution. *Nature* 392: 917–920.
Lambrecht, K. 1929. *Neogaeornis wetzeli* n.g., n.sp., der erste Kreidevögel der südlichen Hemisphaere. *Palaeontologische Zeitschrift* 11: 121–128.
———. 1933. *Handbuch der Palaeornithologie.* Berlin: Gebrüder Borntraeger.
Leanza, H. A., S. Apesteguía, F. E. Novas, and M. de la Fuente. 2004. Cretaceous terrestrial beds from the Neuquén Basin (Argentina) and their tetrapod assemblages. *Cretaceous Research* 25: 61–87.
Leonardi, G. (ed.). 1987. *Glossary and Manual of Tetrapod Footprint Palaeoichnology.* Rio de Janeiro, Brazil: Departamento Nacional da Produção Mineral (DNPM).
Lockley, M. G., S. Y. Yang, M. Matsukawa, F. Fleming, and S. K. Lim. 1992. The track record of Mesozoic birds: evidence and implications. *Philosophical Transactions of the Royal Society of London* B 336: 113–134.
Martin, L. D., and J. Tate Jr. 1976. The skeleton of *Baptornis advenus* (Aves: Hesperornithiformes). In S. L. Olson (ed.), *Collected Papers in Avialian Paleontology Honoring the 90th Birthday of Alexander Wetmore,* 35–66. Smithsonian Contributions to Paleobiology 27.
Noriega, J., and C. P. Tambussi. 1995. A Late Cretaceous Presbyornithidae (Aves: Anseriformes) from Vega Island, Antarctic Peninsula: paleobiogeographic implications. *Ameghiniana* 32: 57–61.
Novas, F. E., and D. Pol. 2002. Alvarezsaurid relationships reconsidered. In L. M. Chiappe and L. Witmer (eds.), *Mesozoic Birds: Above the Heads of Dinosaurs,* 121–125. Berkeley: University of California Press.
Olson, S. L. 1992. *Neogaeornis wetzeli* Lambrecht, a Cretaceous loon from Chile (Aves: Gaviidae). *Journal of Vertebrate Paleontology* 12 (1): 122–124.
Padian, K., and P. E. Olsen. 1984. The fossil trackway *Pteraichnus:* not pterosaurian, but crocodilian. *Journal of Paleontology* 58: 178–184.
Powell, J. E. 1987. Hallazgo de un dinosaurio hadrosáurido (Ornithischia, Ornithopoda) en la Formación Allen (Cretácico Superior) de Salitral Moreno, Provincia de Río Negro, Argentina. *Actas del X Congreso Geológico Argentino* 3: 149–152. San Miguel de Tucumán.
Schweitzer, M. H., F. Jackson, L. M. Chiappe, J. G. Schmitt, J. O. Calvo, and D. E. Rubilar. 2002. Late Cretaceous avian eggs with embryos from Argentina. *Journal of Vertebrate Paleontology* 22 (1): 190–194.

Stinnesbeck, W. 1986. Zu den faunistischen und palökologischen verhältnissen in der Quiriquina Formation (Maastrichtium) Zentral-Chiles. *Palaeontographica Abt. A* 194: 99–237.

Unwin, D. M. 1989. A predictive method for the identification of vertebrate ichnites and its application to pterosaur tracks. In D. D. Gillette and M. G. Lockley (eds.), *Dinosaur Tracks and Traces,* 259–274. Cambridge: Cambridge University Press.

Walker, C. 1981. New subclass of birds from the Cretaceous of South America. *Nature* 292: 51–53.

11. Ichthyosauria

Marta Fernández

Introduction

Different lineages of reptiles successfully colonized marine environments during the Mesozoic. Among these, fish-shaped ichthyosaurs were the group that acquired the highest level of aquatic adaptation. Anatomical changes of this taxon include streamlined bodies with a lunate tail and a dorsal fin. The arrangement of the limb bones is also strongly modified, forming fins. The evolution of finlike limbs comprises drastic changes, including the loss of digit I (Motani 1999a) and the shortening of all limb bones (except the humerus and femur) correlated with the loss of perichondral ossification (Caldwell 1997a,b). On the basis of the morphological modifications of their skeletons, it is assumed that fish-shaped ichthyosaurs have developed good swimming capabilities that allowed them a worldwide distribution (McGowan 1978; Sander 2000; McGowan and Motani 2003). The biochron of ichthyosaurs extends from the Early Triassic (Spathian) to the Late Cretaceous (Cenomanian). Because belemnites were an important item in the ichthyosaur's diet, their ultimate extinction has been correlated with the extinction of marine invertebrates (especially belemnites) during Cenomanian-Turonian boundary events (Bardet 1992). However, the recent finding of turtles and birds in the gut contents of Cretaceous ichthyosaurs challenges this hypothesis (Kear et al. 2003).

The evolutionary history of ichthyosaurs was almost exclusively defined by what was known from the Northern Hemisphere. Long blank zones characterize the global record. Within this context, the Patagonian record is very important, filling in areas that are either otherwise completely unknown (Bajocian) or are very poorly represented (Tithonian) worldwide. The study of Patagonian forms allows us to better understand the evolutionary history of this group.

Institutional abbreviations. MACN, Museo Argentino de Ciencias Naturales "Bernardino Rivadavia," Buenos Aires, Argentina; MCNAM, Museo de Ciencias Naturales y Antropológicas "J. C. Moyano," Mendoza, Argentina; MLP, Museo de La Plata, La Plata, Argentina; MOZP, Museo "Prof. Dr. Juan Olsacher," Zapala, Neuquén, Argentina.

Brief History of Ichthyosaur Discoveries in Patagonia

The history of ichthyosaur discoveries in Argentine land is more than a century old. Despite this long history, the record of the Eastern Pacific margin of South America became significant only in the last half of the 20th century. Unlike what happened in the Northern Hemisphere, the first discoveries were incomplete and/or had dubious stratigraphic provenance. The first description was made by Dames (1893) on a few vertebrae and ribs from Tithonian sediments of La Cienaguita (Mendoza, Argentina). He proposed a new species to enclose that material: *Ichthyosaurus bondenbenderi.* In 1895, Phillipi cited and briefly described 17 posterior dorsal vertebrae, one humerus, one pelvic girdle bone, ribs, and a caudal vertebra from the Cajón del Durazno (south of Mendoza Province). On this basis, Phillipi (1895) named the new species *Ichthyosaurus immanis* and referred another three vertebrae as likely pertaining to *Ichthyosaurus leucopetraeus* (Burmeister and Giebel 1861), but no illustrations were included in his paper. Casamiquela (1970) referred these remains as of Callovian age and pointed out that they belong to a large species comparable to *Ancanamunia* (Rusconi 1942). He noted that the humerus has two articular faces, but he did not illustrate the material either. Because the description is insufficient and no illustrations were published, Gasparini (1985) considered it a nomen vanum. Both the humerus and the pelvic girdle described by Phillipi, although incomplete, could have shown diagnostic features, but unfortunately the repository could not be confirmed in subsequent papers (Z. Gasparini personal communication 2005). Consequently, nothing new may be reported from these first discoveries. In 1939, Cabrera described a rostral fragment of an ichthyosaur from the Bajocian of Curru-Charahuilla, Neuquén, and on this basis, he proposed a new species, *Stenopterygius grandis.*

As all the species names mentioned above are of doubtful diagnosis, I agree with McGowan and Motani (2003) in considering them nomina dubia (including *Ichthyosaurus immanis* [Phillipi 1895], which is not mentioned by McGowan and Motani 2003). Huene (1925) described a few vertebrae and a fragment of an anterior forefin of an ichthyosaur from the Neocomian of Cerro Belgrano (Santa Cruz), and in 1927 he proposed the new genus and species *Myobradypterygius hauthali* for its reception (Huene 1927). McGowan (1972) transferred this species to the genus *Platypterygius* (Huene 1922).

Between 1938 and 1949, Rusconi, as a result of fieldwork mainly in Mendoza Province (Argentina), exhumed several remains of Jurassic marine reptiles, including ichthyosaurs. Rusconi's papers were the most significant on South American ichthyosaurs known at the time. Although most of the taxa proposed by Rusconi are currently considered nomina dubia, his fieldwork and laboratory work were the first to mark the diversity and number of marine reptiles in this land. On the basis of this material, he named seven new species of ichthyosaurs, three of which were assigned to the genus *Ancanamunia: A. mollensis* (Rusconi 1938), *A. mendozana* (Rusconi 1940), and ?*A. espinacitensis* (Rusconi 1949). The type species is *A. mendozana* by original designation (Rusconi 1940, 2). Four other species were then added: *Ichthyosaurus inexpectatus, I. saladensis,* ?*I. sanjuanensis,* and *Macropterygius incognitus* (Rusconi 1948, 1949). Of the species described by Rusconi, *Ichthyosaurus inexpectatus* is not an ichthyosaur but more likely the metriorhynchid *Dakosaurus* (Vignaud and Gasparini 1996); and, except for *A. mendozana,* which is a synonym of *Ophtalmosaurus natans* (McGowan and Motani 2003), the rest of the names are nomina dubia. After this first period, there is a gap during which papers on South American ichthyosaurs were absent in the paleontological literature. In the 1970s, Z. Gasparini initiated a research trend, which is still ongoing, among the main results of which is the creation of the most important collection of Tithonian marine reptiles in the world (e.g., Gasparini and Fernández 1997).

Shultz et al. (2003) described a few vertebrae and dorsal ribs of an ichthyosaur from Torres del Paine National Park (Chile) (Fig. 11.1). This is the southernmost record of an ichthyosaur from Patagonia. The specimen lacks diagnostic features, and because it was found in a large block in glaciofluvial sediments, no stratigraphic position can be precisely assigned.

Systematic Paleontology

The phylogeny of ichthyosaurs and their relationships with other amniotes is still controversial. Three recent phylogenetic analyses of the group are based on large data sets: Motani (1999b), Sander (2000), and Maisch and Matzke (2000). McGowan and Motani (2003) comparatively analyzed these proposals. Of the three data sets, that of Motani (1999b) has the best methodology. I agree with Montoni's opinion, and therefore, in the Systematic section of this chapter, taxa are arranged according to the phylogeny proposed by Motani (1999b). The term Thunnosauria is used for the clade including the last common ancestor of *Stenopterygius quadriscissus* and *Ichthyosaurus communis* and all its descendants (Motani 1999b, 484) (Fig. 11.2).

Triassic

Ichthyosaurs were widely diversified during the Triassic and were fully marine animals (Sander 2000). Although a worldwide

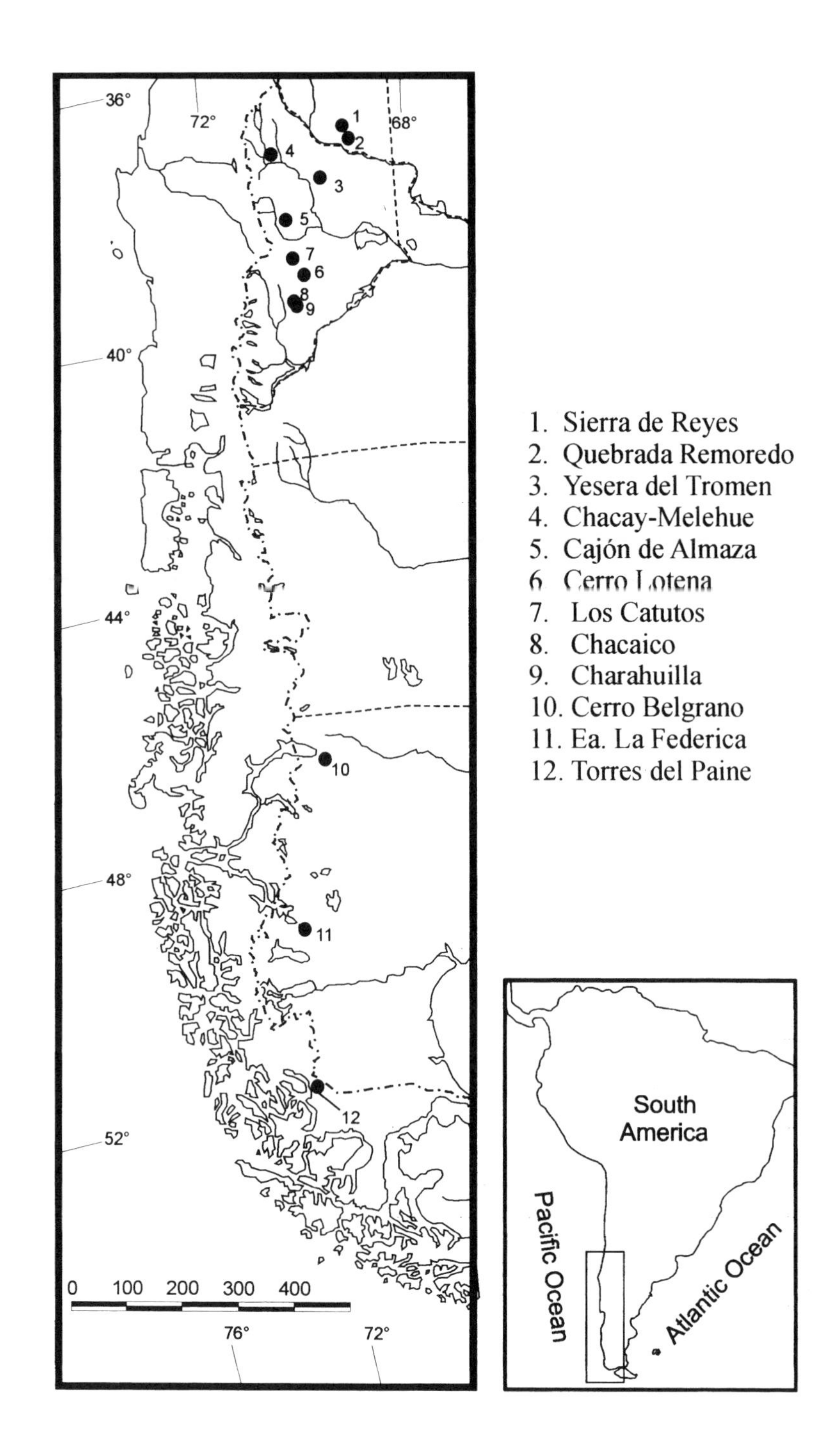

Figure 11.1. Location map of Patagonian ichthyosaurs outcrops.

distribution could be expected for Triassic ichthyosaurs, the most productive sites are all in the Northern Hemisphere. The only Triassic record of South America was found outside Patagonia in the late Triassic of Quebrada Doña Inés (26° 07'S; 69° 20'W), Chile, and comprises one paddle bone, a fragment of scapula or humerus,

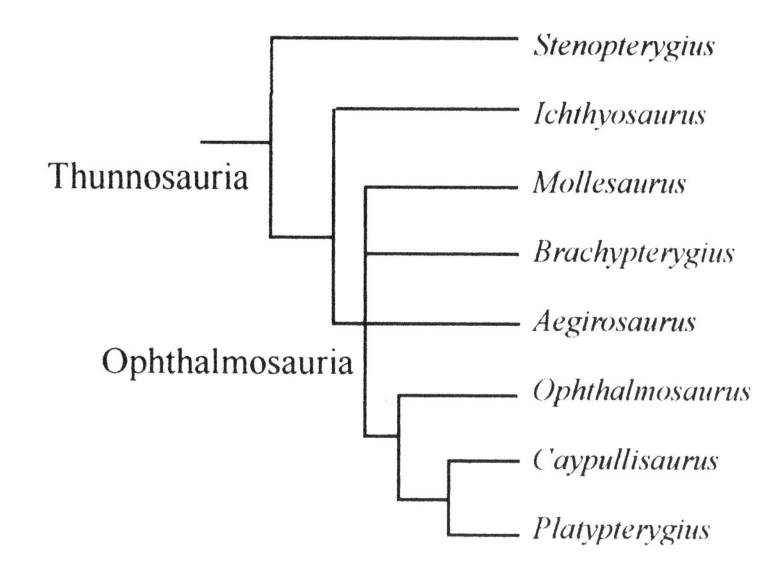

Figure 11.2. Phylogenetic hypothesis of Thunnosauria relationships. Simplified and modified from Motani (1999b).

and five teeth (Suárez and Bell 1992). This situation could be due to the lack of good exposures of Triassic marine sediments in the most prospected areas.

Early Jurassic

Early Jurassic ichthyosaurs are particularly abundant in the European margins of the Tethys. By contrast, the record of Early Jurassic ichthyosaurs in Patagonia is scarce and not identifiable at the familial or lower level. Although extensive exposures of the marine Lower Jurassic occur in the Neuquén Basin, they are mainly known for their invertebrate fossils and have not yet been prospected for marine reptiles. Rusconi (1949) described, on the basis of vertebral centra, the species ?*Ichthyosaurus sanjuanensis* and ?*Ancanamunia espinacitense* from the Liassic of San Juan, but these names are nomina dubia, and the geographic and stratigraphic data are not precise. Up to the present, only some vertebral centra and ribs found in sediments of the Puesto Araya Formation (lower Sinemurian, Río Atuel, Mendoza Province, Argentina; Fernández and Lanés 1999) may be added to the previous records. Although the material is quite fragmentary, it is important because it is the first record of a Lower Jurassic ichthyosaur with precise stratigraphic provenance in the Neuquén Basin.

Middle Jurassic

There is a gap in the ichthyosaur record after the Toarcian until the Callovian, and only fragmentary material has been found from this interval. Until now, this gap has only been interrupted by the Aalenian-Bajocian ichthyosaurs excavated from the Neuquén Basin (Fernández 1994, 1998; Spalletti et al. 1994) (Fig. 11.3A, C–G). Tavera (1981) described ichthyosaur remains of several specimens

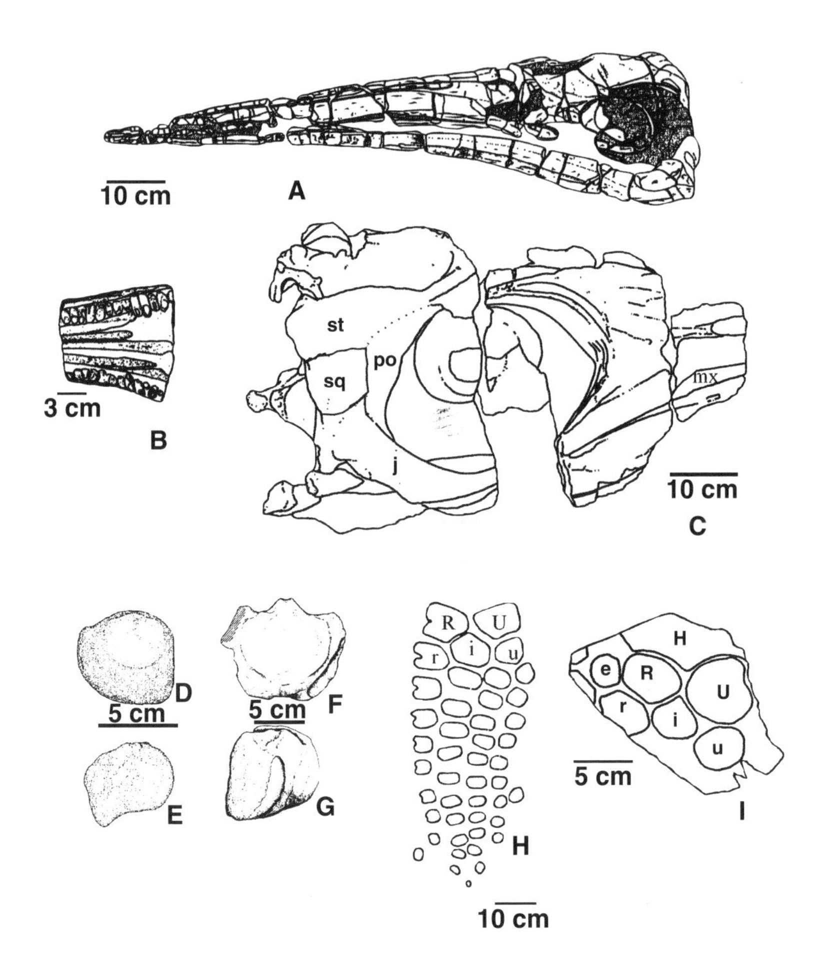

Figure 11.3. Aalenian-Bajocian ichthyosaurs from Patagonia. (A) Stenopterygius cayi *(MOZ 5803), skull in left lateral view. (B)* "Stenopterygius grandis" *(MLP 39-VII-2-2), part of the rostrum in ventral view. (C)* Mollesaurus periallus *(MOZ 2282), skull in right lateral view. (D, E)* Stenopterygius cayi *(MOZ 5803), basioccipital in occipital and lateral view. (F, G)* Mollesaurus periallus *(MOZ 2282), basioccipital in occipital and lateral view. (H)* Stenopterygius cayi *(MOZ 5803), left forefin. (I) Ophthalmosauridae (MLP 92-III-2-1), forelimb.*

found in the upper part of the Lautaro Formation (early Bajocian) (Jensen and Vicente 1976) exposed at Quebrada La Iglesia, Chile (28° 07'S; 69° 58'W). He identified them as *Ichthyosaurus acutirostris* (Owen 1840) and *I. posthumus* (Wagner 1852). However, part of the material identified as *I. acutirostris* (Tavera 1981, pl. I) was reassigned to the metriorhynchid *Metriorhynchus* Gasparini et al. (2000). The specimens T331, T332, and T334 (Tavera 1981, pl. II, figs. 1–5, 7) are badly preserved ichthyosaur vertebrae that cannot be identified at familial or lower level. T330 (Tavera 1981, pl. II, fig. 6) is a skull in occipital view that probably corresponds to a crocodile (Z. Gasparini personal communication 2005), T 338 and 337 (Tavera 1981, pl. IV, figs. 1, 4) are ribs, and T342 (Tavera 1981, pl. IV, figs. 2, 3) is a fragmented rostrum of an ichthyosaur, but no diagnostic features have been preserved. Unfortunately, the repository of the material described by Tavera (1981) cannot be confirmed.

The first ichthyosaur from the Middle Jurassic of Patagonia (Argentina) was described by Cabrera (1939) as a new species: *Stenopterygius grandis* from the early Bajocian of Charahuilla

(Figs. 11.1, locality 10, 11.3B). The material consists of an incomplete snout lacking diagnostic features, so until more complete material becomes available, this name must be considered a nomen dubium as proposed by McGowan and Motani (2003).

In 1984, Dr. C. Gulisano (ex Yacimientos Petrolíferos Fiscales of Argentina) found a fragment of an ichthyosaur forefin in the Middle Jurassic of southern Mendoza. This material was deposited at the Museo de La Plata collection (MLP 92-III-2-1), but it was not described until fairly recently (Fernández 2003). In the 1990s, several field trips were carried out in the locality Chacaico Sur (Neuquén Province). McGowan and Motani (2003, 59) mentioned in their review of the stratigraphic position of the major ichthyosaur-producing localities that the Middle Jurassic localities of Neuquén Province are Bajocian and Bathonian in age. Nevertheless, to date, no ichthyosaur remains have been found in Bathonian sediments, although this could change in the future: crocodile remains have recently been exhumed from Bathonian levels of the Los Molles Fomation (Gasparini et al. 2005). Pliosaurs and the most complete ichthyosaurs were recovered from the top of Los Molles Formation (early Bajocian) (Fernández 1994, 1999; Spalletti et al. 1994; Gasparini 1997).

The following taxa are now recognized for the Aalenian-Bajocian of Patagonia:

Ichthyosauroidea McGowan and Motani 2003
Stenopterygiidae Kuhn 1934
***Stenopterygius* Jaekel 1904**
***Stenopterygius cayi* (Fernández 1994) nov. comb.**
Fig. 11.3A, D, E, H

***Chacaicosaurus cayi* Fernández 1994, 292, figs. 2–5**
***Chacaicosaurus cayi* Maisch and Matzke 2000, 76–77**
***Chacaicosaurus cayi* McGowan and Motani 2003, 90–91, fig. 79**

Holotype. MOZ-P-5803, a complete skull, a forefin, isolated phalanges, cervical vertebrae, and interclavicle, proximal parts of one humerus and femur.

Locality and age. Top of Los Molles Formation (Spalletti et al. 1994), Chacaico Sur, Neuquén, Argentina (Fig. 11.1, locality 9). The holotype was found associated with ammonites identified as *Sonninia* cf. *alsatica* by Dr. A. Riccardi and referred to the zone of *Emileia giebeli* subzone *E. multiformis* (early Bajocian).

Emended diagnosis. Moderate-sized ichthyosaurs with long and slender snout (snout ratio 0.80), forefin with four digits, radius and seven first elements of the first digit notched, radius and ulna long (width/length ratio 1.5), ulnare smaller than the intermedium, no digital bifurcation, maximum number of elements in the longest digit no less than 14. Manus digits distally separated in adults.

Comments. The holotype is the only known individual of the species. This species was originally described as a new genus and species family incertae sedis because at that moment, the relationships among ichthyopterygians were not established, nor were the major clades defined. Papers such as those by Motani (1999a,b), Sander (2000), and Maisch and Matzke (2000) marked a step forward in this topic, and for the first time, phylogenetic hypotheses of most of the known taxa became available. In 2003, McGowan and Motani summarized all the information available about the systematics of the group. According to these authors, the forefin with four to six digits, notching in some elements of the leading edge, humerus with two distal facets, paired fins disproportional in length, skull and jaw length <1 m are among the diagnostic features of *Stenopterygius*. All these characters are also present in MOZ 5803, and on this basis, *Chacaicosaurus cayi* (Fernández 1994) is included in the genus *Stenopterygius*. Therefore, *Chacaicosaurus* (Fernández 1994) should be considered as a junior synonym of *Stenopterygius* (Jaekel 1904).

Ophthalmosauridae Baur 1887
***Mollesaurus* Fernández 1999**
Fig. 11.3C, F, G

Ophthalmosaurus *Maisch and Matzke 2000, 78*

Type species. Mollesaurus periallus (Fernández 1999).

Emended diagnosis. Large ichthyosaur differing from the other members of the family by the following combination of characters: reduced exposition of the maxilla in lateral view; strong crest of the prefrontal in the anterior margin of the orbit; postorbital region of the skull broad; squamosal with a subquadrangular outline and broadly exposed in lateral view; basioccipital condyle not clearly set off from the extracondylar area; extracondylar area reduced.

Comments. This taxon is known only by the holotype of the species (MOZ 2282). McGowan and Motani (2003) mentioned the sclerotic ring occupying only about half of the orbit as a diagnostic character of this genus. However, the reduced sclerotic ring in comparison to the size of the orbit in MOZ 2282 could be better explained in terms of ontogenetic variation, and therefore this feature has no taxonomic value.

***Mollesaurus periallus* Fernández 1999**
Fig 11.3C

Aff. *Ophthalmosaurus* Seeley 1874; Spalletti et al. 1994, 333, pl. 1A
***Ophthalmosaurus periallus* Seeley 1874; Maisch and Matzke 2000, 78**
***Mollesaurus periallus* Fernández 1999; McGowan and Motani 2003, 107–108, fig. 90**

Holotype. MOZ 2282 V, partial skull, 39 articulated vertebrae and associated ribs, and right clavicle.

Type locality and age. Chacaico Sur (39°15'S; 70°18'W), 70 km southwest of Zapala, Neuquén Province, Northwestern

Patagonia, Argentina (Fig. 11.1, locality 9). Top of Los Molles Formation. The holotype was found in association with the ammonites *Sonninia (Papilliceras) espinazitensis* and *Sonninia* cf. *espinazitensis* referable to the *Emileia giebeli* ammonite Zone; early Bajocian; Middle Jurassic (Spalletti et al. 1994).

Diagnosis. As for the genus by monotypy.

Comments. Maisch and Matzke (2000) proposed *Mollesaurus periallus* as a junior synonym of *Ophthalmosaurus icenicus*. Nevertheless, the comparable elements known in both species are too different to support their proposal. Thus, the most conspicuous features to distinguish these two taxa are: the elongate nares; the broad postorbital region of the skull; the reduced extracondylar area of the basioccipital; the condyle not clearly set off from the extracondylar area; the strong ridge in the anterior margin of the orbit; the relative long vertebral centra (in *Ophthalmosaurus* sp. the dorsal vertebrae are short with height/length ratio above 3.0 [Motani 1999b]).

Ophthalmosauridae gen. et sp. indeterminate
Fig. 11.3I

Referred specimens. MLP 92-III-2-1, distal part of the humerus, the zeugopodium, and the radiale, intermedium, and ulnare from the mesopodium, and part of the elements distal to the extrazeugopodial element (Fernández 2003).

Locality and age. MLP 92-III-2-1 was found at Quebrada Remoredo (37°S; 69° 45'W) in the western slope of Sierra de Reyes, Mendoza Province, Argentina (Fig. 11.1, locality 2). Ammonoid Zone of *Puchenquia malarguensis* (Aalenian-Bajocian boundary, A. Riccardi personal communication) of Los Molles Formation.

Comments. Although incomplete, MLP 92-III-2-1 is significant because it represents the oldest record of Ophthalmosauria, extending the geological range of the clade back to the Aalenian-Bajocian boundary. The presence of an extrazeugopodial element anterior to the radius (Fig. 11.3I) and the associated digits distal to it permit MLP 92-III-2-1 to be unequivocally referred to the clade Ophthalmosauria as defined by Motani (1999a,b). Its inclusion within the Ophthalmosauria also shortens a long ghost lineage that links the Liassic *Ichthyosaurus* with the Ophthalmosauria (Fernández 2003).

Significance of Aalenian-Bajocian Ichthyosaurs from Patagonia

Before the discoveries of Patagonian ichthyosaurs, there was a gap in the record from the Toarcian up to the Callovian of approximately 16 Ma (based on Gradstein et al. 1995), which represents most of the Middle Jurassic. The discoveries of Aalenian-Bajocian ichthyosaurs from Patagonia help document the evolution of certain characters through the Jurassic. Thus, most of the Early Jurassic ichthyosaurs have a basioccipital with an extensive extracondylar area, which becomes reduced in different degrees in Callovian and younger forms. The evolutionary pattern of the forefin was extensively analyzed through the stratigraphic record (Motani 1999a), and Early Jurassic ichthyosaurs (as well as some Late Triassic forms) were characterized by the shortening of the zeugopodial elements and by the loss of digit I. These two changes persisted until the extinction of the group, but in Callovian and post-Callovian forms, the unusual addition of an extrazeugopodial element with a digit (or more) distal to it occurred. Until the discoveries from Patagonia, there was no evidence about the timing of these changes. On the basis of Patagonian material, we now know that as early as the Aalenian-Bajocian boundary, an additional extrazeugopodial element and digits distal to it in the forefin were acquired.

When analyzing the Patagonian record as a whole, the most conspicuous feature is its transitional pattern. There are typical "Early Jurassic elements" represented by *Stenopterygius cayi* and "Callovian–post Callovian elements" exemplified by *Mollesaurus* and MLP 92-III-2-1. McGowan (1991, 1995, 1997) described an interesting transitional fauna from the Late Triassic of British Columbia. On the basis of the analysis of this fauna, he proposed that the end of the Triassic was not marked by a major turnover in ichthyosaurian species; rather, there was a more gradual period of transition. Until more specimens become available, it would be premature to discuss whether a similar scenario could be used for the end of the Early Jurassic. However, evidence up to the present strongly suggests this. On the basis of the extensive exposures of the marine Middle Jurassic in the Neuquén Basin and the characters of the facies where marine reptiles were found (Spalletti et al. 1994), more material will probably be found in the future to support this hypothesis.

Late Jurassic

All Callovian and Late Jurassic ichthyosaurs belong to the single clade Ophthalmosauridae (Baur 1887; Motani 1999b). Most of the ichthyosaur material from Patagonia has been exhumed from Tithonian levels of the Vaca Muerta Formation. This is because most of collection effort was concentrated in several localities of the Neuquén Basin where the Vaca Muerta Formation is well exposed. Material described by Rusconi (1940, 1942) as *Ancanamunia mendozanus* (=*Myobradypterygius mendozanus*) was also exhumed from Tithonian levels of this formation. The information to date permits us to recognize the following taxa in the Late Jurassic of Patagonia:

Ophthalmosauridae Baur 1887
***Caypullisaurus* Fernández 1997**
***Caypullisaurus bonapartei* Fernández 1997**
Fig. 11.4A–E

?*Platypterygius* sp. Gasparini and Goñi 1990, 303, pl. II, 4, 5
***Ophthalmosaurus* sp. Gasparini and Fernández 1997, fig. 3E**

Holotype. MACN-N-32, skull and mandible, pectoral girdle, both forefins, and 53 articulated vertebrae and associated ribs (Fig. 11.4A).

Referred specimens. MLP 83-XI-16-1. This partial skeleton

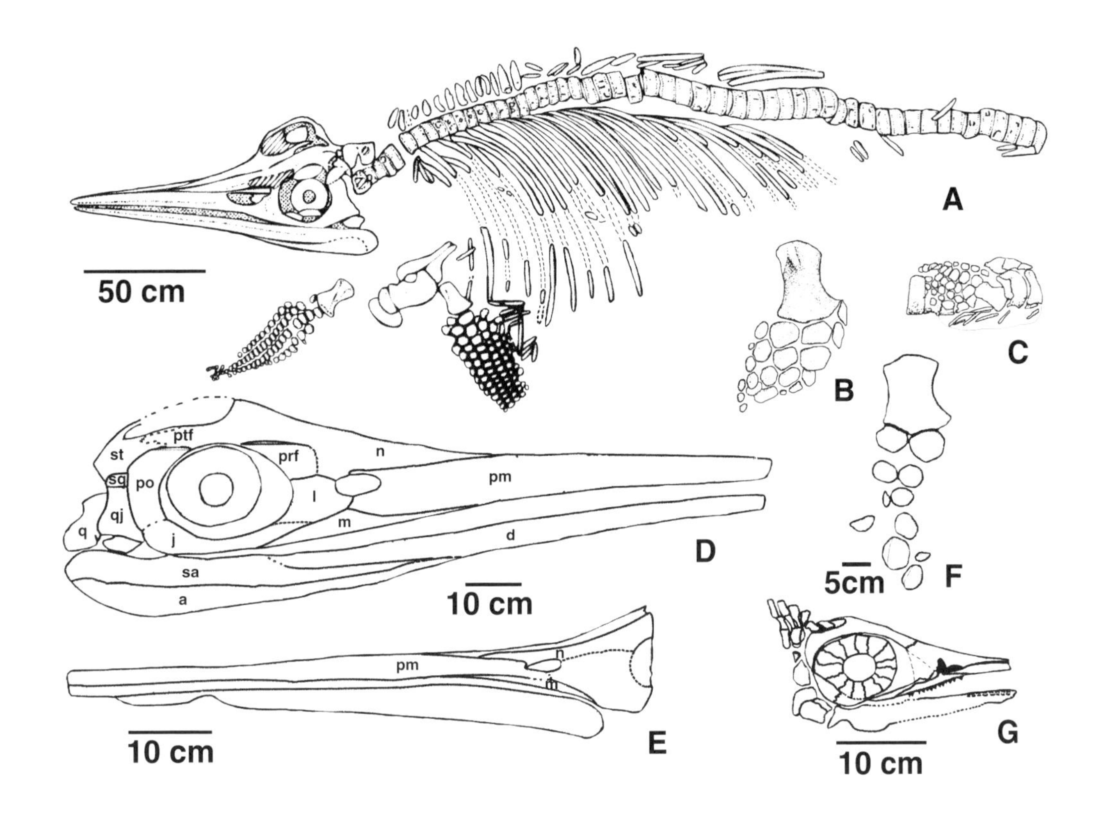

Figure 11.4. Tithonian ichthyosaurs from Patagonia: (A–E) Caypullisaurus bonapartei. *(A) MACN-N-32, general view. (B) MLP 83-XI-16-1, right forefin in dorsal view. (C) Right hindfin of MLP 83-XI-16-1. (D) MOZ-P-6067, skull in right lateral view. (E) MLP 85-I-14-1, skull in left lateral view. (F)* Ophthalmosaurus *sp., MOZ-P-6145, right forefin. (G) Ophthalmosauridae (MOZ-P-1854), skull in right lateral view. (B) and (C) not in scale.*

comprises approximately 48 articulated vertebrae and associated ribs, proximal part of right forefin with humerus (Fig. 11.4B), partial pectoral girdle, and complete right hindfin (Fig.11.4C). It was exhumed from the lower Tithonian levels of the Vaca Muerta Formation exposed at Cerro Lotena (Fig. 11.1, locality 7). Although the skull was not preserved, it is very interesting material because the fin elements are articulated. The proximal part of the forefin is approximately in its anatomical position. This fact helps to identify the dorsal and ventral aspects, and therefore to establish the homology of the fin elements (Fernández 2001). The preservation of the hindfin is also very interesting. Because ichthyosaurs lack the osseous connection between pelvic girdle and the vertebral column, it is very difficult to identify the original orientation of the elements (McGowan and Motani 2003).

MLP 83-XI-16-1. This specimen has its right hindfin articulated and is preserved in the same block with the vertebral centra. It seems not to be displaced from its original position but rotated. The analysis of the homologies of the elements of this hindfin will be presented in the future. This material is up to now the only articulated and three-dimensionally preserved hindfin of an ophthalmosaurid, and therefore, its analysis will help to clarify the homologies of zeugopodium elements.

MLP 85-I-15-1. This specimen was exhumed from the late Tithonian sediments of the Vaca Muerta Formation exposed in

Chacay Melehue (Neuquén Province, Argentina). It comprises the complete skull and mandible exposed in right view, and the proximal portion of a forefin. It was described by Fernández (1998, figs. 1, 2). Unfortunately, it was acid prepared, so most of the skull sutures are not clearly identifiable.

MOZ-P 6097 (Fig. 11.4D). This specimen consists of a complete skull and mandible exposed in right view, with a skull length of 105 cm. It was excavated from late Tithonian sediments of the Vaca Muerta Formation exposed in Cajón de Almaza (Neuquén Province, Argentina). The skull was preserved within a concretion, and its right surface was manually prepared. The upper and lower jaw margins were not prepared in a way that permits features of the dentition to be described. The quality of preservation permits the identification of sutures, especially those of the cheek region, which cannot be seen in the holotype and in MLP 85-I-15-1. The postorbital region is broad, and both postorbital and quadratojugal are well exposed in lateral view; this configuration is similar to that of *Platypterygius* described by Romer (1968) and McGowan (1972), except for the presence of a well-exposed squamosal. In the lower jaw, the contact between the angular, surangular, and dentary can be traced. One of the synapomorphies defining the clade Ophthalmosauria is the angular, largely exposed laterally, reaching as far anteriorly as the surangular (Motani 1999b). With the inclusion of MOZ-P 6097 in *Caypullisaurus,* the presence of this character is confirmed for the genus.

MOZ-P 6139. This is the largest specimen found, with a skull length of 151 cm. It comprises the complete skull and mandible, and cervical and dorsal vertebrae. No forefin elements have been recovered. It was excavated from late Tithonian–early Berriasian sediments of the Vaca Muerta Formation exposed in the Yesera del Tromen (Neuquén Province, Argentina) (Spalletti et al. 1999) (Fig. 11.1, locality 4). Sutures of the cheek region are visible and display the same pattern as in MOZ-P 6097.

MLP 85-I-14-1. This specimen was first described by Gasparini and Goñi (1990) as *?Platypterygius.* It consists of the preorbital portion of a skull (Fig. 11.4E) excavated from lower Tithonian levels of the Vaca Muerta Formation exposed in Cerro Lotena (Fig. 11.1, locality 7). On the basis of the form of the naris and the broad contact between premaxilla and the lachrymal below the naris, this material is referred here to *Caypullisaurus bonapartei.*

Type locality and age. Cerro Lotena (39° 11'S; 69° 40'W), Neuquén Province, Argentina. Ammonite *Virgatosphinctes mendozanus* Zone, early Tithonian, Vaca Muerta Formation, Jurassic (Leanza 1980) (Fig. 11.1, locality 7).

Geographic distribution. Type locality, Cajón de Almaza (38° 08'S; 70° 20'W), Yesera del Tromen (37° 18'S; 69° 53'W), Chacay Melehue (37° 17'S; 70° 20'W); Malargüe, Sierra de Reyes, Mendoza Province (Argentina).

Stratigraphic range. Vaca Muerta Formation (Tithonian-Berriasian) Neuquén Basin, Argentina.

Emended diagnosis. Premaxilla-lachrymal contact below the nare broad; postorbital region of the skull not reduced; humerus with three distal facets, the smallest being the anterior, articulating with the preaxial accessory element and bearing distally two digits; large intermedium placed between radiale and ulnare and not between the radius and ulna; distal border of the radius straight and almost parallel to the anterior border of the intermedium and to the distal facets of the humerus for the radius; phalanges polygonal and tightly packed.

Comments. Caypullisaurus is sister taxon of *Platypterygius* (Fig. 11.2) (Motani 1999b). The close relationships between the Tithonian ichthyosaurs from Patagonia and *Platypterygius* was first noted by Gasparini and Goñi (1990), who identified the MLP 85-I-14-1 as ?*Platypterygius.*

Ophthalmosaurus Seeley 1874
***Ophthalmosaurus* sp.**

Material. Forefin of MCNAM 119 (Rusconi 1948, figs. 65–67, pl. IX); MOZP uncatalogued (Fernández 2000, fig. 3); MOZ 6145, part of a skull and right forefin.

Locality and age. Bardas Blancas, Malargüe, Mendoza Province, Argentina; Cañadón de los Loros, Pampa Tril, Neuquén Province, Argentina. Tithonian, Vaca Muerta Formation.

Comments. The holotype of *Ancanamunia mendozana* (Rusconi 1940) consists of approximately 40 vertebrae, humerus and zeugopodium, phalanges, and part of a pectoral girdle. Although all the material has been described as belonging to a single specimen, the possibility that the specimen was composed of fragments of more than one individual must not be discarded. The vertebral column of MCNAM 119 has no diagnostic features, as in most Jurassic ichthyosaurs. Thus, they may correspond to the most frequent taxon of the Tithonian of the Vaca Muerta Formation, *Caypullisaurus.* However, they are clearly different from the vertebrae of *Ophthalmosaurus* (Seeley 1874) because one diagnostic feature of the genus is the shortening of the dorsal vertebrae (dorsal centrum height/length ratio >3.0) (Motani 1999b). On the contrary, the humerus with three distal facets and the ulnar smaller than radial, the roughly pentagonal outline of the radius, and the phalanges rounded with free rugose margins are diagnostic features of *Ophthalmosarus.*

The presence of *Ophthalmosaurus* in the Tithonian of the Vaca Muerta Formation is also confirmed by the fin (MOZ uncatalogued) from Pampa Tril (Neuquén) (Fernández 2000) and by another specimen recently excavated from the same area (MOZ 6145). This latter specimen consists of the posterior part of the skull and the right forefin. Although most of the skull is poorly preserved, the basioccipital could be removed from the matrix and prepared. The forefin (Fig. 11.4F) has the same pattern as the forefins of the two valid species of *Ophthalmosaurus: O. icenicus* and *O. natans.* McGowan and Motani (2003) pointed out that there are no significant differences between the two species, but they retained the name *O. natans* for the reception of the New World material. However, the finding of MOZ 6145 suggested that all the American material of *Ophthalmosaurus* is probably not referable to a single species. The Patagonian *Ophthalmosaurus* differs from *O. natans* in having an orbit of moderate size, and a basioccipital with a very small extracondylar area (e.g., this area is almost hidden in occipital view).

Ophthalmosauridae gen. et sp. indeterminate
***Ophthalmosaurus monocharactus* Gasparini 1988**
***Aegirosaurus* sp. Maisch and Matzke 2000**
Fig. 8

Material. MOZ 1854 (Fig. 11.4G), an almost complete and well-preserved skeleton.

Locality and age. El Ministerio Quarry, Los Catutos area, Neuquén Province, Argentina (Fig. 11.1, locality 8). Middle Tithonian, Vaca Muerta Formation (Leanza and Zeiss 1990).

Comments. This material represents a new Ophthalmosauridae that will be described elsewhere. Maisch and Matzke (2000, fig. 8) referred this specimen to the genus *Aegirosaurus* erected by Bardet and Fernández (2000), but without any explicit justification. However, at this stage of knowledge, I find no reason for this assignment. Comparable diagnostic elements preserved in *Aegirosaurus* and in MOZ 1854 are clearly different. Thus, in *Aegirosaurus,* the snout is long and slender (snout ratio >0.66), the humerus has a distal facet for the intermedium, and the pre- and postaxial elements do not contact the humerus, whereas in MOZ 1854 (Gasparini 1988, figs. 1, 2) the snout is short, the intermedium does not contact the humerus, and pre- and postaxial elements are in contact with the humerus.

Significance of Tithonian Ichthyosaurs from Patagonia

Traditionally, ichthyosaur diversity during the Late Jurassic has been considered low because most of the specimens were referable to a single genus, *Ophthalmosaurus* (McGowan 1991). Moreover, the abundance of ichthyosaurs within the marine reptile community was considered low (Bardet 1995) because most of the *Ophthalmosaurus* specimens were collected from the Callovian of the Oxford Clay. The Late Jurassic record was mostly restricted to the Oxfordian of the Sundance Formation, USA (Gilmore 1905, 1906), and to the Kimmeridgian of England (McGowan 1976). This landscape changed during the 1990s. The revision of undescribed material and new discoveries in the Tithonian of Patagonia (Fernández 1997; Spalletti et al. 1999), France (Bardet et al. 1997), and Russia (Arkhangelsky 1997, 1998; Efimov 1998, 1999a,b) expanded the knowledge of Tithonian ichthyosaurs. To the list of valid Late Jurassic genera (*Ophthalmosaurus, Brachypterygius,* and *Nannopterygius*), *Caypullisaurus, Aegirosaurus,* and *Undorosaurus* are now added (Mc-

Gowan and Motani 2003). Thus, Late Jurassic ichthyosaurs would be assigned to six genera. This diversity at generic level is not significantly lower than that recognized for the Early Jurassic, for example, which has been characterized as a diverse, ichthyosaur-dominated fauna (Massare 1988). A main difference between the Early and Late Jurassic record is their abundance and our knowledge of them. It is noteworthy that, except for *Ophthalmosaurus* and *Brachypterygius*, the remaining Late Jurassic genera are monotypic. Also, all the genera except *Ophthalmosaurus* are known by a few geographically restricted specimens. Thus, up to now, *Caypullisaurus* has been recorded only in the Neuquén Basin. However, on the basis of the supposed swimming capability of Jurassic ichthyosaurs, a worldwide distribution may be expected, so this fact must be interpreted in terms of a collecting bias and not as endemism.

Cretaceous

Cretaceous ichthyosaur material is very scarce, probably because of the lack of prospecting effort. Two taxa have been identified in the Cretaceous of Patagonia:

Ophthalmosauridae Baur 1887
***Caypullisaurus* Fernández 1997**
***Caypullisaurus bonapartei* Fernández 1997**

Referred specimen. Tril 20-IX-97-1 (Spalletti et al. 1999, 119). Partial skull, humerus, proximal carpus, part of the vertebral column and associated ribs.

Locality and age. Yesera del Tromen, Neuquén Province, Argentina. Vaca Muerta Formation, middle Berriasian (Fig. 11.1, locality 4).

Comments. This specimen is the first record of this taxon in the Early Cretaceous. The discovery of this specimen demonstrated that at least two genera, *Caypullisaurus* and *Brachypterygius* (the latter with a stratigraphic distribution that extends from the Kimmeridgian to the Albian [McGowan and Motani 2003]), crossed the Jurassic-Cretaceous boundary. This fact is significant because it has been proposed that a Tithonian-Berriasian extinction affected certain ichthyosaurs because the taxa recorded in the Tithonian had no record in the Berriasian (Bardet 1995).

***Platypterygius* Huene 1922**
***Platypterygius hauthali* Huene 1927**
Fig. 11.5

***Myobradypterygius hauthali* Huene 1927, 29**
***Platypterygius hauthali* McGowan 1972, 17**
***Platypterygius hauthali* Gasparini and Goñi 1990, 17**

Holotype. MLP 79-I-30-1, left humerus, and part of zeugopodium and autopodium.

Type locality and age. Cerro Belgrano, Santa Cruz Province, Argentina (Fig. 11.1, locality 11). Río Belgrano Formation (Bar-

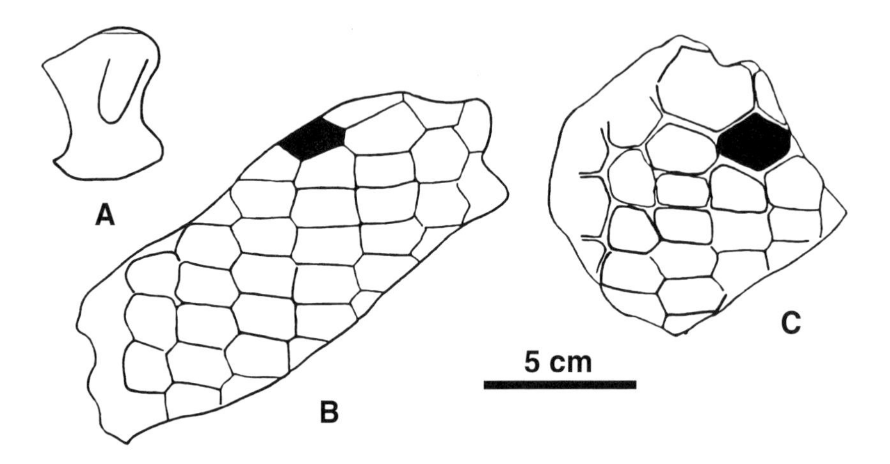

Figure 11.5. Barremian ichthyosaurs from Patagonia. (A–C) Platypterygius hauthali. *(A) MLP 79-I-30-1, left humerus in dorsal view. (B) MLP 79-I-30-1, part of the zeugopodium and autopodium. (C) MLP 79-I-30-2, part of the zeugopodium and autopodium. Shaded area indicates intermedium.* Modified from Fernández and Aguirre-Urreta (in press).

remian) (Aguirre-Urreta 2002; Bonaparte et al. 2002; Fernández and Aguirre-Urreta 2005).

Referred specimens. MLP 79-I-30-2, slab with two different forefins on each side, four blocks with vertebral centra from the type locality. MLP 85-I-18-1, vertebral centra from Estancia La Federica.

Geographical distribution. Type locality and Lago San Martín Formation (late Barremian) Estancia La Federica, southeast of San Martín Lake, Santa Cruz Province, Argentina (Fig. 11.1, locality 12).

Emended diagnosis. This specimen differs from the other species of the genus by the combination of the following characters: humerus with a small facet for the extrazeugopodial element anterior to the radius; hexagonal intermedium articulating distally with two digits (Fernández and Aguirre-Urreta 2005).

Comments. McGowan and Motani (2003, 124) retained *Platypterygius hauthali* for the reception of New World material from the earliest Cretaceous, although they did not recognize any diagnostic features for them. They also suggested that this could be synonymous with the contemporaneous *P. platydactylus* (Broili 1907) from Europe. However, from the earliest Cretaceous of the New World, not only *P. hauthali,* but also *P. sachicarum* (Páramo 1997) from the Barremian-Aptian of Colombia and *Caypullisaurus bonapartei* from the Berriasian of Neuquén have been recorded.

Conclusions

Patagonian ichthyosaur records are important because they document poorly known episodes of the history of the group worldwide, especially those of the Aalenian-Bajocian and Tithonian-Berriasian. The discoveries of Aalenian-Bajocian ichthyosaurs in Patagonia showed that *Stenopterygius,* a typically Early Jurassic form, survived at least in the early Bajocian; and that ophthalmosaurids are represented in the fossil record as early as the Aalenian.

On the basis of *Caypullisaurus* material discovered in the Neuquén Basin, this taxon is one of the best-known Late Jurassic ichthyosaurs. Specimens of *Caypullisaurus* exhumed from Berriasian sediments of the Vaca Muerta Formation demonstrated that at least two genera, *Caypullisaurus* and *Brachypterygius,* crossed the Jurassic-Cretaceous boundary.

Acknowledgments. I thank Z. Gasparini (Museo de La Plata) for her constant support and comments on a draft of this chapter. N. Bardet (Musée National d'Histoire Naturelle, Paris) and J. Massare (SUNY College, Brockport, N.Y.) provided many constructive comments. C. Deschamps helped with the English translation. This research was partially supported by a grant (PICT 8439) from the Agencia de Promoción Científica y Tecnológica de Argentina and Programa de Incentivos UNLP (N388).

References Cited

Aguirre-Urreta, M. B. 2002. Invertebrados del Cretácico Inferior. In M. Haller (ed.), *Geología y Recursos Naturales de Santa Cruz,* 439–459. Relatorio XV Congreso Geológico Argentino, II-6.

Arkhangelsky, M. S. 1997. On a new ichthyosaurian genus from the Lower Volgian Substage of the Saratov, Volga region. *Paleontologicheskii Zhurnal* 31: 87–91.

———. 1998. On the ichthyosaurian fossils from the Volgian Stage of the Saratov region. *Paleontologicheskii Zhurnal* 32: 87–91.

Bardet, N. 1992. Stratigraphic evidence for the extinction of the ichthyosaurs. *Terra Nova* 4: 649–656.

———. 1995. Evolution et extinction des reptiles marins au cours du Mésozoïque. *Palaeovertebrata* 24: 177–283.

Bardet, N., and M. Fernández. 2000. A new ichthyosaur from the Upper Jurassic lithographic limestones of Bavaria. *Journal of Paleontology* 74: 503–511.

Bardet, N., M. Duffaud, M. Martin, and J.-M. Mazin. 1997. Découverte de l'ichthyosaure *Ophthalmosaurus* dans le Tithonien (Jurassique supérieur) du Boulonnais, Nord de la France. *Neues Jahrbuch für Geologie und Paläontologie* 205: 339–354.

Baur, G. 1887. Über den ursprung der extremitäten der Ichthyopterygia. *Jahresberichte und Mitteilungen des Oberrheinischen Geologischen Vereines* 20: 17–20.

Bonaparte J. F., A. M. Báez, A. L. Cione, and J. L. Panza. 2002. Vertebrados Mesozoicos. In M. J. Haller (ed.), *Geología y Recursos Naturales de Santa Cruz,* 421–431. XV Congreso Geológico Argentino, El Calafate, Santa Cruz.

Broili, F. 1907. Ein neuer *Ichthyosaurus* aus der norddeutschen Kreider. *Palaeontographica* 54: 139–152.

Burmeister, H., and G. Giebel. 1861. Die versteinnerungen von Juntas in Tala des Río Copiapó. *Abhandlungen der Naturforschenden Gesellschaft Halle* 6: 111–144.

Cabrera, Á. 1939. Sobre un nuevo ictiosaurio del Neuquén. *Notas del Museo de La Plata* 4: 485–491.

Caldwell, M. A. 1997a. Limb ossification patterns of the ichthyosaur *Stenopterygius,* and a discussion of the proximal tarsal row of ichthyosaurs and other neodiapsid reptiles. *Zoological Journal of the Linnean Society* 120: 1–25.

———. 1997b. Modified perichondral ossification and the evolution of paddle-like limbs in ichthyosaurs and plesiosaurs. *Journal of Vertebrate Paleontology* 17: 534–547.
Casamiquela, R. M. 1970. Los vertebrados jurásicos de la Argentina y Chile. *Actas del IV Congreso Latinoamericano de Zoología,* 873–890. Caracas.
Dames, W. 1893. Ueber das Vorkommen von Ichthyopterygiern im Tithon Argentiniens. *Zeitschrift der Deutschen Geologischen Gesellschaft* 45: 23–33.
Efimov, V. M. 1998. An ichthyosaur, *Otschevia pseudoscythica* gen. et sp. nov. from the Upper Jurassic strata of the Ulyanovsk Region (Volga region). *Paleontologicheskii Zhurnal* 2: 82–86.
———. 1999a. Ichthyosaurs of a new genus *Yasykovia* from the Upper Jurassic strata of European Russia. *Paleontological Journal* 1: 92–100
———. 1999b. A new Family of ichthyosaurs, the Undorosauridae fam. nov. from the Volgian Stage of the European Part of Russia. *Paleontological Journal* 33: 51–58.
Fernández, M. 1994. A new long-snouted ichthyosaur from the Early Bajocian of Neuquen Basin (Argentina). *Ameghiniana* 31: 291–297.
———. 1997. A new ichthyosaur from the Tithonian (Late Jurassic) of the Neuquén Basin, Northwestern Patagonia, Argentina. *Journal of Paleontology* 71: 479–484.
———. 1998. Nuevo material de *Caypullisaurus bonapartei* Fernández (Reptilia: Ichthyosauridae) del Jurásico Superior de la cuenca Neuquina, Argentina. *Ameghiniana* 35 (1): 21–24.
———. 1999. A new ichthyosaur from the Los Molles Formation (Early Bajocian), Neuquén Basin, Argentina. *Journal of Paleontology* 73: 677–681.
———. 2000. Late Jurassic ichthyosaur from the Neuquén Basin, Argentina. *Historical Biology* 14: 133–136.
———. 2001. Dorsal or Ventral? Homologies of the forefin of *Caypullisaurus* (Ichthyosauria: Ophthalmosauria). *Journal of Vertebrate Paleontology* 21 (3): 515–520.
———. 2003. Ophthalmosauria (Ichthyosauria) forefin from the Aalenian-Bajocian boundary of Mendoza province, Argentina. *Journal of Vertebrate Paleontology* 23: 691–694.
Fernández M., and M. B. Aguirre-Urreta. 2005. Revision of *Platypterygius hauthali* von Huene, 1927 (Ichthyosauria: Ophthalmosauridae) from the Early Cretaceous of Patagonia, Argentina. *Journal of Vertebrate Paleontology* 25 (3): 583–587.
Fernández M., and S. Lanés. 1999. Presencia de ictiosaurios en el Sinemuriano del Río Atuel, Cuenca Neuquina, Mendoza. *Ameghiniana* 36: 100.
Gasparini, Z. 1985. Los reptiles marinos jurásicos de América del Sur. *Ameghiniana* 22: 23–34.
———. 1988. *Ophthalmosaurus monocharactus* Appleby (Reptilia, Ichthyopterygia), en las calizas litográficas titonianas del área Los Catutos, Neuquén, Argentina. *Ameghiniana* 25: 3–16.
———. 1997. A new pliosaur from the Bajocian of the Neuquén Basin, Argentina. *Palaeontology* 40: 135–147.
Gasparini, Z., and M. Fernández 1997. Tithonian marine reptiles of the eastern Pacific. In J. Callaway and E. Nicholls (eds.), *Ancient Marine Reptiles,* 435–440. San Diego: Academic Press.
Gasparini, Z., and R. Goñi. 1990. Los ictiosaurios jurásico-cretácicos de la

Argentina. In W. Volkheimer (ed.), *Bioestratigrafía de los Sistemas Regionales del Jurásico y Cretácico de América del Sur,* 299–311. Comité Sudamericano del Jurásico y Cretácico.

Gasparini, Z., P. Vignaud, and G. Chong. 2000. The Jurassic Thalattosuchia (Crocodyliformes) of Chile: a paleobiogeographic approach. *Bulletin de la Société Géologique de France* 171: 657–664.

Gasparini, Z., M. Cichowolski, and D. G. Lazo. 2005. First record of *Metriorhynchus* (Reptilia: Crocodyliformes) in the Bathonian (Middle Jurassic) of the Eastern Pacific. *Journal of Paleontology* 79: 801–805.

Gilmore, C. W. 1905. Osteology of *Baptanodon* Marsh. *Memoirs of the Carnegie Museum* 2: 77–129.

———. 1906. Notes on osteology of *Baptanodon. Memoirs of the Carnegie Museum* 2: 325–337.

Gradstein, F. M., F. S. Agterberg, J. G. Ogg, J. Hardenbol, P. Van Veen, J. Thierry, and Z. Huang. 1995. A Triassic, Jurassic and Cretaceous time scale. In W. A. Berggren, D. V. Kent, and J. Hardenbol (eds.), *Geochronology, Time Scales and Global Stratigraphic Correlation,* 95–126. Special Publication. Society for Sedimentary Geology.

Huene, F. 1922. Die ichthyosaurier des Lias und ihre zusammenhänge. *Monographie zur Geologie und Paläontologie* 1. Berlin: Verlag von Gebrüder Borntraeger.

———. 1925. Ichthyosaurier-reste aus Argentinien. *Centralblatt für Mineralogie, Geologie, und Paläontologie* 1925: 90–95.

———. 1927. Beitrag zur kenntnis mariner mesozoischer Wirbeltiere in Argentinien. *Centralblatt für Mineralogie, Geologie und Paläontologie* B, 1927: 22–29.

Jaekel, O. 1904. Eine neue darstellung von *Ichthyosaurus. Zeitschrift der Deutschen Geologischen Gesellschaft* 56: 26–34.

Jensen, O., and J. C. Vicente. 1976. Estudio geológico del área de "Las Juntas" del río Copiapó, provincia de Atacama, Chile. *Revista de la Asociación Geológica Argentina* 31: 145–173.

Kear, B. P., W. E. Boles, and E. T. Smith. 2003. Unusual gut contents in a Cretaceous ichthyosaur. *Proceedings of the Royal Society of London* B, 270: 206–208.

Leanza, H. 1980. The Lower and Middle ammonite fauna from Cerro Lotena, Province of Neuquén, Argentina. *Zitteliana* 5: 1–49.

Leanza, H., and A. Zeiss. 1990. Upper Jurassic lithographic limestones from Argentina (Neuquén Basin): stratigraphy and fossils. *Facies* 22: 169–186.

Maisch, M. W., and A. T. Matzke. 2000. The Ichthyosauria. *Stuttgarter Beiträge zur Naturkunde, Serie B (Geologie und Paläontologie)* 298: 1–59.

Massare J. A. 1988. Swimming capabilities of Mesozoic marine reptiles: implications for method of predation. *Paleobiology* 14: 187–205.

McGowan, C. 1972. The systematics of Cretaceous ichthyosaurs with particular reference to the material from North America. *Contributions to Geology* 11: 9–29. University of Wyoming.

———. 1976. The description and phenetic relationships of a new ichthyosaur genus from the Upper Jurassic of England. *Canadian Journal of Earth Sciences* 13: 668–683.

———. 1978. Further evidence for the wide geographical distribution of ichthyosaur taxa (Reptilia: Ichthyosauria). *Journal of Paleontology* 52: 1155–1162.

———. 1991. An ichthyosaur forefin from the Triassic of British Columbia exemplifying Jurassic features. *Canadian Journal of Earth Sciences* 28 (10): 1553–1560.

———. 1995. A remarkable small ichthyosaur from the Upper Triassic of British Columbia, representing a new genus and species. *Canadian Journal of Earth Sciences* 32 (3): 292–303.

———. 1997. A transitional ichthyosaur fauna. In J. M. Callaway and E. L. Nicholls (eds.), *Ancient Marine Reptiles,* 61–80. San Diego: Academic Press.

McGowan, C., and R. Motani. 2003. Ichthyopterygia. In H.-D. Sues (ed.), *Handbook of Paleoherpetology.* Part 8, 1–175. Munich: Verlag Dr. Friedrich Pfeil.

Motani, R. 1999a. On the evolution and homology of ichthyosaurian forefins. *Journal of Vertebrate Paleontology* 19: 28–41.

———. 1999b. Phylogeny of the Ichthyopterygia. *Journal of Vertebrate Paleontology* 19: 473–496.

Owen, R. 1840. Report on British fossil reptiles. Part I. *Report of the British Association for the Advancement of Science* 9: 43–126. Plymouth.

Páramo, M. E. 1997. *Platypterygius sachicarum* (Reptilia, Ichthyosauria) nueva especie del Cretácico de Colombia. *Revista Ingeominas* 6: 1–12.

Phillipi, R. A. 1895. *Ichthyosaurus immanis* Ph. una nueva especie sudamericana de este jenero (SIC). *Anales de la Universidad de Chile* 90: 837–941.

Romer, A. S. 1968. An ichthyosaur skull from the Cretaceous of Wyoming. *Contributions to Geology* 7: 27–41. University of Wyoming.

Rusconi, C. 1938. Restos de ictiosaurios del Jurásico Superior de Mendoza. *Boletín Paleontológico de Buenos Aires* 10: 1–4.

———. 1940. Nueva especie de ictiosaurio del Jurásico de Mendoza. *Boletín Paleontológico de Buenos Aires* 11: 1–4.

———. 1942. Nuevo género de ictiosaurio argentino. *Boletín Paleontológico de Buenos Aires* 13: 1–2.

———. 1948. Ictiosaurios del Jurásico de Mendoza (Argentina). *Revista del Museo de Historia Natural de Mendoza* 2: 17–160.

———. 1949. Presencia de ictiosaurios en el Liásico de San Juan. *Revista del Museo de Historia Natural de Mendoza* 3: 89–94.

Sander, P. M. 2000. Ichthyosauria: their diversity, distribution, and phylogeny. *Paläontologische Zeitschrift* 74: 1–35.

Seeley, H. G. 1874. On the pectoral arch and forelimb of *Ophthalmosaurus,* a new ichthyosaurian genus from the Oxford clay. *Quarterly Journal of the Geological Society of London* 30: 696–707.

Shultz, M., A. Fildani, and M. Suarez. 2003. Occurrence of the southernmost South American ichthyosaur (Jurassic–Lower Cretaceous), Parque Nacional Torres del Paine, Patagonia, Southernmost Chile. *Palaios* 18: 69–73.

Spalletti, L., Z. Gasparini, and M. Fernández. 1994. Facies, ambientes y reptiles marinos de la transición entre las formaciones Los molles y Lajas (Jurásico medio), Cuenca Neuquina, Argentina. *Acta Geologica Leopoldensia* 17: 329–344.

Spalletti L., Z. Gasparini, G. Veiga, E. Schwartz, M. Fernández, and S. Matheos. 1999. Facies anóxicas, procesos deposicionales y herpetofauna de la rampa marina titoniano-berriasiana en la Cuenca Neuquina (Yesera del Tromen), Neuquén, Argentina. *Revista Geológica de Chile* 26: 109–123.

Suárez, M., and C. M. Bell. 1992. The oldest South American ichthyosaur from the Late Triassic of Northern Chile. *Geological Magazine* 129 (2): 247–249.

Tavera, J. 1981. *Ichthyosaurus* de la Formación Lautaro, en el área de Manflas, Región de Atacama, Chile. *Comunicaciones, Universidad de Chile, Departamento de Geología* 33: 1–16.

Vignaud, P., and Z. Gasparini. 1996. New *Dakosaurus* (Crocodylomorpha, Thalattosuchia) in the Upper Jurassic of Argentina. *Comptes Rendus de l'Académie des Sciences de Paris* 322: 245–250.

Wagner, J. A. 1852. Neu-aufgefundene saurier-ueberreste aus den lithographischen schiefern und dem obern Jurakalke. *Abhandlungen der Koeniglich Bayerischen Akademie der Wissenschaften, Mathematisch-Physikalische* Classe, München, 6: 663–710.

12. Plesiosauria

ZULMA GASPARINI

Introduction

Plesiosauria are marine reptiles of the Sauropterygia clade (Storrs 1991, 1993) with a wide record, both temporal (from the end of the Triassic to the end of the Cretaceous) and geographic. Plesiosaur remains have been found in all continents, including Antarctica (Welles 1952, 1962; Persson 1963; Brown 1981; Bakker 1993; Brown and Cruickshank 1994; Bardet 1995; Carpenter 1996, 1997, 1999; Maisch and Rücklin 2000; O'Keefe 2001a, 2004; Storrs et al. 2001; Cruickshank and Fordyce 2002; Gasparini et al. 2002a,b, 2003a,b; Kear 2003; Noè et al. 2004; Sachs 2004, 2005). As with other groups of Jurassic-Cretaceous marine reptiles, the knowledge of plesiosaurs exploded between the end of the nineteenth and the beginning of the twentieth centuries (Callaway and Nicholls 1997), followed by a few, although valuable, contributions in the middle of the twentieth century (e.g., Welles 1952, 1962; Tarlo 1960; Persson 1963; Welles and Gregg 1971). During the last 25 years, interest in plesiosaurs has again surged, giving rise to revisions of old collections of the Jurassic (mainly in Europe) and the Cretaceous (mainly in North America), and to the study of new specimens discovered worldwide, particularly in continents where they were almost unknown, such as Africa, Australia, Antarctica, and South America. Despite this progress, studies are skewed because most of the records come from the Northern Hemisphere. This results in some important differences when undertaking phylogenetic studies, in which taxa from the Southern Hemisphere may or may not be included (e.g., Bakker 1993; Carpenter 1997; O'Keefe 2001a,b, 2004; Kear 2002, 2003; Gasparini et al. 2003b; Sachs 2004).

In South America, the record of plesiosaurs is scarce, and most

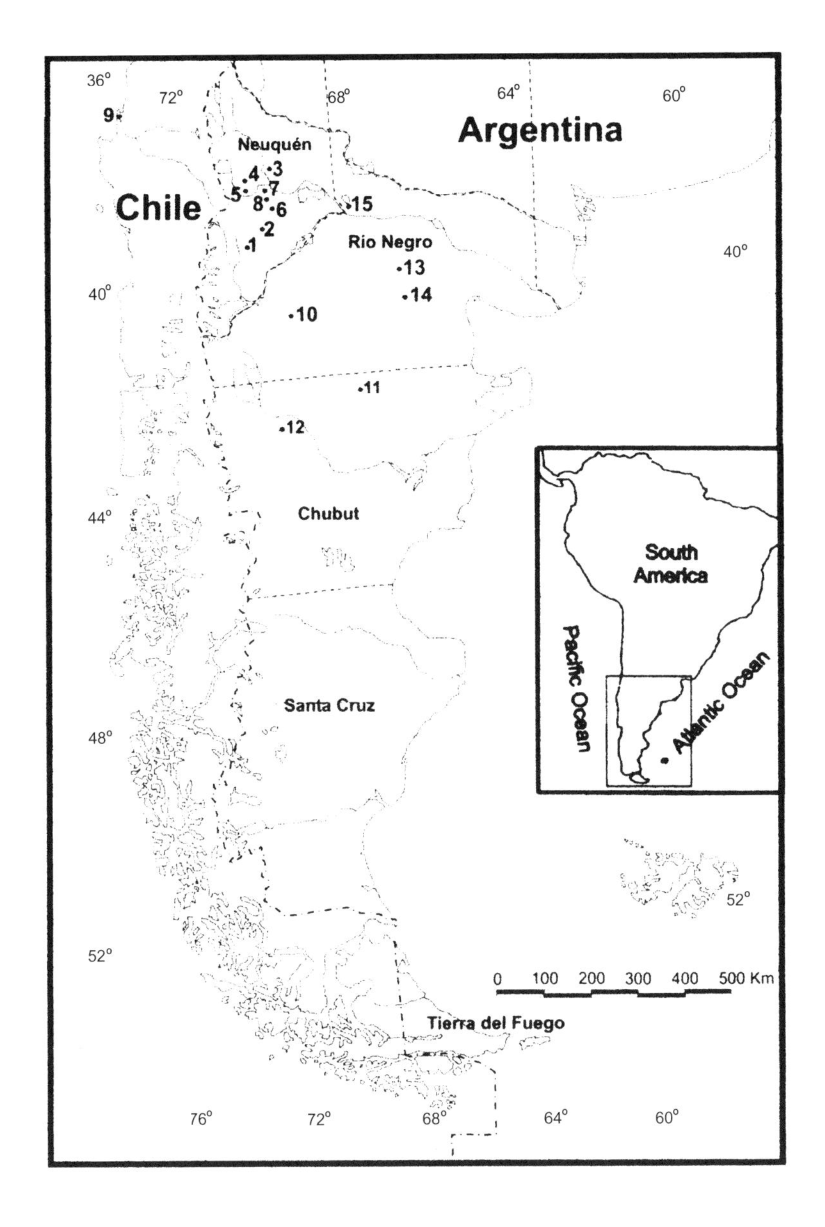

Figure 12.1. Main Patagonian localities with marine Mesozoic reptiles. 1, Chacaico Sur; 2, Cerro Lotena; 3, Yesera del Tromen/ Pampa Tril; 4, Trincajuera Creek; 5, Arroyo Truquicó; 6, Cerro Negro; 7, Agua de la Mula; 8, Bajada del Agrio; 9, Quiriquina Island and neighboring coast; 10, Cari-Laufquen Grande; 11, La Colonia; 12, Cañadón del Loro; 13, Bajos de Trapalcó; 14, Bajos de Santa Rosa; 15, Lago Pellegrini.

of it comes from Argentinean-Chilean Patagonia (Fig. 12.1). The first plesiosaur from the Southern Hemisphere, referred to *Plesiosaurus chilensis* (Gay 1848), was found in upper Maastrichtian rocks in the Chilean Patagonia. A century later, Cabrera (1941) described *Aristonectes parvidens,* the first relatively complete plesiosaur from the Argentinean Patagonia (upper Maastrichtian). Between the middle of the nineteenth century and the beginning of the 1980s, some Jurassic and Cretaceous plesiosaurs were found in Patagonia, generally as a result of fieldwork performed for other purposes. Some of these discoveries deserve to be mentioned or described. Only in the last 25 years has intensive prospecting in the north of the Argentinean Patagonia yielded plesiosaurs from the lower Bajocian–uppermost Maastrichtian lapse, showing the highest taxonomic diversity of these marine reptiles (rhomaleosaurids, pliosaurids, polycotylids, cryptoclidids, and elasmosaurids) in South America.

The objective of this chapter is to provide an update of the Patagonian plesiosaurs. Although most of them were found in the Neuquén Basin (Spalletti and Franzese this volume), Patagonian plesiosaurs belong to two different geologic events. The Jurassic and Early Cretaceous forms correspond to the Pacific embayment in northwestern Patagonia, whereas those of the Late Cretaceous come from the southern Pacific (Chilean Patagonia) and southern Atlantic, which, by the end of the Cretaceous, flooded a large part of the Argentinean Patagonia (Uliana and Biddle 1988; Spalletti and Franzese this volume).

Institutional abbreviations. BMNH R, The Natural History Museum, London, England; MOZ, Museo Prof. Dr. "Juan Olsacher," Zapala, Neuquén, Argentina; MLP, Museo de Ciencias Naturales de La Plata, La Plata, Argentina; MNHN SGO, Museo Nacional de Historia Natural, Santiago, Chile; MML, Museo Municipal de Lamarque, Lamarque, Río Negro, Argentina; NZGS, New Zealand Geological Survey, New Zealand; MPEF, Museo Paleontológico "Egidio Feruglio," Trelew, Chubut, Argentina; MUCPv, Museo de la Universidad Nacional del Comahue, Neuquén, Argentina.

Systematic Paleontology

Sauropterygia Owen 1860

Plesiosauria de Blainville 1835

Since their definition, the Plesiosauria (de Blainville 1835) were accepted as taxonomically divided into the Pliosauroidea (skull large, number of cervical vertebrae less than 28) and the Plesiosauroidea (skull small, number of vertical vertebrae more than 28) (Welles 1943). However, Williston (1907), Bakker (1993), and Carpenter (1997) stated that short necks may have evolved more than once in plesiosaurs (O'Keefe 2002). O'Keefe (2001a, 2002) published a cladistic analysis and taxonomic revision of the Plesiosauria. In this synthesis of the Patagonian Plesiosauria, the terms and taxonomy proposed by O'Keefe (2001a, 2002, 2004) for Pliosauroidea are accepted, but the resolution in Plesiosauroidea is not absolutely clear (O'Keefe 2001a, fig. 20). In this regard, the author agrees with Kear (2003) in that the Cimoliasauridae are not sufficiently supported. Likewise, and according to the systematic frame proposed by Gasparini et al. (2003b), *Aristonectes* Cabrera (=*Morturneria*), is considered an elasmosaurid in this chapter. As stated above, these differences of criteria are unavoidable in this stage of the knowledge of the Plesiosauria. Most taxa are represented by incomplete specimens or noncomparable parts of the skeleton, and several taxa from the Southern Hemisphere have not been considered even in the most inclusive phylogenies (O'Keefe 2001a, 2004). That said, because O'Keefe's (2001a) is the most

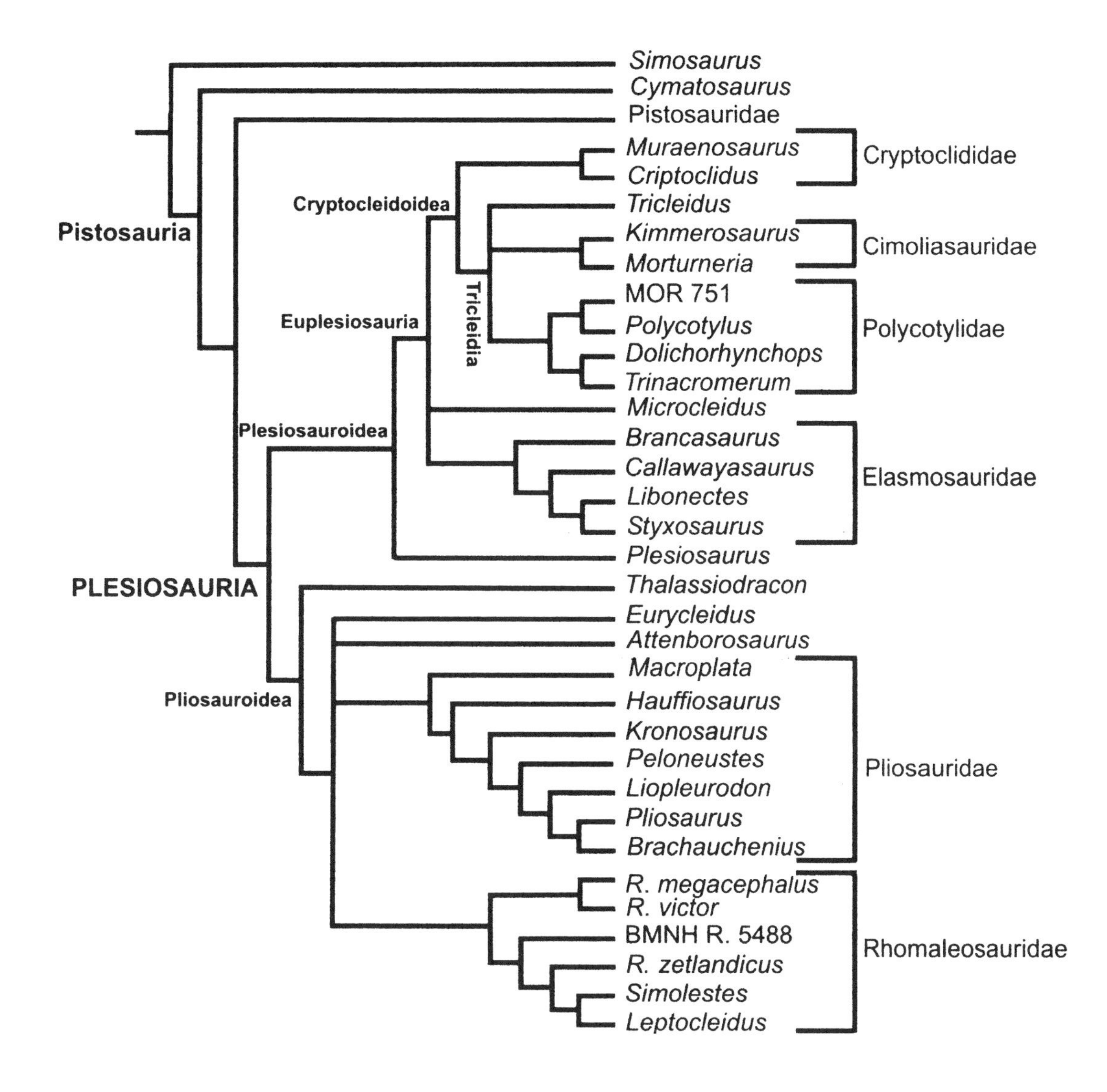

Figure 12.2. Phylogenetic relationships of Plesiosauria, simplified from O'Keefe (2001a).

comprehensive proposal, it is taken as reference in this chapter (Fig. 12.2), including observations that I regard as pertinent.

Middle Jurassic

Pliosauroidea (Seeley) Welles 1943 (sensu O'Keefe 2001a)
Rhomaleosauridae (Kuhn) (sensu O'Keefe 2001a)
***Maresaurus coccai* Gasparini 1997**
Fig. 12.3A, B

Holotype. MOZ 4386P, articulated skull and mandible, fused atlas and axis, and the first cervical vertebrae (Gasparini 1997, text, fig. 1–4; Fig. 12.3A, B).

Locality and age. Chacaico Sur, 70 km southwest of Zapala, Neuquén Province, Argentina (Fig. 12.1, locality 1); upper part of

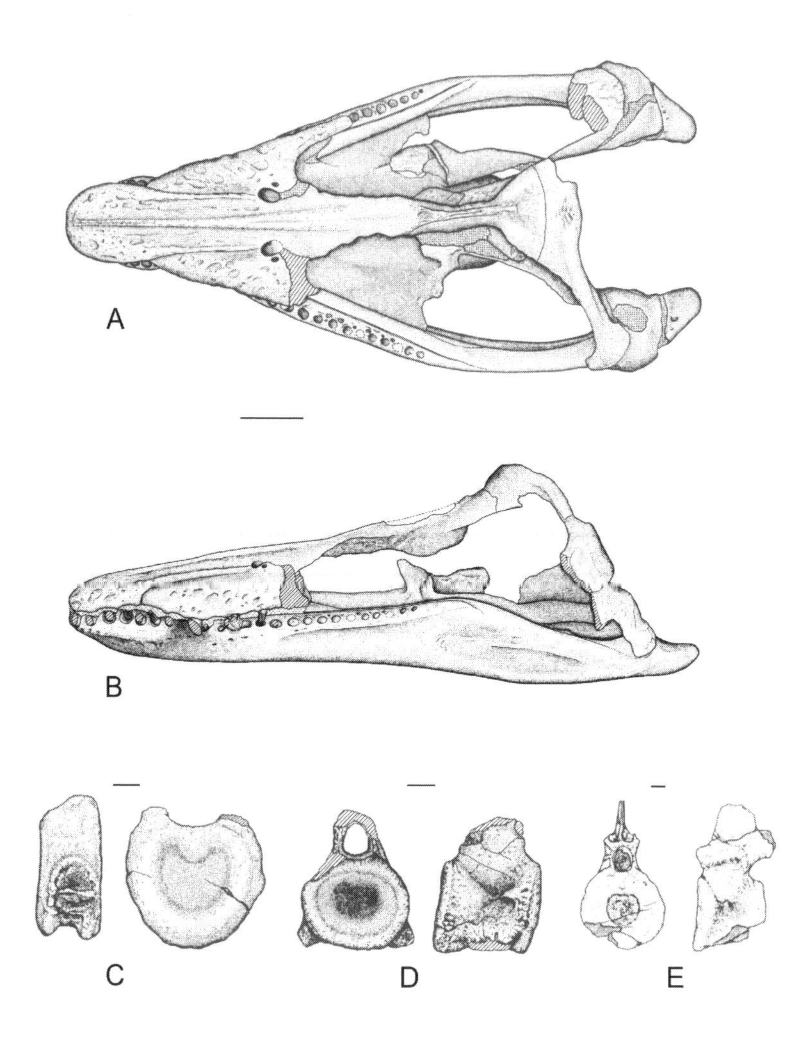

Figure 12.3. Middle Jurassic Patagonian plesiosaurs. Maresaurus coccai *(type, MOZ 4386P): (A) Dorsal view. (B) Lateral view. Scale bar = 10 cm. Simplified from Gasparini (1997). Pliosauroidea (MOZ 6002P) anterior cervical vertebra; cf.* Muraenosaurus *sp. (MOZ 6004) anterior cervical vertebra; cf.* Cryptoclidus *sp. (MOZ 6006P) cervical vertebra. (C–E) Scale bar = 1 cm.* Simplified from Gasparini and Spalletti (1993).

the Los Molles Formation, Cuyo Group; lower Bajocian (Spalletti et al. 1994; Gasparini 1997).

Diagnosis. Rostrum deep, with marked sagittal crest formed by union of premaxillae and two conspicuous parallel crests formed by dorsal union of premaxillae and maxillae. Deep notch between premaxilla and maxilla, and marked anterior wave in maxilla incorporating six alveoli. Spatulate symphysis with six pairs of alveoli. Posterior region of the parietal wide. Lateral palatal fenestration present. Parasphenoid does not completely separate the interpterygoid vacuities. Pterygoid wings very expanded and elevated. Twenty-four teeth in dentary. All teeth with circular section, densely distributed nondichotomized striae, and without carinae. Cervical zygapophyses as wide as centra (Gasparini 1997).

Comments. Maresaurus and *Rhomaleosaurus* share the parietal, which does not reach the occipital edge of the skull roof and does not form the sagittal crest between the squamosal; the occipital condyle, which is not visible in dorsal view; and the squamosal dorsal ramus, which has an elliptical cross section. *Maresaurus* and

Simolestes (Andrews 1913) share the lack of the anterior interpterygoid vacuity and less than 26 alveoli in the dentary. Records of Bajocian marine reptiles are scarce worldwide (Bardet 1995). The only Bajocian plesiosaur of this age published to date is the rhomaleosaurid *Maresaurus coccai* from the early Bajocian of northwestern Patagonia. O'Keefe (2001a) has redefined the Pliosauroidea, including the subclades Pliosauridae and Rhomaleosauridae (Fig. 12.2). He included in the latter *Rhomaleosaurus zetlandicus, Simolestes,* and *Leptocleidus,* although he states that according to his cladogram, the genus *Rhomaleosaurus* is paraphiletic. *Maresaurus* shares characters with the Rhomaleosauridae, and its inclusion in a phylogenetic analysis could resolve the relatively weak support of the nodes (O'Keefe 2001a, fig. 20).

Pliosauroidea gen. et sp. indeterminate

Material. MOZ 6002P, anterior cervical vertebra (Gasparini and Spalletti 1993, lam. 1A–C, fig. 3C; Fig. 12.3C).

Locality and age. Chacaico-Sur (Fig. 1.1), Neuquén Province, Argentina. Uppermost part of the middle Lajas Formation, early Callovian (Gasparini and Spalletti 1993, figs. 1, 2).

Comments. The centrum is very short and bears a small narrowing in the synapophysis, suggesting that the ribs were bicipital. The incomplete vertebra of Chacaico Sur (MOZ 6002P) is the single evidence of Callovian pliosauroids in the Neuquén Basin.

Plesiosauroidea Welles 1943
Cryptoclididae Williston 1925
***Muraenosaurus* Seeley 1874**
cf. *Muraenosaurus* sp.
Fig. 12.3D

Material. MOZ 6003 P, MOZ 6004 P, MOZ 6010, cervical vertebrae; MOZ 6005 P, radius (Gasparini and Spalletti 1993, lam. 1D–G; Fig. 12.3D).

Locality and age. As for the pliosauroid MOZ 6002P (see above).

Comments. According to the proportions of the vertebrae, and following the criterion ratio cervical vertebra height/length less than 1 (Bardet et al. 1991), this material was referred to cf. *Muraenosaurus* (Seeley 1874) (Gasparini and Spalletti 1993).

***Cryptoclidus* Phillips 1871**
cf. *Cryptoclidus* sp.
Fig. 12.3E

Material. MOZ 6006 P; MOZ 1011 P (cervical vertebrae); MOZ 6007, radius (Gasparini and Spalletti 1993, lam. 1H–J; Fig. 12.3E).

Locality and age. As for the pliosauroid MOZ 6002P (see above).

Comments. Likewise, following the criterion height/length

more than 1 (Bardet et al. 1991), the vertebrae were referred to the cryptoclidid cf. *Cryptoclidus* (Phillips 1871) (Gasparini and Spalletti 1993, lam. 1H–J; Fig. 12.3E). O'Keefe (2001a) included *Muraenosaurus* and *Cryptoclidus* in the family Crytoclididae [*sic*]. For Gasparini et al. (2003b), these taxa would not be closely related.

Late Jurassic

Pliosauroidea Welles 1943
Pliosauridae Seeley 1874

Comments. Large plesiosaurs have been found in different stratigraphic levels of the Vaca Muerta Formation (Tithonian-Berriasian). Some specimens were referred to *Pliosaurus* sp. (Gasparini and Fernández 1997) and others to *Liopleurodon* sp. (Gasparini and Fernández 1997; Gasparini et al. 1999; Spalletti et al. 1999). In agreement with Bardet et al. (1994) and O'Keefe (2001a), *Liopleurodon* (Sauvage 1873) is a poorly defined taxon that is based on enormous pliosaurid postcranial bones. Recently, Noè et al. (2004) proposed that the specimens from the Kimmeridge Clay referred to *Liopleurodon* could correspond to *Pliosaurus macromerus* (Phillips 1871). However, and as these authors say, the validity of *Liopleurodon* should be supported on more complete specimens. In the meantime, to preserve the information, the Patagonian specimens are listed in this review as they were originally mentioned.

***Liopleurodon* Sauvage 1873**
***Liopleurodon* sp.**
Fig. 12.4A

Material. MLP 80-V-29-1, incomplete mandibular symphysis and rami, and attached palatal fragments (Fig. 12.4A).

Locality and age. Cerro Lotena (Fig. 12.1, locality 2), Zapala Department, Neuquén Province, Argentina. Lower sector of the Vaca Muerta Formation, early Tithonian (Leanza 1981).

Comments. MLP 80-V-29-1 has a short symphysis, with lateral dentary edges not expanded, with six pairs of alveoli (Fig. 12.4A). These characters are present in *Liopleurodon* (Tarlo 1960). Originally Gasparini et al. (1983) referred this specimen to *Stretosaurus* sp. Currently *Stretosaurus* is thought to be a junior synonym of *Liopleurodon* (Tarlo 1960; O'Keefe 2001a), and even *Liopleurodon* is believed to be a synonym of *Pliosaurus* (Noè et al. 2004). The material from Cerro Lotena is under revision.

***Liopleurodon* sp.**

Material. MOZ 6144P, a skull and articulated mandible (2.50 m) and part of vertebral column that includes the cervicals and part of the dorsals articulated. MOZ 6141P, mandibular symphysis, right mandibular ramus, cervical vertebrae.

Locality and age. In front of Yesera del Tromen, Pampa Tril

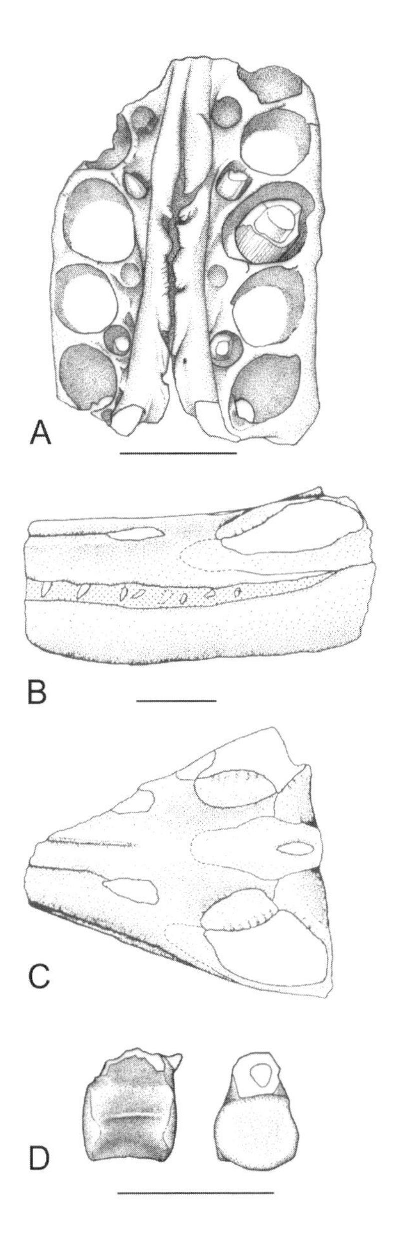

area, northwestern Neuquén province, Argentina (Fig. 12.1, locality 3). Vaca Muerta Formation, late Tithonian (Spalletti et al. 1999).

Comments. MOZ 6144P belongs to the largest pliosaurid recovered in the Southern Hemisphere, but it was much damaged by road machinery. A mandibular symphysis (MOZ 6141P) with subparallel external margins, bearing six pairs of alveoli, was recovered. The right mandibular ramus is 1.70 m long. Both specimens are currently under preparation and study.

Figure 12.4. Late Jurassic–Early Cretaceous Patagonian plesiosaurs. (A) Liopleurodon *sp. (MLP 80-V-29-1), mandibular symphysis. Scale bar = 5 cm. Pliosauridae in (A) lateral and (B) dorsal views. Scale bar = 10 cm. Simplified from Gasparini et al. (1997). (D) Elasmosauridae (MOZ 689/94PV) anterior cervical vertebra. Scale bar = 5 cm.* Modified from Lazo and Cichowolski (2003).

***Pliosaurus* Owen 1841**
***Pliosaurus* sp.**

Material. MOZ 3728 P, mandible and partial palate.

Locality and age. Cajón de Almanza, Neuquén Province (Gasparini and Fernández 1997, fig. 1), Vaca Muerta Formation, late Tithonian (Gasparini and Fernández 1997).

Comments. This material was mentioned by Gasparini and Fernández (1997) and referred to *Pliosaurus* sp. It is not illustrated, and it is currently under study.

Pliosauridae gen. et sp. indeterminate

Material. MOZ 6113P (Gasparini et al. 1997, figs. 5, 6), partial skull and mandible (Fig. 12.4B, C).

Locality and age. Trincajuera Creek, west Neuquén Province, Argentina (Fig. 12.1, locality 4), Vaca Muerta Formation, late Tithonian (Gasparini et al. 1997).

Comments. One of the most distinctive features of the specimen is the prefrontal expansion, an area that is rarely preserved or extremely reduced in pliosauroids, except in the pliosaurid *Peloneustes philarchus* (BMNH R.85574; Gasparini et al. 1997; O'Keefe 2001a, fig. 10). However, it differs from *Peloneustes* in the slitlike anterior interpterygoid vacuity, a character that it shares with *Liopleurodon* (O'Keefe 2001a).

Early Cretaceous

Plesiosauroidea Welles 1943
Elasmosauridae Cope 1869
Gen. et sp. indeterminate

Material. MOZ 6890/94PV, postcranial material, fragmented and isolated, found in five levels of the Agrio Formation, which encompasses from the upper Valanginian to lower Hauterivian (Lazo and Cichowolski 2003, fig. 3).

Locality and age. West of Neuquén Province, Argentina (Fig. 12.1, localities 5–8). Lower Member of the Agrio Formation (Aguirre-Urreta 1998): Arroyo Truquicó (MOZ 6890 PV), upper Valanginian; Cerro Negro (MOZ 6891 PV) upper Valanginian; Agua de la Mula (MOZ 6892 PV), upper Valanginian; Agua de la Mula (MOZ 6893 PV), lower Hauterivian; Bajada del Agrio (MOZ 6894 PV), lower Hauterivian (Lazo and Cichowolski 2003).

Comments. These fragments are the first record of plesiosauroids in the Early Cretaceous of Patagonia. Although the material is fragmentary, some cervical vertebrae have the proportions and the typical lateral keel of the elasmosaurids (Lazo and Cichowolski 2003, fig. 4D).

Late Cretaceous

The first published mention of South American plesiosaurs is that of Gay (1848), who mentioned the presence of *Plesiosaurus chilensis* in rocks of the Quiriquina Formation (late Maastrichtian) outcropping on Quiriquina Island and the neighboring coast, Chile (Fig. 12.1, locality 9). Marine turtles, mosasaurs, and many plesiosauroid plesiosaurs have been found in these levels. On the basis of fragmentary and often isolated material, new taxa have been determined, most of them invalidated by other authors (Welles 1962; Persson 1963; Gasparini and Goñi 1985). Likewise, plesiosaurs in southern Chile have been reported, but not studied (Gasparini 1979, 1985). Numerous records of fragments of Plesiosauria from the Late Cretaceous have been reported from the Argentinean Patagonia, most of them undetermined and even with dubious stratigraphic provenance (Gasparini et al. 2001a,b). Those specimens that were studied and/or revised recently are mentioned in this synthesis.

Plesiosauroidea Welles 1943
Polycotylidae Williston 1908 (O'Keefe 2001a)
***Sulcusuchus* Gasparini and Spalletti 1990**
***Sulcusuchus erraini* Gasparini and Spalletti 1990**
Fig. 12.5A, B

Holotype. MLP 88-IV-10-1, rostrum fragment (Gasparini and Spalletti 1999, lam. 1).

Referred specimen. MPEF 650, rostrum fragment with part of the frontoparietal bridge, incomplete basicranium and two mandibular fragments (Gasparini and de la Fuente 2000, fig. D–F; Fig. 12.5A, B).

Locality and age. MLP 88-IV-10-1, northwest margin of the Cari Laufquen Grande Lake, Río Negro Province, Argentina (Fig. 12.1, locality 10). Basal sector of the Coli-Toro Formation, early Maastrichtian (Gasparini and Spalletti 1990); MPEF 650, Cerro Bosta, La Colonia area, northeast Chubut Province (Fig. 12.1, locality 11); Middle Member of the La Colonia Formation, Campanian-Maastrichtian (Gasparini and de la Fuente 2000).

Diagnosis. Rostrum and mandible long and very narrow. Deep furrow in the external side of the maxilla and dentary. Interalveolar space very short. Pineal foramen absent. Posterior margin of the squamosal vertical. Pterygoids form an anteroposteriorly wide plate below the cranium (Gasparini and de la Fuente 2000).

Comments. The species was determined on the basis of a rostrum fragment (originally described as dentary) of a probable Dyrosauridae crocodile (Gasparini and Spalletti 1990). Later, with the discovery of a skull and articulated mandible, the partially preserved (MPEF 650) mandibular fragment was proved to be a plesiosaur with a long and very narrow rostrum. It shows characters as the long mandibular symphysis, the external margin of the squamosal vertical, and the pterygoids forming a wide plate behind the interpterygoid vacuities that are shared with the Polycotylidae (Gasparini and de la Fuente 2000).

The Coli Toro and La Colonia formations are of coastal environments, although with strong continental influence. This is

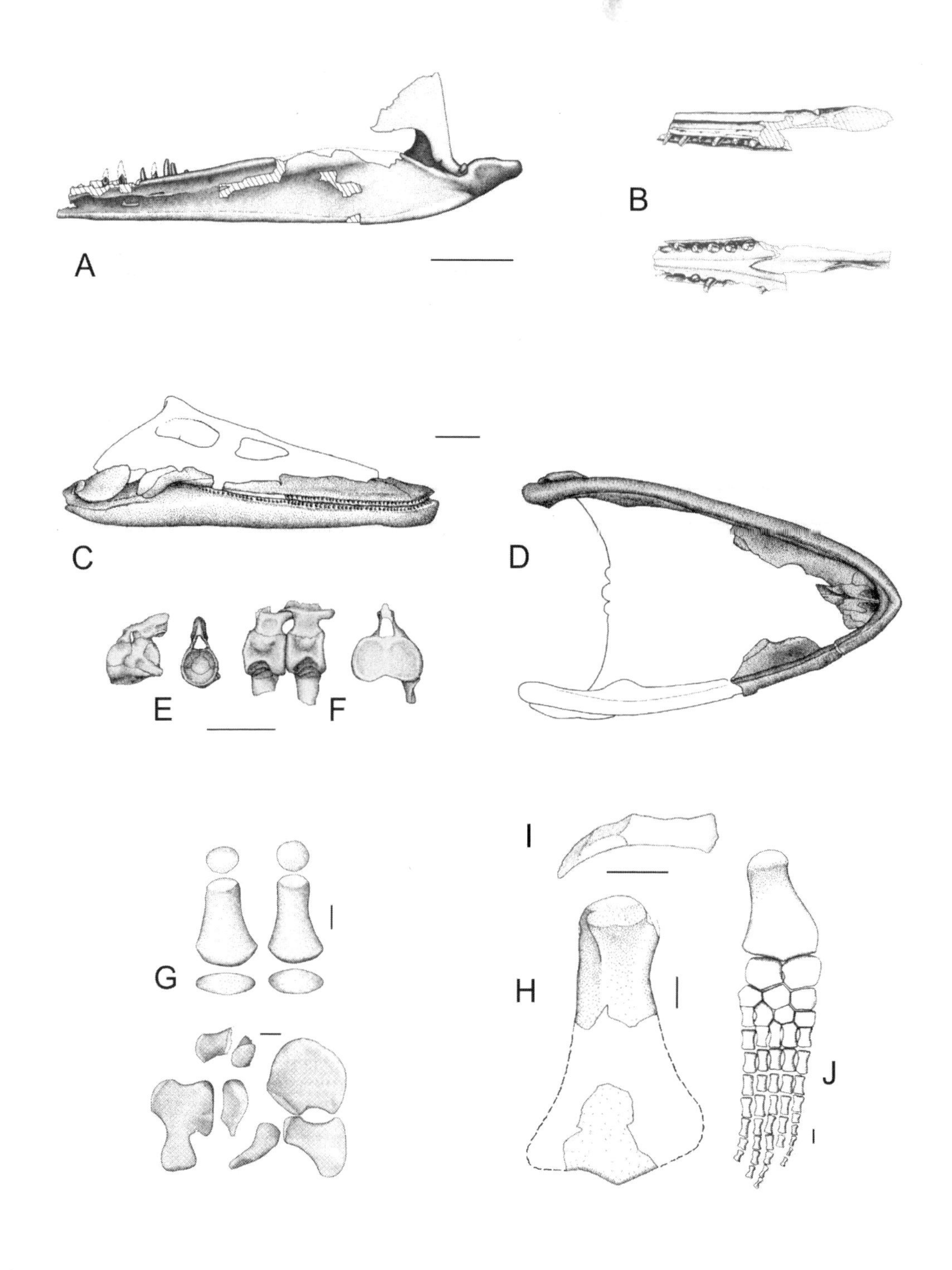
A
B
C
D
E
F
G
H
I
J

demonstrated by both the sedimentary pattern, and the prevailing of the terrestrial and freshwater faunas (Gasparini and Spalletti 1990; Gasparini and de la Fuente 2000).

Plesiosauroidea Welles 1943
Elasmosauridae Cope 1869
***Aristonectes* Cabrera 1941**
***Aristonectes parvidens* Cabrera 1941**
Fig. 12.5C–F

Holotype. MLP 40-XI-14-6, part of a skull attached to the mandible, vertebrae, and an incomplete limb (Gasparini et al. 2003b; figs. 1, 2A–G, 3; Fig. 12.5C–F).

Referred specimens. MNHN Sgo PV-82 fragment of rostrum and mandible in occlusion (Casamiquela 1969; Gasparini et al. 2003b, fig. 2H); MNHN Sgo PV-957 Skull, mandible, and three cervical vertebrae (Suárez and Fritis 2002).

Locality and age. MLP 40-XI-14-6, Cañadón del Loro, middle Chubut River, northwestern Chubut Province, Argentina (Fig. 12.1, locality 12); Paso del Sapo Formation, Lefipan Member, Maastrichtian (Gasparini et al. 2003b), MNHN SGO PV-82, Quiriquina Island (Chile), MNHN SGO PV-957, Quiriquina Formation, late Maastrichtian (Stinnesbeck 1986); TTU P 9219, Seymour Island, Antarctic Peninsula; upper "molluscan units" López de Bertodano Formation, late Maastrichtian (Chatterjee and Small 1989).

Diagnosis. Ogive-shale low and wide skull without premaxillary-maxillary constriction and with anteriorly dorsoventrally flattened maxilla; homodont dentition; strongly outwardly oriented alveoli; dental formula: 10–13 premaxillary; 51–53 maxillary; 60–65 dentary teeth; paired vomeronasal foramina (Gasparini et al. 2003b).

Comments. Since its description by Cabrera (1941), the phylogenetic position of *Aristonectes* has been a subject of debate. It has been dubiously considered as an elasmosaurid (Cabrera 1941; Bardet et al. 1991), an aberrant pliosaur (Welles 1962), a "cimoliasaurid" (Persson 1963), and a cryptoclidid (Chatterjee and Small 1989; Brown 1993). Most recently, in several studies, *Aristonectes* (=*Morturneria*) has been assigned to the Cryptoclididae (Cruickshank and Fordyce 2002), Cimoliasauridae (O'Keefe 2001a; Fig. 12.2), and Elasmosauridae (Gasparini et al. 2003b). These latter authors consider the type of *Morturneria seymourensis* (Chatterjee and Small 1989; Chatterjee and Creisler 1994) as a juvenile of *Aristonectes parvidens.*

***Tuarangisaurus* Wiffen and Moisley 1986**
***Tuarangisaurus*? *cabazai* Gasparini, Salgado, and Casadío 2003**
Fig. 12.5G

Holotype. MML-PV5, an incomplete specimen without skull and neck (Gasparini et al. 2003a, figs. 5–7; Fig. 12.5G).

Figure 12.5. (opposite page) Late Cretaceous Patagonian plesiosaurs. (A, B) Sulcusuchus erraini *(MPEF 650), mandible in lateral view, rostrum in lateral and ventral views. Scale bar = 5 cm. Simplified from Gasparini and de la Fuente (2000). (C–F)* Aristonectes parvidens *(holotype MLP 40-XI-14-6), partial skull and mandible in (C) lateral and (D) ventral views. Scale bar = 10 cm. (E) Atlas axis in lateral and anterior views. Scale bar = 10 cm. (F) Cervical in lateral and anterior views. Scale bar = 10 cm. Simplified from Gasparini et al. (2003a,b). G.* Tuarangisaurus? cabazai *(holotype MML-PV5) postcranial bones: humerus and femur. Scale bar = 5 cm. Pectoral and pelvic bones. Scale bar = 5 cm. Simplified from Gasparini et al. (2003a). (H–J) Cf.* Mauisaurus *sp. (H) Right femur (MML-PV3). Scale bar = 5 cm. (I) Ilium (MML PV4). Scale bar = 5 cm. (J) Anterior paddle (MML-PV4). Scale bar = 5 cm.* Simplified from Gasparini et al. (2003a).

Locality and age. Depressions of Trapalcó, central Río Negro Province, Argentina (Fig. 12.1, locality 15). Upper part of the Jagüel Formation, late Maastrichtian (Gasparini et al. 2003a, fig. 2).

Diagnosis. Elasmosaur with thick and rounded postcranial elements. Relatively high and short vertebral centra, with low and thick neural spines. Coracoid with robust posterior process. Ischium short and narrow (Gasparini et al. 2003a).

Comments. The general morphology of the MML-PV5 coincides with that of the specimens referred by Wiffen and Moisley (1986) to cf. *T. keyesi* (NZGS, CD427), and aff. *T. keyesi* (NZGS, CD428; NZGS, CD429). *Tuarangisaurus*? *cabazai* differs from *Tuarangisaurus keyesi* by the shorter neural spines (in dorsal and caudal vertebrae) and the elongate ischia. Several characters of MML-PV5, such as short neural spines, tuberosity/trochanter unseparated by condylar isthmus, and poor definition of the articular facet of the propodials and mesopodials, are interpreted as indicative of juvenile specimens (Brown 1981). However, Gasparini et al. (2003a) pointed out that some isolated remains similar to MMP-PV5, but larger, support the hypothesis that some characters supposedly indicative of immaturity do not vary significantly during ontogeny and are therefore taxonomically useful. Recently, one other propodial bone (MML-PV 6) was found in the depressions of Trapalcó at 4.30 m below the K/T boundary. This bone corresponds to cf. *T. cabazai* and undoubtedly belongs to a specimen larger than the holotype (Gasparini et al. 2003a, fig. 7). Recently, Sachs (2005) described a new species, *Tuarangisaurus australis,* from the Albian of Australia. This species was founded in an almost complete skull and mandible, with an attached atlas axis. Unfortunately, these elements were not preserved in MML-PV5 from north Patagonia. Isolated vertebrae similar to those of *T*?. *cabazai* were found in the northwestern Río Negro Province (MUCPv-131), in rocks of the Allen Formation referred to the late Campanian–early Maastrichtian. This would support a wider biochron of the species. According to Sachs (2005), *Tuarangisaurus* was present in the Albian of the northwest of the modern Australia.

Mauisaurus Hector 1874
cf. _Mauisaurus_ sp.
Fig. 12.5H–J

Material. MML-PV3, right mandibular fragment, prefrontal, and part of postcranial bones including part of the paddles (Gasparini et al. 2003a, figs. 3A–C, 4B; Fig. 12.5H). MML-PV4, incomplete postcranium (Gasparini et al. 2003a, figs. 3D–G, 4A, C; Fig. 12.5I, J); "*Cimoliosaurus*" *andium,* vertebrae and a paddle illustrated by Broili (1930, fig. 1), of unknown repository.

Locality and age. MML-PV3 and MML-PV4, depressions of Trapalcó and Santa Rosa, Río Negro Province, Argentina (Fig. 12.1, localities 15 and 16), upper part of the Jagüel Formation. The specimen MML-PV4 is only at 0.30 m below the Cretaceous-

Paleogene boundary (Gasparini et al. 2003a, figs. 1, 2). The paddle illustrated by Broili (1930) was found in the Quiriquina Formation, outcropping at Quiriquina Island and adjacent coast (Fig. 12.1, locality 9), referred to the late Maastrichtian (Stinnesbeck 1986).

Comments. MML-PV3 and MML-PV4 are adult specimens with a body length between 9 and 11 m (Gasparini et al. 2003a). Humerus proportions, the shape of radius and ulna, and the pentagonal pedal intermedium match those of *Mauisaurus haasti* from the upper Maastrichtian of New Zealand (Welles and Gregg 1971). Similar paddles were found in specimens from the late Maastrichtian of Chile and from islands northeast of the Antarctic Peninsula (Gasparini et al. 2003a).

Plesiosauria gen. et sp. indeterminate

Reports of Plesiosauria in Chilean Patagonia go back to the middle of the nineteenth century (Gay 1848) and continue until today. Most of them are based on very incomplete fragments found on Quiriquina Island and the neighboring coast (Gasparini 1985). A few remains imply the presence of these reptiles in the Maastrichtian of southern Chile (Chong and Gasparini 1976). Reports of plesiosaurs in the Argentinean Patagonia include those of Ameghino (1893), although the material has never been found, and other, more recent ones (Casamiquela 1980; Gasparini and Salgado 2000; Gasparini et al. 2001a,b). On this regard, it is noteworthy that the species *Trinacromerum lafquenianum* (Gasparini and Goñi 1985), from the Campanian-Maastrichtian of Pellegrini Lake (northwest Patagonia), was invalidated by Gasparini and Salgado (2000), who referred the material as Elasmosauridae indet.

Plesiosaurs of Patagonia

Although the Patagonian plesiosaurs are among the first Mesozoic reptiles known in South America, most of the findings are recent and scarce with respect to those discovered in the Northern Hemisphere. For these reasons, and because some specimens are currently under study and/or in revision, the knowledge of these pelagic reptiles is still limited. In addition, although most Jurassic forms are known by skull and/or mandibles, those of the Late Cretaceous are generally represented by incomplete postcranial skeletons–an important matter when evaluating phylogenetic proposals. Despite these limitations, the record of Patagonian Plesiosauria is the most complete in South America, and the most complete Gondwanan record.

To date, there are no records of Early Jurassic plesiosaurs in Patagonia. This has to be interpreted as a record fault. It coincides with the lack of exploration in rocks of such age. Further north, in Mendoza Province, out of Patagonia, crocodiles and ichthyosaurs are recorded in Early Jurassic rocks (Gasparini 1985). Since the Middle Jurassic (early Bajocian) a top predator, the rhomaleosaurid *Maresaurus coccai,* is recorded in the Chacaico Sur area (Fig. 12.1, locality 1) and is associated with ophthalmosaurian and stenopterygian ichthyosaurs (Fernández this volume). In this area, numerous remains of cryptoclidids (cf. *Cryptoclidus;* cf. *Muraenosaurus*) appear dismembered in early Callovian rocks (when this sector of the basin became more shallow), along with scarce pliosauroid vertebrae. Dismemberment of skeletons coincides with an environment of high energy near the coast (Spalletti et al. 1994).

Tithonian plesiosaur remains are frequent in the Neuquén Basin, and all of them are pliosaurids (*Liopleurodon* sp., *Pliosaurus* sp.). This may suggest a biased record, or it may coincide with the environment in which they were preserved. This environment is interpreted as the basinal part of a marine ramp, not near the coast (Spalletti et al. 1999). When the southwest sector of the Neuquén Basin became shallower, toward the Valanginian-Hauterivian, the action of this area's high environmental energy broke up the plesiosaur remains. Also, the remains are exclusively elasmosaurids (Lazo and Cichowolski 2003), which generally indicates coastal environments.

According to the paleogeographic model proposed by Spalletti et al. (2000) (see Spalletti and Franzese this volume), the Neuquén Basin was separated from the Pacific by island arches. A partially protected basin, with high biodiversity, could have been a favorable environment for feeding and especially for reproduction. The diversity of pelagic predators over millions of years (Gasparini and Fernández 1996, 1997, 2005) would support this hypothesis.

During the Early Cretaceous, and as a consequence of different tectonic events, the Eastern Pacific withdrew from northwest Patagonia (Spalletti and Franzese this volume), together with the marine herpetofauna characteristic for the Jurassic. Only toward the end of the Campanian, and as a consequence of the South Atlantic ingression over a large part of the Argentinean Patagonia (Malumián 1999), this area became a large archipelago, thus favoring the incoming of new marine reptiles, especially by the end of the Maastrichtian, when the ingression reached its maximum extension (Gasparini et al. 2001a,b). Elasmosaurids (*Tuarangisaurus? cabazai,* cf. *Mauisaurus* sp., *Aristonectes parvidens*) were the most abundant, but a polycotylid with narrow and long rostrum (*Sulcusuchus erraini*) was also present. The marine tetrapod richness of the Eastern Pacific during the late Maastrichtian is documented in the Quiriquina Formation outcropping on the homonymous island and the neighboring coast, where plesiosaurs are represented by at least *Aristonectes parvidens* and cf. *Mauisaurus* sp. (Gasparini et al. 2003a).

With their different swimming and prey-capture strategies (Massare 1988; O'Keefe 2002), plesiosaurs were large pelagic reptiles that swam in all the seas of the world, even at high latitudes (Kear 2003). The possibility of long-distance displacements according to biological cycles cannot be discarded. The certainty of Plesiosauria taxonomic determinations, and consequently of biogeographic interpretation, is not very high. However, some evidence may be outlined. The rhomaleosaurid *Maresaurus* of the early Bajocian has close affinities with forms of the European Early Jurassic and Callovian (Gasparini 1997). Likewise, forms referred to cf. *Cryptoclidus* and cf. *Muraenosaurus* entered the Neuquén Basin during the early Callovian, when specimens of these genera lived in the European Tethys (Brown 1981). During the Middle Jurassic, continental masses still formed Pangaea, and thus dispersions (and/or yearly migrations) should have followed the coastlines through the Eastern Tethys, in this case circumnavigating the globe or crossing the Pacific. Or perhaps the reptiles swam through an epicontinental sea that preceded the Caribbean Corridor (Gasparini 1992). This latter hypothesis is the most plausible because it is the shortest way, and it is supported by a record of invertebrates in common in the Western Tethys (Europe) and Eastern Pacific (Damborenea 2000).

Toward the end of the Jurassic, Laurasia and Gondwana were defined with the opening of new seaways that would have facilitated the dispersion of the marine herpetofauna. Large pliosaurids (*Liopleurodon, Pliosaurus*), which were frequent in the Kimmeridge Clay and equivalent basins in France (Bardet 1995; Noè et al. 2004), were also found in the Neuquén Basin, which implies some migratory periods in whatever direction. Recently, a large pliosaur referred to *Liopleurodon* was found in rocks of the Kimmeridgian of East Mexico (Buchy et al. 2003), precisely in what was the Caribbean Corridor (Iturralde-Vinent 2004). Also, the rupture of South Gondwana produced new seaways, enlarging the possibilities of dispersion and/or yearly migration for pelagic forms (Gasparini and Fernández 1997).

Toward the end of the Mesozoic, the continent-sea distribution was more or less similar to today's. However, Patagonia was covered by the Pacific in the southwest, and a large South Atlantic ingression had made a large archipelago of it. The plesiosaurs found in Patagonia have a mainly South Gondwanan distribution. *Aristonectes* (=*Morturneria*) is recorded in the late Maastrichtian of central Patagonia (Chubut) and from the coasts of Quiriquina Island (Chile) and Seymour Island (Antarctic Peninsula). Remains of elasmosaurids referred to cf. *Mauisaurus* sp. have been found in north Argentinean Patagonia, in the area of Quiriquina, and in Seymour Island, whereas *Mauisaurus haasti* has been found in New Zealand. Another example would be *Tuarangisaurus*. Although *Tuarangisaurus keyesi* (Wiffen and Moisley 1986) was found in New Zealand and *Tuarangisaurus australis* (Sachs 2005) in Australia, a specimen referred to *Tuarangisaurus*? *cabazai* (Gasparini et al. 2003a) was found in the north Argentinean Patagonia. No polycotylids are known from other parts of the world that may be referred to the peculiar *Sulcusuchus*.

The plesiosaur diversity toward the end of the Cretaceous in Patagonia was important. A specimen referred to cf. *Mauisaurus* sp. (MML-PV4)—a tetrapod record that is closest to the great extinction in South America (Concheyro et al. 2002; Gasparini et al.

2002b)—was found only 30 cm below the Cretaceous-Paleogene boundary.

Acknowledgments. I am greatly indebted to Sergio Cocca and Rafael Cocca (Museo "Prof. Dr. Juan Olsacher," Zapala) for collaboration in discovery and extraction of most of the Jurassic plesiosaurs of the Argentinean Patagonia that I mention in this chapter. Likewise, I thank Daniel Cabaza and collaborators (Museo Municipal de Lamarque) for their participation in the discovery and extraction of most of the Cretaceous plesiosaurs of Argentina. I extend thanks to Drs. Nathalie Bardet (Museum National d'Histoire Naturelle, Paris) and Sven Sachs (Institut für Paläontologie, Freie Universität Berlin) for their valuable suggestions. Dr. Cecilia Deschamps (Museo de La Plata) helped with the English-language version of this chapter, and Lic. Diego Brandoni and Jorge González helped with the graphics. This chapter was financially supported by Agencia Nacional de Promoción Científica y Tecnológica (PICT 8439) and National Geographic Society (grants 6882-00 and 7757-04).

References Cited

Aguirre-Urreta, M. B. 1998. The ammonites *Karakaschiceras* and *Neohoploceras* (Valanginian Neocomitidae) from the Neuquén Basin. West-Central Argentina. *Journal of Paleontology* 72: 39–59.

Ameghino, F. 1893. Sobre la presencia de vertebrados de aspecto mesozoico, en la formación santacruceña de la Patagonia austral. *Revista del Jardín Zoológico de Buenos Aires* 1 (3): 75–84.

Andrews, C. 1913. *A Descriptive Catalogue of the Marine Reptiles of The Oxford Clay,* Part II. London: British Museum of Natural History.

Bakker, R. 1993. Plesiosaur extinction cycles—Events that mark the beginning, middle and the end of the Cretaceous. In W. Caldwell and E. Kauffman (eds.), *Evolution of the Western Interior Basin,* 641–644. Special Paper 39. Geological Association of Canada.

Bardet, N. 1995. Evolution et extinction des reptiles marins au cours du Mésozoïque. *Palaeovertebrata* 24 (3–4): 177–283.

Bardet, N., G. Lachkar, and F. Escuillé. 1994. Presénce à l'Oxfordien de *Pliosaurus brachyspondylus* (Reptilia, Pliosauridae) déterminée par les données palynologiques. *Revue de Micropaléontologie* 37: 181–188.

Bardet, N., J.-M. Mazin, E. Cariou, R. Enay, and J. Krishna. 1991. Les Plesiosauria du Jurassique supérieur de la province de Kachch (Inde). *Comptes Rendus de l'Académie des Sciences de Paris* 2 (313): 1343–1347.

Broili, F. 1930. Plesiosaurierreste von der Insel Quiriquina. *Neues Jahrbuch für Mineralogie, Geologie und Paläontologie* 63 (8): 497–514.

Brown, D. 1981. The English Upper Jurassic Plesiosauroidea (Reptilia) and a review of the phylogeny and classification of the Plesiosauria. *Bulletin of the British Museum (Natural History): (Geology* 35 (4): 253–347.

———. 1993. A taxonomic reappraisal of the families Elasmosauridae and Cryptoclididae (Reptilia: Plesiosauroidea). *Revue de Paléobiologie* 7: 9–16.

Brown, D., and A. R. I. Cruickshank. 1994. The skull of the Callovian plesiosaur *Cryptoclidus eurymerus*, and the sauropterygian cheek. *Palaeontology* 37 (4): 941–953.

Buchy, M.-C., E. Frey, W. Stinnesbeck, and J. G. López-Oliva. 2003. First occurrence of a gigantic pliosaurid plesiosaur in the Late Jurassic (Kimmeridgian) of Mexico. *Bulletin de la Société Géologique de France* 174 (3): 271–278.

Cabrera, Á. 1941. Un plesiosaurio nuevo del Cretáceo del Chubut. *Revista Museo de La Plata* (n.s.) 2 (8): 113–130.

Callaway, J., and E. Nicholls. 1997. *Ancient Marine Reptiles*. San Diego: Academic Press.

Carpenter, K. 1996. A review of short-necked plesiosaurs from the Cretaceous of the Western Interior, North America. *Neues Jahrbuch für Geologie und Paläontologie, Abhandlungen* 201: 259–287.

———. 1997. Comparative cranial anatomy of two North American Cretaceous plesiosaurs. In J. Callaway and E. Nicholls (eds.), *Ancient Marine Reptiles*, 191–216. San Diego: Academic Press.

———. 1999. Revision of North America elasmosaurs from the Cretaceous of the Western Interior. *Paludicola* 2 (2): 148–173.

Casamiquela, R. 1969. La presencia de *Aristonectes* Cabrera (Plesiosuaria) del Maastrichtiense del Chubut, Argentina. Edad y carácter de la transgresión "rocanense." *Actas de las IV Jornadas Geológicas Argentinas* 1: 199–213.

———. 1980. Considérations écologiques et zoogéographiques sur les vertébrés de la zone litorale de la mer du Maestrichtien dans le Nord de la Patagonie. *Mémoires Société Géologique de France*, n.s., 139: 53–55.

Chatterjee, S., and B. Creisler. 1994. *Alwalkeria* (Theropoda) and *Morturneria* (Plesiosauria), new names for preoccupied *Walkeria* Chatterjee, 1987 and *Turneria* Chatterjee and Small, 1989. *Journal of Vertebrate Paleontology* 14: 142.

Chatterjee, S., and B. Small. 1989. New plesiosaurs from the Upper Cretaceous of Antarctica. In J. A. Crame (ed.), *Origins and Evolution of the Antarctic Biota*, 197–215. Special Publication 47. Geological Society of London.

Chong, G., and Z. Gasparini. 1976. Los vertebrados mesozoicos de Chile y su aporte geopaleontológico. *Actas del IV Congreso Geológico Argentino* 1: 45–67.

Concheyro, A., C. Nañez, and S. Casadío. 2002. El límite Cretácico-Paleógeno en Trapalcó, provincia de Río Negro, Argentina: una localidad clave en América del Sur? *Actas del XV Congreso Geológico Argentino* 1: 590–595.

Cruickshank, A., and R. Fordyce. 2002. A new marine reptile (Sauropterygia) from New Zealand: further evidence for a Late Cretaceous austral radiation of cryptoclidid plesiosaurs. *Palaeontology* 45 (3): 557–575.

Damborenea, S. 2000. Hispanic Corridor: Its evolution and the biogeography of bivalve molluscs. *GeoResearch Forum* 6: 369–380.

de Blainville, H. D. 1835. Description de quelques espèces de reptiles de la Californie, précédée de l'analyse d'un système général d'Erpétologie et d'Amphibiologie. *Nouvelles Annales du Muséum d'Histoire Naturelle de Paris* 3: 233–296.

Gasparini, Z. 1979. Comentarios críticos sobre los vertebrados mesozoicos de Chile. *Actas del II Congreso Geológico Chileno*, Arica 1979, 3: 15–32.

———. 1985. Los reptiles marinos jurásicos de América del Sur. *Ameghiniana* 22: 23–34.
———. 1992. Marine Reptiles. In G. Westermann (ed.), *The Jurassic of the Circum-Pacific*, 361–364. London: Cambridge University Press.
———. 1997. A new pliosaur from the Bajocian of the Neuquén Basin, Argentina. *Palaeontology* 40 (1): 135–147.
Gasparini, Z., and M. de la Fuente. 2000. Tortugas y plesiosaurios de la Formación La Colonia (Cretácico superior) de Patagonia, Argentina. *Revista Española de Paleontología* 15 (1): 23–351.
Gasparini, Z., and M. Fernández. 1996. Biogeographic affinities of the Jurassic marine reptile fauna of South America. IV International Congress Jurassic Stratigraphy and Geology, Mendoza 1994. *GeoResearch Forum* (1–2): 443–450.
———. 1997. Tithonian marine reptiles of the Eastern Pacific. In J. Callaway and E. Nicholls (eds.), *Ancient Marine Reptiles*, 435–440. San Diego: Academic Press.
———. 2005. Jurassic marine reptiles in the Neuquén Basin. In G. Veiga, L. Spalletti, J. Howell, and E. Schwarz (eds.), *The Neuquén Basin: A Case Study in Sequence Stratigraphy and Basin Dynamics*, 279–294. Special Publication 252. Geological Society of London.
Gasparini Z., and R. Goñi. 1985. Los plesiosauros cretácicos de América del Sur y del Continente Antártico. *Coletanea de Trabalhos de Paleontologia* DNPM, Ser. Geologia, Brasilia, 27 (2): 55–56.
Gasparini, Z., and L. Salgado. 2000. Elasmosáuridos (Plesiosauria) del Cretácico del norte de Patagonia. *Revista Española de Paleontología* 15: 13–21.
Gasparini, Z., and L. Spalletti. 1990. Un nuevo cocodrilo en los depósitos mareales maastrichtianos de la Patagonia noroccidental. *Ameghiniana* 27 (1–2): 141–150.
———. 1993. First Callovian plesiosaurs from the Neuquén Basin, Argentina. *Ameghiniana* 30: 245–254.
Gasparini, Z., R. Goñi, and O. Molina. 1983. Un plesiosaurio (Reptilia) tithoniano en Cerro Lotena, Neuquén (Argentina). *Actas del V Congreso Latinoamericano de Geología* 5: 33–47.
Gasparini, Z., L. Spalletti, and M. de la Fuente. 1997. Marine reptiles of a Tithonian transgression, western Neuquén Basin, Argentina. Facies and Paleoenvironments. *Geobios* 30(5): 701–712.
Gasparini, Z., L. Spalletti, M. Fernández, and M. de la Fuente. 1999. Tithonian marine reptiles from the Neuquén Basin: diversity and paleoenvironments. *Revue de Paléontologie, Genève,* 18(1): 333–345.
Gasparini, Z., S. Casadío, M. Fernández, and L. Salgado. 2001a. Marine reptiles from the Late Cretaceous. *Journal of South American Earth Sciences* 14: 51–60.
Gasparini, Z., M. de la Fuente, M. Fernández, and P. Bona. 2001b. Reptiles from Late Cretaceous coastal environments of northern Patagonia. In *VII International Symposium on Mesozoic Terrestrial Ecosystems*, 101–105. Special Publication 7. Asociación Paleontológica Argentina.
Gasparini, Z., L. Spalletti, S. Matheos, and M. Fernández. 2002a. Reptiles marinos y paleoambientes del Jurásico superior–Cretácico inferior en la Yesera del Tromen (Neuquén, Argentina): un caso de estudio. *Actas del XV Congreso Argentino de Geología* 1: 473–478.
Gasparini, Z., S. Casadío, M. de la Fuente, L. Salgado, M. Fernández, and A. Concheyro. 2002b. Reptiles acuáticos en sedimentitas lacustres y

marinas del Cretácico Superior de Patagonia (Río Negro), Argentina. *Actas del XV Congreso Geológico Argentino* 1: 495–499.

Gasparini, Z., L. Salgado, and S. Casadío. 2003a. Maastrichtian plesiosaurs from northern Patagonia. *Cretaceous Research* 24: 157–170.

Gasparini, Z., N. Bardet, J. Martin, and M. Fernández. 2003b. The elasmosaurid plesiosaur *Aristonectes* Cabrera from the latest Cretaceous of South America and Antarctica. *Journal of Vertebrate Paleontology* 23 (1): 104–115.

Gay, C. 1848. *Historia física y política de Chile.* Vol. 2. Zoologia and Atlas. Paris.

Hector, J. 1874. On the fossil Reptilia of New Zealand. *Transactions of the New Zealand Institute* 6: 333–358.

Iturralde-Vinent, M. 2004. The conflicting paleontologic vs stratigraphic record of the formation of the Caribbean Seaway. In C. Bartolini, R. Buffler, and J. Blickwede (eds.), *The Circum-Gulf of Mexico and the Caribbean: Hydrocarbon Habitats, Basin Formation, and Plate Tectonics,* 75–88. Memoir 79. American Association of Petroleum Geologists.

Kear, B. 2002. Darwin Formation (Early Cretaceous, Northern Territory) marine reptiles remains in the South Australian Museum. *Records of South Australian Museum* 35: 33–47.

———. 2003. Cretaceous marine reptiles of Australia: a review of taxonomy and distribution. *Cretaceous Research* 24: 277–303.

Lazo, D., and M. Cichowolski. 2003. First plesiosaur remains from the Lower Cretaceous of the Neuquén Basin, Argentina. *Journal of Paleontology* 77 (4): 784–789.

Leanza, H. 1981. The Jurassic-Cretaceous boundary beds in west-central Argentina and their ammonite zones. *Neues Jahrbuch für Geologie und Paläontologie, Abhandlungen* 161: 62–92.

Maisch, M., and M. Rücklin. 2000. Cranial osteology of the Sauropterygian *Plesiosaurus brachypterygius* from the lower Toarcian of Germany. *Palaeontology* 43 (1): 29–40.

Malumián, N. 1999. La sedimentación y el volcanismo terciarios en la Patagonia extraandina. In R. Caminos (ed.), *Geología Argentina,* 557–612. Anales del Instituto Geológico Rec. Minería, 29 (18).

Massare, J. 1988. Swimming capabilities of Mesozoic marine reptiles: implications for method of predation. *Paleobiology* 14: 187–205.

Noè, L., D. Smith, and D. Walton. 2004. A new species of Kimmeridgian pliosaur (Reptilia; Sauropterygia) and its bearing on the nomenclature of *Liopleurodon macromerus. Proceedings of the Geologist's Association* 115: 13–24.

O'Keefe, F. 2001a. A cladistic analysis and taxonomic revision of the Plesiosauria (Reptilia: Sauropterygia). *Acta Zoologica Fennica* 213: 1–63.

———. 2001b. Ecomorphology of plesiosaur flipper geometry. *Journal of Evolutionary Biology* 14: 987–991.

———. 2002. The evolution of plesiosaur and pliosaur morphotypes in the Plesiosauria (Reptilia: Sauropterygia). *Paleobiology* 28 (1): 101–112.

———. 2004. Preliminary description and phylogenetic position of a new plesiosaur (Reptilia: Sauropterygia) from the Toarcian of Holzmaden, Germany. *Journal of Paleontology* 78 (5): 973–988.

Persson, P. 1963. A revision of the classification of the Plesiosauria with a synopsis of the stratigraphical and geographical distribution of the group. *Lunds Universitets Arsskrift,* N.F. 2, 59 (1): 1–59.

Phillips, J. 1871. *Geology of Oxford and the Valley of the Thames.* Oxford: Oxford University Press.

Sachs, S. 2004. Redescription of *Woolungasaurus glendowerensis* (Plesiosauria: Elasmosauridae) from the Lower Cretaceous of Northeast Queensland. *Memoirs of the Queensland Museum* 49 (2): 713–731.

———. 2005. *Tuarangisaurus australis* sp. nov. (Plesiosauria: Elasmosauridae) from the lower Cretaceous of Northeastern Queensland, with additional notes on the phylogeny of the Elasmosauridae. *Memoirs of the Queensland Museum* 50 (2): 429–444.

Sauvage, H. 1873. Notes sur les reptiles fossiles. *Bulletin de la Societé Geologique de France* 1 (3): 365–380.

Seeley, H. 1874. On *Muraenosaurus leedsi,* a plesiosaurian from the Oxford Clay. *Quarterly Journal of the Geological Society* 30: 197–208.

Spalletti, L., Z. Gasparini, and M. Fernández. 1994. Facies, ambientes y reptiles marinos de la transición entre las Formaciones Los Molles y Lajas (Jurásico medio), Cuenca Neuquina. *Acta Geologica Leopoldensia* 17: 329–344.

Spalletti, L., Z. Gasparini, G. Veiga, E. Scharwz, M. Fernández, and S. Matheos. 1999. Facies anóxicas, procesos deposicionales y herpetofauna de la Rampa marina titoniana-berriasiana en la Cuenca Neuquina. *Revista Geológica de Chile* 26: 109–123.

Spalletti, L., J. Franzese, S. Matheos, and E. Schwarz. 2000. Sequence stratigraphy of a tidally-dominated carbonate-siliciclastic ramp: the Tithonian of the Southern Neuquén Basin, Argentina. *Journal of the Geological Society* 157: 433–446.

Stinnesbeck, W. 1986. Zu den faunischen und palökologischen Verhältnissen in der Quiriquina Formation (Maastrichtium) Zentral-Chile. *Palaeontographica,* Abteilung A, 194: 99–237.

Storrs, G. 1991. Anatomy and relationships of *Corosaurus alcovensis* (Diapsida: Sauropterygia) and the Triassic Alcova Limestone of Wyoming. *Bulletin of the Peabody Museum of Natural History* 44: 1–151.

———. 1993. Function and phylogeny in sauropterygian (Diapsida) evolution. *American Journal of Science* 293A: 63–90.

Storrs, G., M. Arkhangelsky, and V. Efimov. 2001. Fossil marine reptiles from Russia and other former Soviet republics. In M. Benton, M. Shishkin, D. Unwin, and E. Kurochkin (eds.), *The Age of Dinosaurs in Russia and Mongolia,* 187–210. Cambridge: Cambridge University Press.

Suárez, M., and O. Fritis. 2002. Nuevo registro de *Aristonectes* (Plesiosauroidea, incertae sedis) del Cretácico tardío de la Formación Quiriquina, Cocholgüe, Chile. *Boletín de la Sociedad Biológica* 73: 87–93.

Tarlo, L. 1960. A review of Upper Jurassic pliosaurs. *Bulletin of the British Museum of Natural History (Geology)* 4 (5): 147–189.

Uliana, M., and K. Biddle. 1988. Mesozoic-Cenozoic paleogeographic and geodynamic evolution of southern South America. *Revista Brasileira de Geociencias* 18: 172–190.

Welles, S. 1943. Elasmosaurid plesiosaurs with description of new material from California and Colorado. *Memoirs of the University of California* 13: 125–215.

———. 1952. A review of the North American Cretaceous elasmosaurs. *University of California, Publication in Geological Sciences* 29: 47–144.

———. 1962. A new species of elasmosaur from the Aptian of Colombia and a review of the Cretaceous plesiosaurs. *University of California Publications in Geological Sciences* 44 (1): 1–96.

Welles, S., and D. Gregg. 1971. Late Cretaceous marine reptiles of New Zealand. *Records of the Canterbury Museum* 9 (1): 1–111.

Wiffen, J., and W. Moisley. 1986. Late Cretaceous reptiles (families Elasmosauridae and Pliosauridae) from the Mangahouanga Stream, North Island, New Zealand. *New Zealand Geology and Geophysic* 29: 205–252.

Williston, S. W. 1907. The skull of *Brachauchenius,* with observations of the relationships of the plesiosaurs. *Proceedings of the National Museum* 32: 477–493.

———. 1908. North American plesiosaurs: *Trinacromerum. Journal of Geology* 16: 715–736.

———. 1925. *The Osteology of the Reptiles.* Cambridge, Mass.: Harvard University Press.

13. Ichnology

Jorge O. Calvo

Introduction

The first reptilian track from Patagonia to be described comes from Plottier, in the Neuquén Province (Huene 1931) (Fig. 13.1, locality 4). Unfortunately, this material is at present missing, and no more data can be obtained in order to improve its original description. R. M. Casamiquela was a pioneer of the ichnological studies in Patagonia (see Salgado this volume). In the 1960s and 1970s, he published several papers on Mesozoic reptilian and mammal ichnofaunas (Casamiquela 1960, 1962, 1964, 1974). Since then, no more ichnological studies on Patagonia were carried out until 1989, when J. Calvo published the first record of dinosaur tracks at the Ezequiel Ramos Mexía Lake, in Neuquén (Calvo 1991), and R. Coria described a fossil track from the Jurassic of Chubut (Coria 1989). Other new papers on reptilian tracks, mainly on dinosaurs, have been published (Calvo et al. 1990, 2000; Calvo 1991, 1997, 1999; Calvo and Coria 1995; Calvo and Moratalla 1998; Manera de Bianco and Calvo 1999; Calvo and Lockley 2001; Coria et al. 2002).

A very good overview of the South American track record has been made by Leonardi (1994). Leonardi's book includes all museum specimens and field and locality data. Therefore, in this work, I am not going to go into detail about the quantity and quality of track materials. Track descriptions in this chapter are taken from original sources but have been abbreviated and modified as necessary for consistency. Where necessary, the term "Diagnosis" is used.

Two localities in Patagonia have so far yielded the best track record. One is Los Menucos in Río Negro Province (Upper Triassic), and the other locality is Lake Ezequiel Ramos Mexía, in the province

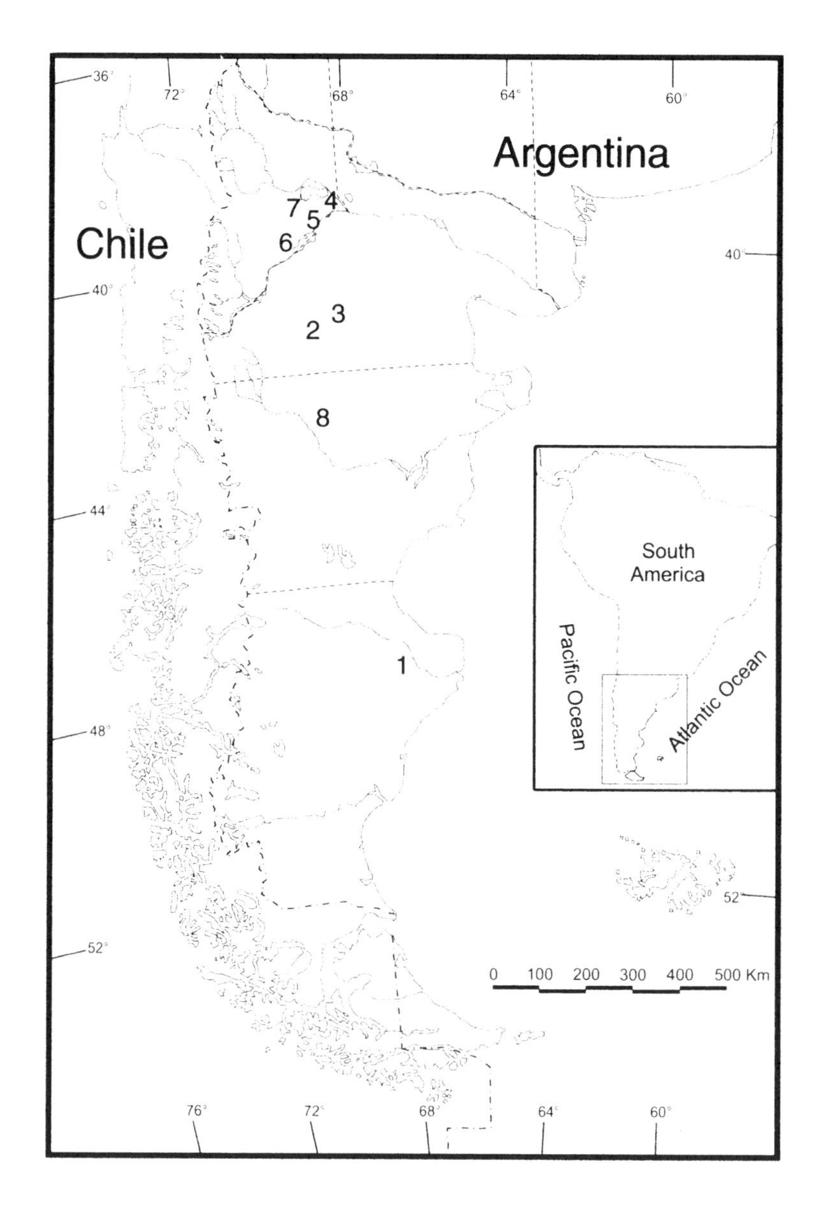

Figure 13.1. Map showing the ichnofossil sites. 1, Laguna Manantiales; 2, "Montón-110" (Ing. Jacobacci); 3, Los Menucos; 4, Plottier; 5, Villa El Chocón; 6, Picún Leufú; 7, Sierra Barrosa; 8, Cerro Cóndor.

of Neuquén (Lower or lower Upper Cretaceous) (Fig. 13.1). Here, I present a summary of the reptilian Mesozoic footprints of Patagonia. An update of the footprint record, as well as some comments, where necessary, are presented. The taxonomic assignment follows the previously published papers mentioned above.

Institutional abbreviations. CePaLB, Centro Paleontológico "Lago Barreales," Universidad Nacional del Comahue, Neuquén, Argentina; MCF-PVPH, Museo "Carmen Funes," Paleontología de Vertebrados, Plaza Huincul, Neuquén, Argentina; MLP, Museo de La Plata, Buenos Aires, Argentina; MUCPv, Museo de la Universidad Nacional del Comahue, Neuquén, Argentina, Paleontología de Vertebrados.

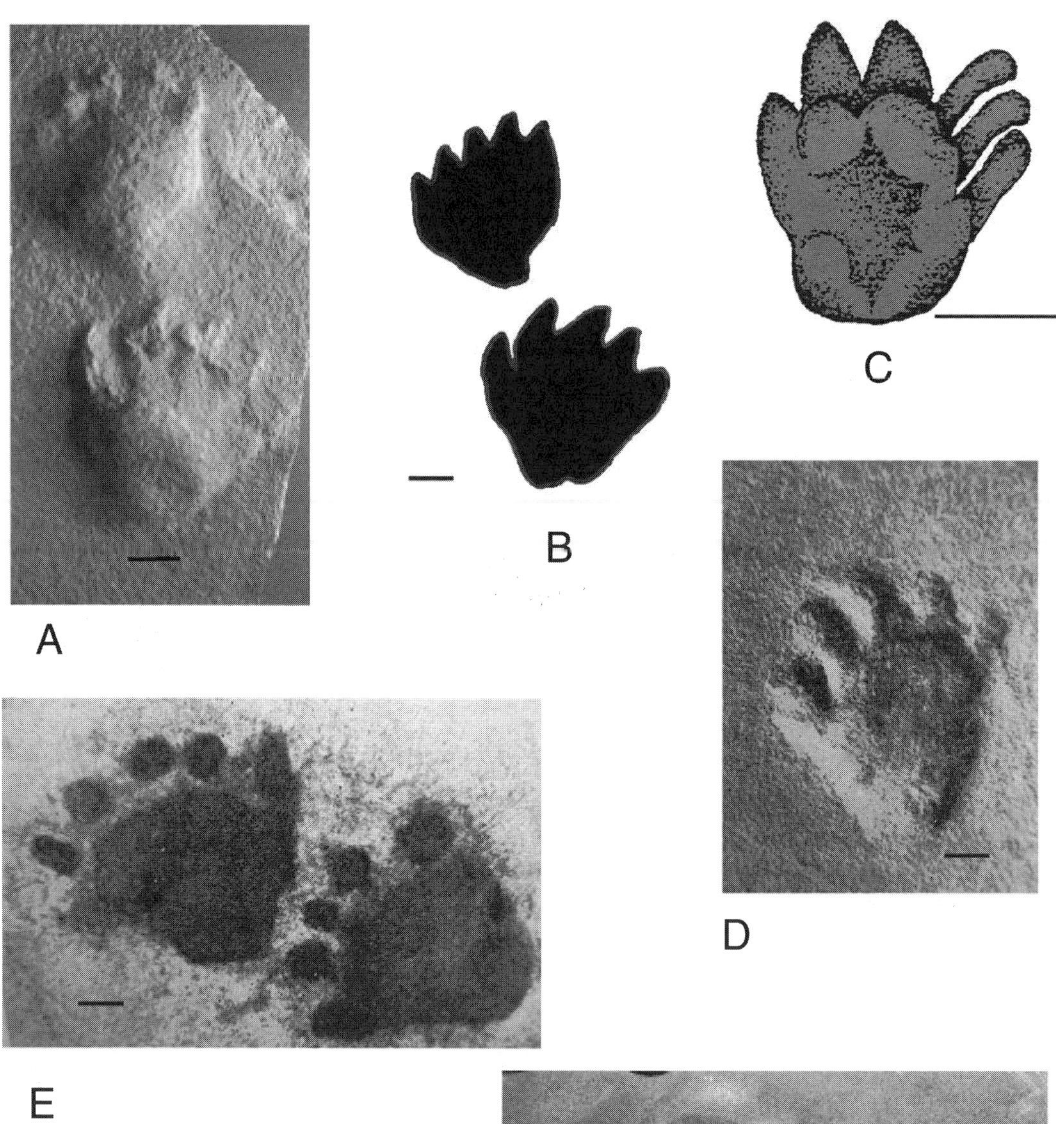

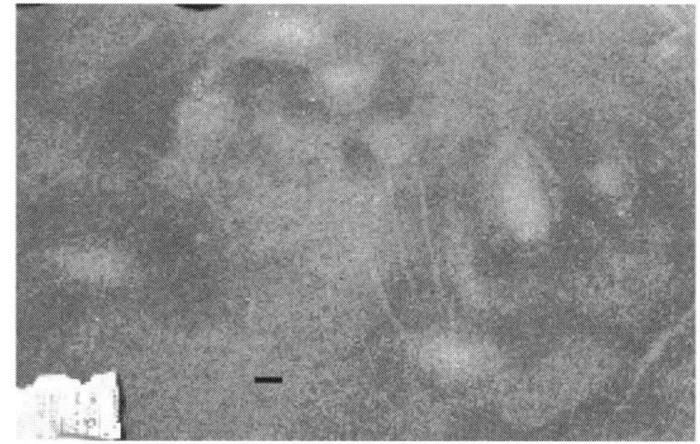

Figure 13.2. (A) Photograph of Calibarichnus ayestarani. *(B) Outline drawing of* Calibarichnus ayestarani. *(C) Outline of anterior footprints of* Palaciosichnus zetti *(modified from Casamiquela 1964). (D) Photograph of the pes track of* Gallegosichnus garridoi. *(E) Photograph of manus and pes tracks of* Gallegosichnus garridoi. *(F) Photograph of manus and pes of* Shimmelia chirotheroides. *Scale bar = 1 cm.*

Systematic Paleontology

Reptilia Linnaeus 1858
Synapsida Osborn 1903
Therapsida Broom 1905
Theriodontia Owen 1867
Calibarichnus Casamiquela 1964
Calibarichnus ayestarani Casamiquela 1964

Fig. 13.2A, B

Holotype. MLP 60-XI-31-4, cast.

Locality and age. La Vieja Quarry, Los Menucos, Río Negro Province, Argentina (Fig. 13.1, locality 3). Vera Formation, Upper Triassic (Kokogian et al. 2001).

Diagnosis. Quadrupedal, plantigrade and pentadigitigrade impressions. Manus directed inward. Pes directed outward. All digits have claw impressions. Digits short, subequal, and arranged in a fan shape. In the pes, digits I and V are slightly more separated from digits II and IV, respectively, than the central ones among them (Casamiquela 1964).

Comments. Casamiquela (1964) had doubts about referring this specimen to a determinate group; he proposed a relationship with therapsids, chelonians, or mammals. However, Leonardi (1994) confirmed its assignation to Therapsida.

Palaciosichnus Casamiquela 1964
Palaciosichnus zettii Casamiquela 1964

Fig. 13.2C

Holotype. MLP 6-XI-31-6.

Locality and age. La Vieja Quarry, Los Menucos, Río Negro Province, Argentina (Fig. 13.1, locality 3). Vera Formation, Upper Triassic (Kokogian et al. 2001).

Diagnosis. Prints corresponding to a quadrupedal and plantigrade form. Anterior footprints diverge 80° outward from the midline. They have an aberrant bifid autopodium that diverges 65°. Three short and massive digits with claw impressions are directed outward. The forward directed digits are slender and longer. Manus is well separated from the pes, and shows five short and massive digit impressions with radial shape. There is no impression of a tail (modified from Casamiquela 1964).

Comments. Casamiquela (1964) described a sixth toe impression present in two left anterior prints. But the material is not very well preserved, and the taxon has been discarded by Leonardi (1994) because of the lack of diagnostic characters. He interpreted it as *Dicynodontipus.*

Anomodontia Owen 1860
Rogerbaletichnus Casamiquela 1964
Rogerbaletichnus aquilerai Casamiquela 1964

Holotype. MLP 60-XI-31-5.

Locality and age. La Nueva Quarry, Los Menucos, Río Negro

Province, Argentina (Fig. 13.1, locality 3). Vera Formation, Upper Triassic (Kokogian et al. 2001).

Diagnosis. Quadrupedal and plantigrade impressions. Pace angulations 120°. Digits I to IV are rounded or polygonal, decreasing in size from I to IV. Digit V is separated from others and sometimes from the palm. It is longer and has a sickle-shaped nail impression. The pes impression outline is round and/or oval, and the toes appear to possess large pads or oval or polygonal hooves. The first toe is slightly abducted (it diverges away from the plane of symmetry). The foot oversteps and partially overlaps the manus print (modified from Casamiquela 1964).

Comments. Leonardi (1994) recognized the presence of another slab with a trackway stored at the Museo "Jorge Gerhold" de Ingeniero Jacobacci. The fossil comes from La Vieja Quarry.

Therapsida incertae sedis
***Gallegosichnus* Casamiquela 1964**
***Gallegosichnus garridoi* Casamiquela 1964**
Fig. 13.2D, E

Holotype. MLP 60-XI-31-7, cast.

Paratypes. MLP 60-XI-31-8 and 9.

Referred specimens. MLP-60-X-31-2A, MLP-66-XI-15-2, 3.

Locality and age. La Vieja Quarry, Los Menucos, Río Negro Province, Argentina (Fig. 13.1, locality 3). Vera Formation, Upper Triassic (Kokogian et al. 2001).

Diagnosis. Quadrupedal, plantigrade, and pentadactyl animal. Subtriangular manus print directed inward. Digit impressions are short; the longest digit is IV. Subtriangular pes print longer and thinner than those of the manus and outward directed. Digit impressions subequal in length. There is no impression of claws. Pace angulations between 95° to 105° (Casamiquela 1964).

Comments. It is the most abundant ichnogenus at the Los Menucos Locality and surrounding area. The material stored at La Plata Museum is the best preserved from the Los Menucos site: minute details, including the morphology of the palm and plant with their pads and wrinkles, are preserved.

Archosauria Cope 1869
Pseudosuchia Zittel 1887/1890
Chirotheridae Abel 1935
***Shimmelia* Casamiquela 1964**
***Shimmelia chirotheroides* Casamiquela 1964**
Fig. 13.2F

Holotype. MLP-60-XI-31-1, cast.

Paratype. MLP 60-XI-31-2, cast.

Locality and age. La Vieja Quarry, Los Menucos, Río Negro Province, Argentina (Fig. 13.1, locality 3). Vera Formation, Upper Triassic (Kokogian et al. 2001).

Diagnosis. Pentadactyle and digitigrades foot and hand. Digit V of large size is directed laterally on foot. Digit III is the longest. Digits II and IV of the same size and digit I very reduced. Digits I to IV are closer and parallel with distal ends rounded. Manus print directed outward with digits similar in shapes and length to that of the pes impressions. Hand and foot overstep a good part of the hand. There are no impressions of claws and tails. The large hand is very different from other chirotherian hands (Casamiquela 1964).

Comments. The poor quality of the material does not permit detailed observations; however, several features that differentiate it from other chirotherian ichnogenera. This justifies preserving the ichnogenus (Leonardi 1994).

Pterosauria Kaup 1834
Pteraichnidae Lockley et al. 1995
***Pteraichnus* Stokes 1957**
***Pteraichnus* sp. (Calvo 1999)**
Fig. 13.3A

Material. MUCPv-196, manus casts; MUCPv-20-23.

Locality and age. West coast of Ezequiel Ramos Mexía, Neuquén Province, Argentina (Fig. 13.1, locality 5). Candeleros Formation, Neuquén Group, Albian (Calvo 1991; Calvo and Salgado 1996) or lower Cenomanian (Leanza et al. 2004).

Description. Wide trackway with prominent manus impressions. Manus is tridactyl, strongly asymmetrical, with three digit impressions showing strong outward rotation. Digit I is the shortest, II is intermediate in length, and IV is the longest. Pes impressions very elongated, functionally tetradactyl, with digit impressions II and III slightly longer than digits IV and I. Pes impressions subrectangular in shape and three times longer than wide (Calvo and Lockley 2001).

Comments. The record of pterosaur tracks in southern continents is scarce; good material has been described only from Villa El Chocón. A short review of these tracks was given by Calvo and Moratalla (1998) and Calvo (1999). A more detailed description and evaluation was carried out by Calvo and Lockley (2001).

Saurischia Seeley 1888
Theropoda Marsh 1881
Carnosauria Huene 1914
***Abelichnus* Calvo 1991**
***Abelichnus astigarrae* Calvo 1991**
Figs. 13.3B, 13.5B

Holotype. MUCPv-74.

Hypodigm. MUCPv-148.

Referred specimens. MUCPv-139.

Locality and age. Peninsula Nueva and Cerrito del Bote Island, Picún Leufú, Neuquén Province, Argentina (Fig. 13.1, locality 6). Candeleros Formation, Neuquén Group, Albian (Calvo 1991; Calvo and Salgado 1996) or lower Cenomanian (Leanza et al. 2004).

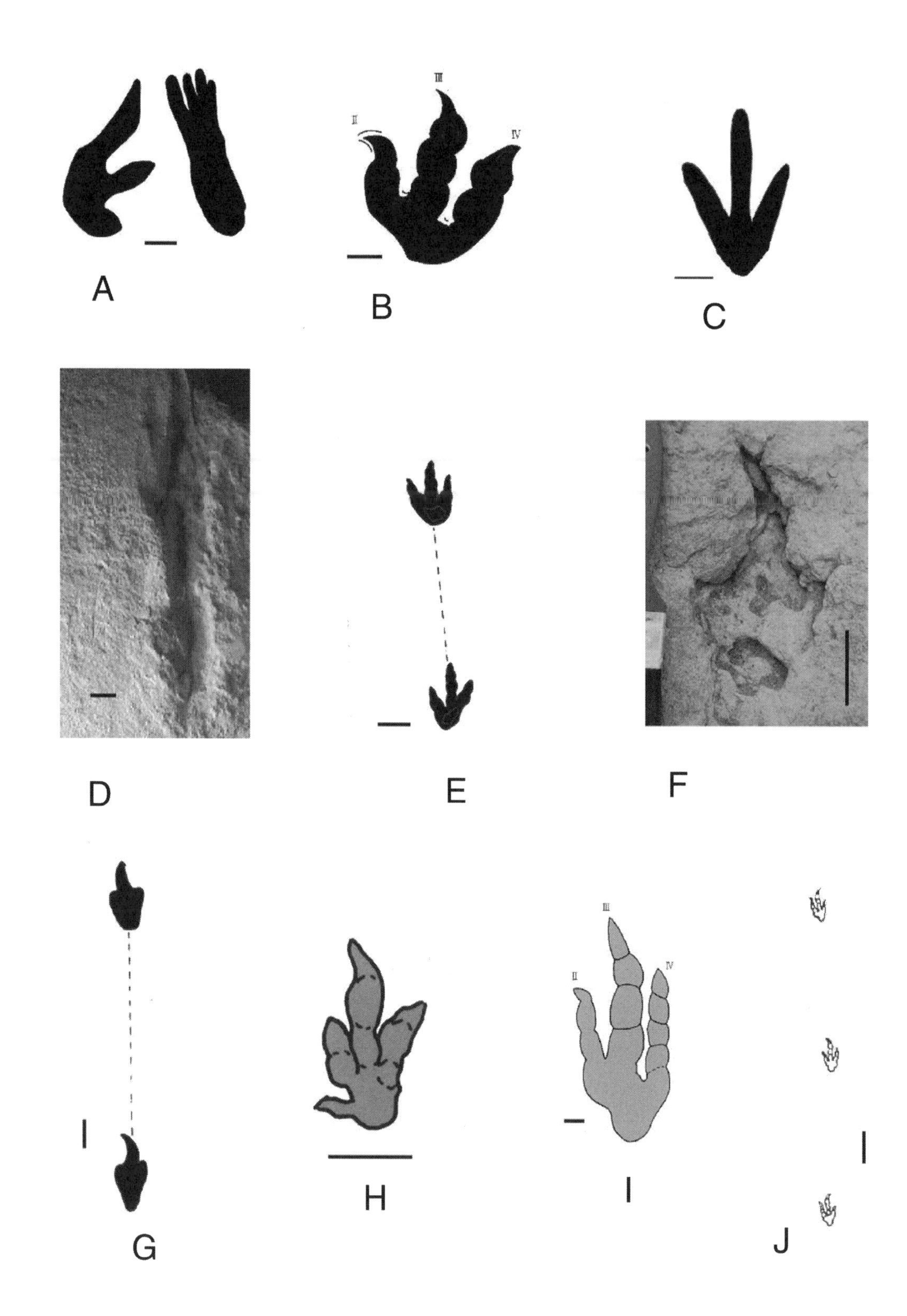
A
III
II
IV
B
C
D
E
F
G
H
III
II
IV
I
J

Diagnosis. Giant biped and tridactyl theropod tracks. Tracks have large and wide digit impressions. Digit III more developed than digits II and IV. Claw impressions are very prominent. Claw on digit III displaced inwardly. Digit IV larger than digit II, with claws displaced outward. The pace angulation is less than 150°. The heel is practically rounded, with a small surface (Calvo 1991).

Comments. These tracks belong to a huge theropod. There are two trackways at Península Nueva. The biggest tracks are 50 cm in length and the pace is 130 cm. The smallest track (MUCPv-148) is 36 cm and the pace is almost 100 cm.

Other tracks have been identified at Cerrito del Bote (Calvo 1989; Calvo and Mazzetta 2004), Cañadón de Coria, and Balneario Villa El Chocón localities (Calvo 1999). Calvo (1999) inferred that these huge theropod tracks could belong to the huge carnivorous dinosaur *Giganotosaurus carolinii* (see Coria this volume).

Coelurosauria Huene 1914
Anchisauripodidae Lull 1904
***Wildeichnus* Casamiquela 1964**
***Wildeichnus navesi* Casamiquela 1964**
Fig. 13.3C

Holotype. MLP 60-X-31-5, cast.

Paratypes. MLP 60-X-31-9 and 11; PVL2302B and 2305.

Referred specimens. PVL 3700-3701-2305, MLP 65-XI-12-1, MLP.60-x-31-11, MLP.65-XI12-½, MLP.65-XI-12-1, MLP.65-XI-12-1.

Locality and age. Laguna Manantiales, Santa Cruz Province (Fig. 13.1, locality 1). La Matilde Formation, upper Middle Jurassic–lower Upper Jurassic (Riccardi and Damborenea 1993).

Diagnosis. Impressions corresponding to a tridactyl, digitigrade, and bipedal dinosaur of small size. Pace angulation of almost 180°. Digit III is the longest; the lateral ones are shorter. Interdigital angle subequal (Casamiquela 1964).

Comments. Digits have deeper impressions at the distal end. The proximal end of digit III has a smooth contact with the divarication of digits II and IV. The posterior border of the heel is not well preserved.

Sauropodidae Haubold 1969
***Sarmientichnus* Casamiquela 1964**
***Sarmientichnus scagliai* Casamiquela 1964**
Fig. 13.3D

Holotype. MLP 60-X-31-A.

Paratypes. MLP 60-X-31-B and -C, MLP 60-X-31-2A and -2C.

Referred specimens. MLP 65-XI-12-1 (Leonardi 1994).

Locality and age. Laguna Manantiales, Santa Cruz Province

Figure 13.3. (opposite page) (A) Outline of manus (left) and pes (right) of Pteraichnus *sp. (modified from Calvo 1999). Scale bar = 1 cm. (B) Outline of footprint of* Abelichnus astigarrae *(modified from Calvo 1991). Scale bar = 10 cm. (C) Outline of footprint of* Wildeichnus navesi. *Scale bar = 1 cm. (D) Photograph of* Sarmientichnus scagliai. *Scale bar = 1 cm. (E) Outline of* Bressanichnus patagonicus *(modified from Calvo 1991). Scale bar = 10 cm. (F) Photograph of* Deferrariischnium mapuchensis *from Picún Leufú locality, in situ. Scale bar = 5 cm. (G) Outline of one step of* Deferrariischnium mapuchensis *(modified from Calvo 1991). Scale bar = 10 cm. (H) Outline of* Picunichnus benedettoi *(modified from Calvo 1991). Scale bar = 10 cm. (I) Outline of a small theropod track from Isla Cerrito del Bote (modified from Calvo 1999). Scale bar = 1 cm. (J) Outline of a trackway of a small theropod from Isla Cerrito del Bote. Scale bar = 10 cm.*

(Fig. 13.1, locality 1). La Matilde Formation, upper Middle Jurassic–lower Upper Jurassic (Riccardi and Damborenea 1993).

Diagnosis. Bipedal and digitigrade dinosaur of small size. Pace angulation closer to 180°. Monodactyl. Digit with acuminate ungual phalanx, but without a true claw impression (Casamiquela 1964).

Comments. The presence of a deep impression of digit III contrasts with the lack of evidence of lateral digits impressions. However, an inverted mold present on slab MLP 60-X-31-1-B shows a second track impression of digit III and part of digit II. The counterpart slab MLP 60-X-31-1-A does not have any impression of digit II. Therefore, I agree with Casamiquela (1964) about the dinosaurian origin of these tracks. All tracks have a small lateral depression placed at one-third of the length. This position coincides with the position of the proximal end of lateral digits.

Coelurosauria incertae sedis
***Delatorrichnus* Casamiquela 1964**
***Delatorrichnus goyenechei* Casamiquela 1964**

Holotype. MLP 60-X-31-6.

Paratypes. MLP 60-X-31-7; MLP 65-XI-12-1/1, -1/2, -1/3.

Referred specimens. PVL 3677, PVL 3690, PVL 3682, MACN 18615.

Locality and age. Laguna Manantiales, Santa Cruz Province, Argentina (Fig. 13.1, locality 1). La Matilde Formation, upper Middle Jurassic–lower Upper Jurassic (Riccardi and Damborenea 1993).

Diagnosis. Very small and digitigrade quadrupedal dinosaur with three functional toes on the foot and one in the hand. Pace angulation almost 180°. The foot has three wide toe impressions. The central toe is longer than the lateral ones. Its distal end is acuminate. Manus impression is digitiform and the size of the foot toes. It is located close to the foot impression but on the external side; therefore, the foot looks like a tetradactyl, although it is clearly tridactyl. No tail impression is visible (Casamiquela 1964).

Comments. The ichnospecies was established by Casamiquela (1964) and later revised by the same author (Casamiquela 1966). The new evidence shows a manus with three toes, with digit III being the longest. The manus is bigger than the foot. *Delatorrichnus* was, without doubt, a quadrupedal animal.

Coelurosauria incertae sedis
***Bressanichnus* Calvo 1991**
***Bressanichnus patagonicus* Calvo 1991**
Fig. 13.3E

Plastotype. MUCPv-60.

Hypodigm. MUCPv-61.

Referred specimens. MUCPv-68 and -69.

Locality and age. Peninsula Nueva, Picún Leufú, Neuquén Province, Argentina (Fig. 13.1, locality 6). Candeleros Formation,

Neuquén Group, Albian (Calvo 1991; Calvo and Salgado 1996) or lower Cenomanian (Leanza et al. 2004).

Diagnosis. Midsized trackway of a biped dinosaur. Digit III is more developed than the lateral ones with claw impressions. The track length is between 20 and 25 cm. The length/width ratio is 1.5. Digit III is curved internally, the heel is subrounded and small. The pace angulations is close to 150°. Tracks are slightly asymmetrical. The hypex formed between digits II and III is posteriorly located with respect to the hypex between digits III and IV (Calvo 1991).

Comments. At Villa El Chocón locality (Fig. 13.1, locality 5), six trackways are referred to this ichnogenus. Some of these tracks are stored at CePaLB.

Coelurosauria incertae sedis
***Deferrariischnium* Calvo 1991**
***Deferrariischnium mapuchensis* Calvo 1991**
Fig. 13.3F, G

Plastotype. MUCPv-62.

Hypodigm. MUCPv-63.

Referred specimens. MUCPv-66.

Locality and age. Península Nueva, Picún Leufú, Neuquén Province, Argentina (Fig. 13.1, locality 6). Candeleros Formation, Neuquén Group, Albian (Calvo 1991; Calvo and Salgado 1996) or lower Cenomanian (Leanza et al. 2004).

Diagnosis. The tracks belong to a small theropod dinosaur characterized by the impression of a long digit III, displaced internally and with a sharp claw. It occupies more than 50% of the track length. Digits II and IV are poorly developed, without claw impressions, and diverge no more than 25° with respect to digit III. The small heel connects with the lateral digits (Calvo 1991).

Comments. At Villa El Chocón, only one poorly preserved track of this ichnogenus has been recognized (Calvo 1999).

Coelurosauria incertae sedis
***Picunichnus* Calvo 1991**
***Picunichnus benedettoi* Calvo 1991**
Fig. 13.3H

Holotype. MUCPv-72.

Locality and age. Península Nueva, Picún Leufú, Neuquén Province, Argentina (Fig. 13.1, locality 6). Candeleros Formation, Neuquén Group, Albian (Calvo 1991; Calvo and Salgado 1996) or lower Cenomanian (Leanza et al. 2004).

Diagnosis. Tetradactyl track with good impressions of claws. Digit I, which is connected to the posterior half of the heel, is directed medially and is very small. Digit III is the longest. Digits II and IV diverge slightly with respect to digit III. Phalanges impressions are well marked and in general longer than wide. Tracks are asymmetrical, with a small heel (Calvo 1991).

Comments: Knowledge on this ichnogenus is based on just one track.

Theropoda ichnogen. et ichnosp. indeterminate (Calvo 1989)
Fig. 13.3I, J

Plastotype. MUCPv-133-134-135.

Locality and age. Cerrito del Bote Island, Picún Leufú, Neuquén Province, Argentina (Fig. 13.1, locality 6). Candeleros Formation, Neuquén Group, Albian (Calvo 1991; Calvo and Salgado 1996) or lower Cenomanian (Leanza et al. 2004).

Description. Small theropod tracks. The pace between two tracks is 45 cm; the length of each track is approximately 10 cm. They are long and thin with sharp claw impressions.

Comments. At Cerrito del Bote Island, five theropod trackways very close to each other were found. All trackways run in the same direction, and they are separated by just 20 to 40 cm. Calvo (1999) supposes that the record of these theropod trackways are evidence that the animals probably lived and moved socially. Paleobiological studies showed that they moved slowly, because the ratio between the pace length/hip height is between 2.0 and 2.9 (Calvo and Mazzetta 2004).

Figure 13.4. (opposite page) (A) Outline of a theropod track from Chubut (modified from Coria 1989). Scale bar = 10 cm. (B) Outline of cf. Ignotornis *(MCF-PVPH-SB 415.20). Scale bar = 1 cm. (C) Photograph of cf.* Ignotornis *(MCF-PVPH-SB 415.14). The white arrow indicates the hallux impression. Scale bar = 1 cm. (D) Outline of cf.* Ignotornis *(modified from Coria et al. 2002, fig. 6, #97) showing the hallux impression (MCF-PVPH-SB 415.17c #97). Scale bar = 1 cm. (E) Photograph of cf.* Ignotornis *(MCF-PVPH-SB 415.16b). Scale bar = 1 cm. (F) Outline of cf.* Ignotornis *trackway and* Barrosopus slobodai *trackway based on the original map of Coria et al. (2002, fig. 6) (MCF-PVPH-SB 415.17). Circle refers to a detail. Scale bar = 10 cm. (G) Outline of cf.* Ignotornis *trackway and* Barrosopus slobodai *trackway based on the original trackways. (MCF-PVPH-SB 415.17). Circle refers to a detail. Scale bar = 10 cm. (H) Outline in detail of cf.* Ignotornis *trackway and* Barrosopus slobodai *trackway based on the original trackways (MCF-PVPH-SB 415.17). Scale bar = 10 cm. (I) Photograph of* Barrosopus slobodai *#87 (MCF-PVPH-SB 415.17c). Scale bar = 1 cm. (J) Photograph of an avian track erroneously assigned to* Barrosopus slobodai *(MCF-PVPH-SB 415-2). Scale bar = 1 cm.*

Theropoda ichnogen. et ichnosp. indeterminate (Coria 1989)
Fig. 13.4A

Locality and age. Cerro Cóndor, Chubut Province, Argentina (Fig. 13.1, locality 8). Cañadón Asfalto Formation, Callovian (Page et al. 1999).

Description. The material consists of just one poorly preserved track, 21.6 mm long and 13.4 mm wide. The tridactyl track shows claw impressions. Digit III is the longest. A small heel is present.

Comments. The material is very poor, and no more data can be obtained.

Aves Linnaeus 1758
***Ignotornis* Mehl 1931 cf. *Ignotornis* sp.**
Fig. 13.4B–H

Coria, Currie, Ebert, and Garrido (2002, 458), fig. 3C, D
"cf. *Aquatilavipes*" Coria, Currie, Ebert, and Garrido (2002, 456–458), figs. 3A, B, 4–6

Materials. MCF-PVPH-SB 415.14/15/20-415.16a/16b.

Locality and age. At 30 km northeast of Plaza Huincul, Sierra Barrosa, Neuquén Province, Argentina (Fig. 13.1, locality 7). Anacleto Formation, Campanian (Leanza 1999).

Description. Separate narrow digital outlines with proximal union of pedal digits II, III, and IV. Footprint wider greater than long. Pedal digits II, III, and IV proximally fused. Hallux impression.

Comments. Coria et al. (2002, 458) described this material as follows: "Bird ichnites similar in shape and size to cf. *Aquatilavipes* but with distinct hallux impressions similar to cf. *Ignotornis* (Mehl

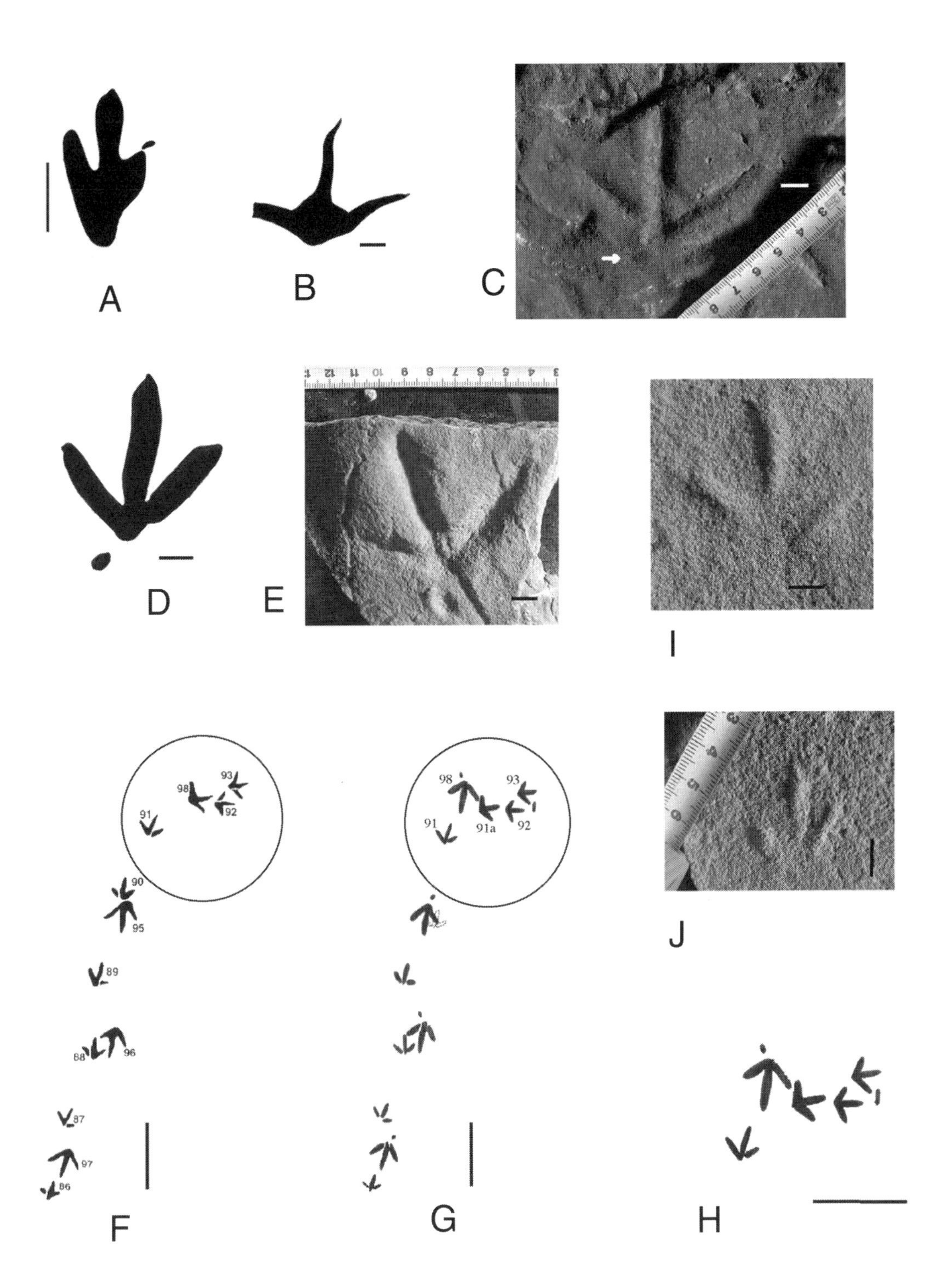
A
B
C
D
E
I
J
91
98
93
92
90
95
89
88
96
87
97
86
F
98
93
91
91a
92
G
H

1931) and *Jindongornipes* (Lockley et al. 1992). The average length is 50 mm (excluding the hallux), an average width 58 mm, and an average divarication between digits II and IV of 132°. The second digit seems to be much thinner that the third and fourth digits." Although more than 50 specimens were found in situ, few specimens are actually available in the collection of the Museo "Carmen Funes."

On the other hand, Coria et al. (2002) described tracks and trackways from Sierra Barrosa as cf. *Aquatilavipes* (Fig. 13.4B, C). *Aquantilavipes* (Currie 1981) was found in the Lower Cretaceous of Canada. This assignation is based on similar divarication size, absence of a hallux, and other morphological characters corresponding with the specimens of Canada, *Aquatilavipes*. The average divarication proposed for the Sierra Barrosa tracks, of 123° (Coria et al. 2002, table 1), is not very clear: it seems to be the wider divarication, but not the average divarication value. On the other hand, the divarication of cf. *Aquatilavipes* sp. (Coria et al. 2002, figs. 3A and 4B, among others) exceeds 130° (Fig. 13.4B). The range of divarication shown for the material from Sierra Barrosa is too variable and is not useful as a diagnostic character. On the other hand, the database given in their table 1 does not correspond with the illustrations. Specimen MCF-PVPH-SB 415.20 (Fig. 13.4B), assigned to cf. *Aquatilavipes* sp., is a counterpart, and the hallux is not preserved. The divarication between digits II and IV is 140° (Fig. 13.4B). Drawings of *Yacoraitichnus avis* (Alonso and Marquillas 1986, fig. 4) from Argentina show a very close similarity with cf. *Aquatilavipes* (Coria et al. 2002, fig. 3A). Moreover, Coria et al. (2002, fig. 3B) show a track with a posterior depression, called a metatarsal pad (Fig. 13.4C), similar to that classified as a hallux impression for *Ignotornis* (Fig. 13.4E) (Coria et al. 2002, figs. 3C, 6). The hallux is clearly seen in tracks 95 and 97 from PVPH-SB-415-17c (Fig. 13.4D, G, H). This trackway was assigned to cf. *Aquatilavipes* (Coria et al. 2002, fig 6, footprints 95 to 97) (Fig. 13.4F); however, it really belongs to cf. *Ignotornis*, which indeed has a hallux. Therefore, I prefer not to consider these tracks as cf. *Aquatilavipes;* instead, these tracks could be assigned to cf. *Ignotornis* (sensu Coria et al. 2002).

Barrosopus slobodai Coria, Currie, Eberth, and Garrido 2002
Fig. 13.4F–J

Holotype: MCF-PVPH-SB 415 17c.

Referred specimens. MCF-PVPH-SB 415–17.

Locality and age. At 30 km northeast of Plaza Huincul, Sierra Barrosa, Neuquén Province, Argentina (Fig. 13.1, locality 7). Anacleto Formation, Campanian (Leanza 1999).

Previous comment. Originally Coria et al. (2002, 458) diagnosed this ichnospecies as follows: "Tridactylous footprint of a small-sized avian theropod characterized by having conspicuous separation of digit II impression from the other two digits. Divarication between digits II and IV ranges between 100° and 120°.

Trackways show a long stride, suggesting the trackmarker was long-legged."

Diagnosis emended. Tridactylous footprint of a small-sized bird. Digit IIs and IV separated from digit III. Digit II directed backward with respect to digit IV. Divarication between digits II and IV ranges between 100° and 120°. There is no impression of a heel (Fig. 13.4I).

Comments. Barrosopus slobodai is assigned with doubt to Aves by Coria et al. (2002, 458); however, later, they assigned this ichnospecies unquestionably to birds (Coria et al. 2002, 461). The assignation of specimen MCF-PVPH-SB 415-2 (Fig. 13.4J) to *Barrosopus slobodai* is not valid because the divarication is less than 90°. The holotype shows the impression of three digits separated from each other and with different depths; in other words, there is no heel impression. This morphology is completely different from the holotype (Coria et al. 2002, fig. 3E) and from the original drawings (Coria et al. 2002, fig. 6), where only digit II is separated from digit III (Fig. 13.4F). There is a small separation between digit II and the rest of the digits. Digit II is placed at the same level as digit IV in Coria et al. (2002, fig. 6). In contrast, the holotype shows that digit II is placed at a lower level than digit IV (Coria et al. 2002, fig. 3E) and digit III is not fused to digit IV (Fig. 13.4I). On slab PVPH-513-415-17-h, track 98, figured by Coria et al. (2002, fig. 6), was misinterpreted because it was drawn using two different tracks, one assigned to the *Barrosospus slobodai* trackway (Fig. 13.4F) and the other assigned to the cf. *Aquatilavipes* trackway. In order to resolve this, a new interpretation is given (Figs. 13.4G, H).

I agree that the trackway of *Barrosopus slobodai* was produced by a bird. The emendation of the diagnosis was necessary in order to maintain this taxon as valid.

Sauropodomorpha Huene 1932
Prosauropoda Huene 1920
Ichnogen. et ichnosp. indeterminate (Manera de Bianco and Calvo 1999)

Material. Several tracks, which could not be collected, observed in the field.

Locality and age. Los Menucos, Río Negro Province, Argentina (Fig. 13.1, locality 3). Vera Formation, Upper Triassic (Kokogian et al. 2001).

Description. The manus track has a subcircular shape (150×120 mm); the pes track has a subtriangular shape (410×300 mm). It presents four striations that could belong to toe impressions. The other nine tracks are larger and elliptical (1000×600 mm). Finally, there is an isolated liver-shaped manus track (510×310 mm).

Comments. The shape and size and the age of the manus and pes impressions indicate that these tracks were left by a prosauropod (Manera de Bianco and Calvo 1999).

Sauropoda Marsh 1878
***Sauropodichnus* Calvo 1991**
***Sauropodichnus giganteus* Calvo 1991**
Fig. 13.5A–C

Holotype. MUCPv-145 and -146.

Locality and age. Península Nueva and Cerrito del Bote Island, Picún Leufú, Neuquén Province, Argentina (Fig. 13.1, locality 6). Candeleros Formation, Neuquén Group, Albian (Calvo 1991; Calvo and Salgado 1996) or lower Cenomanian (Leanza et al. 2004).

Diagnosis emended. Wide-gauge trackway of a huge sauropod. The pace angulation is less than 160°. The internal border of left and right tracks is well separated from the midline, as in *Brontopodus*. Pes tracks are between 60 to 100 cm in length. The shape is subtriangular when well preserved and circular in outline when badly preserved. Manus tracks are wider (40 cm) than long (25 cm), but always smaller than the pes tracks. Manus tracks are crescent-shaped, with the posterior border concave and the anterior border convex (emended from Calvo 1991, 1999).

Comments. Sauropodichnus giganteus was the first ichnospecies attributed to a sauropod discovered in Argentina. Calvo (1989, 1999) recognized the existence of better impressions of sauropod tracks at Cerrito del Bote and Villa El Chocón localities (Calvo and Mazzetta 2004). The trackways belong to a wide-gauge sauropod because the footprints are separated from the midline. The pace angulation is less than 160°. This corrects data erroneously published in the original diagnosis (more than 60° in Calvo [1991, 1999]).

Calvo (1999) interpreted all these tracks and trackways as having been left by the sauropod *Andesaurus delgadoi* according to their width and to the lack of phalanges and claws on the forefoot, resulting in a liver-shaped manus.

Ornithopoda Marsh 1881
Iguanodontidae Cope 1869
***Limayichnus* Calvo 1991**
***Limayichnus major* Calvo 1991**
Figs. 13.5B, D

Plastotype. MUCPv-65.

Hypodigm. MUCPv 64-67-70 and -73.

Locality and age. Península Nueva and Cerrito del Bote Island, Picún Leufú locality (Fig. 13.1, locality 6); Balneario Villa El Chocón, Neuquén Province, Argentina (Fig. 13.1, locality 5). Candeleros Formation, Neuquén Group, Albian (Calvo 1991; Calvo and Salgado 1996) or lower Cenomanian (Leanza et al. 2004).

Diagnosis Bipedal trackmaker with tridactyl large ichnite. The pace angulation is over 150° and less than 170°. Tracks are symmetrical, without impressions of heel and claws. The toe impressions are short, with rounded distal ends. The posterior contour is

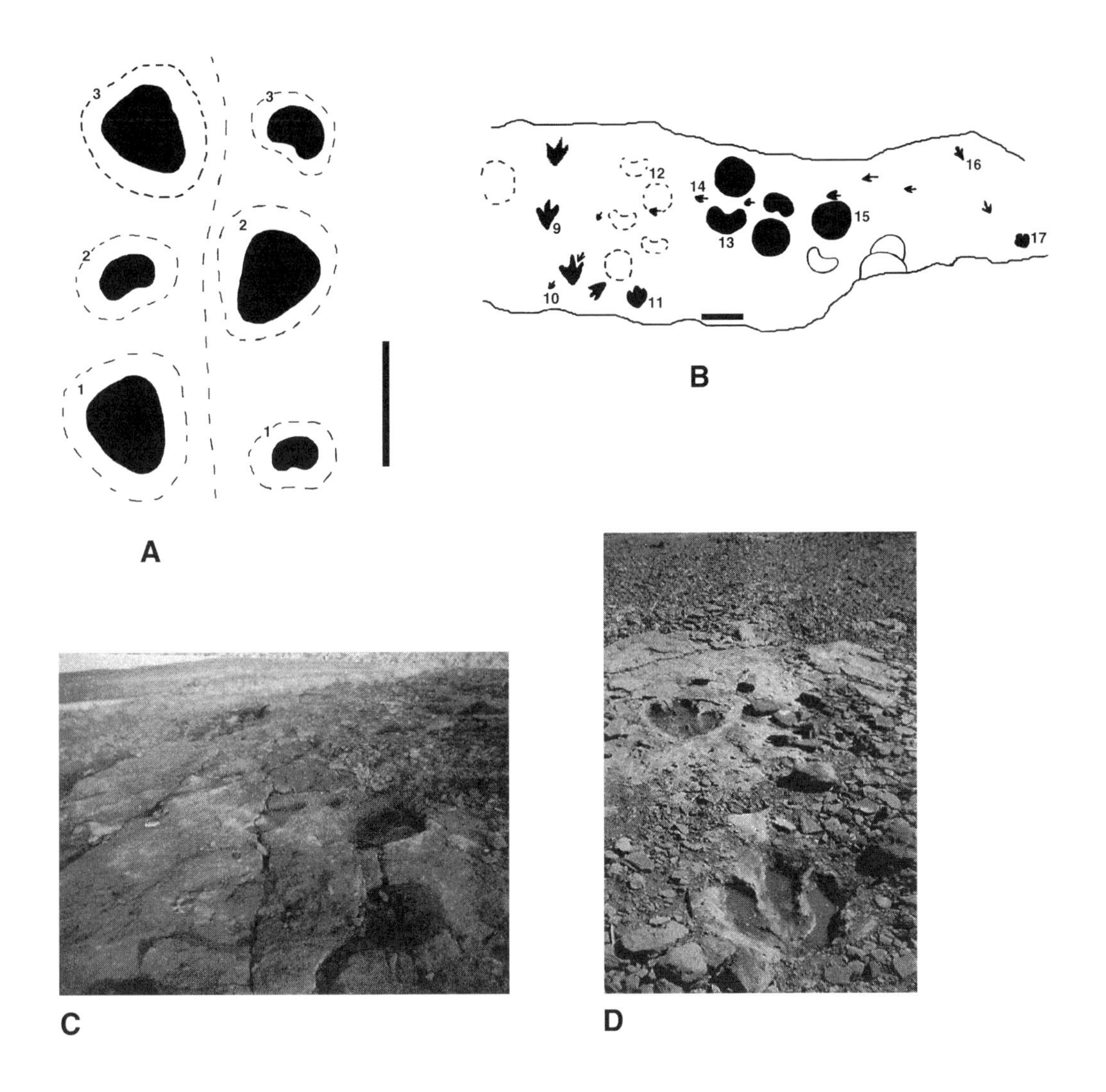

rounded. Toes II and IV diverge slightly from digit III, with an angle of 25° or less. Digit III has oval-shaped digital pads, which are separated from the "plantar" pad. Digit II and IV present an unique U-shaped pad. The preserved trackway shows directional changes in each of the three successive tracks (Calvo 1991).

Comments. It is the most abundant dinosaur track in the area. The holotype and three more trackways (MUCPv-66, -70, and -73) come from Península Nueva. The ichnospecies *Bonaparteichnium tali* (Calvo 1991) was interpreted has a nomen vanum and referred to *Limayichnus major* (Calvo 1999). Meyer (2000) has doubts about the ornithopod origin of these tracks; however, he does not support his alternative hypothesis with data. In fact, theropod and ornithopod trackways occur closely spaced at the Balneario Villa El Chocón, but they still show strong morphological differences. Although *Limayichnus* tracks are almost of the same size as the theropod tracks, the latter shows acute claw im-

Figure 13.5. (A) Drawing of the trackway of Sauropodichnus giganteus. *Scale bar = 100 cm. (B) Drawing of the* Sauropodichnus *#13 and #15, Abelichnus #9, and Limayichnus #11 tracks at Balneario de Villa El Chocón site. Scale bar = 100 cm. (C) Photograph of* Sauropodichnus giganteus *at Isla Cerrito del Bote site. (D) Photograph of* Limayichnus major *at Balneario de Villa El Chocón site.*

pressions, pace angulations of more than 170°, and well-defined, separated toes.

***Sousaichnium* Leonardi 1979**
***Sousaichnium monettae* Calvo 1991**

Holotype. MUCPv-71.

Locality and age. Península Nueva, Picún Leufú, Neuquén Province, Argentina (Fig. 13.1, locality 6). Candeleros Formation, Neuquén Group, Albian (Calvo 1991; Calvo and Salgado 1996) or lower Cenomanian (Leanza et al. 2004).

Diagnosis. Large ichnite, tridactyl, with the central digit a little longer than in *Sousaichnium pricei.* Digit IV longer than digit II. The angulation between digits II and III is greater than that between digits III and IV. Digit III and the "heel" are somewhat oriented internally, different than that of *S. pricei,* where it is oriented centrally. Digit III shows parallel borders and not tapering, as in *S. pricei.* The hypex is more acute than in *S. pricei. Sousaichnium monettae* is similar to *S. pricei* in the distribution of the digits, heel shape, and general contour (Calvo 1991).

Comments. It is the only record of this ichnospecies in the area.

Reptilia incertae sedis
***Ingenierichnus* Casamiquela 1964**
***Ingenierichnus sierrai* Casamiquela 1964**

Holotype. MLP 60-XI-31-3 cast.

Locality and age. La Vieja Quarry, Los Menucos, Río Negro Province, Argentina (Fig. 13.1, locality 3). Vera Formation, Upper Triassic (Kokogian et al. 2001).

Diagnosis. Small-sized animal. Foot and hand impressions poorly preserved. Tail or body impression undulating with a variable width, but wider than those of the footprints. Pace angulations 108° (Casamiquela 1964).

Comments. Hand-foot sets with a continuous, sinuous, and broad impression along the midline, probably left by the body of the animal while walking. The only possible interpretation appears to be for a quadrupedal lizardlike reptile with a sprawling gait (with overlap), perhaps with short but somewhat functional legs (Leonardi 1994).

Problematic Tracks

Casamiquela (1980) described *Neorotodactylus leonardii,* a footprint cast of a tetrapod assigned to Lepidosauria. The material comes from the Lower Jurassic of the Neuquén Province (Piedra Pintada locality). *Neorotodactylus leonardii* is characterized by having a pentadactyl track with finger V totally reverted. There is no claw impression. A reevaluation of the material indicates that it is not a tetrapod footprint (Leonardi 1994). The ichnogenus *Patagonichnornis venetiorum* (Leonardi 1994), found at the "Montón-I10" locality, Rio Negro Province (Fig. 13.1, locality 2), was described as such by Casamiquela in an unpublished paper, although it was illustrated by Leonardi (1994) under the same name. The material comes from rocks dated to the upper Maastrichtian. Casamiquela assigned this print to the Charadriiformes; however, Leonardi (1994) has doubt about such an assignment. However, the material is clearly a bird footprint. The ichnotaxa is not valid because it was not accurately presented in a scientific publication.

Casamiquela (1964, 45) mentioned the presence of some Triassic coelurosaurian footprints at Bajo Caracoles locality (Santa Cruz Province, Argentina). However, the material does not correspond with the strata present in the area (Casamiquela 1964; Leonardi 1994). Therefore, I believe that this material does not come from this locality.

Paleobiology

Recent studies of specimens from Cerrito del Bote Island (Ezequiel Ramos Mexía ichnofauna) (Calvo and Mazzetta 2004) provide more insight into the velocity, locomotive capacity, and body mass of dinosaurs.

All the trackways studied in the area show that the dinosaurs moved slowly, with a maximum speed of 6.5 km/h for a small theropod and a minimum of 3.2 km/h for a middle-sized sauropod (Calvo and Mazzetta 2004; Mazzetta and Calvo 2004). I have used the formula proposed by Alexander (1976): $v = 0.25\ g^{0.5} s^{1.67} h^{-1.17}$, where s and h are expressed in meters; v is expressed as meters per second, and g refers to gravity acceleration. Even though we do not know enough about the behavior of dinosaurs, data taken from the Cerrito del Bote site imply that these velocities are well below their locomotive capacities. This suggests that the dinosaurs moved slowly when they made the tracks—they were not running away from a predator, for example. At the Cerrito del Bote locality, the small theropod trotted instead of running. The close association and the same orientation of trackways shows that small dinosaurs moved in groups, probably indicating gregarious behavior.

The height of the pelvis for the sauropod would be at 2.72 m, and its weight would reach 20.4 tons (Calvo and Mazzetta 2004). In the same way, the estimation given for small theropods suggests that the body mass was 12 to 21 kg (Calvo and Mazzetta 2004). The body masses of the trackmakers at Cerrito del Bote were estimated by applying the concept of geometrical similitude (see Alexander [1985] for a detailed discussion of this).

Acknowledgments. I thank Rodolfo Coria, Zulma Gasparini, and Leonardo Salgado for allowing me to contribute to this volume. I also thank Dr. Christian A. Meyer and Dr. Martin Lockley for their critical reviews. Funding comes from National University of Comahue project T-021, and the companies Chevron-Texaco, Repsol-YPF, Pan American Energy, Duke Energy, Petrobras Energía, Total, and Skanska.

References Cited

Alexander, R. M. 1976. Estimates of speeds of dinosaurs. *Nature* 261: 129–130.

———. 1985. Body support, scaling, and allometry. In M. Hildebrand, D. M. Bramble, K. F. Liem, and D. B. Wake (eds.), *Functional Vertebrate Morphology,* 26–37. Cambridge, Mass.: The Belknap Press of Harvard University Press.

Alonso, R. N., and R. A. Marquillas. 1986. Nueva localidad con huellas de dinosaurios y primer hallazgo de huellas de aves en la Formación Yacoraite (Maastrichtiano) del Norte argentino. *Actas del IV Congreso Argentino de Paleontología y Bioestratigrafía* 2: 33–42.

Calvo, J. O. 1989. Nuevos hallazgos de huellas de dinosaurios en el Albiano-Cenomaniano de la localidad de Picún Leufú. Provincia del Neuquén, Patagonia, Argentina. *Resúmenes de las VI. Jornadas Argentinas de Paleontología de Vertebrados,* 68–70. San Juan.

———. 1991. Huellas Fósiles de dinosaurios en la Formación Río Limay (Albiano-Cenomaniano), Picún Leufú, Provincia del Neuquén, Argentina. (Ornithischia-Saurischia: Sauropoda-Teropoda). *Ameghiniana* 28 (3–4): 241–258.

———. 1997. La fauna de Dinosaurios saurópodos y de huellas de dinosaurios en El Chocón y zonas aledañas. In *Seminario Nacional de Turismo Científico,* 8–13. Editorial Fundación Siglo XXI.

———. 1999. Dinosaurs and other vertebrates of the Lake Ezequiel Ramos Mexía Area, Neuquén-Patagonia, Argentina. In Y. Tomida, T. H. Rich, and P. Vickers-Rich, (eds.), *Proceedings of the Second Gondwanan Dinosaur Symposium,* 13–45. Monograph 15. National Science Museum.

Calvo, J. O., and R. Coria. 1995. Huellas de dinosaurios en Neuquén. *Ciencia Hoy* 5 (29): 22–30.

Calvo, J. O., and M. G. Lockley. 2001. The first pterosaur tracks from Gondwana. *Cretaceous Research* 22: 585–590.

Calvo, J. O., and G. V. Mazzetta. 2004. Nuevos hallazgos de huellas de Dinosaurios en la Formación Río Limay (Albiano-Cenomaniano), Picún Leufú, Neuquén, Argentina. *Ameghiniana* 41 (4): 545–554.

Calvo, J. O., and J. Moratalla. 1998. First record of pterosaur tracks in Southern Continents. *Resúmenes de las III Reunión Argentina de Icnología y I Reunión de Icnología del MERCOSUR,* 8. Mar del Plata.

Calvo, J. O., and L. Salgado. 1996. Sauropod crossing: the Africa/South America connection. *Dinosaur Report* 1 (Summer): 4–6.

Calvo, J. O., R. Coria, and L. Salgado. 1990. Nuevas localidades con huellas de dinosaurios del Miembro Candeleros (Albiano-Cenomaniano) (Formación Río Limay, Grupo Neuquén) Provincia del Neuquén. *Ameghiniana* 26 (3–4): 241.

Calvo, J. O., K. Moreno, and D. Rubilar. 2000. First record of Lower Cretaceous dinosaur tracks (Sauropoda-Theropoda) in Río Negro Province, Patagonia, Argentina. In *31 International Geological Congress,* Río de Janeiro, Brazil. Abstract volume CD.

Casamiquela, R. M. 1960. El hallazgo del primer elenco (icnológico) Jurásico de vertebrados terrestres de Latinoamérica (noticia). *Revista de la Asociación Geológica Argentina* 15 (1–2): 5–14.

———. 1962. Sobre la presencia de un presunto Sauria aberrante en el Liásico del Neuquén (Patagonia). *Ameghiniana* 2(10): 183–186.

———. 1964. *Estudios Icnológicos:. Problemas y Métodos de la Icnología con Aplicación al Estudio de Pisadas Mesozoicas (Reptilia, Mammalia) de la Patagonia.* Colegio Industrial Pío IX. Buenos Aires.

———. 1966. Algunas consideraciones teóricas sobre los andares de los dinosaurios saurisquios: implicancias filogenéticas. *Ameghiniana* 4 (10): 373–385.

———. 1974. Nuevo material y reinterpretación de las icnitas mesozoicas (neotriásicas) de Los Menucos, Provincia de Río Negro (Patagonia). *Actas del 1er Congreso Argentino de Paleontología y Bioestratigrafía* 1: 555–580.

———. 1980. Nuevos argumentos en favor de la transferencia de *Rotodactylus,* icno-género reptiliano triásico, de los crocodiloides (Archosauria) a los lacertoides (insertae sedis). *Ameghiniana* 17 (2): 121–129.

Coria, R. A. 1989. Primer registro icnológico de dinosaurio en el Jurásico medio de la Argentina. In *Resúmenes de las VI Jornadas Argentinas de Paleontología de Vertebrados,* 107–108. San Juan.

Coria, R. A., P. J. Currie, D. Eberth, and A. Garrido. 2002. Bird footprints from the Anacleto Formation (Late Cretaceous), Neuquén, Argentina. *Ameghiniana* 39 (4): 453–464.

Currie, P. J. 1981. Bird footprints from the Gething Formation (Aptian, Lower Cretaceous) of northeastern British Columbia, Canada. *Journal of Vertebrate Paleontology* 1: 257–264.

Huene, F. von. 1931. Verschiedene mesozoische wirbeltierreste aus Südamerika. *Neues Jahrbuch für Mineralogie, Geologie und Paläontologie,* Beilagen-Band, Abt. B, 66 (B): 181–198.

Kokogian, D. A., L. A. Spalletti, E. M. Morel, A. E. Artabe, R. N. Martínez, O. A. Alcober, J. P. Milana, and A. M. Zavattieri. 2001. *Estratigrafía del Triásico Argentino.* In A. E. Artabe, E. M. Morel, and A. B. Zamuner (eds.), *El Sistema Triásico en la Argentina,* 23–54. La Plata: Fundación Museo de la Plata "Francisco Pascasio Moreno."

Leanza, H. A. 1999. *The Jurassic and Cretaceous Terrestrial Beds from Southern Neuquén Basin, Argentina.* Field Guide. INSUGEO 4.

Leanza, H. A., S. Apesteguía, F. E. Novas, and M. S. de la Fuente. 2004. Cretaceous terrestrial beds from the Neuquén Basin (Argentina) and their tetrapod assemblages. *Cretaceous Research* 25: 61–87.

Leonardi, G. 1994. *Annotated Atlas of South America Tetrapod Footprints (Devonian to Holocene).* Brazil: Companhia de Pesquisa de Recursos Minerais.

Lockley, M. G., S. Y. Yang, M. Matsukawa, F. Fleming, and S. K. Lim. 1992. The track record of Mesozoic birds: evidence and implications. *Philosophical Transactions of the Royal Society* 336: 113–134. London.

Manera de Bianco, T., and J. O. Calvo. 1999. Hallazgo de huellas de gran tamaño en el Triásico de Los Menucos, provincia de Río Negro, Argentina. In *Resúmenes de las XV Jornadas Argentinas de Paleontología de Vertebrados,* 16. La Plata.

Mazzetta, G. V., and J. O. Calvo. 2004. On the locomotory implications of the dinosaur ichnofauna at Ezequiel Ramos Mexía lake, Neuquén, Argentina–Patagonia. In *ICHNIA 2004,* 54. Trelew, Chubut, Argentina: First International Congress on Ichnology.

Mehl, M. G. 1931. Additions to the vertebrate record of the Dakota sandstone. *American Journal of Science* 21: 441–452.

Meyer, C. A. 2000. The Río Limay vertebrate ichnofauna (Cretaceous-Patagonia) revisited—evidence for a Cenomanian not Albian age of

Giganotosaurus. In *Abstracts of the 5th European Workshop on Vertebrate Paleontology,* 52–53. Karlsruhe, Germany.

Page, R., A. Ardolino, R. E. Barrio, M. de Franchi, A. Lizuain, S. Page, and D. S. Nieto. 1999. Estratigrafía del Jurásico y Cretácico del Macizo de Somún Curá, Provincias de Río Negro y Chubut. In R. Caminos (ed.), *Geología Argentina,* 460–488. Subsecretaría de Minería de la Nación, Buenos Aires.

Riccardi, A. C. and S. E. Damborenea. 1993. Léxico estratigráfico de la Argentina. *Asociación Geológica Argentina, Serie "B" (Didáctica y Complementaria)* 21.

14. Reptilian Faunal Succession in the Mesozoic of Patagonia

An Updated Overview

Zulma Gasparini, Leonardo Salgado, and Rodolfo A. Coria

Introduction

The first synopsis on Mesozoic tetrapods of South America, with a hypothesis of their origin and geographic distribution, was made by Bonaparte (1978, 1979). He successively enlarged this synthesis on the basis of further discoveries that were mainly from the Cretaceous of Argentina (Bonaparte 1986, 1992, 1996). Other, more recent contributions were mostly made by the authors of this book; those references, together with other complementary ones, are included in the previous chapters. The amount of new information published in the last 25 years, particularly information referring to the Mesozoic reptiles of Patagonia, cannot be passed over when systematic studies, phylogenetic analyses, and biogeographic hypotheses are undertaken.

The record of Mesozoic reptiles in Patagonia, with some 121 taxa recognized at the genus or species level (Appendix), comprises data from the Late Triassic (Calvo this volume; Coria and Cambiaso this volume; Salgado and Bonaparte this volume) to the uppermost Maastrichtian (Gasparini this volume), approximately 150 million years (GSA 1999). During this time, the distribution and physiography of seas and land were greatly modified, first by the breakup of Pangaea and then Gondwana, and next by the different geotectonic processes that affected Patagonia and defined the depocenters in which the studied herpetofauna is recorded. A report of these and

			Los Menucos Basin	El Deseado Basin	P. de Agnia Basin	Cañadón Asfalto Basin	San Jorge Basin	Neuquén Basin
Jurassic	Late	Tith						20 29 91 97 98 1 2 12 92
		Kim				42 61		
		Oxf					110-111-112	
	Middle	Clv				40 41 72 82		95 96
		Bth						18
		Baj						89 90 94
		Aal			39			
	Early	Toa						
		Plb						
		Sin						
		Het						
Triassic	Late	Rht						
		Nor	103-107 121	31 38				
		Crn						
	Middle	Lad						
		Ans						
	Early	Ol						
		In						

Figure 14.1. Distribution of Patagonian Triassic and Jurassic reptiles, according to basins and ages (each taxon is indicated with its respective number, as listed in the appendix).

other large geotectonic events that affected the paleorelief of Patagonia is synthesized and illustrated by Spalletti and Franzese (this volume).

Undoubtedly, such events, as well as other minor ones, acted directly on the environment, climate, and biota. The consequences of these processes on the diverse groups of both continental and marine reptiles were different. A new seaway has double consequences in the geographic distribution of the biota: on the one hand, it favors the dispersion of the marine organisms, and on the other hand, it creates a barrier to the continental forms. The case of the Patagonian reptiles of the Jurassic is illustrative in this sense.

The paleontological record is always both incomplete and strongly biased for several reasons, and this is particularly true of the record of Mesozoic Patagonian reptiles. First, not all the Patagonian Mesozoic basins potentially rich in reptiles have been equally surveyed. This is clearly reflected in the number of taxa recorded in these basins; the Neuquén Basin stands out in this regard (Figs. 14.1,

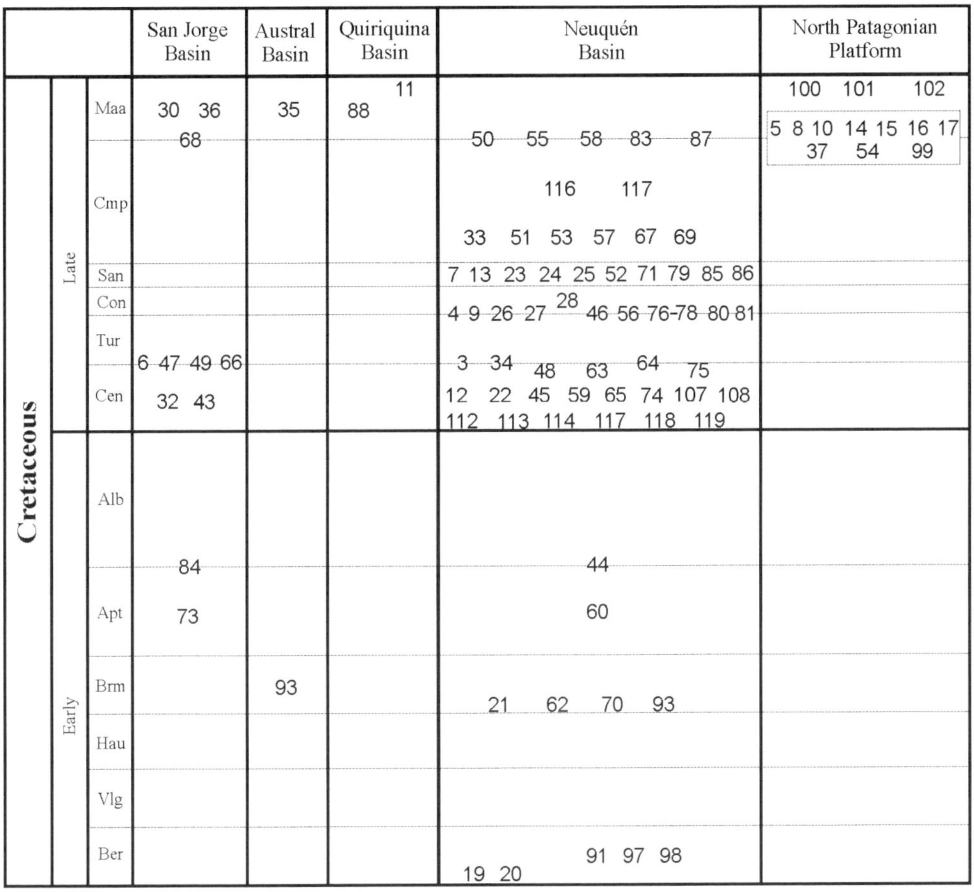

			San Jorge Basin	Austral Basin	Quiriquina Basin	Neuquén Basin	North Patagonian Platform
Cretaceous	Late	Maa	30 36	35	11 88		100 101 102
		Maa/Cmp	68			50 55 58 83 87	5 8 10 14 15 16 17 37 54 99
		Cmp				116 117 33 51 53 57 67 69	
		San				7 13 23 24 25 52 71 79 85 86	
		Con				28	
		Con/Tur				4 9 26 27 46 56 76-78 80 81	
		Tur					
		Tur/Cen	6 47 49 66			3 34 48 63 64 75	
		Cen	32 43			12 22 45 59 65 74 107 108 112 113 114 117 118 119	
	Early	Alb					
		Alb/Apt	84			44	
		Apt	73			60	
		Brm		93		21 62 70 93	
		Hau					
		Vlg					
		Ber				91 97 98 19 20	

Figure 14.2. Distribution of Patagonian Cretaceous reptiles, according to basins and ages. Each taxon is indicated with its respective number, as listed in the appendix.

14.2). In contrast, fieldwork in the Austral Basin has been scarce because of the difficulties of access to the sites and the high costs of exploration, and the result is scarce records compared with those from the Neuquén and San Jorge Basins. Marine reptile remains were mainly found in two areas: at the ingression of the Proto-Pacific over northwest Patagonia (the Neuquén Basin from the end of the Triassic to the Barremian), and at the transgression of the south Atlantic over a large part of this sector (North Patagonian Platform, during the middle Campanian to the Early Paleogene). Between both transgressions, a period of approximately 70 million years passed without marine sediments, and consequently there is no fossil record. In the Chilean Patagonia reptiles, mainly marine forms have been recorded in the Late Cretaceous, suggesting the potential of the findings if systematic fieldwork could be accomplished. Another bias to take into account is the body size of the recorded reptiles. Mid-size to large forms prevail (especially dinosaurs, with almost 50% of the total taxa at generic or specific level; see appendix). This is certainly related to diagenetic processes, but it is also because techniques to obtain microvertebrates are rarely applied.

Even with this bias, Patagonia has an outstanding record of fossil reptiles, and the evolution of Patagonian fauna can be broadly traced during most of the Mesozoic. Here, we aim to organize the information given in previous chapters of this volume in a synthesis ordered by time and by basin (Figs. 14.1, 14.2).

The history of the fauna cannot be separated from the history of the flora, which is most closely related to the paleoenvironmental evolution of the continent and region. For example (and even when there are not enough data to confirm it), some changes of the composition of the fauna of the saurischians along the Cretaceous of Patagonia have been related to certain turnovers in the flora, which in turn would have been produced by climatic and physiographic changes (Coria and Salgado 2005, and references therein).

The first researcher to approach the study of the Mesozoic continental tetrapod faunas of Argentina as a whole was Bonaparte (1979, 1986, 1996, and references therein). Later, Novas (1997) and especially Leanza et al. (2004) developed similar faunistic interpretations, the former for South America and the latter for the Cretaceous of the Neuquén Basin. Leanza et al. (2004, 64), in particular, proposed an arrangement of "assemblages," which were based on the tetrapods recorded in a single lithostratigraphic unit and a restricted area within the Neuquén Basin. Leanza et al. aimed to "provide the basis for developing a useful tool for addressing inter-regional correlations" (2004, 82). However, when the whole continental and marine herpetofauna of Patagonia (Argentina and Chile) is analyzed through the Mesozoic, these "assemblages" do not appear clearly.

Given the diversity of the Patagonian record and its temporal and geographic breadth, the main goal of this chapter is to provide an overview of the reptilian Mesozoic faunal succession. The arrangement is temporal, and the selected major faunal intervals indicate the moments with the highest number of recorded taxa. In this synthesis, some discontinuities can be observed in the record of the different groups, which are interpreted in terms of global and regional extinction, turnover, and diversification.

Major Faunal Intervals

Triassic

Unlike the record of Triassic reptiles from northwest Argentina (Bonaparte 1997; Caselli et al. 2001), that of Patagonia is less diverse, which is undoubtedly related to the comparatively scarce fieldwork (Calvo this volume; Coria and Cambiaso this volume; Salgado this volume) and to the relatively smaller Triassic exposures (Bonaparte 1978, 1997). The greatest reptile richness of the Patagonian Triassic is without doubt its paleoichnological diversity (six of the eight taxa at the generic level for the Triassic are ichnites; see appendix). A series of footprints found in rocks of northern Patagonia, in the area of Los Menucos, Río Negro (Vera Formation, Los Menucos Basin, Fig. 14.1; Spalletti and Franzese this volume, Fig. 2.3B), of the Late Triassic, suggests the presence of theriodonts, anomodonts, and other therapsids with uncertain affinities, in addition to chirotherids and prosauropods (Calvo this volume). The assemblage of ichnological evidence suggests a significant diversity of reptiles in the southwestern end of Pangaea. Likewise, the presence of prosauropods (*Mussaurus patagonicus*) is corroborated by partially complete skeletons both of adults and newborns (Salgado and Bonaparte this volume) found in rocks of the early Late Triassic (Laguna Colorada Formation, Deseado Basin, Fig. 14.1; Spalletti and Franzese this volume, Fig. 2.3B). Heterodontosaurid ornithischian dinosaurs (*Heterodontosaurus* sp.) are also present in the southern tip of Argentina in that same formation (Coria and Cambiaso this volume).

Except for the footprints of coelurosaurs, the clades that the reptilian ichnofauna has been referred to are also present in the Late Triassic of the Argentine northwest and clearly belong to Pangaean groups (Bonaparte 1997). Thus, the heterodontosaurids are also present in the Early Jurassic of South Africa and in the Early Cretaceous of Europe (Báez and Marsicano 2001; Norman et al. 2004; Coria and Cambiaso this volume). Prosauropods also had a worldwide distribution during the Triassic (Galton and Upchurch 2004). In sum, the assemblage of ichnological and bony material evidence suggests that a significant diversity of reptiles with strong Pangaean affinities was present in the southern end of South America.

Early Jurassic

Exploration that sought continental and marine reptiles of the Early Jurassic in Patagonia has been scarce. This has influenced the limited knowledge we currently have of the Patagonian (and South American) herpetofauna of those times, in contrast to those of the Northern Hemisphere, where a spectacular diversification, particularly of diapsids, is observed (see Weishampel et al. 2004).

The single continental reptile of the Lower Jurassic (although it may correspond to the lowest part of the Middle Jurassic) is the basal eusauropod *Amygdalodon patagonicus* (Salgado and Bonaparte this volume), found in Pampa de Agnia, Chubut (Pampa de Agnia Basin, Fig. 14.1; Spalletti and Franzese this volume, Fig. 2.4B). The marine reptiles found in the Neuquén Basin (Spalletti and Franzese this volume, Fig. 2.4B) are limited to vertebrae of a thalattosuchian crocodile of the Lower Jurassic of Mendoza (Gasparini 1981) and some remains of indeterminable ichthyosaurs of Mendoza and Neuquén (Fernández this volume). In contrast, both thalattosuchians and ichthyosaurs of the Early Jurassic of the Northern Hemisphere are numerous and diverse (Bardet 1995; McGowan and Motani 2003).

Middle Jurassic (Aalenian-Callovian)

After the Middle Jurassic, the record of Patagonian reptiles increases significantly, but Pangaean lineages still prevail, at least among dinosaurs (basal eusauropods, basal tetanurans). A single form of continental reptile is recorded in the Middle Jurassic of the Neuquén Basin from the Aalenian of Mendoza (Spalletti and Franzese this volume, Fig. 2.5A). It is a fragment of a sacrum of a sauropod dinosaur similar to *Amygdalodon*, redeposited in marine sedimentites (Salgado and Gasparini 2004). All the other continental reptiles of the Middle Jurassic are from the scarce exposures of Cañadón Asfalto Basin (Fig. 14.1; Spalletti and Franzese this volume, Fig. 2.5B). In the Callovian of Chubut the basal eusauropod *Volkheimeria* and the Cetiosauridae *Patagosaurus* (Salgado and Bonaparte this volume) have been found. In these levels, the oldest tetanuran theropods of Patagonia were also found: *Condorraptor* (Rauhut 2005), and the putative basal spinosauroid *Piatnitzkysaurus* (Holtz et al. 2004a). The large theropod fauna of the Jurassic of Patagonia, which is mainly composed of basal tetanurans, contrasts markedly with the faunas of large nontetanuran theropods that will characterize the uppermost Cretaceous of Patagonia (the Campanian-Maastrichtian interval). The record of tetanurans is one of the outstanding points of the Middle Jurassic of Patagonia and basically coincides with the first records of the group worldwide, because the single Tetanurae before the Middle Jurassic is *Cryolophosaurus*, from the Lower Jurassic of Antarctica (Holtz et al. 2004b). Although the tetanurans could have been present since the Norian (Arcucci and Coria 2003), they certainly spread globally during the Middle Jurassic.

In addition to the dinosaurs, the paleoherpetologic list of the Cañadón Asfalto Basin, which has been intensively explored in recent years, includes pterosaurs (Rauhut et al. 2001; Unwin et al. 2004), pancryptodiran turtles (de la Fuente this volume), and sphenodonts (Albino this volume), all of them currently under study.

All the marine reptiles of the Middle Jurassic of Patagonia were found in the Neuquén Basin (Fig. 14.1) and comprise the Aalenian–early Callovian lapse (Spalletti and Franzese this volume, Figs. 2.5A, 2.5B). Those of the Aalenian–early Bajocian lapse are conspicuous members of Gondwana and have particular significance because, except for scarce remains of Europe (Bardet 1995; McGowan and Motani 2003) and Chile (Gasparini et al. 2000), they are completely unknown in other parts of the world. The oldest is an ophthalmosaurian ichthyosaur from the Aalenian of Mendoza, which is also the oldest record of this clade worldwide (Fernández this volume). Likewise, remains of the early Bajocian show the diversity of the marine herpetofauna during that time in the Southern Hemisphere, represented by the rhomaleosaurid pliosaur *Maresaurus coccai* (Gasparini this volume), the Stenopterygiidae ichthyosaur *Stenopterygius cayi* (Fernández 1994, this volume), and the ophthalmosaurian *Mollesaurus periallus* (Fernández this volume). In early Bajocian rocks in Chile, fragmentary remains of ichthyosaurs and plesiosaurs were found, as well as part of a skull referred to the metriorhynchid *Metriohrynchus* sp., which is the oldest record of the genus (Gasparini et al. 2000).

In the late Bathonian, the most complete metriorhynchid of the Middle Jurassic is recorded in the Neuquén Basin, which was referred to *Metriorhynchus* aff. *M. brachyrhynchus* (Pol and Gasparini this volume). Finally, in early Callovian rocks, also of the Neuquén Basin (Spalletti and Franzese this volume, Fig. 2.5B), fragmentary remains of pliosauroid and plesiosauroid plesiosaurs were found (cf. *Muraenosaurus;* cf. *Cryptoclidus*) (Gasparini this volume).

The marine reptiles of the Middle Jurassic of Patagonia (and north-central Chile) belong to the same genera or to closely related forms recorded in the European Tethys (Bardet 1995; O'Keefe 2001; McGowan and Motani 2003), suggesting paleobiogeographic relationships between both areas. In this regard, the Hispanic Corridor was proposed as the most parsimonious way for this interchange, which since the Oxfordian became the Caribbean Corridor that connected the Western Tethys with the Eastern Pacific (Gasparini 1992; Iturralde-Vinent 2003; Gasparini and Iturralde-Vinent 2006).

Late Jurassic (Oxfordian-Tithonian)

The early Late Jurassic of Santa Cruz (San Jorge Basin; Fig. 14.1; Spalletti and Franzese this volume, Fig. 2.6A) yielded an important association of reptile and mammal footprints (Calvo this volume). Some of these footprints, undoubtedly of small theropods (*Wildeichnus navesi, Sarmientichnus scagliai, Delatorrichnus goyenechei*), have been referred to coelurosaurs (Calvo this volume). However, most of those assignments have to be confirmed, because according to current phylogenetic patterns, not all the small theropods belong to this group. If they are indeed coelurosaurs, they would be the oldest ones of Gondwana and would increase the tetanuran diversity of this lapse. Such paleoichnological association, even with putative coelurosaurs, has been described for the Jurassic-Cretaceous of São Paulo State (Bonaparte 1979, Fernándes 2005).

From the Late Jurassic to the pre-Turonian Cretaceous, elements mostly of the late Pangaean lineages are recorded (among sauropod dinosaurs, dicraeosaurids, basal titanosauriforms, and basal diplodocoids), as well as groups that originated and diversified in Gondwana (early Gondwanan lineages, such as the rebbachisaurids, basal titanosaurians, basal allosauroids [carcharodontosaurids], and basal abelisauroids among dinosaurs; and basal mesoeucrocodylians and notosuchians among crocodiles) (Leanza et al. 2004). Although in the Neuquén Basin the continental Jurassic levels are scarce and not well surveyed relative to marine ones, some important remains have been found. For example, in the Tordillo Formation, north of Neuquén, fragments of femur, tibia, and fibula that belong to a lineage of basal eusauropods that would have lasted in Gondwana until the Cretaceous (on the basis of the record of the sauropod *Jobaria* in the Lower Cretaceous of Niger) have been found (García et al. 2003).

In Chubut, Kimmeridgian-Tithonian rocks (Cañadón Asfalto Basin; Spalletti and Franzese this volume, Fig. 2.6A) yielded the neosauropod dinosaurs *Brachytrachelopan mesai* and *Tehuelchesaurus benitezi* (Salgado and Bonaparte this volume). The latter are the single Patagonian representatives of the neosauropod radiation of the Late Jurassic (Wilson and Sereno 1998). The records of *Brachytrachelopan* (Upper Jurassic, Cañadón Asfalto Basin) and *Amargasaurus* (Lower Cretaceous, Neuquén Basin) show the persistence in Patagonia of a late-Pangaean lineage of diplodocoids: the Dicraeosauridae. These sauropods are not recorded in levels after the Barremian, and they probably became extinct before the end of the following interval (Salgado and Bonaparte this volume).

The Patagonian marine reptiles of the Late Jurassic were found mainly in Tithonian rocks of the Neuquén Basin (Fig. 14.1; Spalletti and Franzese this volume, Fig. 2.6A), although in some localities of this basin these same genera are recorded in the Berriasian, demonstrating that they cross the J-K boundary (Spalletti et al. 1999; Gasparini 2003; Gasparini and Fernández 2005). Numerically, ophthalmosaurian ichthyosaurs, such as the gigantic *Caypullisaurus bonapartei* and *Ophthalmosaurus* sp., prevailed (Fernández this volume), but other offshore predators were also present. Such is the case of the pliosaurids *Pliosaurus* sp. and *Liopleurodon* sp. (Gasparini this volume) and the metriorhynchids *Geosaurus araucanensis* and *Dakosaurus andiniensis* (Gasparini et al. 2006; Pol and Gasparini this volume). Likewise, the most ancient marine turtles were present in the Proto–Eastern Pacific, such as the eucryptodiran *Neusticemys neuquina* and the panpleurodiran *Notoemys laticentralis* (de la Fuente this volume). Over the Tithonian seas of the Neuquén Basin flew different pterodactyloids, one of them referred to *Herbstosaurus pigmaeus,* and other assigned to a Ctenochasmatoidea (Codorniú and Gasparini this volume; Codorniú et al. 2006).

Many of the genera of Tithonian marine reptiles of Patagonia are also represented in the Northern Hemisphere, mainly in the European Tethys (*Ophthalmosaurus, Pliosaurus, Liopleurodon, Geosaurus, Dakosaurus*), suggesting close biogeographic relationships that were probably favored by the opening of new seaways such as the Caribbean Corridor (Iturralde-Vinent 2003) and the Weddell Sea (Spalletti and Franzese this volume), which connected the south of the Eastern Pacific with the Indian Ocean.

Early Cretaceous–Early Late Cretaceous (Berriasian-Cenomanian)

The oldest reptiles of the Patagonian Cretaceous are marine and were found in the Neuquén Basin. In the Berriasian, the genera *Geosaurus, Dakosaurus* (crocodyliforms), *Liopleurodon* (pliosaurid), and *Caypullisaurus* (ohpthamosaurian ichthyosaur) are recorded, the same as in the Tithonian; consequently, there is no evidence of extinction of pelagic offshore forms in the Jurassic-Cretaceous transition. Later, in coincidence with the shallowing of the Neuquén Basin (Spalletti and Franzese this volume, Fig. 2.6B) in the southern sector, the only marine reptiles of the Valanginian-Hauterivian are the elasmosaurid plesiosaurs (Lazo and Cichowolski 2003). In the Austral Basin (Spalletti and Franzese this volume, Fig. 2.6B), there are records of Barremian ichthyosaurs in Chile (Fernández and Aguirre-Urreta 2005) and pteranodontoid pterodactyloids in the lower Barremian of western Santa Cruz (Codorniú and Gasparini this volume).

Since the Barremian, when the Proto-Pacific definitively withdrew from the Neuquén Basin (Spalletti and Franzese this volume, Fig. 2.7A), the basin yielded the most complete example of the continental Cretaceous herpetofauna of South America. Only since this moment can the continental faunistic lists of the Neuquén Basin be compared with other Patagonian basins (Austral and San Jorge Basins).

As we said before, Bonaparte was the first to accomplish a systematic study of the tetrapod successions for the Mesozoic of Patagonia (1979, 1986, 1996, and references therein). Recently, Leanza et al. (2004) recognized a series of assemblages for the Cretaceous tetrapods of the Neuquén Basin. Some of the elements these authors report as part of those local faunas or assemblages are recorded also in other basins of Patagonia. Thus, since the Lower Cretaceous, especially since the Aptian, the associations recorded in the Neuquén Basin can be compared with other depocenters of Patagonia on the basis of the presence of common elements.

The main localities that yielded remains of continental reptiles of this interval Berrisasian-Cenomanian are La Amarga (Barremian of Neuquén), Lohan Cura (Aptian-Albian, Neuquén), Ocho Hermanos (Cenomanian, Chubut), Picún Leufú-El Chocón, Plaza Huincul, Cortaderas (Cenomanian of Neuquén), and La Buitrera (Cenomanian, Río Negro).

In the Neuquén Basin, the Berriasian-Cenomanian period includes the Amargan (La Amarga Formation), Lohancuran (Lohan Cura Formation), and Limayan (Río Limay Subgroup) assemblages of Leanza et al. (2004). In the San Jorge Basin, Cretaceous elements of this lapse are recorded in the Matasiete Formation (Martínez et al. 1989) and Cerro Barcino Formation (Apesteguía and Giménez 2001), and in the Lower Member of the Bajo Barreal Formation (Powell et al. 1989; Lamanna et al. 2001; Martínez et al. 2004a,b), where some taxa similar to those of the Limayan assemblage of the Neuquén Basin are recorded (with rebbachisaurid dinosaurs and other basal diplodocoids, titanosauriforms and basal titanosaurs, and carcharodontosaurs).

The end of the Berriasian-Cenomanian period in Patagonia was characterized by a marked faunistic turnover that affected many continental forms. This is relatively well documented for the saurischians of the Neuquén Basin and probably occurred globally for this group (Coria and Salgado 2005). Diplodocoids (rebbachisaurids and other forms of basal diplodocoids) are recorded in all the pre-Turonian Cretaceous, but are absent in the next period (Turonian-Santonian). The record of the sauropod *Amargasaurus* in the Barremian of the Neuquén Basin shows the persistence in the Lower Cretaceous of Patagonia, at least until the Barremian, of a lineage that was already present in the Tithonian of the Cañadón Asfalto Basin (*Brachytrachelopan*). In this sense, just as with the marine forms, a continuity between the Late Jurassic and the Early Cretaceous in Patagonia is seen, taking into account different depocenters, although such continuity is not as well documented as in the case of the marine reptiles.

Among dinosaurs, the basal diplodocoids are recorded in the San Jorge Basin (Sciutto and Martínez 1994) and in the Neuquén Basin (Calvo and Salgado 1995; Leanza et al. 2004). This period attested also the appearance and basal diversification of the titanosaurs. Basal forms of this group, such as *Chubutisaurus,* the sauropod of the Matasiete Formation, and *Epachthosaurus* in the San Jorge Basin (del Corro 1975; Martínez et al. 1989, 2004b), *Andesaurus* and the sauropod of the La Antena Quarry (Villa El Chocón) in the Neuquén Basin (Calvo and Bonaparte 1991; Calvo 1999; Simón and Calvo 2002), will survive up to the end of the Berriasian-Cenomanian interval.

Within the theropod dinosaurs, the first Abelisauria, a group that probably originated in Gondwana (Coria this volume), are recorded. In the Neuquén Basin, remains of abelisaurs are recorded in the La Amarga Formation (the noasaurid *Ligabueino andesi,* Bonaparte 1996), the Candeleros Formation (*Ekrixinatosaurus,* Calvo et al. 2004), and the Huincul Formation (the neoceratosaur *Ilokelesia,* Coria and Salgado 1998). In Chubut, in turn, in levels of the Bajo Barreal Formation, the abelisaur *Xenotarsosaurus bonapartei* is recorded together with other remains of indeterminate species (Martínez et al. 1986, 2004a,b; Lamanna et al. 2002).

Another Gondwanan group of large theropods, the carcharodontosaurids, is recorded, in the San Jorge Basin, *Tyrannotitan* (Aptian) (Novas et al. 2005b), and in the Neuquén Basin, *Giganotosaurus* (early Cenomanian) (Coria and Salgado 1995) and *Mapusaurus* (Coria and Currie 2006). All the carcharodontosaurids of the Neuquén Basin belong to the clade of the giganotosaurines. These are not likely to have spread over Africa, even when a connection was possible (Sereno et al. 2004). The giganotosaurs, as well as the remaining carcharodontosaurids, apparently became extinct by the end of this interval. Although carcharodontosaurid-type teeth are recorded in the next interval (Veralli and Calvo 2004), it is not certain whether they actually belong to carcharodontosaurids (Coria this volume).

Other groups of continental reptiles were present in Patagonia in this lapse, besides the dinosaurs. The oldest ones are the basal crocodyliform *Amargasuchus* and the pterodactyloids of the Barremian in the Neuquén Basin. It is not until the end of the period (Cenomanian) that the first notosuchians (*Araripesuchus*) (Pol and Gasparini this volume), chelid turtles (*Prochelidella*) (de la Fuente this volume), iguanids (Apesteguía et al. 2005), and the sphenodontid *Kaikaifilusaurus* (Apesteguía and Novas 2003; Simón and Kellner 2003) are recorded.

Turonian-Santonian

This interval is represented in the Neuquén Basin in the associations of the Neuquenian and part of the Coloradan (only the forms recorded in the Bajo de la Carpa Formation) of Leanza et al. (2004). In the remaining Patagonian basins, it is not yet clear which formations represent this interval–probably the Barreal Formation in some sectors (J. Rodríguez personal communication 2005). But the knowledge of the herpetofauna of this interval comes almost exclusively from the Neuquén Basin (Fig. 14.2). In the Austral Basin, south of Sehuén River (Santa Cruz Province) at Mata Amarilla locality, and south of Lago Viedma, at Cerro de los Hornos, a rich fossil reptile fauna has been recorded that is probably Coniacian or Santonian in age, and that is currently under study (Goin et al. 2002; Novas et al. 2002).

Several localities of this period, especially the Turonian-Coniacian of the Neuquén Basin (Portezuelo Formation), have been intensively studied in recent years, the most outstanding of which are Sierra del Portezuelo (Novas and Puerta 1997; Novas 1998; Novas and Pol 2005) and the Futalognko Site at Lago Los Barreales (Calvo et al. 2002). Some of the most spectacular findings of dinosaurs and other reptiles of recent years were found here. The last stages of this interval are undoubtedly represented by the Bajo de la Carpa Formation, Neuquén Province. Remains of this unit have been known since the first discoveries of Santiago Roth by the end of the nineteenth century (Salgado this volume).

During the interval Turonian-Santonian, South America became progressively isolated from other continents. Accordingly, the forms recorded in Patagonia belong to Late Gondwanan lineages, groups that have evolved and diversified after the breakup of Gondwana (Bonaparte 1996; Novas 1997; Leanza et al. 2004).

Concerning dinosaurs, the derived titanosaurs (or titanosauroids sensu Salgado 2003) diversified after the extinction of the basal titanosaurs (andesauroids, epachthosaurines, and like forms) and the diplodocoids. The latter occurred by the end of the Cenomanian; in the Neuquén Basin, the time frame is represented by the top of the Huincul Formation, and in the San Jorge Basin by the base of the Upper Member of the Bajo Barreal Formation. In other continents (clearly North America), and unlike what is observed in Patagonia, the decline of sauropods by the end of the Early Cretaceous resulted in the diversification of ornithischians (Bakker 1978; Coria and Salgado 2005).

Bonaparte (1979) was the first to see that in Patagonia, sauropods with amphiplatyan caudal vertebrae (recorded in the Lower Cretaceous) are replaced (in the Upper Cretaceous) by others with procoelous caudals (or titanosauroids sensu Salgado 2003). Bonaparte attributed the flourishing of Titanosauridae (sensu Powell 2003) to the extinction of those "pre-Senonian sauropods" of amphiplatyan caudals. Powell et al. (1989), in turn, recorded an assemblage of elements of both associations (sauropods with amphiplatyan caudals on the one hand, and procoelous on the other) in a single lithostratigraphic unit (the Bajo Barreal Formation in the San Jorge Basin), although they were certainly found in different levels. Today, we know that those sauropods with procoelous caudals—"pre-Senonians," as named by Bonaparte (1979, 234), or "non-titanosaurid" according to Powell et al. (1989, 170)—in fact belong to two different lineages of basal diplodocoid and basal titanosaur sauropods, respectively (Lamanna et al. 2001; Salgado and Coria 2005).

Examples of titanosauroid dinosaurs of this interval are *Mendozasaurus, Rinconsaurus,* and *"Antarctosaurus" giganteus.* Salgado and Bonaparte (this volume) recognize in addition the persistence of a lineage of titanosaurs with amphiplatyan caudals.

Within the theropod dinosaurs, the basal abelisauroids seem to have lasted during the Turonian-Santonian interval as the single members of the large continental predators of the Lower Cretaceous, facing the extinction of carcharodontosaurs of the previous interval. Although a few remains of this age are known (the abelisauroid *Bayosaurus pubica,* for example; Coria this volume), the abelisaurs are abundant in the following interval (Campanian-Maastrichtian). Abelisaurs from the Turonian-Santonian lapse probably shared the role of large predators with the large tetanuran theropods: coelurosaurs and the middle-sized dromaeosaurid-related forms (*Unenlagia* and *Neuquenraptor*). Small alvarezsaurid theropods present in this period are likely to have lasted until the following interval (Coria et al. personal communication).

During the Turonian-Santonian lapse of Patagonia, other group of reptiles were diversified in addition to the dinosaurians. Such is the case of the chelid turtles (*Prochelydella, Lomalatachelys*) and Podocnemidoidae (*Portezueloemys*) (de la Fuente this volume), notosuchians (*Notosuchus, Comahuesuchus, Pehuenchesuchus, Cynodontosuchus*), and peirosaurids (*Lomasuchus, Peirosaurus*) (Pol and Gasparini this volume) and the first ophidians (*Dinilysia*) (Albino this volume). The oldest birds of Patagonia are also recorded, with two lineages already well differentiated: the flying enantiornithines (*Neuquenornis*) and the flightless basal ornithuromorphs (*Patagopteryx*).

There is no record of marine herpetofauna during the Turonian-Santonian lapse in Patagonia.

Campanian-Maastrichtian

In the Neuquén Basin, this span of time encompasses the Anacleto Formation (part of the continental tetrapod assemblage included in the Coloradan) and the Allen Formation (Allenian assemblages) (Leanza et al. 2004), besides the marine Jagüel Formation. In the San Jorge Basin, the continental levels of the Campanian-Maastrichtian were deposited in the upper part of the Bajo Barreal Formation and the Laguna Palacios Formation. In the Austral

Basin, the deposits of this time belong to the continental Pari Aike Formation (Novas et al. 2004, 2005a). Outside the area of these basins, the taxa of this interval are those recorded in the Allen and Jagüel Formations (central and south Río Negro Province), Los Alamitos Formation (southeast of this province), and Angostura Colorada and Coli-Toro Formations (south-southwest of the province).

According to Leanza et al. (2004), during this period, in the Neuquén Basin, the taxa of continental tetrapods recorded in the lower Campanian (Anacleto Formation) would form a single assemblage (Coloradan) with the taxa recorded in the underlying Bajo de la Carpa Formation, Santonian in age. However, in this chapter, we gather all the Campanian-Maastrichtian continental reptiles in a single interval, because they basically belong to the same taxa, as well as forms that would have entered from North America since the Middle Campanian.

The events characterizing this interval are, in the case of sauropod dinosaurs, the appearance of the Saltasaurini (only in the Neuquén Basin) and the expansion and diversification of the Aeolosaurini (Neuquén and San Jorge Basins, Salgado and Coria 1993; Casal et al. 2002; Powell 2003). These latter are also recorded in the Campanian-Maastrichtian of Brazil (Kellner and Azevedo 1999), suggesting an expansion of their area of distribution during the Campanian. Among the theropods, both in Northern and Central Patagonia, the abelisaurs seem to definitively dominate as large predators (the megaraptors disappear from the record), with the Abelisaurinae and Carnotaurinae recorded first. According to the record, only the tetanuran *Quilmesaurus* (Allen Formation in the Neuquén Basin) would have coexisted with large abelisaurs during this period. But in Laurasia, the large tetanurae do not disappear but rather multiply, as do the ornithischians in the case of herbivorous forms.

Other dinosaur groups appear in the Patagonian record as the hadrosaurs and ankilosaurs (Coria and Cambiaso this volume). The former undoubtedly come from the Northern Hemisphere, probably North America, through an array of islands emerged in the present Caribbean Sea. In this regard, Iturralde-Vinent and McPhee (1999) and Iturralde-Vinent (2003) point out that this connection occurred between the middle Campanian and the early Maastrichtian, approximately 5 Ma, enough time to permit an important faunistic interchange. Hadrosaurs are recorded in the San Jorge Basin (with *Secernosaurus*) and in the North Platform (with *Kritosaurus* and an indeterminate genus of Lambeosaurinae) (Coria and Cambiaso this volume). There are no records in the Austral Basin to date, although the remains of this group of ornithopods found in Antarctica clearly indicate that by the end of the Cretaceous, these forms already inhabited Patagonia at least north of the Antarctic Peninsula. No certain records of ceratopsids have been reported so far for Patagonia; however, they were important in the faunas of the Late Cretaceous of North America (Coria and Cambiaso this volume, where a discussion on the status of the alleged ceratopsian *Notoceratops bonarelli* is presented). Another group recorded in the Campanian-Maastrichtian of Patagonia is that of the dromeosaurid theropods (Novas et al. 2003), which were very abundant in the Cretaceous of North America. However, they are also recorded in levels that correspond to the previous interval (Novas and Pol 2005). Concerning the Patagonian and Antarctic ankylosaurs, it is not certain that they have Laurasian lineages (Salgado and Gasparini 2006). In sum, the importance of the incoming North American reptiles, beyond the abundant record of several different species of hadrosaurs, is still uncertain.

As stated above, the forms recorded in the early Campanian—at least those for the Neuquén Basin—before the settlement of the connection with North America (found in the Anacleto Formation) seem to be the same (even at the genus level, as the case of the sauropod *Neuquensaurus* [Salgado unpublished data]) as the following continental formations (Allen Formation of the Neuquén Basin) because a turnover or extinction caused by the entering of northern forms was not seen.

From the middle Campanian to the Maastrichtian-Paleogene, a South Atlantic transgression related to the subsidence of wide areas of the austral portion of South America partially covered Patagonia, forming a large archipelago with lagoons, open embayments, and marine environments. This ingression of the Atlantic covered several continental depocenters such as the Neuquén, San Jorge, and Austral Basins (Spalletti and Franzese this volume, Fig. 2.8B), but their deposits surpassed the limits of the previous basins. (In Fig. 14.2, when taxa of this interval have been recorded in sedimentite outcroppings in the mentioned basins, they are included in the column of the corresponding basin, but when the record is from a locality outside these limits, it is included in the column of North Patagonian Platform.)

Most of the reptiles of the Campanian-Maastrichtian transgression were found in the first three Mesozoic basins—Neuquén, San Jorge, and Austral—and were the continental forms (although littoral) that prevailed in both amount and taxonomic diversity (Fig. 14.2). Likewise, it is noteworthy that all these continental forms, as well as other formerly associated marine ones, belong to the Campanian and Campanian–early Maastrichtian (Bonaparte 1987; González Riga 1999; Gasparini et al. 2001; Martinelli and Forasiepi 2004), whereas those found in uppermost Maastrichtian rocks are, to date, exclusively marine reptiles (Gasparini et al. 2003). Accordingly, in the Quiriquina Basin, in south-central Chile, and also in the late Maastrichtian (Stinnesbeck 1986), plesiosaurs, mosasaurs, and marine turtles and birds have been found.

The cheloniofauna of the lowlands of Trapalcó and Santa Rosa is similar to the fauna described for the Los Alamitos and La Colonia Formations, and to other lithostratigraphic units of the Upper Cretaceous of Patagonia that are exposed in areas outside of previous continental basins (de la Fuente this volume).

With regard to the marine reptiles of the Campanian and Campanian–early Maastrichtian, only plesiosaurs and mosasaurs have been found in the Argentinean Patagonia, along with marine turtle remains found above the K-P boundary in the Province of La Pampa (de la Fuente and Casadío 2000). Generally, disarticulated remains are most common, in accordance with coastal or mixed environments, and policotylids (*Sulcusuchus erraini*) and elasmosaurids have been found (Gasparini this volume).

The reptiles of the uppermost Maastrichtian that have been preserved in strictly marine environments are generally more complete, and they belong to elasmosaurids (*Aristonectes parvidens, Mauisarus* sp., *Tuarangisaurus? cabazai*) (Gasparini this volume) and mosasaurs of at least three different species: *Plioplatecarpus* sp., *Mosasaurus* cf. *M. hoffmani,* and an indeterminate species of Mosasaurinae (Gasparini et al. 2005b; M. Fernández personal communication 2005). In Chilean Patagonia, also in late Maastrichtian sedimentites, the following taxa have been recorded: the elasmosaurids *Aristonectes* sp., cf. *Mauisaurus;* mosasaurs; eucryptodiran pancheloniid turtles (de la Fuente this volume); and modern birds of the Family Gaviidae, or loons (Chiappe this volume). The distribution of elasmosaurids suggests that these pelagic reptiles moved both along the southwest of the Proto-Pacific and the south of the South Atlantic, reaching the north of the Antarctic Peninsula through the Weddell Sea (Gasparini et al. 2003).

Conclusions

Evidence yielded by the analysis of the systematic chapters of this volume (chapters 3–12) suggests a rough correspondence among some geodynamic macroevents that affected the distribution of seas and lands during the Mesozoic, particularly in Patagonia (Spalletti and Franzese this volume), with the succession of their herpetofauna. Likewise, local extinction seems to be coincident with the evolution of each basin, as well as some faunistic turnovers–particularly the one occurred during the Middle-Late Cretaceous in the Neuquén Basin, where currently the largest number of Mesozoic Patagonian taxa is recorded.

1. Pangaea predominated during the Triassic and most part of the Jurassic. Accordingly, the Late Triassic herpetofauna of Patagonia is composed of clades of Pangaean distribution (see Holtz et al. 2004b, figs. 27.6, 27.7). This is the case of the theriodonts, anomodonts, chirotherids (based on their ichnological record), prosauropods, and heterontosaurid ornithischians (bony record). This assemblage demonstrated a high diversity by the end of the Triassic in the southwestern end of Pangaea. Interestingly, in Patagonia and the rest of South America, none of these groups crosses the Triassic-Jurassic boundary as they do in other continents (Galton and Upchurch 2004; Norman et al. 2004; Marsicano et al. 2005). However, as already discussed, it has to be recognized that little has been explored in the Lower Jurassic of South America; hence, at first sight, this can appear biased.

2. According to recent paleogeographic reconstruction (Blakey 2001), the definitive break of Pangaea into two supercontinents, Laurasia and Gondwana, occurred in the Late Jurassic. However, some events anticipated that separation, such as the Hispanic Corridor that intermittently connected the Central Atlantic with the Eastern Pacific up to its definitive opening in the Oxfordian (Iturralde-Vinent 2003). In accordance with the distribution of macro- and microinvertebrates (Damborenea 2000), the Hispanic Corridor played a main role in the interchange between the western Tethys and the Eastern Pacific. Precisely, the marine herpetofauna of the Middle Jurassic of the Neuquén Basin (and north-central of Chile) has strong affinities with that recorded in the Lower-Middle Jurassic of Europe, sharing genera (*Stenopterygius, Metriorhynchus*) and like forms (*Maresaurus*, cf. *Muraenosaurus*, cf. *Cryptoclidus*).

3. In continental Patagonia, some Pangaean elements (basal eusauropods, basal tetanurans, pancryptodiran turtles) persisted during the Early and Middle Jurassic. One of the outstanding points of the Middle Jurassic of Patagonia is the tetanuran dinosaurs, whose record approximately matches with the first records of the clade worldwide (Holtz et al. 2004b).

4. In agreement with the separation of Gondwana, the continental herpetofauna of Patagonia is represented from the end of the Jurassic up to the Cenomanian by some late-Pangaean lineages, together with other mainly Gondwanan lineages. The former are represented among the dinosaurs by dicraeosaurids, basal titanosauriforms, and basal diplodocoids. The Gondwanan lineages of dinosaurs are rebbachisaurids, basal titanosaurians, basal allosauroids (carcharodontosaurids), and basal abelisauroids; and among other groups, chelids and mesoeucrocodylians. The oldest coelurosaurian dinosaurs would be represented by a series of footprints in the Late Jurassic of Patagonia and south Brazil (Fernandes 2005), although this assignment has to be restudied. Unresolved phylogenetic relationships of the sphenodontids of Patagonia preclude determining whether they were related to Pangaean or Gondwanan groups.

5. Concerning the marine herpetofauna of the Late Jurassic–Early Cretaceous, its composition seems to be more closely related to the regional evolution of the Neuquén Basin (Spalletti and Franzese this volume). During the Tithonian-Berriasian, almost all the reptiles were offshore pelagic forms (*Caypullisaurus, Dakosaurus, Pliosaurus, Liopleurodon, Neusticemys*). However, when the south of the basin shallowed, the Proto-Pacific began to withdraw to the northwest of Patagonia (toward the Valangian-Hauterivian), only elasmosaurids are recorded. Since the Barremian, there are no records of marine reptiles in Patagonia, except in the southwest of the Austral Basin facing the Pacific (*Platypterygius*).

6. No extinction is observed by the end of the Jurassic. On the contrary, most pelagic reptiles (*Dakosaurus, Geosaurus, Liopleurodon, Caypullisaurus*) cross the Jurassic-Cretaceous boundary.

Likewise, but in the continental environment, the sauropod *Amargasaurus* (Barremian of the Neuquén Basin) indicates the persistence in the Lower Cretaceous of a lineage already present in the Tithonian of the Cañadón Asfalto Basin (*Brachytrachelopan*).

7. Their isolation continued with the separation of South America from South Africa in the Aptian-Albian (Blakey 2001), and in the Turonian-Santonian lapse, the continental reptiles still belonged to late Gondwanan lineages (Bonaparte 1996; Novas 1997; Leanza et al. 2004). This is the case for the derived titanosaurs (titanosauroids sensu Salgado 2003), and possibly for abelisaurs. The record of titanosauroids begins in Patagonia in levels deposited immediately after the opening of the South Atlantic (Calvo and Salgado 1998). Although titanosauroids are also recorded in the Upper Cretaceous of continental Africa (Smith et al. 2001), the evolutionary history of the group in this continent is still mostly unknown. The record of abelisaurs in the Cenomanian of Niger suggests that the South America–Africa connection could have been working at least up to that point (Sereno et al. 2004). Chelid turtles and podocnemidoids (*Portezueloemys*), the basal *Dinilysia,* the madtsoiids, the diverse class Notosuchia (*Araripesuchus, Notosuchus, Comahuesuchus, Cynodontosuchus*), and basal neosuchians (*Lomasuchus, Peirosaurus*) are also outstanding Gondwanan elements of Patagonia during this interval.

8. During the Campanian-Maastrichtian, several large geodynamic events affected the faunistic composition of Patagonia. On the one hand, the connection between the Americas (Iturralde-Vinent 2003) should have favored the faunistic interchange, although it is currently impossible to determine how they could have affected it. In the Patagonian continental herpetofauna, there are some northern elements, such as the hadrosaurs and probably the ankylosaurs, although in the latter, unsolved phylogenetic relationships prevent us from recognizing the extent of their relationships to those of North America (Salgado and Gasparini 2006).

Other important events affected Patagonia regionally. A major ingression of the South Atlantic (North Patagonian Platform and Austral Basin) turned Patagonia into a wide archipelago. Because of this, frequent associations of mixed marine and continental fauna are recorded, essentially in late Campanian–early Maastrichtian levels.

9. During the Campanian-Maastrichtian lapse, the Saltasaurini sauropod dinosaurs (recorded exclusively in the Neuquén Basin) appeared, and the Aeolosaurini (recorded in the Neuquén and San Jorge Basins, and in the Bauru Basin in Brazil) diversified. Among theropods, the abelisaurs dominate as large predators. The large megaraptors, characteristic of the Turonian-Coniacian interval, disappeared from the record.

10. In marine areas of the interval Campanian–early Maastrichtian, elasmosaurid and polycotylid (*Sulcusuchus*) plesiosaurs are recorded. The South Atlantic ingression over Patagonia accelerated during the Late Cretaceous–Danian (Spalletti and Franzese this

volume), and the fossil-bearing levels belong to the last stages of the Mesozoic, including the record of several species of mosasaurs (*Plioplatecarpus* sp., *Mosasaurus* cf. *M. hoffmani,* Mosasaurinae indet.) and elasmosaurid plesiosaurs (*Aristonectes, Tuarangisaurus? cabazai,* cf. *Mauisaurus* sp.). In the area of Quiriquina, Chile, in late Maastrichtian rocks, marine herpetofauna is recorded, such as mosasaurs, plesiosaurs (*Aristonectes,* cf. *Mauisarus*), pancheloniid turtles, and gaviid birds. The record of marine reptiles continues in the Paleocene as the osteopygine of La Pampa.

11. Although the history of Patagonian marine reptiles is documented–although preliminary up to the end of the Mesozoic—there is no record of the last stages of the successions in continental forms. Their record is truncated after the middle Maastrichtian, a lapse of approximately 4 Ma—long enough to hide the reasons for the end. The K-P boundary marks the great extinction, and this is also seen in Patagonia (Gasparini et al. 2003). From the great taxonomic diversity of the Mesozoic, only the chelids, meiolanids, madtsoine boids, and sebecosuchians survived in the Paleogene.

References Cited

Apesteguía, S., and O. Giménez. 2001. The Late Jurassic–Early Cretaceous worldwide record of basal titanosauriforms and the origin of titanosaurians (Sauropoda): new evidence from the Aptian (Lower Cretaceous) of Chubut Province, Argentina. *Journal of Vertebrate Paleontology* 21 (Suppl. to 3): 29A.

Apesteguía, S., and F. E. Novas. 2003. Late Cretaceous sphenodontian from Patagonia provides insight into lepidosaur evolution in Gondwana. *Nature* 425: 609–612.

Apesteguía, S., F. L. Agnolin, and G. L. Lio. 2005. An early Late Cretaceous lizard from Patagonia, Argentina. *Comptes Rendus Palevol* 4: 311–315.

Arcucci, A. B., and R. A. Coria. 2003. A new Triassic carnivorous dinosaur from Argentina. *Ameghiniana* 40 (2): 217–228.

Báez, A. M., and C. A. Marsicano. 2001. A heterodontosaurid ornithischian dinosaur from the Upper Triassic of Patagonia. *Ameghiniana* 38 (3): 271–279.

Bakker, R. T. 1978. Dinosaur feeding behavior and the origin of the flowering plants. *Nature* 274: 661–663.

Bardet, N. 1995. Evolution et extinction des reptiles marins au cours du Mésozoïque. *Palaeovertebrata* 24: 177–283.

Blakey, R. C. 2001. Regional paleogeographic views of earth history. Available at http://jan.ucc.nau.edu/~rcb7/globaltext.html. Accessed June 19, 2006.

Bonaparte, J. F. 1978. *El Mesozoico de América del Sur y sus Tetrápodos.* Opera Lilloana 26.

———. 1979. Faunas y Paleobiogeografía de los tetrápodos Mesozoicos de América del Sur. *Ameghiniana* 16 (3–4): 217–246.

———. 1986. History of the terrestrial Cretaceous vertebrates of Gondwana. *Actas del IV Congreso Argentino de Paleontología y Bioestratigrafía* 2: 63–95.

———. 1987. The Late Cretaceous fauna of Los Alamitos, Patagonia, Ar-

gentina. *Revista del Museo Argentino de Ciencias Naturales "Bernardino Rivadavia," Paleontología* 3 (3): 103–179.
———. 1992. Una nueva especie de Triconodonta (Mammalia), de la Formación Los Alamitos, Provincia de Río Negro y comentarios sobre su fauna de mamíferos. *Ameghiniana* 29: 99–110.
———. 1996. Cretaceous tetrapods of Argentina. In F. Pfeil and G. Arratia (eds.), *Contributions of Southern South America to Vertebrate Paleontology. Münchner Geowissenschaftliche Abhandlungen, Rehihe A, Geologie und Paläontologie* 30: 73–130.
———. 1997. El Triásico de San Juan–La Rioja y sus Dinosaurios. *Museo Argentino de Ciencias Naturales,* 1–190. Buenos Aires.
Calvo, J. O. 1999. Dinosaurs and other vertebrates of the Lake Ezequiel Ramos Mexía area, Neuquén-Patagonia, Argentina. In Y. Tomida, T. H. Rich, and P. Vickers-Rich (eds.), *Proceedings of the Second Gondwanan Dinosaur Symposium,* 13–45. National Science Monograph 15.
Calvo, J. O., and J. F. Bonaparte. 1991. *Andesaurus delgadoi* gen. et sp. nov. (Saurischia, Sauropoda), dinosaurio Titanosauridae de la Formación Río Limay (Albiano-Cenomaniano), Neuquén, Argentina. *Ameghiniana* 28: 303–310.
Calvo, J. O., and L. Salgado. 1995. *Rebbachisaurus tessonei* sp. nov.: a new Sauropoda from the Albian-Cenomanian of Argentina–new evidence on the origin of Diplodocidae. *Gaia* 11: 13–33.
———. 1998. Nuevos restos de Titanosauridae (Sauropoda) en el Cretácico Inferior de Neuquén, Argentina. In *Resúmenes VII Congreso Argentino de Paleontología y Bioestratigrafía,* 59. Bahía Blanca, Buenos Aires: Asociación Paleonológica Argentina.
Calvo, J. O., D. Rubilar-Rogers, and K. Moreno. 2004. A new Abelisauridae (Dinosauria: Theropoda) from Northwestern Patagonia. *Ameghiniana* 41 (4): 555–563.
Calvo, J. O., J. Porfiri, C. Veralli, F. Poblete, and A. Kellner. 2002. Futalognko Paleontological Site, one of the most amazing Continental Cretaceous Environments of Patagonia, Argentina. *Resúmenes del 1° Congreso Latinoamericano de Paleontología de Vertebrados,* p. 19.
Casal, G., M. Luna, L. Ibiricu, E. Ivany, R. D. Martínez, M. Lamanna, and A. Koprowsky. 2002. Hallazgo de una serie caudal articulada de Sauropoda de la Formación Bajo Barreal, Cretácico Superior del sur de Chubut. *Ameghiniana* 39 (Suppl.): 8R.
Caselli, A. T., C. A. Marsicano, and A. Arcucci. 2001. Sedimentología y Paleontología de la Formación Los Colorados, Triásico Superior (Provincias de La Rioja y San Juan, Argentina). *Revista Asociación Geológica Argentina* 56 (2): 173–188.
Codorniú, L., Z. Gasparini, and A. A. Paulina-Carabajal. 2006. A Late Jurassic pterosaur (Reptilia, Pterodactyloidea) from northwestern Patagonia, Argentina. *Journal of South American Earth Sciences* 20: 383–389.
Coria, R. A., and P. J. Currie. 2006. A new carcharodontosaurid (Dinosauria, Theropoda) from the Upper Cretaceous of Argentina. *Geodiversitas* 28 (1): 71–118.
Coria, R. A., and L. Salgado. 1995. A new giant carnivorous dinosaur from the Cretaceous of Patagonia. *Nature* 377 (6546): 224–226.
———. 1998. A basal Abelisauria Novas 1992 (Theropoda-Ceratosauria) from the Cretaceous of Patagonia, Argentina. *Gaia* 15: 88–102.
———. 2005. Mid-Cretaceous turnover of saurischian dinosaur communities: evidence from the Neuquén Basin. In G. D. Veiga, L. A. Spalletti,

J. A. Howell, and E. Schwarz (eds.), *The Neuquén Basin, Argentina: A Case Study in Sequence Stratigraphy and Basin Dynamics,* 317–327. Special Publication 252. Geological Society, London.

Damborenea, S. 2000. Hispanic Corridor: Its evolution and the biogeography of bivalve molluscs. *GeoResearch Forum* 6: 369–380.

De la Fuente, M. S., and S. Casadío. 2000. Un nuevo osteopigino (Chelonii: Cryptodira) de la Formación Roca (Paleoceno inferior) de Cerros Bayos, provincia de La Pampa, Argentina. *Ameghiniana* 37 (2): 235–246.

Del Corro, G. 1975. Un nuevo saurópodo del Cretácico Superior, *Chubutisaurus insignis* gen. et sp. nov. (Saurischia, Chubutisauridae, nov.) del Cretácico Superior (Chubutiano), Chubut, Argentina. *Actas del Primer Congreso Argentino de Paleontología y Bioestratigrafía* 2: 229–240.

Fernandes, A. C. S. 2005. Paleoicnología en ambientes desérticos: analise da icnocenose de Vertebrados da Pedreira Sao Bento (Formaçao Botucatu, Jurassico Superior–Cretáceo Inferior, Bacia do Parana), Araraquara, SP. Thesis, Universidade Federal de Rio de Janeiro.

Fernández, M. 1994. A new long-snouted ichthyosaur from the Early Bajocian of Neuquén Basin (Argentina). *Ameghiniana* 31: 291–297.

Fernández, M., and M. B. Aguirre-Urreta. 2005. Revision of *Platypterygius hauthali* von Huene, 1927 (Ichthyosauria: Ophthalmosauridae) from the Early Cretaceous of Patagonia, Argentina. *Journal of Vertebrate Paleontology* 25 (3): 583–587.

Galton, P. M., and P. Upchurch. 2004. Prosauropoda. In D. B. Weishampel, P. Dodson, and H. Osmólska (eds.), *The Dinosauria,* 2nd ed., 232–258. Berkeley: University of California Press.

García, R. A., L. Salgado, and R. A. Coria. 2003. Primeros restos de dinosaurios saurópodos en el Jurásico de la Cuenca Neuquina, Patagonia, Argentina. *Ameghiniana* 40 (1): 123–126.

Gasparini, Z. 1981. Los Crocodylia fósiles de Argentina. *Ameghiniana* 18 (3–4): 177–205.

———. 1992. Marine reptiles. In G. Westermann (ed.), *The Jurassic of the Circum-Pacific,* 361–364. London: Cambridge University Press.

———. 2003. Los reptiles marinos jurásicos en la Cuenca Neuquina, Argentina. *Anales Academia Nacional de Ciencias Exactas, Físicas y Naturales* 55: 121–138.

Gasparini, Z., and M. Fernández. 2005. Jurassic marine reptiles in the Neuquén Basin. In G. Veiga, L. Spalletti, J. Howell, and E. Schwarz (eds.), *The Neuquén Basin: A Case Study in Sequence Stratigraphy and Basin Dynamics,* 279–294. Special Publication 252. Geological Society, London.

Gasparini, Z., and M. Iturralde-Vinent. 2006. The Cuban Oxfordian herpetofauna in the Caribbean Seaway. *Neues Jahrbuch für Geologie und Paläontologie* 240 (3): 343–371.

Gasparini, Z., D. Pol, and L. Spalletti. 2006. An unusual marine crocodyliform from the Jurassic-Cretaceous boundary of Patagonia. *Science* 311 (5757): 70–73.

Gasparini, Z., L. Salgado, and S. Casadío. 2003. Maastrichtian plesiosaurs from northern Patagonia. *Cretaceous Research* 24: 157–170.

Gasparini, Z., P. Vignaud, and G. Chong. 2000. The Jurassic Thalattosuchia (Crocodyliformes) of Chile: a paleobiogeographic approach. *Bulletin de la Societé Géologique, France* 171: 657–664.

Gasparini, Z., S. Casadío, M. Fernández, and L. Salgado. 2001. Marine reptiles from the Late Cretaceous of northern Patagonia. *Journal of South American Earth Sciences* 14: 51–60.

Gasparini, Z., L. Salgado, M. Fernández, S. Casadío, A. Concheyro, A. Parras, and S. Ballent. 2005b. The last marine reptiles over Patagonia: signs of South Gondwanan distribution. In R. J. Pankhurst and G. D. Veiga (eds.), *Gondwana XII: Geological and Biological Heritage of Gondwana, Abstracts,* 166. Córdoba, Argentina: Academia Nacional de Ciencias.

González Riga, B. 1999. Hallazgo de vertebrados fósiles en la Formación Loncoche, Cretácico Superior de la Provincia de Mendoza, Argentina. *Ameghiniana* 36: 401–410.

Goin, F. J., D. G. Poiré, M. S. de la Fuente, A. L. Cione, F. E. Novas, E. S. Bellosi, A. Ambrosio, O. Ferrer, N. D. Canessa, A. Carlini, J. Ferigolo, A. M. Ribeiro, M. S. Sales Viana, M. A. Reguero, M. G. Vucetich, S. Marenssi, M. F. de Lima Filho, and S. Agostinho. 2002. Paleontología y geología de los sedimentos del Cretácico Superior aflorantes al sur del río Shehuen (Mata Amarilla, Provincia de Santa Cruz, Argentina). *Actas del XV Congreso Geológico Argentino,* El Calafate, 1: 603–608.

GSA (Geological Society of America). 1999. 1999 geologic time scale. Available at *http://www.geosociety*.org/science/timescale/timescl.pdf. Accessed June 19, 2006.

Holtz, T. R., R. E. Molnar, and P. J. Currie. 2004a. Basal Tetanurae. In D. B. Weishampel, P. Dodson, and H. Osmólska (eds.), *The Dinosauria,* 2nd ed., 71–110. Berkeley: University of California Press.

Holtz, T. R., R. E. Molnar, and M. C. Lamanna. 2004b. Mesozoic biogeography of Dinosauria. In D. B. Weishampel, P. Dodson, and H. Osmólska (eds.), *The Dinosauria,* 2nd ed., 627–642. Berkeley: University of California Press.

Iturralde-Vinent, M. A. 2003. A brief account of the evolution of the Caribbean seaway: Jurassic to present. In D. R. Prothero, J. D. Ivany, and E. Nesbitt (eds.), *From Greenhouse to Icehouse: The Marine Eocene-Oligocene Transition,* 386–396. New York: Columbia University Press.

Iturralde-Vinent, M. A., and R. D. E. McPhee. 1999. Paleogeography of the Caribbean region: implications for Cenozoic biogeography. *Bulletin of the American Museum of Natural History* 328: 1–95.

Kellner, A. W. A., and S. A. K. Azevedo. 1999. A new sauropod dinosaur (Titanosauria) from the Late Cretaceous of Brasil. In Y. Tomida, T. H. Rich, and P. Vickers-Rich (eds.), *Proceedings of the Second Gondwanan Dinosaur Symposium, National Science Monographs* 15: 111–142.

Lamanna, M. C., R. D. Martínez, M. Luna, G. Casal, P. Dodson, and J. Smith. 2001. Sauropod faunal transition through the Cretaceous Chubut Group of Central Patagonia. *Journal of Vertebrate Paleontology* 21 (Suppl. to 3): 70A.

Lamanna, M. C., R. D. Martínez, and J. B. Smith. 2002. A definitive abelisaurid theropod dinosaur from the early Late Cretaceous of Patagonia. *Journal of Vertebrate Paleontology* 22 (1): 58–69.

Lazo, D., and M. Cichowolski. 2003. First plesiosaur remains from the Lower Cretaceous of the Neuquén Basin, Argentina. *Journal of Paleontology* 77 (4): 784–789.

Leanza, H. A., S. Apesteguía, F. E. Novas, and M. S. de la Fuente. 2004. Cretaceous terrestrial beds from the Neuquén Basin (Argentina) and their tetrapod assemblages. *Cretaceous Research* 25: 61–87.

Marsicano, C. A., R. M. H. Smith, and C. A. Sidor. 2005. Tracking the Triassic-Jurassic boundary in the roof of Africa. In R. J. Pankhurst and G. D. Veiga (eds.), *Gondwana 12: Geological and Biological Heritage of Gondwana,* Abstracts. Córdoba, Argentina: Academia Nacional de Ciencias.

Martinelli, A. G., and A. M. Forasiepi. 2004. Late Cretaceous vertebrates from Bajo de Santa Rosa (Allen Formation), Río Negro Province, Argentina, with the description of a new sauropod dinosaur (Titanosauridae). *Revista del Museo Argentino de Ciencias Naturales* 6 (2): 257–305.

Martínez, R., O. Giménez, J. Rodríguez, and G. Bochatey. 1986. *Xenotarsosaurus bonapartei* gen. et sp. nov. (Carnosauria, Abelisauridae), un nuevo Theropoda de la Formación Bajo Barreal, Chubut, Argentina. *Actas IV Congreso Argentino de Paleontología y Bioestratigrafía* 2: 23–31.

Martínez, R. D., O. Giménez, J. Rodríguez, and M. Luna. 1989. Hallazgo de restos de saurópodos en cañadón Las Horquetas, Formación Matasiete (Aptiano), Chubut. *Resúmenes de las VI Jornadas Argentinas de Paleontología de Vertebrados,* 49–51. San Juan.

Martínez, R. D., F. E. Novas, and A. Ambrosio. 2004a. Abelisaurid remains (Theropoda, Ceratosauria) from southern Patagonia. *Ameghiniana* 41 (4): 577–585.

Martínez, R. D., O. Giménez, J. Rodríguez, M. Luna, and M. C. Lamanna. 2004b. An articulated specimen of the basal Titanosaurian (Dinosauria: Sauropoda) *Epachthosaurus sciuttoi* from the early Late Cretaceous Bajo Barreal Formation of Chubut Province, Argentina. *Journal of Vertebrate Paleontology* 24 (1): 107–120.

McGowan, C., and R. Motani. 2003. Ichthyopterygia. In H.-D. Sues (ed.), *Handbook of Paleoherpetology,* Part 8, 1–175. Munich: Verlag Dr. Friedrich Pfeil.

Norman, D. B., H.-D. Sues, L. M. Witmer, and R. A. Coria. 2004. Basal Ornithopoda. In D. B. Weishampel, P. Dodson, and H. Osmólska (eds.), *The Dinosauria,* 2nd ed., 393–412. Berkeley: University of California Press.

Novas, F. E. 1997. South American dinosaurs. In P. J. Currie and K. Padian (eds.), *Encyclopedia of Dinosaurs,* 678–689. San Diego: Academic Press.

———. 1998. *Megaraptor namunhuaiquii,* gen. et sp. nov., a large-clawed, Late Cretaceous theropod from Patagonia. *Journal of Vertebrate Paleontology* 18: 4–9.

Novas, F. E., and D. Pol. 2005. New evidence on deinonychosaurian dinosaurs from the Late Cretaceous of Patagonia. *Nature* 3285: 1–3.

Novas, F. E., and P. F. Puerta. 1997. New evidence concerning avian origins from the Late Cretaceous of Patagonia. *Nature* 387: 390–392.

Novas, F., E. Bellosi, and A. Ambrosio. 2002. Los “Estratos con Dinosaurios” del lago Viedma y río La Leona (Cretácico, Santa Cruz): sedimentología y contenido fosilífero. In N. Cabaleri, C. A. Cingolani, E. Linares, M. G. López de Luchi, H. A. Ostera, and H. O. Panarello (eds.), *Actas del XV Congreso Geológico Argentino,* article 315. CD-ROM.

Novas, F. E., A. Cambiaso, and A. Ambrosio. 2004. A new basal iguanodontian (Dinosauria, Ornithischia) from the Upper Cretaceous of Patagonia. *Ameghiniana* 41: 75–82.

Novas, F. E., J. I. Canale, and M. P. Isasi. 2003. Un terópodo maniraptor del Campaniano-Maastricthiano del norte Patagónico. *Ameghiniana* 40 (4, Suppl.): 63R.

Novas, F. E., L. Salgado, J. Calvo, and F. Agnolin. 2005a. A giant titanosaur (Dinosauria, Sauropoda) from the Late Cretaceous of Patagonia. *Revista del Museo Argentino de Ciencias Naturales* (n.s.), 7 (1): 37–41.
Novas, F. E., S. de Valais, P. Vickers-Rich, and T. Rich. 2005b. A large Cretaceous theropod from Patagonia, Argentina, and the evolution of carcharodontosaurids. *Naturwissenschaften* 92: 226–230.
O'Keefe, F. 2001. A cladistic analysis and taxonomic revision of the Plesiosauria (Reptilia: Sauropterygia). *Acta Zoologica Fennica* 213: 1–63.
Powell, J. E. 2003. Revision of South American titanosaurid dinosaurs: palaeobiological, palaeobiogeographical and phylogenetic aspects. *Records of the Queen Victoria Museum* 111: 1–173.
Powell, J. E., O. Giménez, R. D. Martínez, and J. Rodríguez. 1989. Hallazgo de saurópodos en la Formación Bajo Barreal de Ocho Hermanos, Sierra de San Bernardo, Provincia de Chubut (Argentina) y su significado cronológico. *Anais do XI Congresso Brasileiro de Paleontología,* Curitiba, 165–176.
Rauhut, O. W. M. 2005. Osteology and relationships of a new theropod dinosaur from the Middle Jurassic of Patagonia. *Palaeontology* 48 (Part 1): 87–110.
Rauhut, O., A. López Arbarello, P. Puerta, and T. Martín. 2001. Jurassic vertebrates from Patagonia. *Journal of Vertebrate Paleontology* 21 (Suppl. to 3): 91A.
Salgado, L. 2003. Should we abandon the name Titanosauridae? Some comments on the taxonomy of titanosaurian sauropods (Dinosauria). *Revista Española de Paleontología* 18: 15–21.
Salgado, L., and R. A. Coria. 2005. Sauropods of Patagonia: systematic update and notes on sauropod global evolution. In V. Tidwell and K. Carpenter (eds.), *Thunder-Lizards: The Sauropodomorph Dinosaurs,* 430–453. Bloomington: Indiana University Press.
Salgado, L., and Z. Gasparini. 2004. El registro más antiguo de Dinosauria en la Cuenca Neuquina (Aaleniano, Jurásico Medio). *Ameghiniana* 41 (3): 505–508.
Salgado, L., and Z. Gasparini. 2006. Reappraisal of the ankylosaurian dinosaur from the Upper Cretaceous of James Ross Island (Antarctica). *Geodiversitas* 28: 119–135.
Sereno, P. C., J. A. Wilson, and J. L. Conrad. 2004. New dinosaurs link southern landmasses in the Mid-Cretaceous. *Proceedings of the Royal Society of London* 271 (1546): 1325–1330.
Sciutto, J. C., and R. D. Martínez. 1994. Un nuevo yacimiento fosilífero de la Formación Bajo Barreal (Cretácico Tardío) y su fauna de saurópodos. *Naturalia Patagonica, Ciencias de la Tierra* 2: 27–47.
Simón, M. E., and J. O. Calvo. 2002. Un primitivo titanosaurio (Sauropoda) del Chocón, Formación Candeleros (Cenomaniano temprano), Neuquén, Argentina. *Ameghiniana* 39: 17R.
Simón, M. E., and A. W. A. Kellner. 2003. New sphenodontid (Lepidosauria, Rhynchocephalia, Eilenodontinae) from the Candeleros Formation, Cenomanian of Patagonia, Argentina. *Boletim do Museu Nacional, Nova Série, Geologia* 68: 1–12.
Smith, J. B., M. C. Lamanna, K. J. Lacovara, P. Dodson, J. R. Smith, J. C. Poole, R. Giegengack, and Y. Attia. 2001. A giant sauropod dinosaur from an Upper Cretaceous Mongrove deposit in Egypt. *Science* 292: 1704–1706.
Spalletti L., Z. Gasparini, G. Veiga, E. Schwartz, M. Fernández, and S.

Matheos. 1999. Facies anóxicas, procesos deposicionales y herpetofauna de la rampa marina titoniano-berriasiana en la Cuenca Neuquina (Yesera del Tromen), Neuquén, Argentina. *Revista Geológica de Chile* 26: 109–123.

Stinnesbeck, W. 1986. Zu den faunischen und Palökologischen Verhältnissen in der Quiriquina Formation (Maastrichtium) Zentral-Chile. *Palaeontographica*, Abteilung A, 194: 99–237.

Unwin, D. M., O. W. M. Rauhut, and A. Haluza. 2004. The first "Rhamphorhynchoid" from South America and the early history of pterosaurs. *74th Annual Meeting of the Paläontologische Gesellschaft*, 235–237A.

Veralli, C., and J. O. Calvo. 2004. Dientes de terópodos carcharodontosáuridos del Turoniano superior–Coniaciano inferior del Neuquén, Patagonia, Argentina. *Ameghiniana* 41 (4): 587–590.

Weishampel, D. B., P. M. Barret, R. A. Coria, J. Le Loeuff, X. Xing, Z. Xijin, A. Sahni, E. M. P. Gomani, and C. R. Noto. 2004. Dinosaur distribution. In D. B. Weishampel, P. Dodson, and H. Osmólska (eds.), *The Dinosauria*, 2nd ed., 517–606. Berkeley: University of California Press.

Wilson, J. A., and P. C. Sereno. 1998. Early evolution and higher-level phylogeny of sauropod dinosaurs. Memoir 5. Society of Vertebrate Paleontology. Supplement to *Journal of Vertebrate Paleontology*, 1–68.

Appendix: Patagonian Mesozoic Reptilian Taxa Included in the Preceding Chapters and Their Respective Ages

Testudines

1. *Notoemys laticentralis*	early and middle Tithonian
2. *Neusticemys neuquina*	early-late Tithonian
3. *Prochelidella argentinae*	Cenomanian-Turonian
4. *Prochelidella portezuelae*	late Turonian–early Coniacian
5. *Palaeophrynops patagonicus*	late Campanian–early Maastrichtian
6. *Bonapartemys bajobarrealis*	Cenomanian-Turonian
7. *Lomalatachelys neuquina*	Santonian
8. *Yaminuechelys gasparinii*	late Campanian–early Maastrichtian
9. *Portezuelomys patagonica*	late Turonian–early Coniacian
10. *Niolamia argentina*	late Campanian–early Maastrichtian
11. *Euclastes* sp.	late Maastrichtian

Lepidosauromorpha

12. *Kaikaifilusaurus calvoi*	early Cenomanian
13. *Dinilysia patagonica*	Santonian
14. *Alamitophis argentinus*	early Campanian–late Maastrichtian
15. *Alamitophis elongates*	early Campanian–late Maastrichtian
16. *Patagoniophis parvus*	early Campanian–late Maastrichtian
17. *Rionegrophis madtsoioides*	early Campanian–late Maastrichtian

Crocodyliformes

18. *Metriorhynchus aff. M. brachyrhynchus*	late Bathonian
19. *Dakosaurus andiniensis*	Tithonian–Berriasian
20. *Geosaurus araucanensis*	Tithonian–Berriasian
21. *Amargasuchus minor*	early Barremian
22. *Araripesuchus patagonicus*	early Cenomanian
23. *Notosuchus terrestris*	Santonian
24. *Comahuesuchus brachybuccalis*	Santonian
25. *Cynodontosuchus rothi*	Santonian
26. *Pehuenchesaurus enderi*	Turonian–Coniacian
27. *Lomasuchus palpebrosus*	late Turonian–early Coniacian
28. *Peirosaurus tormini*	Coniacian

Pterosauria

29. *Herbstosaurus pigmaeus*	late Tithonian

Ornithischia

30. *Notoceratops bonarelli*	Maastrichtian
31. *Heterodontosaurus* sp.	Norian
32. *Notohypsilophodon comodorensis*	Cenomanian?
33. *Gasparinisaura cincosaltensis*	early Campanian
34. *Anabisetia saldiviai*	late Cenomanian–early Turonian
35. *Talenkauen santacrucensis*	Maastrichtian
36. *Secernosaurus koerneri*	Maastrichtian?
37. *Kritosaurus australis*	early Campanian–early Maastrichtian

Sauropodomorpha

38. *Mussaurus patagonicus*	Norian
39. *Amygdalodon patagonicus*	Toarcian-Bajocian
40. *Volkheimeria chubutensis*	Callovian
41. *Patagosaurus fariasi*	Callovian
42. *Tehuelchesaurus chubutensis*	Kimmeridgian-Tithonian
43. *Chubutisaurus insignis*	early Late Cretaceous
44. *Agustinia ligabuei*	Aptian-Albian
45. *Andesaurus delgadoi*	early Cenomanian
46. *Mendozasaurus neguyelap*	late Turonian–late Coniacian
47. *Epachthosaurus sciuttoi*	Cenomanian-Turonian
48. *Argentinosaurus huinculensis*	late Cenomanian
49. *Argyrosaurus superbus*	Cenomanian-Turonian
50. *Antarctosaurus wichmannianus*	Campanian-Maastrichtian
51. *Laplatasaurus araukanicus*	early Campanian
52. *Bonitasaura salgadoi*	Santonian
53. *Pellegrinisaurus powelli*	lower Campanian
54. *Bonatitan reigi*	middle Campanian–early Maastricthian
55. *Aeolosaurus rionegrinus*	Campanian-Maastrichtian
56. *Rincosaurus caudamirus*	late Turonian–Coniacian
57. *Neuquensaurus australis*	early Campanian
58. *Rocasaurus muniozi*	middle Campanian–early Maastrichtian
59. *Limaysaurus tessonei*	early Cenomanian
60. *Rayososaurus agrioensis*	Aptian
61. *Brachytrachelopan mesai*	Tithonian
62. *Amargasaurus cazaui*	Barremian

Non-avian Theropods

63. *Ilokelesia aguadagrandensis*	late Cenomanian
64. *Bayosaurus pubica*	late Cenomanian–early Turonian
65. *Ekrixinatosaurus novasi*	early Cenomanian
66. *Xenotarsosaurus bonapartei*	Cenomanian-Turonian
67. *Abelisaurus comahuensis*	early Campanian
68. *Carnotaurus sastrei*	Campanian-Maastrichtian
69. *Aucasaurus garridoi*	early Campanian
70. *Ligabueino andesi*	early Barremian
71. *Velocisaurus unicus*	Santonian

72. *Piatnitzkysaurus floresi*	Callovian
73. *Tyrannotitan chubutensis*	Aptian
74. *Giganotosaurus carolinii*	early Cenomanian
75. *Mapusaurus roseae*	late Cenomanian
76. *Buitreraraptor gonzalezorum*	Turonian–early Coniacian
77. *Neuquenraptor argentinus*	Turonian–early Coniacian
78. *Unenlagia comahuensis*	Turonian–early Coniacian
79. *Alvarezsaurus calvoi*	Santonian
80. *Patagonykus puertai*	Turonian–early Coniacian
81. *Megaraptor namunhuaiquii*	Turonian–early Coniacian
82. *Condorraptor currumili*	Callovian
83. *Quilmesaurus curriei*	middle Campanian–early Maastrichtian
84. *Genyodectes serus*	Aptian-Albian

Aves

85. *Neuquenornis volans*	middle Santonian
86. *Patagopteryx deferrariisi*	middle Santonian
87. *Limenavis patagonica*	Campanian-Maastrichtian
88. *Neogaeornis wetzeli*	Maastrichtian

Ichthyosauria

89. *Stenopterygius cayi*	early Bajocian
90. *Mollesaurus periallus*	early Bajocian
91. *Caypullisaurus bonapartei*	Tithonian–middle Berriasian
92. *Ophthalmosaurus* sp.	Tithonian
93. *Platypterygius hauthali*	Barremian

Plesiosauria

94. *Maresaurus coccai*	early Bajocian
95. cf. *Muraenosaurus* sp.	early Callovian
96. cf. *Cryptoclidus* sp.	early Callovian
97. *Liopleurodon* sp.	Tithonian–middle Berriasian
98. *Pliosaurus* sp.	late Tithonian–Berriasian
99. *Sulcusuchus erraini*	early Campanian–late Maastrichtian
100. *Aristonectes parvidens*	late Maastrichtian
101. *Tuarangisaurus? cabazai*	late Maastrichtian
102. cf. *Mauisaurus* sp.	late Maastrichtian

Ichnites

103. *Calibarichnus ayestarani*	Late Triassic
104. *Palaciosichnus zettii*	Late Triassic
105. *Rogerbaletichnus aquilerai*	Late Triassic
106. *Gallegosichnus garridoi*	Late Triassic
107. *Shimmelia chirotheroides*	Late Triassic
108. *Pteraichnus* sp.	early Cenomanian
109. *Abelichnus astigarrae*	early Cenomanian
110. *Wildeichnus navesi*	late Middle Jurassic–early Late Jurassic
111. *Sarmientichnus scagliai*	late Middle Jurassic–early Late Jurassic
112. *Delatorrichnus goyenechei*	late Middle Jurassic–early Late Jurassic

113. *Bressanichnus patagonicus*	early Cenomanian
114. Deferrariischnium mapuchensis	early Cenomanian
115. *Picunichnus benedettoi*	early Cenomanian
116. cf. *Ignotornis*	Campanian
117. *Barrosopus slobodai*	Campanian
118. *Sauropodichnus giganteus*	early Cenomanian
119. *Limayichnus major*	early Cenomanian
120. *Sousaichnium monettae*	early Cenomanian
121. *Ingenierichnus sierrai*	Late Triassic

General Index

Page numbers in italics indicate illustrations.

Taxonomic Index